COMPUTERS FOR ARTIFICIAL INTELLIGENCE PROCESSING

COMPUTERS FOR ARTIFICIAL INTELLIGENCE PROCESSING

Edited by Benjamin W. Wah and C. V. Ramamoorthy

A WILEY-INTERSCIENCE PUBLICATION

John Wiley & Sons, Inc.

NEW YORK / CHICHESTER / BRISBANE / TORONTO / SINGAPORE

Library of Congress Cataloging in Publication Data:

Computers for artificial intelligence processing / edited by
Benjamin W. Wah and C. V. Ramamoorthy.
     p.   cm.
"A Wiley-Interscience publication."
Includes bibliographical references.

  1. Electronic digital computers. 2. Computer architecture.
3. Artificial intelligence.   I. Wah, Benjamin W.   II. Ramamoorthy,
C. V. (Chitoor V.), 192

QA76.5.C6144   1990
006.-dc20   89-77799
ISBN 0-471-84811-5   CIP

CONTENTS

SECTION II.   LANGUAGE-BASED AI ARCHITECTURES

7. "Design Decisions in SPUR"   **273**

by Mark Hill, Randy Katz, John Ousterhout,
David Patterson, et al.

CONTRIBUTORS

G. T. Alley
Instrumentation & Controls
  Division
Oak Ridge National Laboratory
P.O. Box 2008
Oak Ridge, TN 37831-6006
Chapter 9

Farokh B. Bastani
Department of Computer Science
University of Houston
Houston, TX 77025
Chapter 16

Professor P. Bruce Berra
Department of Electrical and
  Computer Engineering
Syracuse University
111 Lind Hall
Syracuse, NY 13210
Chapter 11

D. W. Bouldin
Electrical & Computer Engineering
University of Tennessee
Knoxville, TN 37996-2100
Chapter 9

W. L. Bryan
Instrumentation & Controls
Division
Oak Ridge National Laboratory
P.O. Box 2008
Oak Ridge, TN 37831-6006
Chapter 9

Dr. Soon Myoung Chung
Department of Computer Science
  and Engineering
Wright State University
Dayton, OH 45435
Chapter 11

R. O. Eason
Electrical Engineering
University of Maine
Orono, ME 04469
Chapter 9

Scott E. Fahlman
School of Computer Science
Carnegie-Mellon University
Pittsburgh, PA 15213
Chapter 12

Dr. Vijay Garg
Computer Science Division
University of California/Berkeley
Berkeley, CA 94720
Chapter 15

Jayantha Herath
Department of Electrical
 & Computer Engineering
Drexel University
Philadelphia, PA 19104
Chapter 6

Susantha Herath
Department of Electrical
 & Computer Engineering
Keio University
Yokohama, JAPAN
Chapter 6

Geoffrey E. Hinton
Computer Science Department
University of Toronto
Toronto M5S 1A4 CANADA
Chapter 12

Kai Hwang
Department of Electrical
 Engineering
University of Southern California
Los Angeles, CA 90089-0781
Chapter 5

Ravi K. Iyer
Coordinated Science Laboratory
University of Illinois
1101 W. Springfield Avenue
Urbana, IL 61801
Chapter 4

Guo-Jie Li
Institute of Computing Technology
Academia Sinica
P.O. Box 2704-1
Beijing, PROC
Chapter 1

Rene L. Llames
Coordinated Science Laboratory
University of Illinois
1101 W. Springfield Avenue
Urbana, IL 61801
Chapter 4

Matthew B. Lowrie
926 North Scott
Wheaton, IL 60187
Chapter 1

Pankaj Mehra
Coordinated Science Laboratory
University of Illinois
1101 W. Springfield Avenue
Urbana, IL 61801
Chapter 13

David A. Moon
Symbolics Inc.
8 New England Executive Park East
Burlington, MA 01803
Chapter 3

D. F. Newport
Electrical & Computer Engineering
University of Tennessee
Knoxville, TN 37996-2100
Chapter 9

SPUR Project
C/O Professor David Patterson
Computer Science Division
University of California
Berkeley, CA 94720
Chapters 7 & 8

Andrew R. Pleszkun
Department of Electrical
 & Computer Engineering
University of Colorado
Campus Box 425
Boulder, CO 80309
Chapter 2

Professor C. V. Ramamoorthy
Computer Science Division
University of California/Berkeley
Berkeley, CA 94720
Chapter 15

Nobuo Saito
Department of Computer Science
Keio University
Yokohama, JAPAN
Chapter 6

Professor Shashi Shekhar
Department of Computer Science
University of Minnesota
Minneapolis, MN 55455
Chapter 15

Matthew J. Thazhuthaveetil
Department of Electrical Engineering
Pennsylvania State University
University Park, PA 16802
Chapter 2

Dr. Wei-Tek Tsai
Department of Computer Science
University of Minnesota
4-192 EE/CSCI Bldg.,
200 Union St. SE
Minneapolis, MN 55455
Chapter 14

David Ungar
Department of Computer Science
Stanford University
Palo Alto, CA 94305
Chapter 8

Benjamin W. Wah
Coordinated Science Laboratory
University of Illinois
1101 W. Springfield Avenue
Urbana, IL 61801
Chapters 1 & 13

David L. Waltz
Thinking Machines Corporation
245 First Street
Cambridge, MA 02142-1214
Chapter 10

Yoshinori Yamaguchi
Electrotechnical Lab
Sakuramura Niiharigun
Tsukubu 305 JAPAN
Chapter 6

Toshitsugu Yuba
Electrotechnical Lab
Sakuramura Niiharigun
Tsukubu 305 JAPAN
Chapter 6

PREFACE

This book addresses the increasing complexity and the growing need for computational power of artificial intelligence (AI) algorithms and programs. These algorithms and software, which share many common features with symbolic processing, are not supported efficiently by conventional von Neumann computers, which are oriented towards numeric processing. Their efficient evaluation requires new architectural designs, languages, algorithms, and representation schemes to be developed.

This book presents fundamentals in architectures, languages, and software designs for supporting AI applications. It provides a comprehensive treatment of the design issues and current state-of-the-art research efforts in this area, and illustrates these solutions with example designs. The discussion spans from hardware architectures to software engineering methods to meta-level strategy designs.

This book represents a collective effort of fifty-one authors, all recognized experts in areas of computer architecture, parallel processing, artificial intelligence, and software engineering. It was developed over a period of three years and reflects some of the leading efforts in this area.

This book can serve as a reference text for researchers and developers working in the area, as well as an introductory text for beginners. It can also serve as a reference text to accompany an advanced course on computer architecture. The topics selected for presentation provide an overview of the area as well as an in-depth discussion of some of the important and difficult problems in the area. The material presented assumes a basic knowledge on computer system design, computer architecture, artificial intelligence, and software design methods. A senior in Computer Science will possess the necessary background for understanding the material presented.

This book is organized into five major sections. Each section delineates a specific aspect of the problem and may have one or more chapters.

Section 1 presents a comprehensive survey on the design issues and examples of computers oriented towards symbolic processing. An extensive bibliography accompanies the discussion.

Section 2 discusses the design and implementation of special-purpose language-oriented computers for supporting AI processing. Special-purpose languages studied include functional languages, Lisp, production systems, and Smalltalk. Three chapters are devoted to sequential Lisp processing, with discussions on the design issues, memory management, performance evaluation, and an example illustrated with the Symbolics Lisp computer. Three chapters are devoted to multiprocessing and parallel processing of Lisp programs and, in general, functional programs. The last two chapters in this section present architectures for supporting Smalltalk-80 and production systems.

Section 3 examines multiprocessor systems for general AI processing. The Connection Machine is drawn as an example of a symbolic multiprocessor with data-level parallelism. Design of large data/knowledge base machines for AI processing is also studied.

Section 4 discusses connectionist architectures and applications. One chapter is devoted to illustrating the benefits and design issues of connectionist systems. A second chapter presents an extensive survey on connectionist architectures, as well as other computing architectures designed for learning strategies.

The last section addresses software architectures for AI applications and the design of AI software as a software engineering project, two important issues that are largely neglected in the literature. Three aspects are examined: AI and software engineering, development tools for AI programs, and reliability of AI programs.

We would like to thank all the authors who participated in this project for their dedication and patience. We are also grateful to the reviewers who provided many constructive criticisms on this work. We would like to acknowledge the partial support of this project by the National Aeronautics and Space Administration under contract NCC 2-481. Lastly, we are indebted to Miss Vickie DeMoss, who spent many late evenings to enter the text and draw the figures using Interleaf's University Publishing System.

Benjamin W. Wah   *Urbana-Champaign, Illinois*

C. V. Ramamoorthy   *Berkeley, California*

COMPUTERS FOR ARTIFICIAL INTELLIGENCE PROCESSING

CHAPTER 1

COMPUTERS FOR SYMBOLIC PROCESSING

Benjamin W. Wah, Matthew B. Lowrie, and Guo-Jie Li

In this chapter, we provide a detailed survey of the motivations, design, applications, current status, and limitations of computers designed for symbolic processing. Symbolic processing applications are computations performed on words, relations, or meanings. A major difference between symbolic and conventional numeric applications is that the knowledge used in symbolic applications may initially be fuzzy, uncertain, indeterminate, and ill-represented. Hence, the collection, representation, and management of knowledge is more difficult in symbolic applications than in conventional numeric applications.

We first survey various techniques for knowledge representation and processing, from the points of view of both designer and user. Next, we examine the design and choice of a suitable language for symbolic processing and the mapping of applications into a software architecture. Finally, we examine the design process of refining the application requirements into hardware and software architectures and present a discussion of state-of-the art sequential and parallel computers designed for symbolic processing.

1 SYMBOLIC PROCESSING

1.1 Introduction

The development in the 1950s of the programming language IPL by Newell, Shaw, and Simon was a pioneering effort in symbolic processing by computers [172]. Data structures of unpredictable shape and size could be manipulated conveniently by programs written in IPL. Many of the early symbolic

©1989 IEEE. Reprinted, with permission, from *Proceedings of the IEEE,* vol. 77, no. 4, pp. 507–540, April 1989.

programs, including the Logic Theorist and the General Problem Solver, were written in IPL. The invention of Lisp in 1958 by John McCarthy made symbolic programming yet easier. The language featured the recursive use of conditional expressions, the representation of symbolic information externally by lists and internally by linked lists, and representation of program and data using the same data structures [152].

Recent advances in applications of computers suggest that the processing of symbols rather than numbers will be the basis for the next generation of computers. This is highlighted by the numerous research efforts in Japan, Europe, and the United States [1, 196, 228]. Symbolic processing has been applied in a wide spectrum of areas; among them are pattern recognition, natural language processing, speech understanding, theorem proving, robotics, computer vision, and expert systems. Researchers in artificial intelligence, databases, programming languages, cognitive science, psychology, and many other fields have addressed overlapping issues in symbolic processing.

Conventional computers have been designed with numeric processing power as their focus. The disparity between symbolic and numeric operations, therefore, calls for innovative research in alternative architectures for symbolic processing. In this chapter we present a review of the state of the art in computers for symbolic processing. The discussion proceeds in a top-down fashion. First, we present the relevant features and characteristics of symbolic processing. Next, we discuss the role of techniques and methodologies in the design process and classify hardware and software architectures at different levels of design. The general view of computers designed or used for symbolic processing is depicted in Figure 1.1; the section discussing each portion is indicated in the figure.

Section 1 develops a classification of general computations. From this classification, a definition of symbolic processing is derived in Section 1.1. Typical symbolic processing applications and their characteristics are discussed in Section 1.2.

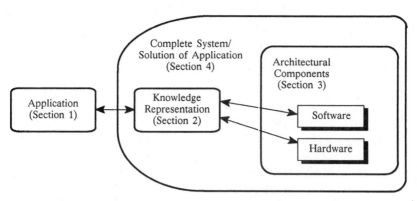

Figure 1.1 Overview of the chapter.

Knowledge representation and knowledge processing are two important characteristics of solutions to a symbolic processing problem. Knowledge representation is the technique for representing data and information in a computer; it is discussed in Section 2.1. Knowledge processing is the technique for controlling the manipulation of knowledge in the system and is the topic of Section 2.2.

The design of a computer relies on various concepts and strategies for implementing knowledge processing techniques. Section 3 examines the architectural concepts behind the design of symbolic processing systems. Software architectures are studied in Section 3.1, and hardware architectures in Section 3.2.

A complete system for symbolic computation is the result of the application of a design philosophy, architectural components, and available technology. Complete systems are the topic of Section 4, in which many existing and experimental systems are discussed and compared.

Symbolic processing systems will evolve as new concepts and technologies develop. Section 5 outlines some recent research that is likely to impact the design of symbolic processing systems in the future.

1.2 Classification of Computations

One of the fundamental debates on intelligent behavior is about the nature of symbols. A number of scientists view human beings and computers as physical symbolic systems that produce an evolving collection of symbolic structures. In their 1975 Turing award lecture, A. Newell and H. Simon stated a general scientific hypothesis—The Physical Symbol System Hypothesis [174]:

> A physical symbol system has the necessary and sufficient means for general intelligent action.

By "necessary," they mean that any system that exhibits general intelligence will prove upon analysis to be a physical symbol system. By "sufficient," they mean that any physical symbol system of sufficient size can be organized to exhibit general intelligence. Research on artificial intelligence (AI) addresses the sufficiency of a physical symbol system for producing intelligent action, while investigators in cognitive psychology attempt to demonstrate that a physical symbol system necessarily exists wherever intelligence is exhibited. Although empirical in nature, the continuous accumulation of empirical evidence on the above hypothesis in the last 30 years has formed the basis of much research on AI and cognitive science.

Since our focus is on computers for symbolic processing, we will first classify computations performed on computers. From this classification is derived the definition of symbolic processing used in this chapter. There are five classes of computations: analog, numeric, word, relational, and meaning. These classes are based on the primary unit of data in the computation.

Analog: The analog class of computation encompasses those computations whose functions have parameters that are continuous variables. This is not the primary area of computation in a digital computer, as digital computers use digital memory. In digital computers, this kind of computation primarily measures environmental parameters.

Numeric: In this class of computation, the primary unit upon which functions are performed represents magnitude. Many applications of computers fall into this category; functions on memory elements containing integers, floating point numbers, and the like are numeric.

Word: In this class of computation, the parameters of functions are words that do not necessarily have quantitative value. Text processing is an example of this kind of computation.

Relational: In relational computations, functions operate on relations among words; that is, the primary units of data are groups of words that have some relational interpretation.

Meaning: Very little research has been done on techniques for automated computation at the meaning level. The primary unit of data is an interrelated series of relations that represent semantics and meaning.

A few examples to illustrate the various classes of computations are shown in Figure 1.2. Figure 1.2(*a*) presents weather forecasting, a standard supercomputer application. The computation begins with analog measurements of the atmosphere (arc 1). These measurements are then converted to numeric entities (arc 2). Most of the computation occurs in the numeric stage (arc 3, representing the conversion of numbers into different sets of numbers), and conversion to meaning is done at the very end (arc 4, which may be done by humans instead of the computer). Story comprehension from speech input is an example that uses the full spectrum of the classes of computations. Figure 1.2(*b*) shows the flow of data. Computation may also flow in the opposite direction; robot control, depicted in Figure 1.2(*c*) is an example of this.

The design of a computer system can be viewed as the maximization of performance with respect to the problem to be solved, subject to cost constraints. Computations in a higher level of abstraction are usually carried out by transforming them into more definite computations. For example, database queries function primarily at the word and relation level of computation. However, if we wish to know the average salary of employees in a database, numeric computations would be required. The design of this system may not, however, benefit from the inclusion of fast arithmetic units, because the performance gain may not be large enough to counteract the loss of other features necessitated by the cost constraint. The difference between this database system and a system that would include fast arithmetic units is in the emphasis of the computations.

A computer that is intended to perform more abstract computations should also be able to perform more definite computations. For instance, the afore-

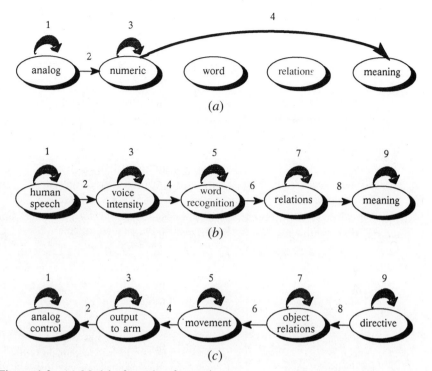

Figure 1.2 (*a*) Model of weather forecasting on a supercomputer; (*b*) model of speech understanding; (*c*) model of robot control.

mentioned database computer should be able to perform fluid dynamics calculations. It may, however, be less efficient than a computer of comparable cost that is intended for numeric calculations.

Using this classification, it is possible to describe concisely what is meant by "symbolic processing". *Symbolic processing is defined as those computations that are performed at the same or a more abstract level than the word level. Symbolic processing computers* are computers that are designed to carry out operations at the same or a more abstract level than the word level.

1.3 Characteristics of Symbolic Processing Applications

This section presents symbolic processing applications and their overall features. A few applications and their characteristics are presented in Table 1.1. The rest of the section lists features that characterize general symbolic processing applications [239]. These features do not apply to every instantiation of a symbolic processing problem and are intended as design guidelines that target general symbolic processing techniques.

TABLE 1.1   Some Symbolic Processing Applications.

Application	Characteristics
Problem Solving general specific accounting programming/compilation text processing human interface	User inputs problem, system attempts to solve User encodes solution of problem Metaknowledge for specific problems is well understood Small-grain parallelism is predominant
Database Management variety of applications often an integral part of larger systems	Organization of information for retrieval Efficient algorithms to consider all run-time possibilities are too complex Metaknowledge is application dependent Large potential for parallelism, both small-grain and large-grain

Expert Systems
 diagnosis
  medical
  plant disease
  computer system errors
 design assistance
  architecture
  computer architecture
  computer chips
 personal systems
  business
  finance
  wine tasting
 others
	Ill-structured collection of facts, inferences, and so on are represented as knowledge-intensive program in specific domain [96]
	Knowledge and metaknowledge are usually provided by designers
	Large potential for parallelism

Natural Language Processing understanding generation translation	Translate natural language to machine representation Translate machine representation to natural language Translate between two forms of natural language [2, 13, 229]
Computer Vision signal processing pattern recognition image understanding	Primarily numeric Patterns viewed as sentences—symbolic Higher-level reasoning at the image-understanding level [3, 76]
Learning expert systems deduction improve system efficiently knowledge acquisition	Ability to adapt to environment to increase efficiency Fundamental to symbolic processing [157, 190]

Incomplete Knowledge. Many applications are nondeterministic; it is not possible to predict the flow of computation in advance. This is due to incomplete knowledge and understanding of the application. The lack of complete knowledge may also lead to a need for dynamic execution, which allows the creation of new data structures and functions during the solution of the problem. Furthermore, data structures used in the solution may be arbitrarily large, necessitating dynamic allocation of memory, tasks, and other resources. System design should meet the need to cope with dynamic and nondeterministic execution. An architecture that can adapt to the more efficient performance of computations not anticipated at design time is referred to as an open system.

Knowledge Processing. A computation can be viewed as manipulations of a set of data. In Section 1.1, computations are classified by the data on which they operated. The nature of the operations performed on the data depends on the application and the way the data is stored. As illustrated in Figure 1.3, it is processing knowledge about the computation, whether that be algorithms, techniques for evaluation, or some other reasoning technique, that controls and dictates the manipulation of data. At a level above this is metaknowledge — knowledge about knowledge [7, 49, 80]. Metaknowledge includes the extent and origin of the domain knowledge of a particular object, the reliability of certain information, and the possibility that an event will occur. Meta-level knowledge can be considered to exist at a single level or in a hierarchy [25] containing an arbitrary number of levels, each serving to direct the use of knowledge at the lower levels.

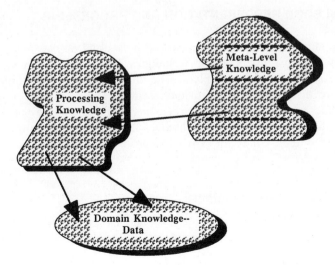

Figure 1.3 Knowledge processing.

Metaknowledge can be classified as deterministic or statistical according to correctness and performance considerations [139]. Deterministic metaknowledge refers to the knowledge about precedence relationships that results from a better understanding of the problem and helps to reduce its resource and time complexity. Statistical metaknowledge can be used to order object-level actions before their performance for greater efficiency.

Rather than adding more heuristics to improve the performance of a system, metaknowledge about the effective use of existing heuristics can be collected and developed. Metaknowledge can also be used in the formalization of belief, default reasoning, inference in changing situations, and other problems [7].

Symbolic Primitives. A general symbolic application may contain primitive symbolic operations. Typical operations are comparison, sorting, selection, matching, and logical operations such as union, negation, intersection, transitive closure, and pattern retrieval and recognition. These operations may be performed at more than one level of computation (for instance, at both word and relation levels). Higher levels of computation may also contain complicated primitive operations such as unification.

Parallel and Distributed Processing. Many symbolic applications exhibit a large potential for parallelism. Parallelism may be categorized into AND-parallelism and OR-parallelism. In AND-parallelism, a number of independent tasks are executed concurrently. OR-parallelism is a technique used to shorten processing time in nondeterministic computations by simultaneously evaluating alternatives at a decision point.

2 KNOWLEDGE REPRESENTATION AND PROCESSING

A symbolic application is a problem whose inputs and outputs are symbolic. *Symbolic processing* refers to the techniques employed by the system for finding the solutions of the problem. The characteristics of symbolic applications were discussed in the previous section. The emphasis of this section is on the *features of symbolic processing* as they relate to the design of computers. Section 2.1 discusses techniques for representing knowledge. The issues involved in the control of knowledge processing are presented in Section 2.2.

2.1 Knowledge Representation

To design an efficient computer for a given application, it is first necessary to characterize the programs that will run on the computer. Of primary importance in the solution of symbolic processing problems is the choice of a knowledge representation [149]. Appropriate symbolic structures to represent knowledge, and appropriate reasoning mechanisms to answer questions and assimilate new information must both be selected [120]. Four criteria can be used to evaluate a knowledge representation scheme: flexibility, user friendliness, expressiveness, and efficiency of processing. Flexibility, user friendliness, and expressiveness are required to simplify the tasks of programming and comprehension. The efficiency or tractability of a knowledge representation scheme dictates the efficiency of the application. Much of the research in this area represents a tradeoff between expressiveness and tractability.

Despite a great deal of effort devoted to research in knowledge representation, very little scientific theory is available either to guide the selection of an appropriate representation scheme for a given application or to show how to transform a representation into a more efficient one. Although a number of knowledge representation schemes have been proposed, none is clearly superior to the others for all applications.

The following sections present two attributes for consideration when comparing knowledge representations: the *local versus distributed* and *declarative versus procedural* attributes. The classical knowledge representation schemes are then evaluated on the basis of these features.

2.1.1 Features of Knowledge Representations

Local Versus Distributed Representations. In a local representation, each conceptual datum is stored in a separate hardware unit. A word or data item stored in a register is an example of local storage for that data item. When stored locally, data are simple to read, update, and understand. Unfortunately, if any hardware unit fails, all data contained in that unit is lost to the system. Most current systems, both symbolic and numeric, use local representations for individual pieces of data.

In a distributed representation, a piece of knowledge is represented by many storage units, and each unit may be part of several pieces of knowledge that correspond to features from different concepts. The advantage of such a representation is that it is fault tolerant. If a small proportion of units fails, the integrity of the distributed data undergoes little change. This property is very attractive for practical implementations. Distributed representations also allow a great deal of parallelism in computation [191]. However, they are usually harder for a user to understand and modify.

Table 1.2 summarizes the salient characteristics of local and distributed representations. It should be noted that there is no concrete boundary between local and distributed features. Some features of a knowledge representation scheme are local, whereas others are distributed. At one extreme is a standard implementation of a simple Lisp program, for instance, which can be thought of as a hierarchy of local representations. The program is stored as one unit of information; the data structures used by the program are also stored as a single entity; finally, each piece of data within a data structure is stored in one memory location. At the other extreme is a standard implementation of a neural network. A predicate logic program, however, is neither fully local nor

TABLE 1.2   Attributes of Local and Distributed Representations.

Attribute	Local	Distributed
Storage technique	Each datum stored in dedicated hardware	Data represented over multiple units
Ease of understanding	Easy for humans to comprehend	Difficult for humans to interpret
Modification of stored data	Simple	More difficult
Fault tolerance	Loss of hardware results in loss of all stored data in that unit	Loss of small proportion of units does not seriously damage integrity of data

fully distributed. The complete program is not a single entity but a set of logic statements. Each statement, however, is an example of a local representation. Predicate logic is considered further in Section 2.1.2.

Declarative Versus Procedural Representations.   The issue of distributed versus local representations concerns methods of representing information in the computer hardware. In contrast, procedural and declarative representations involve techniques for representing processing knowledge and processing methodologies in computer programs.

A program written in a declarative representation consists of a set of domain-specific facts or statements and a technique for inferring knowledge from these statements. It is, therefore, characterized as a set of statements of knowledge *about* the problem domain. Pure predicate logic and production systems (to be discussed in Section 2.1.2) are examples of declarative representations.

In a procedural representation, program statements consist of steps to be taken in the solution of the problem—statements of knowledge about *how* to solve the problem. Examples of procedural program representations include the C language and Lisp.

Declarative representations are user-oriented and emphasize correctness and user friendliness of programs. They are referentially transparent: the meaning of the whole can be derived solely from the meaning of the parts, independent of the sequence of operations that leads to these parts. This may increase programmer productivity [245] and result in tremendous potential for parallelism [88, 128].

Unfortunately, programs written in declarative representations are often inefficient due to nondeterminism, implicit control aspects, and inconsistent knowledge. It is hard to add domain-specific knowledge and meta-level knowledge to declarative programs. The difficulty with using declarative repre-

TABLE 1.3 **Attributes of Declarative and Procedural Representations.**

Attribute	Declarative	Procedural
Emphasis	Knowledge of domain	Knowledge of solution
Technique	Domain-specific statements	Solution techniques
Orientation	User friendliness; ease of understanding	Efficiency of solution; ease of representing control knowledge
Parallelism	Natural, but countered by unnecessary search	Constrained and often user specified
Control	Transparent to the user	Specified by the user

sentations to solve symbolic problems lies in determining how to *use* the facts stored in the program's data structures, not in deciding how to store them.

Procedural programs are not as user friendly as declarative ones because the programmer must specify all control knowledge. In addition, the validity of a procedural statement often relies heavily on other procedural statements in the program, and this complicates both the creation and the modification of software. The loss of flexibility in a procedural programming environment is counteracted by the gain in ease of representing control knowledge. Procedural schemes allow the specification and direct interaction of facts and heuristic information, thereby eliminating wasteful search. Metaknowledge can also be included easily in procedures. Overall, procedural representations are as much concerned with the technique and efficiency of the computation as with the ease of representing the domain knowledge.

The salient features of declarative and procedural representations are summarized in Table 1.3. Practical knowledge representation schemes may have both procedural and declarative features.

2.1.2 Classical Knowledge Representation Schemes.
In this section, the classical knowledge representation schemes are described and evaluated. Those that have received the greatest attention include predicate logic, production systems, semantic networks, frames, procedural languages, and fully distributed representations.

Predicate Logic.   Predicate logic studies the relationship of implication between assumptions and conclusions. Logic often seems a natural way to express certain notions, and there are standard methods of determining the meaning of an expression in a logic formalism [125]. Logic is useful for exploring the epistemological problems that determine how the observed facts can be represented in the memory of a computer independent of how the knowledge is used. The major disadvantage of logic stems from the separation of representation and processing.

Production Systems. Production systems use collections of rules to solve problems. The rules consist of condition and action parts, or antecedent and consequent parts [173]. Production systems provide a useful mechanism for controlling interactions between statements of declarative and procedural knowledge. For this reason, production systems have been used extensively in expert systems and knowledge engineering. Unfortunately, the expressive power of production systems is limited. Some researchers have argued that rule-based expert systems cannot achieve expert-level behavior [59]. Production systems are also inefficient because of their high control overhead.

Semantic Networks. A semantic network is a directed graph whose nodes represent objects, concepts, or situations, and whose arcs represent relationships between nodes [183]. The basic inference mechanism in semantic networks is *spreading activation*. The idea has a clear neural inspiration: certain concepts in memory become a source of activation, and activation spreads in parallel to related concepts. This graphical representation allows certain kinds of inference to be performed by simple graph-search techniques. Although simple semantic networks can only express variable-free assertions, several authors have shown that the expressive power of semantic networks can be extended to equal that of predicate logic [126]. Frequently, semantic networks are used as data structures for manipulation by other knowledge representation schemes (such as Lisp, a procedural representation).

Frame Representations. Frame representations employ a data structure called a *frame* for representing stereotypical situations [158]. The frame description form is an elaboration of the semantic network form. Its emphasis is on the structuring of types (called frames) in terms of their attributes (called slots). A frame contains declarative and procedural information in predefined internal relations. Attached to each frame is heuristic information, such as a procedure on how to use the information in the frame. Although many frame implementation issues are unresolved, the framelike structuring of knowledge appears promising and has appeared in various forms in many conventional languages.

Procedural Representations. In a procedural representation, a knowledge base is viewed as a collection of modules expressed in a procedural language, such as Lisp or C. The procedural scheme is capable of representing heuristic knowledge and performing extended logical inferences, such as plausible reasoning. Because it eliminates wasteful search, this representation scheme can be carried out efficiently. However, it is often limited by the available constructs. Conventional FORTRAN or Pascal programs, for example, have been found to be inadequate for the support of efficient symbolic processing.

Connectionist Representations. A connectionist representation is a form of distributed representation: concept representations are spread over a number of modules or units. When presented with input, units having a positive correlation with an input feature are activated, whereas those with a negative correlation produce inhibitory signals. In this fashion, input can be recognized as a function of connection strengths among units (see also Section 3.2.2) [191]. Distributed representations allow automated procedures for learning concepts and representations, and have great potential for parallel computation. Their major drawback is the difficulty of interpreting system states and internal representations. Furthermore, the learning of connection strengths often requires a lengthy training period.

A given representation may exhibit local or distributed and declarative or procedural aspects at different levels of the representation. Table 1.4 summarizes the characteristics of the representations introduced here and categorizes them by the hierarchy of knowledge representations inherent in the technique.

2.2 Knowledge Processing

Different reasoning methods are associated with different knowledge representation schemes and require different architectural supports. Table 1.5 shows the classical knowledge representation paradigms and their respective reasoning techniques.

It is argued that humans use logic-like reasoning in the domain of rational knowledge and apply memory-based reasoning to perceptual actions. For over 30 years, logic-like deduction has been the dominant paradigm in AI research. This paradigm has been applied to a wide range of problems, especially expert systems. Although intelligent behavior often resembles logic-like reasoning with limited search, the intensive use of memory to recall specific episodes from the past (rather than rules) could be another foundation of machine reasoning [209]. Memory-based reasoning (or case-based reasoning) does not use rules; instead it attempts to solve the problem by direct reference to memory. The Connection Machine is an example of a machine designed for memory-based reasoning, although it can also be programmed to perform logic-like reasoning [103].

A knowledge processing technique must be tailored to cope with the application's requirements. The greatest need in symbolic processing applications is the ability to deal with uncertain, incomplete, or conflicting information. Techniques for dealing with these problems are discussed in Section 2.2.1. Section 2.2.2 discusses methods for exploiting parallelism.

2.2.1 Uncertain, Incomplete, and Inconsistent Knowledge Processing. The techniques for dealing with these problems in knowledge processing are detailed in Figure 1.4. The rest of the section is devoted to a brief discussion of the entries in the figure.

TABLE 1.4 **Characterization of Knowledge Representation Schemes.**

Representation	Level of Representation	Characterization
Logic	Variable	Local/declarative
	Statement/relation	Local/declarative
	Program	Distributed/declarative
Production System	Variable	Local/declarative
	Statement/relation	Local/either
	Program	Distributed/declarative
Semantic Network	Node	Local/declarative
	Arc/relation	Local/declarative
	Network	Local/procedural
	Program	Distributed/declarative
Frames	Variable	Local/declarative
	Statement	Local/either
	Slot	Local/either
	Frame	Local/declarative
	Program	Distributed/declarative
Procedural	Variable	Local/either
	Statement	Local/procedural
	Program	Local/procedural
Connectionist	Connection strength	Local/declarative
	Propagation technique	Local/procedural
	Data and knowledge	Distributed/declarative

TABLE 1.5 **Reasoning Techniques.**

Representation	Typical Reasoning Technique
Logic	Resolution (unification)
Production system	Forward/backward chaining
Semantic network	Spreading activation
Frames	Procedure attachments
Procedural	Control flow
Connectionist	Propagation of excitation

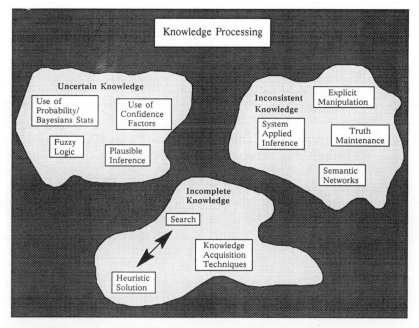

Figure 1.4 Indeterminate knowledge processing: some techniques.

Uncertain Knowledge.   Methods such as conventional knowledge representation techniques based on predicate calculus are not well suited for representing common-sense knowledge. Explicit and implicit quantifiers are fuzzy, and the standard inference methods mentioned earlier make no provision for dealing with uncertainty. Two types of uncertainty have been studied. One type is caused by noisy data and the fuzzy meaning of symbols; the other is associated with uncertain inference rules.

A number of methods and theories of capturing uncertainty have been examined in recent years. Probability and Bayesian statistics form the basis of most approaches to this problem, which include fuzzy logic [250, 251], confidence factors [30], Dempster and Shafer's theory of plausible inference [199], odds [60], and endorsements [38].

Uncertain knowledge is most frequently handled by two principal components. The first is a translation system for representing the meaning of propositions and other semantic entities. The second is an inferential system for deriving answers to questions about information resident in a knowledge base. The application of Bayesian statistics to expert systems follows this approach [60]. A confidence factor (CF), or certainty factor, is used to decide among alternative answers during a consultation session. A CF of a rule is a measurement of the strength of the association between premises and conclusions. A positive CF indicates that the evidence confirms the hypothesis, whereas a negative CF disconfirms the hypothesis.

Dempster and Shafer's theory of plausible inference provides a natural and powerful methodology for representing and combining evidence. Ignorance and uncertainty are directly represented in belief functions and remain throughout the combination process.

Endorsements are records of information that affect a hypothesis' certainty. They can be propagated over inferences, but in a manner that is sensitive to the context of the inference.

Incomplete and Inaccurate Knowledge. A key feature of symbolic computations is nondeterminism arising from the fact that almost any intelligent activity is likely to be poorly understood. This implies that no systematic, direct algorithms are available for performing the activity. When a problem becomes well understood and can be solved by a deterministic algorithm, the solution of the problem is no longer considered to be intelligent [206].

The starting point of any conventional computation is a deterministic algorithm. Since most symbolic processing applications are knowledge intensive, such a deterministic algorithm may not exist. Efficient solution of the problem, therefore, requires continual refinement of the computation technique and may employ various knowledge acquisition techniques. When knowledge of the application domain is incomplete or uncertain, heuristic methods are used [133, 134, 180, 181, 205]. A heuristic suggests plausible actions to take, or implausible ones to avoid. It is desirable to use concise and accurate domain-specific knowledge and metaknowledge. Unfortunately, this body of information is difficult to acquire in practice and may be fallible if incomplete or very large if complete.

The nondeterministic nature of computations and the fallibility of heuristics may lead to anomalies of parallelism. For instance, when multiple processors are used, one or more of the processors may be guided by the heuristic function into a part of the search tree that is not explored in the same order as that in sequential processing. This out-of-order exploration of the search tree, coupled with the pruning of undesirable nodes, may result in a speed-up (as compared to sequential processing) by a factor less than one or greater than the number of processors. Some results on the occurrence of anomalies and how to cope with them can be found in the literature [129, 136, 141, 238].

In addition to heuristics, several new forms of logic for belief and knowledge have been introduced. Traditional reasoning methods suffer from the problem of *logical omniscience* [105]. Logical omniscience refers to the assumption that agents are sufficiently intelligent that they know all valid formulas. Thus, if an agent knows proposition p, and knows that p implies q, then the agent must know q. In real life, people are certainly not omniscient. The newly introduced forms of logic are more suitable than traditional logic for modeling beliefs of humans (or computers) with limited reasoning capabilities [63, 87].

Processing Inconsistent Knowledge. Traditional logic is *monotonic*. Monotonicity implies that new axioms may be added to the list of provable theorems only when they are consistent. Nonmonotonic reasoning provides a more flexible and complete logical system, as well as a more accurate model of human thought processes. The motivations for nonmonotonic reasoning can be classified into two general areas: default reasoning, and reasoning in a changing environment [24].

Default reasoning can be broken into two distinct areas: exceptions to the rule, and autoepistemic logic. As the name implies, exceptions to the rule allows relations that contradict more general relations. A statement with *most* as a relation will not add any information to a monotonic logic system. Nonmonotonic systems allow a representation that includes exceptions to the general rule, without eliminating the validity of the computing environment.

Autoepistemic logic (also known as circumscription or the closed-world assumption) allows the system to reach conclusions about relations about which no facts exist in the database. This requires the assumption that all relevant knowledge is in the database (the closed-world assumption).

In a monotonic system, existing knowledge and data cannot be modified without restarting all inference processes and withdrawing all conclusions. In a world where new discoveries and revisions of previous beliefs are the norm, this is a poor model for a large knowledge-based system. Accommodating a changing environment is particularly important when default reasoning is used. A statement inferred by default may be corrected in light of additional evidence.

The distinguishing feature of the various techniques for dealing with inconsistent knowledge is their method of handling correction of the knowledge base. These methods include explicit encoding, system applied inference, semantic networks, and truth maintenance. In explicit encoding, the programmer is responsible for writing code that will update the database when a new statement is added which may conflict with other statements [72]. A system employing system applied inference contains user-encoded functions that automatically search for inconsistencies in the knowledge base [101]. For example, McCarthy and Hayes have indicated how actions might be described using modal operators like "normally" and "consistent" [151], and Sandewall used a deductive representation of nonmonotonic rules based on a primitive called UNLESS [192]. Reasoning with semantic networks is another technique for the ordering of inferences and default reasoning, although semantic networks have been criticized as insufficiently formal logical systems, lacking a clear inference technique. Doyle's truth maintenance system (TMS) records and maintains the reasons for the system's beliefs. These beliefs can be revised when discoveries contradict assumptions [58]. Improvements have been explored by de Kleer in his assumption-based truth maintenance systems (ATMS) [121-123]. In the IBM YES/MVS expert system, inconsistent deductions are automatically removed and new consequences computed in accordance with the changed facts [197].

2.2.2 Parallel Knowledge Processing.
Humans are often thought of as the most efficient symbolic processing engines. Some researchers claim that symbolic problems can, therefore, be most effectively solved using techniques

similar to those employed in the human brain. Observations of human intelligence suggest that human cognition can be divided into perceptual and rational cognition, each of which may involve a different degree of parallelism. In the perceptual form of cognition, which includes such activities as vision and speech understanding, massive parallel processing is possible due to the large number of independent data and the simple flow of control. In rational cognition, on the other hand, only limited parallelism is possible. Therefore, the problems of high-level reasoning should be solved by trying to accumulate heuristics rather than by trying to exploit parallelism.

Unfortunately, early experiences with symbolic multiprocessor architectures such as Hearsay-II [70], Eurisko [135], and multiprocessor implementation of forward chaining rule based expert systems [75] have shown that parallel symbolic programs exhibit small speed-ups only [119]. This has led to the possibly incorrect conclusion that symbolic programs written for sequential execution have low potential for parallelism.

Methods of parallel knowledge processing are distinguished by four features: deterministic and nondeterministic parallelism, granularity of parallelism, data- and control-level parallelism, and user- and system-defined parallelism. These features are summarized in Table 1.6. Designs of parallel symbolic processors are presented in detail in Sections 3 and 4.

3   ARCHITECTURAL CONCEPTS FOR SYMBOLIC PROCESSING

Once a symbolic processing application has been characterized and a suitable representation technique selected, it is possible to choose appropriate features and attributes for a computer system for that application. An architectural component of a processing system is defined as a hardware or software structure which supports an application. In this section, we discuss current software and hardware architectures that are useful for symbolic processing systems, focusing on architectures specific to symbolic processing and the way that they relate to fundamental design concepts.

Software architectures are comprised primarily of software languages and environments suitable for encoding particular applications. The selection of a software environment imposes certain features that the software and hardware must support. The design process decides among software and hardware implementations of the required features and is driven by a tradeoff between cost and expected performance improvement.

3.1   Software Architectures

The area of software architectures for symbolic processing has two important facets. The first is the design of appropriate software facilities, tools, and languages for the symbolic application; the second concerns the tools used for mapping a symbolic application into software.

TABLE 1.6   Issues in Parallel Processing.

Issue	Definition	Comments
Deterministic and nondeterministic	Concurrent execution of multiple units of computation, all of which are necessary for job completion	Low overhead guarantees speed-up Tasks must be independent Pure functional programming is deterministic
	Multiple potential solutions evaluated in parallel Parallelism used to replace or augment backtracking	Easy to implement—always independent Nondeterministic nature may lead to anomalies in parallelism
Granularity	Size of units of computation to be executed by a single functional unit	Difficult to determine Is a function of knowledge representation, problem complexity, the shape of the search tree, distribution of processing times, and the dynamic nature of the problem [140]
Data level and control level	Data stored one element per processor Executed in single-instruction stream, multiple-data stream (SIMD) fashion	Can be used for large database operations, sort, set operations, statistical analysis, etc. [240] Can be implemented in memory—referred to as an *active memory* [67]
	Independent control for parallel tasks	Major type of parallelism used Multiple-instruction stream, multiple-data stream (MIMD) systems Detection of parallelism can be more difficult than in numeric programs. Nondeterministic nature requires dynamic mapping
User defined and system defined	User specifies portions of program which can execute in parallel	In numeric processing, DOACROSS is a typical example [247]. The FUTURE construct in Multilisp is a symbolic construct [90].
	Parallelism is detected and exploited automatically by the compiler or run-time software and hardware	Fully distributed representations allow massive system-level parallelism. Some systems may employ both—user only aids in indicating available parallelism.

The discussion of software architectures is organized in the following manner. First, the process of designing software facilities and languages is analyzed. Next follows an overview of the most frequently researched and utilized programming paradigms for symbolic processing. Finally, the problem of mapping applications into software is examined.

The Design of Software Languages.   The objective of a software language is to provide software support and implementation of the chosen knowledge representation(s). As when choosing the knowledge representation scheme, the major goals of the language designer are ease and ability to represent the application and the efficient execution of the algorithm. Once the designer has selected a technique or techniques for representing knowledge, the major features of the language become apparent. For example, the choice of a logic representation dominates the characteristics of the Prolog language.

The software technique for implementing these features may not be as clear, however. For instance, Lisp is a procedural language developed for symbolic processing. In its design, functional programming with recursion and list-structured data were selected. These are not obvious choices, but they arise out of the use of a procedural representation. These features were also selected for ease of representation and efficiency of processing. Another example of implementation is the additional "impure" features that may be added to the implementation of a knowledge representation scheme to support efficient computation and provide flexibility to the user. Cuts and side effects in Prolog are well known examples. They can be seen as a procedural addition to a logic representation to support the efficient *implementation* of logic programming.

The extension of conventional von Neumann computer languages for symbolic processing is an issue that has been explored extensively. By their nature, conventional computer languages are based on procedural representations. Examination of the characteristics of symbolic processing, as discussed in Section 3.1.1, reveals the desirable features to incorporate into a conventional programming language. Possible features include data structures, symbolic primitives, and recursion. Conventional languages designed for numeric processing, such as FORTRAN, do not provide adequate support for symbolic processing. In particular, the languages are not sufficiently flexible to enable simple encoding of very complex symbolic operations. For this reason, this section emphasizes the new and less conventional languages for symbolic processing.

As noted in the preceding section, the emphasis in the design of new representations for symbolic processing problems has been on the addition of declarative and distributed features to existing representation schemes. Part of the motivation for this emphasis is referential transparency, or freedom from side effects. This relieves the users of some of the programming burden and allows easier programming of complex applications. The following sections discuss three paradigms for the design of software languages that promote referential transparency: *functional languages* [16, 44, 97], *rule-based languages* [36, 52, 126], and *object-oriented languages* [81, 208, 210, 242].

3.1.1 Functional Programming Languages.

The functional programming approach does not employ states, program counters, or other sequence-related computational constructs. A program is a function in the mathematical sense. The program, or function, is applied to the input, and the function is evaluated to produce the desired output. A functional language can be thought of as a language based on the lambda calculus; operators are applied to data or to the results of other function evaluations. John McCarthy's conception of list processing is a pioneering effort in this area [150]. Examples of functional languages include pure Lisp [152], Backus' FP [16], Hope [32], Val [155], and Id [14].

In a functional language, the meaning of an expression is independent of the history of any computation performed prior to the expression's evaluation (referential transparency). Precedence restrictions occur only as a result of function application. Notions such as side effects and shared memory do not exist in functional programs. The lack of side effects results in the determinacy property that is so valuable in parallel processing. A function has the same value (assuming its evaluation terminates) regardless of the order in which its arguments are computed. Hence, all arguments and distinct elements in dynamically created structures in a functional program can be evaluated concurrently. For example, consider a simple program for computing the average of the numbers in a list *s*.

```
average(s) = div (sum(s), count (s))
```

If we attempt to evaluate `average (1.(2.(3.nil)))`, the computation of `sum (1.(2.(3.nil)))` can clearly proceed independently of the computation of `count (1.(2.(3.nil)))`. The key point is that parallelism in functional languages is implicit and is supported by the underlying semantics. There is no need for special message passing constructs, synchronization primitives, or constructs for specifying parallelism. It has been reported that implementation of functional languages seems easier on a parallel computer than on a sequential computer [46].

Programming in functional languages facilitates specification or prototyping prior to the development of efficient programs. Given a satisfactory specification, it is possible to develop an efficient program through *program transformation* — the systematic refinement of the program specification. Because functional languages are referentially transparent, they can be refined as familiar mathematical forms. Another advantage of functional programming is that it can represent higher-order functions; a function can be passed as an argument. A comparison of functional programming with von Neumann programming is presented in Table 1.7.

TABLE 1.7 Functional Versus von Neumann Programming.

Functional Programming	von Neumann Programming
Programs are built only from other programs.	Programs contain programs, expressions, and variables.
Programs can be freely built from other programs.	Programs are composed only of common data storage.
Same program can treat objects of different structure and size.	Changing size or structure of data means changing the program.
Programs have a strong theoretical background. Programs may be proven correct, as in mathematics.	Few general practical theorems exist about programs. Proving correctness of a program is extremely difficult.

Pure Lisp is a functional language. Many dialects of Lisp, however, are not purely functional. Operations on global variables, property lists, input/output, and other features of these dialects create side effects. Side effects support efficient computation by obviating the recomputation of functions whose results are required in more than one place; they also support convenient input/output. Unfortunately, the property of referential transparency is lost in most practical Lisp languages. Moreover, precedence restrictions are represented not only by functional calls, but also in procedures.

It is not as straightforward to identify the parallel tasks in a language with side effects as it is in a purely functional language. Users must use special primitives to identify independent tasks. Several parallel Lisp languages have been proposed and implemented. Multilisp, developed by R. Halstead at MIT and implemented on a 128-processor Butterfly parallel processor, includes the usual Lisp side effect primitives for altering data structures and changing the values of variables [89]. Parallel symbolic concurrency is introduced by means of the pcall and future constructs [41]. Both utilize an implicit fork-join. For example, (pcall A B C) will result in the concurrent evaluation of expressions A, B, and C, and (future X) immediately returns a pseudo location for the value of X and creates a task to concurrently evaluate X. The use of future allows concurrency between the computation of a value and the use of that value. The primitive future was introduced because pcall alone did not provide a great deal of parallelism [90].

Proponents of functional languages believe that their simplicity and elegance will promote more orderly, more rigorous, more verifiable, and ultimately more efficient programming. Opponents worry about loss of expressiveness caused by the expression-evaluation–only model. The crucial disadvantage of functional programming is that it is difficult to represent the inherent nondeterminism of AI problems in a functional program. The recursive formulation and leftmost outermost reduction feature of functional programs enable a natural formulation of depth-first search, but it is difficult to write a heuristic search program in a purely functional language, since heuristic search is inherently history sensitive. In fact, heuristic search programs written in Lisp include many setq and do statements that are not pure functional primitives [246]. Because of their inability to represent nondeterminism and their inefficiency in dealing with large data structures, pure functional languages are often unsuitable for general symbolic applications. Nonetheless, they are very useful for deterministic symbolic applications.

3.1.2 *Rule-Based Languages.*

Rule-based languages are associated with two major forms of knowledge representation: logic and production systems. The languages associated with these representations are described as rule-based because they emphasize the relation between a condition and an inference or rule.

Logic. In its modest form, a logic program is the procedural interpretation of Horn clauses or predicate logic [125, 126]. Some logic programming ideas, such as automatic backtracking, were used in the early AI languages **QA3**, **PLANNER**, and **MICROPLANNER** [18, 218]. The newer language Prolog is based on logic programming [37, 241]. A logic programming environment is a reasoning-oriented or deductive programming environment. Logic programming has recently received considerable attention because of its choice as the core computer language for the Fifth Generation Computer System Project in Japan [164].

The motivation of logic programming is to separate knowledge from control. However, logic programming implementations often include extralogical primitives to improve their run-time efficiency and provide additional methods for specifying control information; for example, in Prolog the Cut predicate is an extralogical control mechanism used to define a construct similar to the *if-then* of conventional languages. Furthermore, variables in a logic program are often non-directional—that is, they do not have to be defined as input or output variables at compile time, and their mode can be changed at run time according to the context. Thus, dependencies among subgoals are not defined at compile time, and static detection of parallelism is very difficult. The solution is to require users to specify the tasks to be processed in parallel. In Parlog [35], every argument is annotated with a mode declaration that states whether the argument is input (?) or output (^). In Concurrent Prolog [200], a read-only annotation (?) is used. Users can also distinguish between parallel AND and sequential AND by annotating them with "**,**" and "**&**" respectively.

Constructs can also be introduced to restrict parallelism until certain preconditions are satisfied. An example is Parlog's guarded clause. A guarded clause has the format h :- g | b.; where g is the *guard* of the clause and b is its *body*. Subgoals in the body can only be evaluated when all subgoals in the guard have succeeded and bound values have been committed to the body.

User specification of parallelism certainly detracts from the objective of declarative programming. This is a problem even in the restricted AND-parallelism (RAP) model [51], in which the user, although not required to specify parallelism explicitly, must remain aware of the underlying computational model. Both mode declarations in Parlog and read-only annotations in Concurrent Prolog impose a fixed execution order on subgoals, an order which may be inefficient. Choosing the proper subgoals for a guard is sometimes difficult and is not guided by any general principle. The distinction between sequential AND and parallel AND—a linear order—is not sufficient to specify all precedence relationships, which form a partial order. Because of the non-deterministic nature of AI applications, users cannot identify all tasks that can be processed in parallel. A better symbolic processing language should detect parallelism at both compile-time and run-time.

Production Systems.   The other major form of rule-based language that separates knowledge from control is based on the production system representation [28, 175]. A production system consists of a set of data and a set of rules that manipulate the data. A rule is composed of a left hand side (LHS) and a right hand side (RHS). The LHS is the antecedent or situation and represents the conditions necessary for applying the rule. The conditions are in the form of a Boolean combination of clauses [28]. The RHS is called the consequent and indicates a set of changes to be made to the data when the conditions of its LHS are met. Thus, a production system functions procedurally by first matching a logical condition and then modifying the data. Strategies are required for matching data conditions with the conditions of the LHS of rules and for resolving conflicts when more than one rule is matched. The conflict set is the set of antecedents and their bindings that match elements in the working memory. Production systems operate in a recognize-act cycle; first the recognize cycle computes the conflict set, and then the act cycle selects one matching production and makes the changes indicated in the RHS.

A popular programming environment for the implementation of production systems is the OPS5 system [28]. The OPS5 system highlights production system design issues. OPS5 employs data typing, and the working memory (data) is viewed as a separate entity from the production memory, where the rules are stored. OPS5 employs a Rete match algorithm which computes the conflict set but does not select a production on which to act. The algorithm stores the matching condition in the form of a tree. After the recognize-act cycle, the entire conflict set is not recomputed; instead the tree is updated via tokens that reflect the addition or deletion of elements from the working memory [28, 74]. OPS5 has two conflict resolution strategies: LEX and MEA. LEX orders the conflict set by recency of the time tags of the working memory elements that match the condition elements of the production rule. In contrast, MEA orders the conflict set by recency of the working memory element that matches the first condition of the matching conditions, even if it is not a maximum.

Production systems provide a natural programming paradigm for *if-then* programming environments such as those employed in expert systems. Unfortunately, algorithms using iteration or recursion are difficult to encode. Furthermore, rules are independent, and the structural organization of programs requires special attention. Thus, it is difficult to develop large programs using production systems [186].

On Functional Versus Rule-Based Languages. The advantages and disadvantages of functional languages stem from the procedural and formal mathematical nature of the lambda calculus. Rule-based languages have complementary advantages and disadvantages that stem from their declarative nature.

Logic languages are more expressive because of their properties of non-directionality of inputs and outputs, dynamic binding of variables, and nondeterminism. On the other hand, functional programming enables programmers to write more concise programs since higher-order functions permit quantification over individual data items as well as predicates and functions. Programs containing higher-order functions are also easier to understand and reason about.

Although logic languages are more expressive, their implementation in a parallel environment is more difficult because of the non-directionality of their variables. This flexibility complicates the detection of parallelism at compile time and results in the dynamic execution of logic programs. Current Prolog systems also lack a means to describe the termination of computations on conceptually infinite data structures and the concept of function evaluation, which makes the logic base nontransparent. The run-time behavior of pure functional programs is much simpler to control than that of first-order logic programs, particularly in a parallel context. Techniques such as graph reduction and data flow have been studied for use in parallel evaluation of pure functional languages.

It would be advantageous if the simple control mechanism of functional languages could be applied to languages with the great expressive power of logic languages. Considerable efforts have been devoted to the combining of functional and logic programming [52]. Some researchers are trying to simplify logic languages by introducing directionality of information in logic programs [187]; however, this approach will reduce the expressive power of logic programs to that of first-order functional programs.

The alternative approach is to extend functional languages so that they have the expressive power of logic languages while retaining their underlying functional simplicity. The addition of unification to the Hope language is such an example [47]. Subrahmanyam and Yau have proposed FUNLOG, a language that integrates functional and logic programming. FUNLOG provides the programmer with the flexibility to choose between a backtrack-free computational framework and a logic computational framework. *Semantic unification* is the basis for the integration of function and logic, and can be used to replace conventional unification in logic programming [216]. TABLOG, a new approach to logic programming designed by Malachi, Manna, and Waldinger, is based on first-order predicate logic with equality and combines rule-based and functional programming. The use of this richer and more flexible syntax overcomes some of the shortcomings of Prolog syntax [146]. Other languages that combine features of Prolog and Lisp include LOGLISP, QLOG, POPLOG, Qute, and Lambda Prolog [187].

3.1.3 Object-Oriented Languages.   New languages and programming systems are being developed to simplify AI programming radically. *Object-oriented programming* holds promise as a programming framework and can be extended to concurrent systems, databases, and knowledge bases.

In conventional software, data and procedures are the main focus of the representation and are treated as separate entities. The choice of procedures and data is made by the programmer. In an object-oriented system, there is only one entity: the *object*. Objects may be manipulated like data, or describe procedure-like manipulation, or both. Processing is performed by sending and receiving messages to and from the object that possesses the appropriate information. A selector in the object specifies the kind of operation to be performed. Message sending is uniform; a message specifies only the result that is required by the sender, with no information on how it should be accomplished. Objects respond to messages by using their own procedures, called *methods*, to perform operations. Since all communication is via messages, one method may not "call" another method. Any knowledge-representation scheme can be used to represent the procedural knowledge, although most implementations of object-oriented languages have employed procedural representations of control knowledge stored within the object.

In addition to objects and messages, object-oriented languages may employ the concepts of class and instance. A class is a description of similar types of objects; classes allow attributes of objects to be shared by providing a mechanism for inheritance or implicit sharing. Inheritance is used to define objects that are almost like other objects. Thus, classes provide an interface for the programmer to interact with the definitions of objects.

Data abstraction is an important concept that is supported by message sending. Object-oriented languages support both the management and collection of data through *abstract data types*, and the composition of abstract data types through an *inheritance* mechanism.

The requirement of typed data abstraction with inheritance is explicit and definitive, suggesting that object-oriented programming should be characterized by its type mechanisms rather than its communication mechanisms. In a sense, object-oriented programming can be defined as

object-oriented = data abstraction + data types + type inheritance

The object-oriented programming paradigm is used mainly for organizing knowledge domains but is permissive in its methodology for communication. The message/object model provides no new leverage for expressing concurrent problems. Concurrent models, operating systems, and coordination tools can be built from such lower-level objects as processes, queues, and semaphores.

Early exploration of object-oriented programming was found in Simula [42]. A more contemporary object-oriented language which has received a great deal of attention is Smalltalk [81], and the object-oriented languages also include LOOPS [210], Actor [6], CommonObjects [208], OIL [48], and others [242]. Recently, CommonLoops has been suggested as a standard for object-oriented extensions to Common Lisp [23].

The *Actor* model, developed by Hewitt at MIT, combines a formalization of object-oriented language ideas with the effects of parallelism [98]. An actor is the analog of a class or type instance. Computations in the Actor model are partial orderings of events, inherently parallel and having no assignment commands. The language Act3, based on the Actor model, combines the advantages of object-oriented programming with those of functional programming [5]. The Apiary network architecture has been proposed to support the Actor model [99, 100].

3.1.4 *Mapping Applications into Software.* Software development, an active area of research in software engineering, is the process of mapping an application into a language. The process begins with the selection of a solution technique, a selection that includes the choice of a knowledge representation and a method for solving the problem within that representation. Naturally, the available languages impact this choice. In life cycle models of software development [185, 201] the choices may be referred to as requirement analysis.

Software development environments can be classified into four generations [186]: discrete tools, toolboxes, life cycle support and knowledge-based tools, and intelligent life cycle support. Discrete tools, typical in the 1960s and 1970s, are individual debuggers. Toolboxes are integrated packages of tools; Interlisp [224] is the most prevalent example. Life cycle support and knowledge-based tools were developed in the 1980s. Life cycle support refers to software-development environments suitable for each stage in the design cycle; knowledge-based tools incorporate domain knowledge to provide interactive assistance to the programmer. Finally, intelligent life cycle support, a topic for future research [186], provides integrated knowledge-based support for all stages of the software-development cycle. Software engineering environments for distributed software development are another important area of research [201].

As software complexity increases, verification, validation, and the enforcement of a structured discipline become more important to the production of reliable software. Although a number of symbolic languages, such as Prolog, have been criticized for lack of structure, a programming style of hierarchical development can be followed as in conventional structured programming languages such as Pascal. Verification and validation of software written in symbolic processing languages are controversial. If programs are treated as algorithms with well specified requirements, then verification and validation procedures are similar to those of programs written in traditional languages,

TABLE 1.8 Features of Two Example Languages that Can Be Supported by Hardware.

Lisp	Prolog
Data typing Function calls Recursion List structures Garbage collection Individual commands CAR, CDR, etc. Parallelism FUTURE, etc. Application support Application dependent Database support Transitive closure Others	Condition matching Database functions Search Search strategy Backtracking mechanism Unification Parallelism Modes, guards, etc. Application support

and techniques such as test case generation and path testing can be applied. It is, of course, difficult to test the validity of the knowledge used in a program, since it may be heuristic and fuzzy in nature. The same criticism, however, can be made of programs written in traditional languages, whose validity will depend to a great extent on the experience of experts and the procedures used in deriving the knowledge. Systematic knowledge-capture tools will help, but cannot guarantee proper collection and maintenance of consistent knowledge from multiple experts.

3.2 Hardware Architectural Support for Symbolic Processing

The choice of knowledge representation schemes and software languages largely dictates the desirable hardware architectures. This section discusses the desirable intermediate hardware designs of a symbolic processing architecture. These hardware designs can support language-specific features or primitive symbolic operations such as sorting and pattern matching. Table 1.8 shows some features of Lisp and Prolog that require hardware or software support.

Hardware features, like languages, can incorporate many well established design philosophies: for example pipelining, parallel processing, micropro-gramming, and redundancy. Consider, for example, the hardware support of type checking in Lisp. Figure 1.5 illustrates the role that design concepts and design requirements play in a hardware architecture, though hardware design-ers may not utilize such an approach in a real design.

Once a set of hardware and software alternatives has been enumerated, a subset of them must be selected for incorporation into the system. The

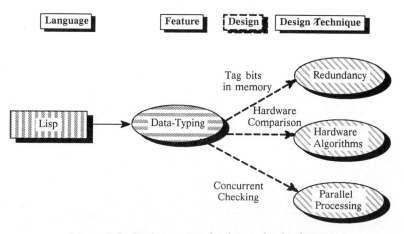

Figure 1.5 Design process for data-typing hardware.

choice will depend on the design philosophy of the system (see Section 4). The selection among competing structures (alternative designs that perform the same function) is made on the basis of anticipated gain in performance versus anticipated cost; both may be difficult to estimate, especially when the structures are not commercially available components. An approximate model is often used to guide the selection.

The hardware architectures currently used to support symbolic computation can be classified into microlevel hardware features, subsystem-level hardware features, and system-level designs. A microlevel architecture is generally a piece of hardware designed to support a feature of the language or primitive operations in the knowlege representation scheme. It is specialized and does not, in general, perform useful computations outside of its role in the system. A subsystem-level architecture is an architecture that performs a complete and useful function, but is often included in a larger system. A system-level design provides a complete hardware/software solution to a symbolic processing application. Microlevel and subsystem-level designs are discussed in Sections 3.2.1 and 3.2.2; system-level designs are presented in Section 4.

3.2.1 Microlevel Hardware Features.

A microlevel hardware architecture is a unit dedicated to the support of a specific symbolic processing technique. As the number of features in symbolic applications and their languages is large, the number of corresponding possibilities for microlevel architectures is very large. Table 1.9 presents some microlevel architectures and the features that they support.

Five microlevel architectures are discussed in detail in this section: stacks, data tags, garbage collection, pattern matching, and unification. The remainder

TABLE 1.9 **Functions that Can Be Supported by Microlevel Architectures.**

Function	Example Architectures
Function calls and recursion	Hardware stacks Register windows Fast memory techniques
Data typing	Memory tags Concurrent tag-checking hardware
Sorting	VLSI sorter
Set intersection	Marker-passing systems such as NETL
Pattern matching	Systolic arrays Content addressable memories Finite state automata
Best-matching	Value-passing system such as Thistle, neural networks
Garbage collection	Hardware pointers Reference counters

of the section is devoted to a discussion of emerging technologies employed in the construction of microlevel hardware.

Hardware Stacks and Fast Stack-Access Techniques. Stack architectures support function calls. This is especially useful for Lisp and other functional programming languages. The Symbolics 3600™ computer [161] contains three stacks: the control stack, the binding stack, and the data stack, used to support tail recursion and shallow binding. To speed access to the control stack, up to four virtual-memory pages of the stack are held in a dedicated fast-access 1K-word memory referred to as the stack buffer. The stack buffer contains all of the current function environment or frame plus as many of the older frames as fit. A second stack buffer contains an auxiliary stack for servicing page faults and interrupts without disturbing the primary buffer.

In ALPHA, a commercially available Lisp computer produced in Japan [94], a hardware stack is divided into four 2K-word physical blocks. Variables or arguments of a Lisp function are stored in different locations of cell space or stack. A great deal of time is required to search for a free variable, especially in the case of deep binding. To speed the evaluation of functions, the hardware stack is designed to support value caching. The value of a variable is fetched from the environment stack when it is first accessed during a function evaluation, and is stored in the value cache. Subsequent accesses will refer to values in the value cache. When the function exits, the variables it used in the value cache are marked invalid. Virtual stacks are also used to avoid the overhead of using a single stack and having to swap the entire stack when process switching.

Fast stack operations are also useful for implementing Prolog. Data is often pushed on and pulled from stacks in backtracking operations. However, the frames of a caller clause may be deeply buried in the stacks; hence, a stack architecture may not be adequate. In the Personal Sequential Inference Machine (PSI, a product of the Japanese Fifth Generation Computer Systems Project), a cache memory is employed. Several operations suited to stack access are carried out in the cache memory.

Tagged Memory.   A conventional von Neumann computer does not distinguish between data and program. Both are stored as fixed-size binary words. Meaning is not inherent in the contents of storage but defined by the program manipulating the storage. A tagged architecture, however, relies on self-identifying representations at all levels of storage. Although tagging has been employed since the 1960s, early design philosophy considered tagging a relatively unimportant and expensive peripheral concept. Tagged memory is now a key feature in many symbolic processing computers. Symbolic architectures often require identification of different physical and abstract data types: for example, integer, character, event, and garbage. During processing, the different operands employed in the computation must be identified. The tagging of data improves real-time type checking.

The most common hardware support for data tagging is the allocation of extra bits in each word to represent its data type. Data type checking at run time may be supported by additional hardware and overlapped with regular processing. The speed of a symbolic computer often depends on how effectively it emulates a tagged-memory architecture [50]. Special hardware for data type tagging has been estimated to improve the performance of Lisp computers by as much as an order of magnitude. Data tagging also supports garbage collection, facilitates better register utilization, reduces memory traffic, and simplifies the design of cooperating parallel processors and specialized functional units [71]. Data tagging is also essential in untyped languages, since the programmer does not specify the data types of instructions. For instance, the programmer need not specify the type of an add instruction as integer, long, real, or double. The type of adder used at run time depends on the types of the operands, which are specified by tags and detected at run time.

Data tags can be used to represent information other than data types. In the Classifier Machine [31], no addresses are used at all; instead tags are used to connect classifiers to each other. This no-addressing technique makes the Classifier Machine startlingly different from classical von Neumann architectures.

Garbage Collection.   Garbage collection is the process of identifying memory cells whose contents are no longer needed by the computation in progress.

In this case, memory cells are contiguous groups of at least one memory word. The process marks cells as available for future use and compacts free memory into contiguous blocks. It has been estimated that 10 to 30 percent of the execution time of large Lisp programs is spent on garbage collection. As garbage collection often requires large continuous segments of time from the CPU, the impact on interactive or real-time systems is great.

Early techniques for automatic garbage collection centered around the use of reference counts [150]. Each cell had an extra field that indicated the number of times the cell had been referenced. The reference count was updated each time a pointer to the cell was created or destroyed. When the reference count reached zero, the cell could be reclaimed as garbage. These techniques were intuitively simple and distributed the processing overhead evenly throughout the task processing time. However, large amounts of space and time may be required for the use of reference counts, although some of the overhead can be shifted to compile time [19]. Furthermore, cyclic structures cannot be reclaimed. Generation scavenging is an important technique which reduces the overall rate of garbage collection by using different garbage collection rates for memory areas of different ages (or generations) [233].

More recent research has focused on parallel garbage collection methods. Parallel garbage collection is garbage collection that is performed concurrently with program execution. Two processing entities are involved: the *mutator* and the *collector*. The mutator is responsible for program execution, and the collector for garbage collection. The techniques often use coloring cells [57, 127], or divide memory space into (two) distinct regions [17, 142]. In the first case, tagging can be a useful microlevel hardware feature. Parallel garbage collection processors can be designed with very simple and fast components and hence are not a bottleneck in the system. Hibino [102] proposed a design whose collector processor cycle time was 200 ns—6 times faster than typical processor cycle times.

Pattern Matching Hardware Support. Addition and multiplication are the mainstay of scientific computations. Similarly, pattern matching is the basic operation of symbolic processing. Two of the major tasks for which a pattern matcher may be employed are finding entries in a database and choosing the next operation to execute. For example, determining the applicable rule in a production system is a pattern matching problem. Empirical results show that 90 percent of execution time in a production system used for an expert system can be spent in the matching phase [75]. Therefore, hardware pattern matching support can simplify the programming task and improve run-time efficiency.

In most symbolic representations, symbols are represented by strings. Conventional string-matching hardware can be classified into four categories. The first category is associative memory, which although straightforward does not easily handle strings of variable lengths. The second category is the cascaded logic memory array, also called a cellular array [137]; each character in a pattern-string is stored in a cell and is compared with a character in the input string. The third category is the finite state automata (FSA) method that uses a transition table to perform complex string matching. Finally, there is the dynamic programming technique that uses statistical characteristics of the general pattern to determine the parameter table for proximity matching. Mukhopadhyay [165] presents a survey of techniques for hardware support of pattern matching, and more recently further techniques have been proposed [69, 220].

Pattern matching in symbolic processors differs from conventional database retrievals in that many symbolic applications contain widely varying field lengths and that uncertainties in data forbid exact matching of patterns. The pattern matching hardware for a symbolic processor must be tailored to the representation(s) for which it will be used.

When matching under uncertainties, *best-matching* is required. Best-matching structures search for the pattern which best matches the defined objective. Best-matching using associative memories has been explored [184], and neural networks also provide potential for performing best-matching (see Section 3.2.2). Kanerva's Sparse Distributed Memory (SDM) is a system designed for best-matching [116]. The proposed prototype consists of a virtual memory that is addressed by a 1024-bit address, and a small physical memory. Each word in the physical memory has a 1024-bit address field and multiple data fields. When a memory address is given, all locations in the physical memory with addresses that differ by less than 450 bits of the given address are accessed, and the corresponding data fields are combined together into a single response. The 450 bits are chosen so that approximately 0.1 percent of the memory words will respond on the average for a physical memory of 4 Mwords.

In a semantic network representation, pattern matching and other functions can be performed in parallel using a marker passing operation [68]. A high-bandwidth communication channel is important for this type of pattern matching.

In a forward chaining rule-based production system, the objects to be matched are constants, and multipattern multiobject pattern matching is required. The Rete Match Algorithm is an efficient solution to this problem [74], and a number of hardware implementations of this technique have been proposed, including tree architectures [202, 212], an SIMD Cellular Array Processor (CAP) [26], and a tagged token data-driven multiprocessor [78]. The key architectural requirements to support the Rete Match Algorithm in parallel production systems are a memory to maintain information across multiple recognize-act cycles, and the proper granularity of parallelism [86].

Unification Hardware.   Unification, a form of pattern matching, is the fundamental technique in logic programming. It determines if two terms can be made textually identical by finding a set of substitutions for variables in the terms, and replaces all occurrences of each variable by that variable's substitution. In general, both terms to be resolved in unification are allowed to contain variables; hence, unification can be thought of as a bidirectional pattern matching operation [126]. Since unification is applied extensively and is known to consume over 60 percent of the execution time in sequential executions, additional hardware or firmware support is desirable.

The primitive operations in unification are: (1) search for the called clause, (2) fetching of arguments of the caller and called predicates, and (3) examination of equivalence of arguments. Fast memory access is required to carry out unification in hardware. To support dynamic memory allocation, an efficient garbage collection technique is also required. Finally, hardware support for data type checking can also speed performance.

Research in this area has concentrated on string-matching hardware [204], uniprocessor machines [170], special unification chips [188], and pipelined unification [169]. To reduce the required memory space and improve performance, structure sharing [29] and techniques for structure copying [156] have also been explored.

The Parallel Inference Engine (PIE) developed at the University of Tokyo employs special hardware, referred to as UNIRED, for unification and reduction. UNIRED is characterized by the following features: tagged memory, high-speed local memory that can be accessed in parallel, parallel hardware stacks, and dedicated internal busses. The unify processor fetches a goal from a memory module and candidate clauses from definition memory, attempts to unify them, generates new goals, and returns the goals to the memory module.

Parallel unification is also an area of great interest. Parallel unification can be performed either by simultaneously unifying each term pair in two atoms, or by finding many possible unifications concurrently [164]. Unfortunately, the unification problem is proven to be log-space-complete in the number of processors. This means that it is not possible to perform parallel unification in $O(\log_k n)$ time using a polynomial number of processors, where k is any constant and n is the total number of nodes and edges of the directed acyclic graph representation of the clauses. It has also been shown, however, that near-linear speed-up can be achieved in parallel unification. Thus, from a practical perspecive, unification algorithms are parallelizable [237]. Array architectures for parallel unification have been proposed, such as the Cellular Array Processor (CAP) [26] and others [204]. A mesh-connected array of unifiers has been proposed to exploit AND-parallelism in unification and may achieve superlinear speed-up [203].

VLSI and Emerging Technologies. Very Large Scale Integrated circuit or VLSI technology has been a major factor in the cost reduction and increased functionality of symbolic processing systems. The high degree of space-time complexity in AI and symbolic computations has necessitated the use of both parallel processing and VLSI technology. The development of specialized microelectronic functional units is among the major objectives of the Japanese Fifth Generation Computing project [77], MCC [73], and DARPA's Strategic Computing projects [45].

VLSI technologies allow a single chip computer to be realized. Although many functions can be implemented on a single chip, the size of the chip and the number of input/output pins are usually limited, and chip area has to be carefully allocated to achieve the highest performance. The Reduced Instruction Set Computer (RISC) is a highly popular design approach that carries out only the most frequently used instructions in hardware and the rest in software [179]. The small chip area required by the control unit of a RISC computer enables it to be placed on a single chip. It has been found that a large number of registers on a single chip is a good design tradeoff that reduces the overhead of swapping registers in context changes.

SOAR, or Smalltalk on a RISC, is a project to develop a RISC chip for Smalltalk-80 [235]. SOAR design details a 32-bit NMOS microprocessor containing 35,700 transistors that runs roughly 400 ns per instruction. Cycle time may be decreased to 290 ns if 3-micron lines are used. FAIM-1 is another project that designs RISC chips to carry out specialized functions in the system [12].

Specialized symbolic processing functions can also be carried out in hardware. An example is the Texas Instruments Lisp chip with over 500,000 transistors on a 1 cm^2 chip, implementing approximately 60 percent of the functions of a Texas Instruments Explorer [147].

Many existing computers for symbolic computation employ VLSI technology. The major building block of the Connection Machine CM-1 and CM-2 is a custom VLSI chip containing 16 processor cells [103]. The chip is implemented on a CMOS die about 1-12cm212 in area. There are approximately 50,000 active devices. Although each addition takes approximately 21 μs, an aggregate maximum rate of 2500 MIPS or 5000 MFLOPS can be achieved with 64K processors implemented in 4K processor chips and 4K floating-point chips.

Cellular array structures are a widely studied technique that can take advantage of the available VLSI technology and exploit data-level parallelism in many symbolic processing problems. The Cellular Array Processor (CAP) is an example in this class [26], and systolic cellular hardware design has been explored for performing unification [204].

In the near future, other emerging technologies may become cost effective for implementing computers for symbolic processing. They include GaAs circuits, wafer-scale integration (WSI), and optical computing techniques.

GaAs circuits have design requirements similar to those of conventional semiconductor circuits, but are much faster. Switching speeds on the order of 10 ps have been reported in High Electron Mobility Transistor (HEMT) GaAs circuits [130]. Gate propagation delays are typically on the order of 200 ps. Unfortunately, the scale of fabrication of GaAs circuits is subject to limited

TABLE 1.10 Microlevel Architectures and Their Significance.

Architecture	Significance
Stacks	For function calls and value binding; more than one stack may be used
Tagging	Data type checking; up to an order of magnitude speed improvement for Lisp
Garbage collection	Reclamation of usable storage; accounts for 10 to 30 percent of run time in Lisp
Pattern matching	Fundamental operation; up to 90 percent of execution time in production systems
Unification	Type of pattern matching for logic; over 60 percent of execution time in sequential logic programs

size and greater numbers of defects; these problems increase costs, already high [154].

WSI is the integration of multiple circuits on the same wafer in order to avoid the high performance and cost penalty of off-chip connections. Unfortunately chip yield in manufacturing is low; the yield of complete wafers would be negligible, although techniques such as Focused Ion Beam (FIB) repair increase yield. Low yields make WSI more suitable for implementing distributed knowledge representation schemes, such as neural networks, in which the loss of a small fraction of the distributed knowledge may not be critical. When GaAs and WSI are combined, it is possible to implement a 32-bit GaAs processor on a single wafer [154]. Design of such a processor for production systems has been explored [131].

Optical processing can be used in the same way as silicon gating. Switching speeds on the order of 5–10 ps are possible in optical gates. Optical circuits do not have the penalty of capacitance [244] and can communicate with low propagation delay and no interference. Two forms of optical processing have been developed: arrays of light rays [21, 79] and optical crossbars [148]. The use of optics in storage media may greatly improve performance in symbolic systems with erratic memory behavior. It has been proposed that optical techniques have the potential for improving data- and knowledge-base processing speeds by two orders of magnitude [21].

Table 1.10 summarizes the microlevel architectures presented in this section. Each type of architecture is accompanied by an example application, and remarks about its significance in supporting that application.

3.2.2 Subsystem-Level Architectures. The subsystem-level architecture represents an intermediate level between microlevel and complete-system designs. In this section, we identify three techniques for classifying the subsystem-level architectures that support different types of knowledge processing.

Three different types of subsystems will emerge from this analysis: data- and knowledge-base machines, inference engines, and neural networks.

Control flow, data flow and demand flow are the names of three important approaches that are used in the design of subsystem-level architectures [236]. Their definitions and relative advantages and disadvantages are summarized in Table 1.11.

Knowledge representation plays an important role in the way that a complete system integrates its components. The Japanese FGCS project emphasizes logic representations and stresses the separate development of knowledge-base and inference engine, which are integrated only after development [9]. This approach to the system architecture is illustrated in Figure 1.6a.

The Connection Machine [103] is designed with an alternative perspective and reflects a semantic-network knowledge representation and memory-based reasoning. All the knowledge in the system is embodied in a collection of facts; no intentional knowledge or rules are employed. The database is implemented directly on the architecture, and inferences are carried out in software and

TABLE 1.11   Control-Driven, Data-Driven, and Demand-Driven Computations.

	Control Flow (control-driven)	Data Flow (data-driven)	Production (demand-driven)
Definition	Conventional computation; token of control indicates when a statement should be executed	Eager evaluation; statements are executed when all of their operands are available	Lazy evaluation; statements are only executed when their result is required for another computation
Advantages	Full control	Very high potential parallelism	Only required instructions are executed
	Complex data and control structures are easily implemented	High throughput	High degree of parallelism
		Free from side effects	Easy manipulation of data structures
Disadvantages	Less efficient	Time lost waiting for unneeded arguments	Does not support sharing of objects with changing local state
	Difficulty in programming	High control overhead	Time needed to propagate demand tokens
	Difficulty in preventing run-time error	Difficulty in manipulating data structures	

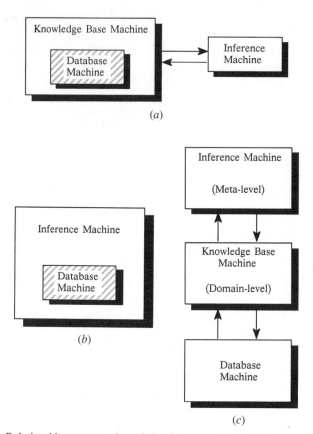

Figure 1.6 Relationships among knowledge-base machine, inference machine, and database machine: (*a*) Japanese FGCS model; (*b*) connection machine model; (*c*) hierarchical model.

message exchanges. This perspective on system design is illustrated in Figure 1.6b.

The last design perspective is hierarchical in nature, as shown in Figure 1.6c. An inference machine handles all meta-level inferences, while a knowledge-base machine uses rules to manipulate domain knowledge. A separate database machine carries out search and selection operations on the domain knowledge.

Database Architectures.   Early studies of specialized database architectures emphasized the use of parallelism; such designs include CASSM [215], RAP [178], and DIRECT [56]. Some later systems, such as the Connection Machine, are designed with massive parallelism for symbolic applications and can be used for a number of specialized database functions [240]. Hawthorne and Dewitt [93] present a comparison of several early parallel database comput-

TABLE 1.12 Differences between Databases and Knowledge-Bases.

Issue	Database	Knowledge-Base
Contents	Collection of data facts	Higher level of abstraction
		Classes of objects
Complexity	Stored items are simple	Stored items are complex relations
Time and Dependence	Data changes over time	Knowledge changes less frequently, except for situation knowledge
Size	Large number of facts	Fewer relations on classes of objects
Use	Operational purposes	Analysis, planning, etc.

ers, and Malabarba [145] gives a survey of commercially available database computers. Commercial manufacturers of database computers include Britton-Lee, Hitachi [227], International Computers Limited [15], and Teradata [171].

These systems may not function well for some applications because the bottleneck in database retrieval is disk input/output and not processor cycles [214]; hence, intelligent database processing using application-dependent knowledge and indexing may be preferable to massive parallelism [214].

Knowledge-Base Architectures. The objectives and requirements of a knowledge-base computer are different from those of a database architecture. The most prominent differences are noted in Table 1.12. An evolving knowledge-base subsystem should include a mechanism that either rejects inconsistent data or rules or performs truth maintenance when such data or rules are inserted. Support for the inference mechanism is also desirable and may take the form of an automatic rule-selection mechanism, logic support, or special hardware for operations such as joins and projections of relations. Finally, the interface to the host computer should be intelligent and may draw on the resources of the knowledge-base.

The issues in the design of a knowlege-base computer include:

Storage and Manipulation of Intentional and Extensional Data.

Extensional data are data that represent facts, that is, statements with no quantified variables. Intentional data are general facts or rules. In

Figure 1.6c, the extensional and intentional data are stored and manipulated separately. In Figures 1.6a and 1.6b, they are processed by the same physical entity.

Relational Operations.

As already mentioned, hardware and software support must be provided.

Hierarchical Storage.

Knowledge can be classified into categories by its degree of generality: from fact through the most general metaknowledge. Access characteristics are highly dependent on the type of knowledge. A hierarchical storage for metaknowledge may be used to exploit the knowledge structure efficiently.

Access-Control Algorithms.

A knowledge-base subsystem may be required to control access to its contents for security, integrity, and concurrency control.

Parallel and Distributed Processing.

Database updates are history-sensitive. The choice of the best technique(s) for exploiting parallel and distributed processing in knowledge-base systems remains an open issue. The data flow approach may be helpful.

In many applications, a database computer may be integrated with an existing host to form a knowledge-base computer: for example, the Intelligent Information Resource Assistant developed at System Development Corporation [118]. The system consists of a Britton-Lee IDM 600 backend database computer, a Xerox 1100 workstation acting as a logic-based deductive engine, and a VAX 11/780 computer as a file and print server. This prototype has demonstrated that a knowledge-base computer can be easily constructed from existing hardware components.

The Japanese FGCS project has developed Delta, a combined knowledge- and database computer. The motivation behind a specialized design is that the integrated system calls for a performance that cannot be met by commercial components. The system consists of a control processor, a relational database engine, and a hierarchical memory. The control processor translates commands (received from an interface processor communicating with a parallel inference machine) into subcommands that the relational database engine can obey. The relational database engine communicates through a data path with the hierarchical memory, which is composed of semiconductor and magnetic disk storage [166].

Hardware Support for Inference Engines.   Inference engines are key components of knowledge processing architectures. Their structure is highly dependent on the knowledge representation and programming language employed.

An important consideration when designing hardware support for inference engines is the architectural support for searching the knowledge base. Deduction and search have been the dominant paradigms for machine inference over the last 30 years. As discussed in Section 2, the best approach to search is the development of superior heuristics combined with efficient hardware. There follow some of the key reasearch issues in search architectures. (More general issues on parallel processing were presented in Table 1.6.)

Prediction of Performance.

A major difficulty in developing search-based inference engines is our inability to estimate their performance without execution of the search. This is because of both the nondeterministic nature of searches and anomalies in parallel search algorithms [238].

Space-Time Tradeoff.

There is a space-time tradeoff in using heuristic knowledge. Very accurate heuristic functions may require greater amounts of space and computation time than less-accurate ones. This relationship must be understood for the design of effective search subsystems.

Architectural Support for Machine Learning.

Heuristic functions used in search algorithms should be improved over time by automatic learning methods. Architectural support for nonmonotonic processing may be helpful.

Management of Large Memory Space.

Heuristic search strategies may require large amounts of memory space to store intermediate results and the heuristic information. The tradeoff between the effectiveness of a search strategy and its overhead must be considered. Techniques for efficient memory management tuned to the search behavior are vital in such a system.

Granularity of Parallelism.

The proper choice of granularity is difficult to determine at design time because of the dynamic nature of the problem. Granularity may have to be varied at run time when more is known about the application.

Scheduling and Load Balancing.

Because many search problems are nondeterministic, direct mapping of a sequential search strategy into a parallel system may not produce the best performance. The key to effective scheduling is the proper order of execution; it is not enough to keep the available processors busy. Counter to intuition, depth-first search is sometimes preferable to best-first search in real systems when memory constraints are considered [249]. Conditions for the sequential search strategy may also have to be relaxed to accommodate the architectural constraints [238]. For instance, when performing a heuristic search, it may not be desirable to select the subproblem with the minimum heuristic value when the overhead of selecting subproblems distributed in local memories of multiple processors is high. Selecting the local subproblem with the minimum heuristic value will suffice in most cases.

Communication of Pruning Information.

When the search space is explored in parallel, excess computation may be performed if pruning information cannot be shared among the processors. In general, a tradeoff exists between search efficiency and communication overhead.

Microlevel hardware features discussed in Section 3.2.2 are often components of the inference subsystem. Data flow, control flow, and demand flow techniques (see Table 1.11) have been employed in designing hardware supports for inference engines for Lisp and logic [10, 114, 167]. An example of a combined data-driven and demand-driven approach is demonstrated in ALICE [43] and Rediflow [117]. Although the theory behind data-driven and demand-driven computations appears very promising, it offers no clear solution for design problems such as the selection of the granularity of parallelism. Furthermore, the proper tradeoff among demand flow, data flow, and control flow in an inference computer remains an open problem.

Artificial Neural Networks.   Automated computation based on a neural network design philosophy originated many years ago. One pioneering contributor was Rosenblatt, who developed the concept of perceptrons [189]. Recently, a great deal of interest has revived in this area. Neural networks, besides capturing the imagination by their resemblance to the human brain, offer great potential for automated learning. Because information is stored in the connections between neurons, neural networks may also be called connectionist systems.

A neural network can be viewed as a collection of eight components: a set of processing units, a state of activation, an output function, a pattern of connectivity, a propagation rule, an activation rule, a learning rule, and an environment [190]. These components are described in Table 1.13. The first

TABLE 1.13 Neural Network Components.

Component	Description
Processing units	Three types: input, output, and hidden
State of activation	Vector of the activation levels of the units in the system
Output function	Function, on the activation level of a unit, which produces the unit's output; may vary between units, but most systems are homogeneous
Pattern of connectivity	The connections determine the performance and function of the system
Propagation rule	A way of combining outputs of units and pattern of connectivity into an input for each unit; usually is a weighted sum of the inputs and the excitatory $(+)$ and inhibitory $(-)$ connection strengths
Activation rule	A function for determining the new activation level of a unit on the basis of current activation and inputs to the unit
Learning rule	Three types: develop new connections, abandon old connections, modify weights; only last has been pursued; almost all learning rules based on Hebbian learning rule
Environment	In which computing engine functions

design consideration is how knowledge should be represented in the set of processing units. There are two possible approaches: local and distributed. A local representation allocates each concept to its own unit. Hopfield's network for solving the traveling salesman problem is an example of a local representation in a neural network [110]; however, although it may solve the problem efficiently, it does not perform learning. In contrast, each unit in a distributed representation participates in the storage of many concepts [106]; in this way, a processing unit in a distributed representation stores an abstract feature at the microlevel [190].

Each unit in a neural network has an associated activation level that may be analog, analog and bounded, or discrete. A unit's output is a function of its activation level—usually a threshold or sigmoidal function. The output of each unit is distributed to a set of processing units. The influence of a unit's output on another unit is determined by the weight of the connection between the units. Finally, a node's activation is modified by the activation rule, a function of inputs, their relative connection strengths, and the current activation level.

Classes of neural networks may be distinguished by their learning paradigms [190]. Networks in the associative learning class learn associations between inputs and their desired outputs, whereas regularity detectors learn to recognize interesting patterns in the input. The literature on this topic includes overviews of designs for neural networks of both classes [4, 144] and an extensive review of learning techniques in alternative neural network strategies [106, 107].

The use of neural networks has been proposed for a variety of applications, although they are not yet widely used in practice. A neural network can easily be adapted to form an associative memory with capability for inexact matching with speed and accuracy in the presence of noise [53]. Neural networks can also be used for speech recognition [198] and for vision tasks such as letter recognition [153]. Neural networks have been applied to combinatorial search problems such as the traveling salesman problem [110, 111], although the solutions of large search problems by neural networks are not yet of as high a quality as those produced by good digital algorithms [20]. Neural networks are also useful for some strategy-learning tasks, and experiments have been performed with a balance-control system [11].

The design of neural networks is still plagued by a number of difficult problems. Firstly, a neural network must be trained for a given application and must be retrained when the system parameters change; there is no systematic method for taking a neural network trained for one application and generalizing it for another. Secondly, all known learning algorithms require extensive training for good performance: for example, over 9,000 learning sweeps of 40 phases each are required to train a shifter network using the Boltzmann Machine learning algorithm [107]. Moreover, learning speed depends on the configuration of the neural network, which cannot be selected systematically. Dedicated hardware to emulate various configurations of neural networks and map the inner-loop operations into analog instead of digital circuits can improve the learning speed significantly, however. An improvement in speed by a factor of a million has been demonstrated using hardware emulation [8]. Thirdly, extremely large neural networks cannot be built with the current technology, in spite of numerous industrial efforts (for example, AT&T's R3 chip has a word size of 1048 bits and performs learning by back propagation). It has been estimated that by the year 2010 it will be possible to build 4.5 million neurons onto one wafer using 0.25 micron lithography and 250 million transistors [62]. It is therefore unlikely that a complete symbolic processor as intelligent as the human brain (with over 10 billion biological neurons) can be built with neural networks alone in the near future. Finally, because neural networks are limited in size, the problem must be partitioned so that part of it can be learned by traditional method and the other part by neural processing. However, knowledge representations in neural networks are drastically different from procedural and declarative representations in conventional symbolic processors, and systematic methods to integrate them are still missing.

4 COMPLETE SYSTEMS

Up to this point, we have focused on individual techniques for effective symbolic computations. In this section, *complete systems* for symbolic applications are classified into single-processor systems, parallel computers, and connectionist systems, and, as before, we analyze the role of individual components in the design of complete systems.

There are two prevailing trends in designs. With the increasing complexity of many symbolic processing applications, the emphasis of knowledge representations, software languages, and support is shifting away from strongly procedural techniques and toward distributed and declarative computing environments. Parallel processing is another obvious trend in the design of computer systems for symbolic processing.

Depending on its starting point, system design can be classified as following either a top-down or a bottom-up approach. Top-down design begins with the specification and analysis of the application. A knowledge representation is then designed and tailored to the needs of the application. Finally a language is designed, and the system is mapped into software and microlevel and subsystem-level structures. This process may be iterated many times if the functional requirements of the design cannot be implemented using the current technology. The FGCS project used a top-down design approach.

In contrast, for a bottom-up design, the designer first selects the technology and design options, such as data- or demand-driven calculation. A language and application suitable for implementation with these options are then sought. Like the top-down design process, this process may be iterated many times if the selected applications cannot be suitably supported by the original design. ZMOB [243] and the Butterfly Multiprocessor [27] were designed using the bottom-up technique.

The designer may also compromise between the top-down and bottom-up approaches, resulting in a middle-out design approach [239]. The middle-out approach begins with the selection of an appropriate and well established knowledge-representation scheme that is suitable for the application. The representation scheme should already have a well developed programming environment that can be modified later according to the needs of a specific system. Primitives for concurrent execution may be added to allow users to annotate concurrent tasks in the application. The hardware and software architectures are then designed. When selecting features to incorporate, the designer considers structures previously designed and used. The middle-out approach can be further subdivided into top-first and bottom-first approaches. In a top-first

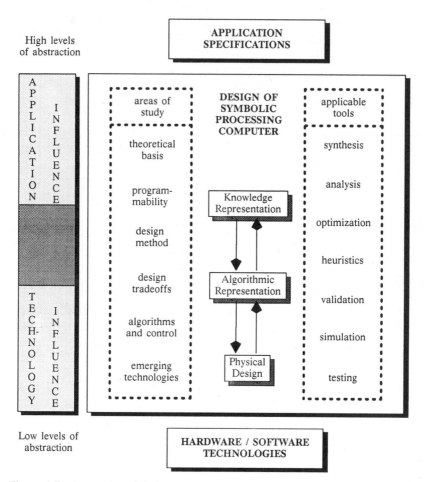

Figure 1.7 Perspective of design of a computer for symbolic processing applications.

middle-out approach, the designers start with a well defined knowledge representation scheme and tailor it to the given application. The architectures of ALICE [45] and FAIM-1 [12] were designed according to this philosophy. In a bottom-first middle-out approach, the designers first develop the architecture to support a well established representation scheme, then map the application to the chosen scheme. DADO [212] was developed using this approach.

The process of designing a computer for symbolic processing is iterative (see Figure 1.7). In mapping from application specifications to hardware and software technologies, the designers repeatedly propose and select knowledge representation schemes, algorithms, and physical design until a feasible and cost-effective mapping is found. Applicable design tools include methods for synthesis, analysis, optimization, heuristic design, validation, simulation, and testing. During the design process, the designers must address issues related to theory, programmability, design methodology, design tradeoffs, effective control, and emerging technologies.

4.1 Single-Processor Symbolic Computers

Lisp has enjoyed the longest tenure in the mainstream of languages for symbolic processing, and has led to the greatest number of computers devoted to its execution. The earliest Lisp computers were PDP-6s; they were followed by the PDP-10s and PDP-20s of Digital Equipment Corporation [152]. The half-word instructions and stack operations of these computers were particularly well suited to Lisp. A great deal of work has also been done to improve garbage collection on the PDP-10s and PDP-20s.

The MIT AI Laboratory introduced a Lisp computer called CONS in 1976 [124], to be followed in 1978 by CADR, their second generation Lisp computer. This computer was the basis for many commercial Lisp computers—the Symbolics LM2, the Xerox 1100 Interlisp computer, and the Lisp Machines Incorporated Series III CADR, all introduced in 1981. Table 1.14 shows some notable and more recent commercial and research-oriented Lisp computers.

In addition to special hardware for improved efficiency, commercial Lisp computers also provide an integrated software development environment, such as KEE or ART, that allows programmers to develop, debug, and maintain large Lisp programs. A recent trend is to implement these development environments efficiently in software (rather than in microcode) on high-speed general purpose workstations. It is likely that in the future special purpose Lisp computers will be used to execute Lisp programs rather than to develop them.

The design of special purpose Lisp computers continues to be a popular research area. Many experimental computer designs have been reported [55, 83–85, 168, 182, 193–195, 222]. A single chip design to support Scheme, a dialect of Lisp, has been demonstrated in the Scheme-79 and Scheme-81 chips [219]. Scheme-79 was limited by its implementation of a register file and slow programmable array logic, but Scheme-81, a redesign, has proved to be much faster.

The popularity of rule-based systems arose at a time when parallel processing was highly popular. As a result, most architectures designed for supporting ruled-based languages have been parallel computers. However, the study of single-processor rule-based systems is still important because a parallel system is limited by the speed of its inferences, and the building blocks of a parallel system are likely to be single-processor symbolic computers.

The three most notable designs for executing logic languages are SRI's pipelined Prolog processor [226], and the Personal Sequential Inference (PSI) Machine and the Cooperative High-Speed Inference Machine (CHI), both developed by the Japanese FGCS Project [223, 230]. PSI, an integrated workstation with an execution speed of 30 KLIPS, was intended to be a software development tool for the project; it has since been redesigned into PSI-II, which

TABLE 1.14 Notable Sequential Symbolic Processing Machines.

Machine	Year/ Status[a]	Primary Language	Features
Scheme-79, Scheme-81	1981/PO	Scheme	Single hip; tail recursion Lexical scoping
Lisp Machines Lambda	1983/CA	Zetalisp, LMLisp	NuBus–multiprocessor capability Stack orientation
Symbolics 3600	1983/CA	Zetalisp, Flavors	Tagged memory Stack buffer Hardware garbage collection Single-address instruction
Tektronix 4400	1984/CA	Smalltalk, Franz Lisp	Lower-end AI workstation Motorola 68010/20

TI Explorer	1984/CA	Common Lisp	NuBus; tagged memory
			Microprogrammed
			Mega-chip version contains 60%
			  of processor in 1 chip with
			  550,000 transistors [147]

Fujitsu ALPHA	1983/CA	Utilisp	Value cache; hardware stack Virtual stack
FGCS PSI-I, PSI-II	1985/CA	KL0 (Logic)	PSI-II; Cmos-GaAs; TTL 200 ns cycle time; cache; stacks Copying for structure data Hardware unification Tagged data Interpretive execution 150 KLIPS average speed
FGCS CHI-I, CHI-II	1986/PO	Current Mode Logic (CML)	CHI-II; CMOS-GaAs; TTL 170 ns Cycle time; Cache About 400 KLIPS average speed for append
SRI's Pipelined Prolog Processor	1984/SI	Prolog	Pipelined execution Instruction execution units Microprogrammed controller Interleaved memory FCFS module queues
SOAR	1984/PO	Smalltalk-80	RISC Expensive procedure calls in Smalltalk-80 Tagged and untagged instructions Large number of registers Automatic storage reclamation Direct object addressing Fast type checking

[a] Status   CA: Commercially available   PO: Prototype operational
          HS: Hardware simulated   UC: Under construction
          SI: Simulations completed   PD: Paper design

has an average performance of 150 KLIPS [230]. CHI was designed with speed in mind, using a less constrained technique. Like PSI the original CHI was redesigned, becoming CHI-II, with an estimated improvement in performance from 280 to 400 KLIPS for the *append* operation [230].

Single-processor support for production systems has focused on additional data memories [132] and RISC architectures [75].

Object-oriented languages have been implemented on sequential processors. The computers of the Xerox 1100 family were among the first workstations on which Smalltalk-80 was installed. Smalltalk-80 has also been implemented on a single chip using the RISC approach in the Smalltalk-On-A-RISC (SOAR) project [234]. SOAR has no microcode or fine-grained addressing hardware, and few multicycle instructions.

4.2 Parallel Symbolic Processors

In this section, we classify parallel symbolic processors in terms of their representations or programming techniques. Section 4.2.1 discusses the methods of communication and synchronization used in parallel symbolic processors. The rest of this section is devoted to discussions of the use of parallel symbolic processors for functional, rule-based, and object-oriented representations.

4.2.1 Communication and Synchronization

Communication.   Message passing, marker passing, and value passing are the three predominant communication methods used by parallel symbolic processors.

Message passing is the conventional method of communication: information is formulated into a message and sent over the interconnection network. The computing elements are generally complex, and communication costs are high. Nonetheless, message passing is popular and has been used in many parallel symbolic processors.

Marker passing is the transfer of single-bit markers from one processor to another. A marker indicates the presence of a given property, and a set of markers indicates the conjunction of a set of properties. Each processor is simple and can store a few distinct marker bits. There is never any conflict: if two markers arrive at the same destination, they are simply ORed together. The basic inference operation is, therefore, set intersection. Marker passing is especially suitable for implementing semantic networks and for recognition problems in hardware. One such system is NETL [64, 66]. The Connection Machine was originally designed as a marker-passing system, but was later modified to use more powerful processing and communication methods and to support virtual processors [68].

The third method of communicating information is value passing, in which information is passed as continuous quantities. Only simple operations are performed on the numbers. The salient feature of this approach is that if several values arrive simultaneously at a single point, they are combined into a single value by a mathematical function, eliminating conflict in information transfer.

Examples of value-passing systems include the Boltzmann machine [66] and other neural computation systems [109]. Iterative relaxation techniques for problems such as low-level vision, speech understanding, and optimization all seem to be suited to value-passing architectures.

Synchronization.   Synchronization is the control of concurrent access to shared items in a parallel processing system. It is important in message-passing systems, because such systems do not prevent competition for shared resources. Synchronization is not critical in marker- or value-passing systems because their predefined methods of passing markers and combining values prevent conflict.

Synchronization is important when data items are shared. In a program written in a procedural language, the order of statements dictates the order of execution; if two statements share a variable, the first must be executed before the second. Hence, when data sharing is necessary, synchronization control is implicitly defined by the order of statements. In contrast, when a program is written in a pure declarative language, the order of execution is not defined; when two statements share a variable, the order of execution is indeterminate. Thus explicit specification of synchronization control is needed when data is shared. Unfortunately, synchronization control cannot be specified explicitly in most declarative languages.

Synchronization can be carried out by shared memory or message passing. Shared memory is popular and has been used in systems such as Aquarius [54], the Concurrent Lisp Machine [217], the Concert MultiLisp Multiprocessor [92], and the Parallel Inference Engine [82]. Blackboard architectures and shared variables are two techniques that can be used for shared-memory synchronization.

The blackboard model was originally developed for abstracting features of the HEARSAY-II speech understanding system [61, 176]. There are three components: a set of knowledge sources, a blackboard, and control. The knowledge for processing the application is partitioned into separate knowledge sources. The data, including input/output and partial solutions, are stored in the blackboard. The blackboard may be partitioned into smaller blackboards, forming a hierarchy of solution spaces. To reach a solution, the knowledge sources manipulate the data in the blackboard, which is the only communication mechanism between them. A monitor is present to ensure that only one knowledge source is changing the blackboard at any time.

A more powerful blackboard architecture has been proposed in which control information (or metaknowledge) is allocated a separate blackboard [95]. This approach is more flexible and suits the nondeterministic nature of symbolic processing.

Synchronization may also be achieved through shared variables. Lisp languages that have been modified for parallel processing often contain shared variables for synchronization. Multilisp provides a mechanism that allows a process to wait for needed values to be generated; as in other languages, procedure activations may not be well nested, and a process can terminate prior to an activation that it began. This problem must be addressed by the programming system [91].

Single assignment languages such as pure Prolog and pure data flow languages do not require careful synchronization of shared variables since a variable may be written to only once [143]. Prolog delays process reduction until enough information is available to make an effective decision. In Guarded Horn Clauses (GHC) [231], the current kernel language of the Japanese FGCS project, OR-parallelism is not exploited, and strict synchronization rule suspends a subgoal if it tries to modify its parent environment. This simplifies implementation, but the resulting language is less expressive [221].

Synchronization in a message-passing system is accomplished through a protocol implemented in hardware or software. In a standard message-passing environment—in an object-oriented programming system, for example—messages may be of arbitrary complexity. Actor is a paradigm for systems with message passing of this nature [98, 100]. When an actor receives a message, it performs predefined primitive actions; in this sense, actor systems are inherently parallel. The Apiary architecture is based on actors [99]. Other message-passing systems for symbolic applications include the Contract Net system [207] and the Rand Distributed Air Traffic Control System [33].

4.2.2 *Parallel Functional Programming Computers.*
The majority of special purpose parallel processors designed to support functional languages are oriented toward Lisp: for example Concert [92], EM-3 [248], and a multi-microprocessor Concurrent Lisp system developed at Kyoto University [217]. All these systems require users to specify, to some extent, the tasks to be decomposed. Compilers for automatic detection of parallelism in sequential Lisp programs are an area of active research. Table 1.15 presents some of the more highly publicized parallel systems for functional programming.

The majority of computers designed for general purpose applications have only a few features specifically appropriate to symbolic processing. In these computers the inference engine and knowledge base are not separated and are almost exclusively implemented by sophisticated software structures. Lisp is added as one of the several languages to accompany a general purpose parallel computer. A Lisp compiler is used for decomposing tasks for parallel processing, and users must perform some task annotation. Examples of commercial multiprocessors include the Butterfly [41], the Connection Machine [39], and the Intel iPSC concurrent computer [22].

The Connection Machine has special chips containing 16-bit serial processors and router circuits. *Lisp in CM-2 allows users to specify a parallel variable (*pvar*), a first-class object with a value for each processor in the computer [40]. The primitive *pvar* can be accessed concurrently (with possible masks) by all local or remote processors in SIMD or multiple SIMD mode. CM-Lisp is a dialect of Common Lisp extended to allow fine-grained, data-oriented parallel processing; it provides higher-level data abstractions called *zappings*, which are similar in structure to arrays or hash tables. Broadcasts, reductions, and combinations can be specified.

TABLE 1.15 Notable Parallel Functional Programming Computers.

Machine	Year/ Status[a]	Primary Language	Intercon- nection	Commu- nication	Features
Butterfly	1985/CA	Multilisp	Butterfly switch	Shared memory	256 MC68000-series PEs Homogeneous, tightly coupled General purpose multiprocessor
iPSC	1986/CA	Common Lisp	Hypercube	Message passing	256 Intel 80286 and 80386 processors No shared environment User decomposes program into concurrent processes that communicate by messages General purpose multiprocessor
Connection Machine	1986/CA	*Lisp CM-Lisp	Hypercube	Message passing	Model CM-2 has 4096 bit serial processors Users annotate Lisp programs from SIMD or multiple SIMD parallel processing C*, Fortran, and Paris (CM-2 assembly language) are also supported
Concurrent Lisp Machine	1983/PO	C-Lisp	Multiple buses	Shared memory	17 MC68000-series PEs Special cell interface Control stack Garbage collector
EM-3	1984/PO	EMLISP	Modified delta network	Message passing	Listlike data-driven language 16 MC68000-series PEs Special router chip Control for function evaluation
Concert	1986/PO	Multilisp	Ringbus	Shared memory	32-64 MC68000-series PEs Network is segmented bus in shape of ring
Rediflow	1984/SI	Functional Equation Language (FEL)	Mesh or richer connections	Message passing	Demand/data-driven Loosely coupled Hardware support for load balancing Distributed garbage collection
Alice	–/UC	Hope Lisp,, Prolog	Cluster of processors, ring buffer	Message passing	Transputer as basic processor Reference counter for garbage collection

[a] Refer to Table 1.14 for explanation of status codes.

Other computers with limited support for parallel symbolic processing, such as ZMOB, are being developed at universities [243]. These computers are not just symbolic processing computers, but also general purpose computers, appropriate for both numeric and symbolic computations.

4.2.3 Parallel Logic Architectures.

This section presents parallel systems suitable for evaluating logic programs. Table 1.16 summarizes notable projects.

Unification and search are two key features by which to evaluate logic programs. Architectures that emphasize efficient search of logic programs include the BAGOF architecture [34] and MANIP-2 [138]. The MANIP-2 architecture is particularly interesting because of its emphasis on heuristic parallel search strategies.

Two significant parallel logic systems have been developed at universities. The Aquarius multiprocessor, developed at the University of California, Berkeley, couples intensive numeric calculations with symbolic manipulations [54]. Its designers used parallelism at all levels of computation and considered cost to be secondary to performance considerations. The Parallel Inference Engine (PIE), under development at the University of Tokyo, has a target of 1,000 processors, and a speed-up of 170 has already been estimated for 256 processors [163]. PIE utilizes only OR-parallelism.

Probably the most massive effort in the development of parallel logic systems has been made by the Japanese Fifth Generation Computing System project (FGCS). The project distinguishes three major development areas: problem solving and inference machines (hardware), knowledge-base management systems (software and algorithms), and an intelligent man-machine interface [162]. The project is divided into three stages. The initial stage, now complete, explored basic computer technology and processing techniques. The middle stage is for the development and construction of experimental subsystems. The final stage is devoted to the development of the complete system.

The initial-stage designs of the Parallel Inference Machines (PIM) were based on two concepts: reduction and data flow [115, 166]. The architectures for the two computers (PIM-R and PIM-D, respectively) were similar, but the evaluation techniques reflected two different philosophies. The hardware of PIM-R and PIM-D was simulated.

PIM-I is a hardware design for the intermediate stage of the FGCS project. The target speed for the 100-processor PIM-I is 10–20 MLIPS, with a target speed of 200–500 KLIPS for the individual processors [230]. The machine language for this computer will be KL1-B, which is based on Guarded Horn Clauses. The software will be developed on a network of PSI systems (multi-PSI) [113].

4.2.4 Parallel Systems for Production System Computations.

The exploration of computers for production systems has been carried out primarily at universities. Table 1.17 presents a summary of these projects.

TABLE 1.16   Notable Parallel Computers for Logic Representations.

Machine	Year/ Status[a]	Interconnection	Communication	Features
BAGOF	1984/PD	Bus	Shared memory	OR-parallelism Separate static and dynamic memory Token pool

MANIP-2	1985/PD	Global broadcast bus	Message passing	Cluster of PEs with
				   local memory
				Distributed selection
				Heuristic guiding
				   and pruning

Aquarius	–/UC	Bus and crossbar	Shared memory	Heterogeneous MIMD
				16 PEs
				Synchronization through
				   Goodman Cache
				Crossbar to shared
				   memory modules
				Special Prolog,
				   floating point, and
				   I/O processors

Parallel Inference Engine (PIE)	1984/SI	Switching network	Shared memory	100s to 1000s of
				   inference units
				Goal rewriting model
				OR-parallelism
				Sequential AND
				   processing
				Activity controllers
				   to control inference
				   tree
				Unify processors
				   connected to
				   definition memory
				   containing program

Parallel Inference Machine-Reduction (PIM-R)	1986/HS	Multistage network	Shared memory	Many inference
				   modules connected
				   to structure
				   memory units
				   through network
				Structure copying

Parallel Inference Machine Dataflow (PID-M)	1986/HS	Multistage network	Shared memory	Multiple PEs connected
				   to structure memory
				Unfolding interpreter
				Asynchronous
				   communication
				Streams for nondeter-
				   ministic control

PIM-1	–/UC	Hierarchy	Shared memory, message passing	100 PEs
				8-PE clusters intercon-
				   nected with shared
				   memory and parallel
				   cache

[a] Refer to Table 1.14 for explanation of status codes.

TABLE 1.17 Notable Parallel Machines for Production Systems.

Machine	Year/ Status[a]	Interconnection	Communication	Features
DADO1 DADO2	1986/PO	Binary tree	Message passing	DADO2; 1023 8-bit processors; 16K of user memory Two modes: MIMD and multiple SIMD Special I/O circuits
Non-Von	1985/PO	Binary tree with leaf connections Connections to LPEs	Message passing	Binary SIMD tree of small PEs Leaves are connected in mesh Large PEs connected by network, with connections to high-level nodes in tree Intelligent disk drives connected to LPEs
PSM	1986/SI	Shared buses	Shared memory	32–64 processors Parallel Rete Match algorithm PEs connected to memory modules through cache Local memory Hardware task scheduler

[a] Refer to Table 1.14 for explanation of status codes.

The DADO1 and DADO2 projects [211, 213] at Columbia University are developing a class of computers based on tree architectures. The upper-level nodes of the tree synchronize and select rules, intermediate nodes match and store rules, and the leaves act as working memory.

Another project in progress at Columbia University is the Non-Von computer—Non-Von-1 was an early version [104, 112]. Unlike DADO2, Non-Von connects smaller processing elements, which are subject to the control of large processing elements into a binary tree. Most of the pattern matching tasks that are done in the working memory have small granularity and are more suitable to be executed on a large number of small processing elements.

Finally, the PSM computer is a large-grain machine that is specifically designed to support the OPS5 system and a parallel Rete Match algorithm. Simulations have shown promising speed-ups, and that 32 processors are sufficient to exploit most of the parallelism in this system [86].

Numerous studies have been made of strategies for mapping production systems to multiprocessors [132, 177, 225, 232].

4.2.5 *Parallel Object-Oriented Architectures.*

Most development work in object-oriented programming has been done on computers such as the Intel iPSC that are not specifically designed for object-oriented computation. Two notable multiprocessors, however, are designed specifically for object-oriented computations. FAIM-1 [12] is a multiprocessor with special RISC processors

TABLE 1.18   Notable Parallel Computers for Object Oriented Computation.

Machine	Year/ Status[a]	Interconnection	Communication	Features
FAIM-1	–/UC	Hexagonal mesh	Message passing	PEs = Hectagons (hexagonal configuration) Heterogeneous shared memory multiprocessor Instruction stream memory, post office communication processor, evaluation processor, and others Three-port switch at edge of array for I/O and wrapping of connections Up to 10 32-bit workstation/processors
Dragon	–/UC	Bus	Shared memory	Tightly coupled Associative cache at each processor
Apiary	1980/PD	Single-stage network	Message passing	Implements Actor model Computations in the Actor model are partial orders of events with no assignment commands

[a] Refer to Table 1.14 for explanation of status codes.

connected by a hexagonal mesh. At first an intermediate language, OIL, was developed for interfacing between its modules. However, the scope of the project has recently been changed to exclude the development of OIL and program the computer in MultiScheme. The second specialized multiprocessor was developed by the Dragon project; it supports only 10 processors [160]. Table 1.18 summarizes notable projects.

4.3   Connectionist Processing

Many connectionist implementations focus on correlations between nodes in a graph and have been designed primarily for semantic networks. Artificial neural networks are also of interest, but unfortunately, current technology for their design has precluded their development in a role greater than that of subsystem.

The four connectionist designs for the implementation of semantic networks correspond to the three types of message-passing environments. NETL uses the most elementary processing elements in a marker-passing system [65, 66]. THISTLE is a similar design, but employs value passing instead of marker passing [66]. The Connection Machine can be programmed to simulate marker

TABLE 1.19   Parallel Connectionist Systems.

Machine	Year/ Status[a]	Interconnection	Communication	Features
NETL	1979/PD	Multilevel switching network	Marker passing	For semantic networks A million processors, each can store 16 markers Simple PEs Only Boolean functions
THISTLE	1983/PD	Multilevel switching	Value passing	For semantic networks Similar to NETL, but with 8-bit value passing
Connection Machine	1986/CA	Hypercube	Message passing	General purpose SIMD and multiple SIMD processing Can be programmed for marker passing operations in semantic networks and simulating neural networks
SNAP	1985/PD	Mesh with global bus	Message passing	Square array of identical processors CACM in each PE for relationships between nodes Communication unit Processing unit

[a] Refer to Table 1.14 for explanation of status codes.

passing in semantic networks and value passing in artificial neural networks by using massive data-level parallelism [39]. Finally, SNAP relies on message passing [159]. The details of these machines are summarized in Table 1.19.

4.4 Summary

The purpose of this section was to give a high-level perspective of existing special purpose computers designed for symbolic processing. Complete systems can be classified by two attributes. The first attribute is the fundamental approach to processing taken by the system: sequential, parallel, or connectionist.

The second attribute is the overall processing technique. For current systems the processing technique can be defined in one of two ways. The first way is by knowledge representation: for example, computers for parallel logic computation. The second definition is based on the programming paradigm: for example, functional or object-oriented.

Systems can be further classified by various design decisions, and by the micro/macro-level architectures employed. From a high-level perspective, however, it is not necessary to do this in order to understand the state of the art in symbolic processing systems.

Figure 1.8 classifies the computers presented in this section on the basis of the two suggested attributes. However, a number of systems have been designed for diverse symbolic applications and cannot be classified into a unique category. The Connection Machine is a notable example of this class.

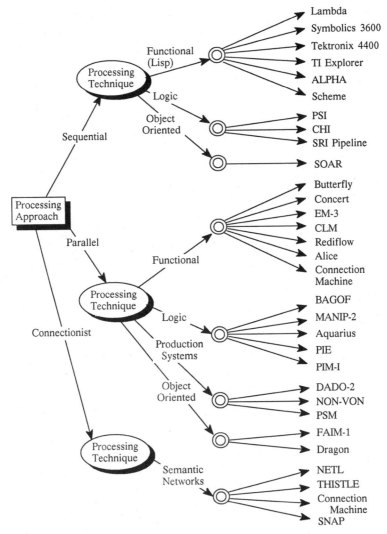

Figure 1.8 Complete symbolic processing systems.

5 RESEARCH DIRECTIONS

We have presented in this paper an extensive discussion and analysis of the state of the art in computer solutions of symbolic processing problems. We conclude this section by indicating some of the research areas where advances will most likely bring about fast and efficient computer solution of these problems.

Technologies.   The design of a system is often driven by its cost; hence, while the fastest technologies are preferable, their selection is subject to cost constraints. New and emerging technologies may give higher performance but are often prohibitively expensive. The candidate technologies that will likely become cost effective in the near future include GaAs circuits [130, 131], Wafer-Scale Integrated circuits (WSI) [154], analog–digital VLSI circuits, and optical computing techniques [21, 79, 244].

These emerging technologies offer tremendous potential for increasing the processing speeds of current computers; the extension of processing power is especially valuable for real-time systems. However, the most that can be expected from these technologies is an improvement in speed of about one to two orders of magnitude in the next ten years. They will not greatly impact the size or type of symbolic applications that are addressed today, since many of these applications involve huge search spaces and an increase in computational speed by one to two orders of magnitude will do little to increase the size of a solvable instance of such a problem [238].

Algorithms.   Research in the area of application-specific algorithms has the greatest potential for speeding applications. New and improved algorithms may be developed by finding alternative ways to incorporate knowledge about the application domain into the computer solution. In this way, advancement of symbolic processing capabilities in the area of application-specific algorithms is tightly linked to advancement in the area of knowledge representations.

Knowledge Representations.   Most new knowledge representations for symbolic processing have emphasized declarative and distributed features, reducing programming complexity. These representation schemes may have to be modified or extended to fit the applications and the computational environment. The addition of temporal features and nonmonotonicity would also be helpful.

A major problem in the area of knowledge representations is the lack of technique to guide the evaluation and selection of a knowledge representation scheme. Research in this area could prove extremely valuable. Learning techniques for incorporating new knowledge about application domains into current knowledge intensive applications may also have a great impact on symbolic processing. Artificial neural networks and connectionist representations already incorporate automated learning techniques into their design at the knowledge representation level.

Software Architecture.   Software architectures are highly dependent on research in the area of knowledge representation. The generation of new software environments, tools, and languages will probably rely on amalgamation of known knowledge representation techniques. Software development systems and automated intelligent programming assistants are prime areas for the advancement of symbolic programming. The problems of program verification and validation and continuous maintenance of symbolic programs are important related topics.

Hardware Architectures.   As with software, hardware architectures are often based on known design techniques such as parallel processing and pipelining. Architectural innovation may be encouraged, however, by the availability of new and emerging technologies.

New hardware architectures are best utilized for operations that the computer performs frequently. Counter to intuition, identification of these tasks is very difficult. Operations may be instructions, parts of instructions, groups of instructions, or frequently recurring tasks. Identification of new and valid areas for development of new hardware architectures is an important area of research.

System Design.   System-level design is often based on an overall design philosophy; for example, systems may contain a mix of data- and control-flow computation. The proper mix of control, data, and demand flow is one area of research that may impact systems for symbolic processing. New systems for symbolic processing may also greatly benefit from the integration of new hardware subsystems and microlevel architectures, and of new and emerging technologies. The major difficulty lies in integrating designs with radically different knowledge representations. The combination of distributed representation (offered by artificial neural networks) and procedural or declarative representation (offered by standard computers) is an interesting area for development.

6 ACKNOWLEDGMENTS

This research was partially supported by National Aeronautics and Space Administration Grant NCC 2-481 and National Science Foundation Grant MIP 88-10584. The research of M. B. Lowrie was supported by a Ph. D Scholarship from AT&T Bell Laboratories.

REFERENCES

1. "ESPRIT: Europe Challenges U.S. and Japanese Competitors," *Future Generation Computer Systems*, vol. 1, no. 1, pp. 61–69, North-Holland, New York, 1984.

2. "Special Issue on Natural Language Processing," *Proc. IEEE*, July 1986.

3. "Special Issue on Computer Vision," *Proc. IEEE*, 1987.

4. Defense Advanced Research Project Agency, *DARPA Neural Network Study*, Lincoln Laboratory, Massachusetts Institute of Technology, Lexington, July 1988.

5. G. Agha and C. Hewitt, "Concurrent Programming Using Actors: Exploiting Large-Scale Parallelism," *Lecture Notes in Computer Science*, no. 206, pp. 19–41, Springer-Verlag, New York, December 1985.

6. G. Agha, *Actor: A Model of Concurrent Computation in Distributed Systems*, MIT Press, Cambridge, 1986.

7. L. Aiello, C. Cecchi, and D. Sartini, "Representation and Use of Metaknowledge," *Proc. IEEE*, pp. 1304–1321, October 1986.

8. J. Alspector and R. B. Allen, "A Neuromorphic VLSI Learning System," *Advanced Research in VLSI: Proc. 1987 Stanford Conference*, ed. P. Loseleben, MIT Press, Cambridge, 1987.

9. M. Amamiya et. al., "New Architecture for Knowledge Based Mechanisms," *Proc. Int'l Conf. on Fifth Generation Computer Systems*, pp. 179–188, Japan, 1981.

10. M. Amamiya et. al., "Implementation and Evaluation of List-Processing-Oriented Data Flow Machine," *Proc. 13th Int'l Symp. on Computer Architecture*, pp. 10–19, IEEE/ACM, 1986.

11. C. W. Anderson, "Strategy Learning with Multilayer Connectionist Representations," *Proc. 4th Int'l Workshop on Machine Learning*, pp. 103–114, June, 1987.

12. J. M. Anderson, W. S. Coates, A. L. Davis, R. W. Hon, I. N. Robinson, S. V. Robison, and K. S. Stevens, "The Architecture of FAIM-1," *Computer*, vol. 20, no. 1, pp. 55–65, IEEE, January 1987.

13. H. L. Andrews, "Speech Processing," *Computer*, vol. 17, no. 10, pp. 315–324, IEEE, October 1984.

14. Arvind, K. Gostelow and W. Plouffe, *An Asynchronous Programming Language and Computing Machine*, Techical Report 114a, University of California, Irvine, December 1978.

15. E. Babb, "Implementing a Relational Database by Means of Specialized Hardware," *Trans. on Database Systems*, vol. 4, no. 1, pp. 1–29, ACM, March 1979.

16. J. Backus, "Can Programming be Liberated From the von Neumann Style? A Functional Style and Algebra of Programs," *Comm. of the ACM*, vol. 21, no. 8, pp. 613–641, 1978.

17. H. G. Baker, Jr., "Optimizing Allocation and Garbage Collection of Spaces," *Artificial Intelligence: An MIT Perspective*, ed. P. H. Winston and R. H. Brown, vol. 1, pp. 391–396, MIT Press, 1979.

18. A. Barr and E. A. Feigenbaum, *The Handbook of Artificial Intelligence*, vol. 1, 2, and 3, William Kaufmann, Los Altos, CA, 1981, 1982.

19. J. M. Barth, "Shifting Garbage Collection Overhead to Compile Time," *Comm. of the ACM*, vol. 20, no. 7, pp. 513–518, July 1977.

20. E. B. Baum, "Towards Practical 'Neural' Computation for Combinatorial Optimization Problems," *AIP Conf. Proc. on Neural Networks for Computing*, pp. 53–58, 1986.

21. P. B. Berra and N. B. Troullinos, Optical Techniques and Data/Knowledge Base Machines," *Computer*, vol. 20, no. 10, pp. 59–70, IEEE, October 1987.

22. D. Billstrom, J. Brandenburg, and J. Teeter, "CCLISP on the iPSC Concurrent Computer," *Proc. 6th Int'l Conf. on Artificial Intelligence*, pp. 7–12, AAAI, Seattle, WA, 1987.

23. D. G. Bobrow, et al., *CommonLoops: Merging Common Lisp and Object-Oriented Programming*, Technical Report ISL-85-8, Xerox Palo Alto Research Center, August 1985.

24. D. G. Bobrow and P. J. Hayes, (ed.), "Special Issue on Nonmonotonic Logic," *Artificial Intelligence*, vol. 13, no. 1 & 2, North-Holland, New York, April 1980.

25. K. Bowen, "Meta-Level Programming and Knowledge Representation," *New Generation Computing*, vol. 3, no. 4, pp. 359–383, 1985.

26. R. Brooks and R. Lum, "Yes, An SIMD Machine Can be Used for AI," *Proc. Int'l Joint Conf. on Artificial Intelligence*, pp. 73–79, 1985.

27. C. M. Brown, C. S. Ellis, J. A. Feldman, T. J. LeBlank, and G. L. Peterson, Research with the Butterfly Multicomputer," *Rochester Research Review,* pp. 3–23, 1984.

28. L. Brownston, R. Farrell, E. Kant, and N. Martin, *Programming Expert Systems in OPS5*, Addison-Wesley, Reading, MA, 1985.

29. M. Bruynooghe, "The Memory Management of PROLOG Implementations," in *Logic Programming*, ed. K. Clark and S. A. Tarnlund, pp. 83–89, Academic Press, New York, 1982.

30. B. G. Buchanan and E. H. Shortliffe, *Rule-Based Experts Programs: The MYCIN Experiments of the Stanford Heuristic Programming Project*, Addison-Wesley, Reading, MA, 1984.

31. A. W. Burks, "Keynote of CONPAR86," *Lecture Notes in Computer Science No. 237*, pp. 1–17, Springer-Verlag, New York, 1986.

32. R. M. Burstall, D. B. MacQueen, and D. T. Sannella, "HOPE: An Experimental Applicative Language," *Conf. Record of Lisp Conf.*, pp. 136–143, Stanford Univ., Menlo Park, CA, 1980.

33. S. Cammarata et al., "Strategies of Cooperation in Distributed Problem Solving," *Proc. Int'l Joint Conf. on Artificial Intelligence*, pp. 767–770, IJCAI, Inc., 1983.

34. A. Ciepielewski and S. Haridi, "Execution of Bagof on the OR-Parallel Token Machine," *Proc. Int'l Conf. Fifth Generation Computer Systems*, pp. 551–560, ICOT and North-Holland, New York, 1984.

35. K. Clark and S. Gregory, "Note on System Programming in PARLOG," *Proc. Int'l Conf. Fifth Generation Computer Systems*, pp. 299–306, ICOT and North-Holland, New York, 1984.

36. K. L. Clark and S. A. Tarnlund, (eds.), *Logic Programming*, Academic Press, New York, 1982.

37. W. F. Clocksin and C. S. Mellish, *Programming in Prolog*, Springer-Verlag, New York, 1981.

38. P. R. Cohen and M. R. Grinberg, "A Theory of Heuristic Reasoning About Uncertainty," *AI Magazine*, pp. 17–24, AAAI, Summer 1983.

39. Thinking Machines Corporation, *Connection Machine Model CM-2 Technical Summary*, Technical Report HA87-4, Cambridge, MA, April 1987.

40. Thinking Machines Corporation, **Lisp Reference Manual*, Version 4.0, Cambridge, MA, October 1987.

41. A. S. Cromarty, What Are Current Expert System Tools Missing?" *Proc. COMPCON Spring*, pp. 411–418, IEEE, 1985.

42. O. J. Dahl and K. Nygaard, "SYMULA—An Algol-Based Simulation Language," *Communications of the ACM*, vol. 9, no. 9, pp. 671–8, September 1966.

43. J. Darlington and M. Reeve, "ALICE: A Multi-Processor Reduction Machine for the Parallel Evaluation of Applicative Languages," *Proc. Conf. on Functional Programming Languages and Computer Architecture*, pp. 65–74, ACM, 1981.

44. J. Darlington, P. Henderson, and D. Turner, *Functional Programming and Its Applications*, Cambridge University Press, Cambridge, UK, 1982.

45. J. Darlington and M. Reeve, *ALICE and the Parallel Evaluation of Logic Programs*, Preliminary Draft, Department of Computing, Imperial College of Science and Technology, London, June 1983.

46. J. Darlington, "Functional Programming," *Distributed Computing*, eds. F. B. Chambers, D. A. Duce, and G. P. Jones, Academic Press, London, 1984.

47. J. Darlington, A. J. Field, and H. Pull, *The Unification of Functional and Logic Languages*, Technical Report, Imperial College, London, February 1985.

48. A. L. Davis and S. V. Robison, "The FAIM-1 Symbolic Multiprocessing System," *Proc. COMPCON Spring*, pp. 370–375, IEEE, 1985.

49. R. Davis and B. Buchanan, "Meta-level Knowledge: Overview and Applications," *Proc. 5th Int'l Joint Conf. on Artificial Intelligence*, pp. 920–928, William Kaufmann, Los Altos, CA, 1977.

50. M. F. Deering, "Architectures for AI," *Byte*, pp. 193–206, McGraw-Hill, New York, April 1985.

51. D. DeGroot, "Restricted AND-Parallelism," *Proc. Int'l Conf. on Fifth Generation Computers*, pp. 471–478, ICOT and North-Holland, New York, November 1984.

52. D. DeGroot and G. Lindstrom (eds.), *Logic Programming*, Prentice-Hall, Englewood Cliffs, NJ, 1985.

53. J. S. Denker, "Neural Network Models of Learning and Adaptation," *Physica*, pp. 216–232, 1986.

54. A. M. Despain and Y. N. Patt, "Aquarius—A High Performance Computing System for Symbolic/Numeric Applications," *Proc. COMPCON Spring*, pp. 376–382, IEEE, February 1985.

55. P. Deutsch, "Experience with a Microprogrammed Interlisp System," *Proc. MICRO*, vol. 11, ACM/IEEE, November 1978.

56. D. J. DeWitt, "DIRECT—A Multiprocessor Organization for Supporting Relational Database Management Systems," *Trans. on Computers*, pp. 395–406, IEEE, June 1979.

57. E. W. Dijkstra, L. Lamport, A. J. Martin, C. S. Scholten, and E. F. M. Steffens, "On-the-Fly Garbage Collection: An Exercise in Cooperation," *Comm. of the ACM*, vol. 21, no. 11, pp. 966–975, November 1978.

58. J. Doyle, "A Truth Maintenance System," *Artificial Intelligence*, vol. 12, no. 3, pp. 231–272, North-Holland, New York, 1979.

59. H. Dreyfus and S. Dreyfus, "Why Expert Systems Do Not Exhibit Expertise," *Expert*, vol. 1, no. 2, IEEE, Summer 1986.

60. R. O. Duda, P. E. Hart, and N. J. Nilsson, "Subjective Bayesian Methods for Rule-Based Inference Systems," *Proc. Nat'l Computer Conf.*, pp. 1075–1082, AFIPS Press, 1976.

61. L. D. Erman, F. Hayes-Roth, V. R. Lesser, and D. R. Reddy, "The Hearsay-II Speech-Understanding System: Integrating Knowledge to Resolve Uncertainty," *Computing Surveys*, vol. 12, no. 2, pp. 213–253, ACM, June 1980.

62. F. Fagin, *How Far Can We Go With Proven Technology?* Special Session on Neural Networks, AAAI Conference, Seattle, WA, July 1987.

63. R. Fagin and J. Halpern, "Belief, Awareness, and Limited Reasoning: Preliminary Report," *Proc. Int'l Joint Conf. on Artificial Intelligence*, pp. 491–501, IJCAI Inc., 1985.

64. S. E. Fahlman, *NETL: A System for Representing and Using Real-World Knowledge*, MIT Press, Cambridge, 1979.

65. S. E. Fahlman, "Design Sketch for a Million-Element NETL Machine," *Proc. 1st Annual Nat'l Conf. on Artificial Intelligence*, pp. 249–252, AAAI, August 1980.

66. S. E. Fahlman and G. E. Hinton, "Massively Parallel Architectures for AI: NETL, THISTLE, and Boltzmann Machines," *Proc. Nat'l Conf. on Artificial Intelligence*, pp. 109–113, AAAI, 1983.

67. S. E. Fahlman, "Parallel Processing in Artificial Intelligence," *Parallel Computing*, vol. 2, pp. 283–286, 1985.

68. S. E. Fahlman and G. E. Hinton, "Connectionist Architecture for Artificial Intelligence," *Computer*, vol. 20, no. 1, pp. 100–109, IEEE, January 1987.

69. C. Faloutos, "Access Method for Text," *Computing Surveys*, ACM, March 1985.

70. R. D. Fennell and V. R. Lesser, "Parallelism in Artificial Intelligence Problem Solving: A Case Study of Hearsay-II," *Trans. on Computers*, vol. C-26, no. 2, pp. 98–111, IEEE, February 1977.

71. E. A. Feustel, "On the Advantages of Tagged Architecture," *Trans. on Computers*, vol. C-22, no. 7, pp. 644–656, IEEE, 1973.

72. R. E. Fikes and N. J. Nilsson, "STRIPS: A New Approach to the Application of Theorem Proving to Problem Solving," *Artificial Intelligence*, vol. 2, no. 3 & 4, pp. 189–208, North-Holland, New York, 1971.

73. M. A. Fischetti, "A Review of Progress at MCC," *Spectrum*, IEEE, March 1986.

74. C. L. Forgy, "Rete: A Fast Algorithm for the Many Pattern/Many Object Pattern Match Problem," *Artificial Intelligence*, vol. 19, no. 1, pp. 17–37, North-Holland, New York, September 1982.

75. C. L. Forgy, A. Gupta, A. Newell, and R. Wedig, Initial Assessment of Architectures for Production Systems," *Proc. Nat'l Conf. on Artificial Intelligence*, pp. 116–120, AAAI, August 1984.

76. K. S. Fu, *Syntactic Methods in Pattern Recognition*, Academic Press, New York, 1974.

77. K. Fuchi, "The Direction the FGCS Project Will Take," *New Generation Computing*, vol. 1, no. 1, pp. 3–9, OHMSHA Ltd. and Springer-Verlag, New York, 1983.

78. J. L. Gaudiot, S. Lee, and A. Sohn, *Data-Driven Multiprocessor Implementation of the Rete Match Algorithm*, Technical Report, Department of Computer Science, University of Southern California, Los Angeles, 1987.

79. T. K. Gaylord and E. I. Verriest, "Matrix Triangularization Using Arrays of Integrated Optical Givens Rotation Devices," *Computer*, vol. 20, no. 12, pp. 59–67, IEEE, December 1987.

80. M. R. Genesereth, "An Overview of Meta-Level Architecture," *Proc. Nat'l Conf. on Artificial Intelligence*, pp. 119–124, AAAI, 1983.

81. A. J. Goldberg and D. Robson, *Smalltalk-80: The Language and Its Implementation*, Addison-Wesley, Reading, MA, 1983.

82. A. Goto, H. Tanaka, and T. Moto-oka, "Highly Parallel Inference Engine PIE—Goal Rewriting Model and Machine Architecture," *New Generation Computing*, vol. 2, no. 1, pp. 37–58, OHMSHA Ltd. and Springer-Verlag, New York, 1984.

83. E. Goto, T. Ida, K. Hiraki, M. Suzuki, and N. Inada, "FLATS, A Machine for Numerical, Symbolic and Associative Computing," *Proc. 6th Int'l Joint Conf. on Artificial Intelligence*, pp. 1058–1066, William Kaufmann, Los Altos, CA, August 1979.

84. N. Greenfeld and A. Jericho, "A Professional's Personal Computer System," *Proc. 8th Int'l Symp. on Comp. Architecture*, pp. 217–226, IEEE/ACM, 1981.

85. M. Griss and M. Swanson, "MBALM/1700: A Microprogrammed Lisp Machine for the Burroughs B1726," *Proc. MICRO-10*, ACM/IEEE, 1977.

86. A. Gupta, C. L. Forgy, A. Newell, and R. Wedig, "Parallel Algorithms and Architectures for Rule-Based Systems," *Proc. 13th Int'l Symp. on Computer Architecture*, pp. 28–37, IEEE/ACM, 1986.

87. J. Halpern and Y. Moses, "A Guide to the Modal Logics of Knowledge and Belief: Preliminary Draft," *Proc. Int'l Joint Conf. on Artificial Intelligence*, pp. 480–490, IJCAI Inc., 1985.

88. R. H. Halstead, Jr., "Implementation of MULTILISP: LISP on a Multiprocessor," *Proc. Symp. on LISP and Functional Programming*, ACM, 1984.

89. R. Halstead, "Parallel Symbolic Computing," *Computer*, vol. 19, no. 8, pp. 35–43, IEEE, August 1986.

90. R. Halstead, Jr., "An Assessment of Multilisp: Lessons from Experience," *Int'l J. Parallel Programming*, vol. 15, no. 6, pp. 459–501, December 1986.

91. R. Halstead, Jr. and J. Loaiza, "Exception Handling in Multilisp," *Proc. Int'l Conf. on Parallel Processing*, pp. 822–830, August 1985.

92. R. Halstead, Jr., T. Anderson, R. Osborne, and T. Sterlig, "Concert: Design of a Multiprocessor Development System," *Proc. Int'l Symp. on Computer Architecture*, pp. 40–48, IEEE/ACM, June 1986.

93. P. B. Hawthorn and D. J. DeWitt, "Performance Analysis of Alternative Database Machine Architectures," *Trans. on Software Engineering*, vol. SE-8, no. 1, pp. 61–75, IEEE, January 1982.

94. H. Hayashi, A. Hattori, and H. Akimoto, ALPHA: "A High-Performance Lisp Machine Equipped with a New Stack Structure and Garbage Collection System," *Proc. 10th Annual Int'l Symp. on Computer Architecture*, pp. 342–348, IEEE/ACM, June 1983.

95. B. Hayes-Roth, "A Blackboard Architecture for Control," *Artificial Intelligence*, vol. 26, no. 3, pp. 251–321, North-Holland, New York, July 1985.

96. F. Hayes-Roth, D. A. Waterman, and D. B. Lenat, *Building Expert Systems*, Addison-Wesley, Reading, MA, 1983.

97. P. Henderson, *Function Programming, Application and Implementation*, Prentice-Hall, Englewood Cliffs, NJ, 1980.

98. C. Hewitt, "Viewing Control Structure as Patterns of Passing Messages," *Artificial Intelligence*, vol. 8, no. 3, pp. 323–364, North-Holland, New York, 1977.

99. C. Hewitt, "The Apiary Network Architecture for Knowledgeable Systems," *Conf. Record of Lisp Conf.*, pp. 107–117, Stanford University, Menlo Park, CA, 1980.

100. C. Hewitt and H. Lieberman, "Design Issues in Parallel Architectures for Artificial Intelligence," *Proc. COMPCON Spring*, pp. 418–423, IEEE, February 1984.

101. C. E. Hewitt, *Description and Theoretical Analysis of PLANNER: A Language for Proving Theorems and Manipulating Models in Robots*, MIT AI Lab. TR-258, 1972.

102. Y. Hibino, "A Practical Parallel Garbage Collection Algorithm and Its Implementations," *Proc. 7th Annual Symp. on Computer Architecture*, pp. 113–120, IEEE/ACM, May 1980.

103. W. D. Hillis, *The Connection Machine*, MIT Press, Cambridge, 1985.

104. B. K. Hillyer and D. E. Shaw, "Execution of OPS5 Production Systems on a Massively Parallel Machine," *J. Parallel and Distributed Computing*, vol. 3, no. 2, pp. 236–268, Academic Press, New York, 1986.

105. J. Hintikka, "Impossible Possible World Vindicated," *J. Philosophical Logic*, pp. 475–484, 1975.

106. G. Hinton, J. L. McClelland, and D. D. Rumelhart, "Distributed Representations," *Parallel Distributed Processing: Explorations in the Microstructure of Cognition*, eds. D. E. Rumelhart, J. L. McClelland, and the PDP Research Group, vol. 1, MIT Press, Cambridge, 1986.

107. G. Hinton and T. J. Sejnowski, "Learning and Relearning in Boltzmann Machines," *Parallel Distributed Processing: Explorations in the Microstructure of Cognition*, eds. D. E. Rumelhart, J. L. McClelland, and the PDP Research Group, vol. 1, MIT Press, Cambridge, 1986.

108. G. E. Hinton, *Connectionist Learning Procedures*, Technical Report CMU-CS-87-115, Carnegie Mellon University, Pittsburgh, PA, June 1987.

109. J. J. Hopfield and D. W. Tank, "Neural Computation of Decisions in Optimization Problems," *Biological Cybernetics*, vol. 52, no. 3, pp. 1–25, July 1985.

110. J. J. Hopfield and D. W. Tank, *Disordered Systems and Biological Organization*, Springer-Verlag, New York, 1986.

111. J. J. Hopfield and D. W. Tank, "Computing with Neural Circuits: A Model," *Science*, pp. 625–633, August 1986.

112. H. A. H. Ibrahim, J. R. Kender, and D. E. Shaw, "Low-Level Image Analysis Tasks on Fine-Grained Tree-Structured SIMD Machines," *J. Parallel and Distributed Computing*, pp. 546–574, Academic Press, New York, December 1987.

113. N. Ichiyoshi, T. Miyazaki, and K. Taki, "A Distributed Implementation of Flat GHC on the Multi-PSI," *Int'l Conf. on Logic Programming*, 1987.

114. N. Ito et al., "The Architecture and Preliminary Evaluation Results of the Experimental Parallel Inference Machine PIM-D," *Proc. 13th Int'l Symp. on Computer Architecture*, pp. 149–156, IEEE/ACM, 1986.

115. N. Ito, H. Shimizu, M. Kishi, E. Kuno, and K. Rokusawa, "Data-Flow Based Execution Mechanisms of Parallel and Concurrent Prolog," *New Generation Computing*, vol. 3, pp. 15–41, OHMSHA Ltd. and Springer-Verlag, New York, 1985.

116. P. Kanerva, *Parallel Structures in Human and Computer Memory*, RIACS Technical Report TR-86.2, NASA Ames Research Center, Moffett Field, CA, January 1986.

117. R. M. Keller, F. C. H. Lin, and J. Tanaka, "Rediflow Multiprocessing," *Proc. COMPCON Spring*, pp. 410–417, IEEE, 1984.

118. C. Kellogg, "Intelligent Assistants for Knowledge and Information Resources Management," *Proc. 8th Int'l Joint Conf. on Artificial Intelligence*, pp. 170–172, William Kaufmann, Los Altos, CA, 1983.

119. D. F. Kibler, "Parallelism in AI Programs," *Proc. Int'l Joint Conf. on Artificial Intelligence*, pp. 53–56, IJCAI, 1985.

120. M. King and M. Rosner, "Scanning the Issue: The Special Issue on Knowledge Representation," *Proc. IEEE*, vol. 74, no. 10, pp. 1299–1303, October, 1986.

121. J. de Kleer, "An Assumption-Based TMS," *Artificial Intelligence*, vol. 28, pp. 127–161, North-Holland, New York, 1986.

122. J. de Kleer, "Extending the ATMS," *Artificial Intelligence*, vol. 28, pp. 163–196, North-Holland, New York, 1986.

123. J. de Kleer, "Problem Solving with the ATMS," *Artificial Intelligence*, vol. 28, pp. 197–224, North-Holland, New York, 1986.

124. T. Knight, *The CONS Microprocessor*, AI Working Paper 80, MIT, Cambridge, November 1974.

125. R. Kowalski, "Predicate Logic as a Programming Language," *IFIP Information Processing*, pp. 569–574, North-Holland, New York, 1974.

126. R. Kowalski, *Logic for Problem Solving*, North-Holland, New York, 1979.

127. H. Kung and S. Song, *An Efficient Parallel Garbage Collection System and Its Correctness Proof*, Technical Report, Department of Computer Science, Carnegie-Mellon University, Pittsburgh, PA, September 1977.

128. R. Kurzweil, "What is Artificial Intelligence Anyway?" *American Scientist*, vol. 73, no. 3, pp. 258–264, May–June, 1985.

129. T. H. Lai and S. Sahni, "Anomalies in Parallel Branch-and-Bound Algorithms," *Comm. of the ACM*, vol. 27, no. 6, pp. 594–602, June 1984.

130. L. E. Larson, J. F. Jensen, and P. T. Greiling, "GaAs High-Speed Digital IC Technology: An Overview," *Computer*, vol. 19, No. 10, pp. 21–28, IEEE, October 1986.

131. T. F. Lehr and R. G. Wedig, "Toward a GaAs Realization of a Production-System Machine," *Computer*, vol. 20, no. 4, pp. 36–49, IEEE, April 1987.

132. D. B. Lenat and J. McDermott, "Less Than General Production System Architectures," *Proc. 5th Int'l Joint Conf. on Artificial Intelligence*, pp. 923–932, William Kaufmann, Los Altos, CA, 1977.

133. D. B. Lenat, "The Ubiquity of Discovery," *Proc. 5th Int'l Joint Conf. on Artificial Intelligence*, pp. 1093-1105, William Kaufmann, Los Altos, CA, 1977.

134. D. B. Lenat, "The Nature of Heuristics," *Artificial Intelligence*, vol. 19, no. 2, pp. 189–249, North-Holland, New York, 1982.

135. D. B. Lenat, "Computer Software for Intelligent Systems," *Scientific American*, vol. 251, no. 3, pp. 204–213, Scientific American Inc., September 1984.

136. G.-J. Li and B. W. Wah, "Computational Efficiency of Parallel Approximate Branch-and-Bound Algorithms," *Proc. Int'l Conf. on Parallel Processing*, pp. 473–480, IEEE, August 1984.

137. G.-J. Li and B. W. Wah, "The Design of Optimal Systolic Algorithms," *Trans. on Computers*, vol. C-34, no. 1, pp. 66–77, IEEE, January 1985.

138. G.-J. Li and Benjamin W. Wah, "MANIP-2: A Multicomputer Architecture for Evaluating Logic Programs," *ICPP*, pp. 122-130, IEEE, June 1985.

139. G.-J. Li and B. W. Wah, "Multiprocessing of Logic Programs," *Proc. Int'l Conf. on Systems, Man and Cybernetics*, pp. 563–567, IEEE, October 1986.

140. G.-J. Li and B. W. Wah, "Optimal Granularity of Parallel Evaluation of AND-Trees," *Proc. Fall Joint Computer Conf.*, pp. 297–306, ACM/IEEE, November 1986.

141. G.-J. Li and B. W. Wah, "How Good are Parallel and Ordered Depth-First Searches?" *Proc. Int'l Conf. on Parallel Processing*, pp. 992–999, IEEE, August 1986.

142. H. Lieberman and C. Hewitt, "A Real-Time Garbage Collector Based on the Lifetimes of Objects," *Comm. of the ACM*, vol. 26, no. 6, pp. 419–429, June 1983.

143. G. Lindstrom and P. Panangaden, "Stream-Based Execution of Logic Programs," *Proc. Int'l Symp. on Logic Programming*, pp. 168–176, IEEE, February 1984.

144. R. P. Lippmann, "An Introduction to Computing with Neural Nets," *IEEE ASSP Magazine*, pp. 4–22, April, 1987.

145. F. J. Malabarba, "Review of Available Database Machine Technology," *Proc. Trends and Applications*, pp. 14–17, IEEE, 1984.

146. Y. Malachi, Z. Manna, and R. Waldinger, "TABLOG: A New Approach to Logic Programming," *Logic Programming*, ed. D. DeGroot and G. Lindstrom, Prentice-Hall, Englewood Cliffs, NJ, 1985.

147. G. Matthews, R. Hewes, and S. Krueger, "Single-Chip Processor Runs Lisp Environments," *Computer Design*, pp. 69–76, PennWell, May 1, 1987.

148. A. D. McAulay, "Spatial-Light-Modulator Interconnected Computers," *IEEE Computer*, vol. 20, no. 10, pp. 45–58, October 1987.

149. G. McCalla and N. Cercone, (eds.), "Special Issue on Knowledge Representation," *Computer*, vol. 16, no. 10, IEEE, October 1983.

150. J. McCarthy, "Recursive Functions of Symbolic Expressions and Their Computation by Machine, Part I," *Comm. of the ACM*, vol. 3, no. 4, pp. 184–195, 1960.

151. J. McCarthy and P. Hayes, "Some Philosophical Problems from the Standpoint of Artificial Intelligence," *Machine Intelligence 4*, pp. 463–502, Elsevier, NY, 1969.

152. J. McCarthy, "History of Lisp," *SIGPLAN Notices*, vol. 13, no. 8, pp. 217–223, ACM, 1978.

153. J. L. McClelland, D. D. Rumelhart, and G. Hinton, "The Appeal of Parallel Distributed Processing," *Parallel Distributed Processing: Explorations in the Microstructure of Cognition*, eds. D. E. Rumelhart, J. L McClelland, and the PDP Research Group, vol. 1, MIT Press, Cambridge, 1986.

154. J. F. McDonald, H. J. Greub, R. H. Steinvorth, B. J. Donlan, and A. S. Bergendahl, "Wafer Scale Interconnections for GaAs Packaging Applications to RISC Architecture," *Computer*, vol. 20, no. 4, pp. 21–35, IEEE, April 1987.

155. J. R. McGraw, "Data Flow Computing: Software Development," *Trans. on Computers*, vol. C-29, no. 12, pp. 1095–1103, IEEE, 1980.

156. C. S. Mellish, "An Alternative to Structure Sharing in the Implementation of a Prolog Interpreter," *Logic Programming*, ed. K. Clark and S. A. Tarnlund, pp. 99–106, Academic Press, New York, 1982.

157. R. S. Michalski, J. G. Carbonell, and T. M. Mitchell, *Machine Learning: An Artificial Intelligence Approach*, Tioga, Palo Alto, CA, 1983.

158. M. Minsky, "A Framework for Representing Knowledge," *The Psychology of Computer Vision*, ed. P. H. Winston, McGraw-Hill, New York, 1975.

159. D. I. Moldovan and Y. W. Tung, "SNAP: A VLSI Architecture for Artificial Intelligence Processing," *J. Parallel and Distributed Computing*, pp. 109–131, Academic Press, New York, May 1985.

160. L. Monier and P. Sidhu, "The Architecture of the Dragon," *Proc. COMPCON*, pp. 118–121, Spring 1985.

161. D. A. Moon, "Symbolics Architecture," *Computer*, vol. 20, no. 1, pp. 43–52, IEEE, 1987.

162. T. Moto-oka, "Overview to the Fifth Generation Computer System Project," *Proc. 10th Annual Int'l Symposium on Computer Architecture*, pp. 417–422, IEEE/ACM, June 1983.

163. T. Moto-oka, H. Tanaka, H. Aida, K. Hirata, and T. Maruyama, "The Architecture of a Parallel Inference Engine (PIE)," *Proc. Int'l Conf. on Fifth Generation Computer Systems*, pp. 479–488, ICOT and North-Holland, New York, 1984.

164. T. Moto-oka and H. S. Stone, "Fifth-Generation Computer Systems: A Japanese Project," *Computer*, vol. 17, no. 3, pp. 6–13, IEEE, March 1984.

165. A. Mukhopadhyay, "Hardware Algorithms for Nonnumeric Computation," *Transactions on Computers*, vol. C-28, no. 6, pp. 384–394, IEEE, June 1979.

166. K. Murakami, T. Kakuta, R. Onai, and N. Ito, "Research on Parallel Machine Architecture for Fifth-Generation Computer Systems," *Computer*, vol. 18, no. 6, pp. 76–92, IEEE, June 1985.

167. K. Murakami, "Research on Parallel Machine Architecture for Fifth Generation Computing Systems," *Computer*, p. 92, IEEE, vol. 18, no. 6, 1985.

168. M. Nagao, J. I. Tsujii, K. Nakajima, K. Mitamura, and H. Ito, "Lisp Machine NK3 and Measurement of Its Performance," *Proc. 6th Int'l Joint Conf. on Artificial Intelligence*, pp. 625–627, William Kaufmann, Los Altos, CA, August 1979.

169. H. Nakagawa, "AND Parallel Prolog with Divided Assertion Set," *Proc. Symp. on Logic Programming*, pp. 22–28, 1984.

170. R. Nakazaki et al., "Design of a High Speed Prolog Machine (HPM)," *Proc. 12th Int'l Symp. on Computer Architecture*, pp. 191–197, IEEE/ACM, 1985.

171. P. M. Neches, "Hardware Support for Advanced Data Management Systems," *Computer*, vol. 17, no. 11, pp. 29–40, IEEE, November 1984.

172. A. Newell, J. C. Shaw, and H. A. Simon, "Programming the Logic Theory Machine," *Proc. 1957 Western Joint Computer Conf.*, pp. 230–240, IRE, 1957.

173. A. Newell and H. A. Simon, in *Human Problem Solving*, Prentice-Hall, Englewood Cliffs, NJ, 1972.

174. A. Newell and H. Simon, "ACM Turing Award Lecture: Computer Science as an Empirical Inquiry: Symbols and Search," *Communications of the ACM*, vol. 19, no. 3, March 1975.

175. A. Newell, "Production Systems: Models of Control Structures," *Visual Information Processing*, ed. W. G. Chase, Academic Press, New York, 1975.

176. H. P. Nii, "Blackboard Systems, Blackboard Application Systems, Blackboard Systems from a Knowledge Engineering Perspective," *AI Magazine*, pp. 82–106. AAAI, August 1986.

177. K. Oflazer, "Partitioning in Parallel Processing of Production Systems," *Proc. Int'l Conf. on Parallel Processing*, pp. 92–100, IEEE, 1984.

178. E. A. Ozkarahan, S. A. Schuster, and K. C. Smith, "RAP—An Associative Processor for Database Management," *Proc. Nat'l Computer Conf.*, pp. 379–388, AFIPS Press, 1975.

179. D. A. Patterson, "Reduced Instruction Set Computers," *Comm. of the ACM*, vol. 28, no. 1, pp. 8–21, January 1985.

180. J. Pearl, *Heuristics—Intelligent Search Strategies for Computer Problem Solving*, Addison-Wesley, Reading, MA, 1984.

181. J. Pearl, "Some Recent Results in Heuristic Search Theory," *Trans. on Pattern Analysis and Machine Intelligence*, vol. PAMI-6, no. 1, pp. 1–13, IEEE, January 1984.

182. E. von Puttkamer, "A Microprogrammed Lisp Machine," *Microprocessing and Microprogramming*, vol. 11, no. 1, pp. 9–14, North-Holland, New York, January 1983.

183. M. R. Quillian, "Word Concepts: A Theory and Simulation of Some Basic Semantic Capabilities," *Behavioral Science*, pp. 410–430, 1967.

184. C. V. Ramamoorthy, J. L. Turner, and B. W. Wah, "A Design of a Fast Cellular Associative Memory for Ordered Retrieval," *Transactions on Computers*, vol. C-27, no. 9, pp. 800–815, IEEE, September 1978.

185. C. V. Ramamoorthy, A. Prakash, W. T. Tsai, and Y. Usuda, Software Engineering, *Computer*, vol. 17, pp. 191–210, IEEE, October 1984.

186. C. V. Ramamoorthy, S. Shekhar, and V. Garg, "Software Development Support for AI Programs," *Computer*, pp. 30–42, IEEE, January 1987.

187. U. S. Reddy, "On the Relationship Between Logic and Functional Languages," *Logic Programming*, ed. D. DeGroot and E. G. Lindstrom, Prentice-Hall, Englewood Cliffs, NJ, 1985.

188. P. Robinson, "The SUM: An AI Co-Processor," *Byte*, pp. 169–180, McGraw-Hill, New York, June 1985.

189. F. Rosenblatt, *Principles of Neurodynamics*, Spartan Books, New York, 1962.

190. D. D. Rumelhart, G. Hinton, and J. L. McClelland, "A General Framework for Parallel Distributed Processing," *Parallel Distributed Processing: Explorations in the Microstructure of Cognition*, ed. D. E. Rumelhart, J. L. McClelland, and the PDP Research Group, vol. 1, MIT Press, Cambridge, 1986.

191. D. E. Rumelhart, J. L. McClelland, and the PDP Research Group (eds.), *Parallel Distributed Processing: Explorations in the Microstructure of Cognition*, MIT Press, Cambridge, 1986.

192. E. Sandewall, "An Approach to the Frame Problem, and its Implementation," *Machine Intelligence 7*, pp. 195–204, John Wiley and Sons, New York, 1972.

193. J. Sansonnet, D. Botella, and J. Perez, "Function Distribution in a List-Directed Architecture," *Microprocessing and Microprogramming*, vol. 9, no. 3, pp. 143–153, North-Holland, New York, 1982.

194. J. P. Sansonnet, M. Castan, and C. Percebois, "M3L: A List-Directed Architecture," *Proc. 7th Annual Symp. on Computer Architecture*, pp. 105–112, IEEE/ACM, May 1980.

195. J. P. Sansonnet, M. Castan, C. Percebois, D. Botella, and J. Perez, "Direct Execution of Lisp on a List-Directed Architecture," *Proc. Symp. on Architectural Support for Programming Languages and Operating Systems*, pp. 132–139, ACM, March 1982.

196. D. Schaefer and J. Fischer, "Beyond the Supercomputer," *Spectrum*, vol. 19, no. 3, pp. 32–37, IEEE, March 1982.

197. M. Schor, "Declarative Knowledge Programming: Better Than Procedural," *Expert*, vol. 1, no. 1, pp. 36–43, IEEE, Spring 1986.

198. T. J. Sejnowski and C. R. Rosenberg, *NETtalk: A Parallel Network That Learns to Read Aloud*, Tech. Rep., Johns Hopkins University, Baltimore, January 1986.

199. G. Shafer, *A Mathematical Theory of Evidence*, Princeton Univ. Press, Princeton, NJ, 1976.

200. E. Shapiro and A. Takeuchi, "Object Oriented Programming in Concurrent Prolog," *New Generation Computing*, vol. 1, no. 1, pp. 25–48, OHMSHA Ltd. and Springer-Verlag, New York, 1983.

201. S. M. Shatz and J. P. Wang, "Introduction to Distributed-Software Engineering," *Computer*, pp. 23–32, IEEE, October 1987.

202. D. E. Shaw, *On the Range of Applicability of an Artificial Intelligence Machine*, Technical Report, Columbia University, New York, January 1985.

203. Y. Shih and K .B. Irani, "Large Scale Unification Using a Mesh-Connected Array of Hardware Unifiers," *Proc. Int'l Conf. on Parallel Processing*, pp. 787–794, Pennsylvania State University Press, 1987.

204. Y. Shobatake and H. Asio, "A Unification Processor Based on Uniformly Structured Cellular Hardware," *Proc. 13th Int'l Symp. on Computer Architecture*, pp. 140–148, IEEE/ACM, 1986.

205. H. A. Simon, "Search and Reasoning in Problem Solving," *Artificial Intelligence*, vol. 21, pp. 7–29, North-Holland, New York, 1983.

206. H. A. Simon, "Whether Software Engineering Needs to Be Artificially Intelligent," *Trans. on Software Engineering*, vol. SE-12, no. 7, IEEE, July 1986.

207. R. G. Smith, "The Contract Net Protocol: High-Level Communication and Control in a Distributed Problem Solver," *Trans. on Computers*, vol. C-29, no. 12, IEEE, 1980.

208. A. Synder, *Object-Oriented Programming for Common Lisp*, Report ATC-85-1, Software Technology Lab., Hewlett-Packard Lab., Palo Alto, CA, 1985.

209. C. Stanfill and D. Waltz, "Toward Memory-Based Reasoning," *Comm. of the ACM*, pp. 1213–1228, December 1986.

210. M. Stefik and D. G. Bobrow, "Object-Oriented Programming: Themes and Variations," *AI Magazine*, pp. 40–62, AAAI, Spring 1986.

211. S. J. Stolfo and D. E. Shaw, *DADO: A Tree-Structured Machine Architecture for Production Systems*, Technical Report, Columbia University, New York, March 1982.

212. S. J. Stolfo and D. P. Miranker, "The DADO Production System Machine," *J. Parallel and Distributed Computing*, vol. 3, no. 2, pp. 269–296, Academic Press, New York, 1986.

213. S. J. Stolfo, "Initial Performance of the DADO-2 Prototype," *Computer*, pp. 75–84, IEEE, January 1987.

214. H. S. Stone, "Parallel Querying of Large Databases: A Case Study," *Computer*, pp. 75–84, IEEE, January 1987.

215. S. Y. W. Su, "Associative Programming in CASSM and Its Applications," *Proc. 3rd Int'l Conf. on Very Large Databases*, pp. 213–228, Morgan Kaufmann, Los Altos, CA, 1977.

216. P. A. Subrahmanyam and J. H. You, "FUNLOG: A Computational Model Integrating Logic Programming and Functional Programming," *Logic Programming*, ed. D. DeGroot and G. Lindstrom, Prentice-Hall, Englewood Cliffs, NJ, 1985.

217. S. Sugimoto, K. Agusa, K. Tabata, and Y. Ohno, "A Multi-Microprocessor System for Concurrent Lisp," *Proc. Int'l Conf. on Parallel Processing*, pp. 135–143, IEEE, 1983.

218. G. J. Sussman and D. V. McDermott, "From PLANNER to CONNIVER–A Genetic Approach," *Fall Joint Computer Conf.*, vol. 41, pp. 129–137, AFIPS Press, 1972.

219. G. J. Sussman, J. Holloway, G. L. Steel, Jr., and A. Bell, "Scheme-79–Lisp on a Chip," *Computer*, vol. 14, no. 7, pp. 10–21, IEEE, July 1981.

220. K. Takahasi, H. Yamada, H. Hagai, and K. Matsumi, "A New String Search Hardware Architecture for VLSI," *Proc. 13th Int'l Symp. on Computer Architecture*, pp. 20–27, IEEE/ACM, 1986.

221. A. Takeuci and K. Fukukuwa, "Parallel Logic Programming Languages," *Proc. 3rd Int'l Conf. on Logic Programming*, Springer-Verlag, New York, 1986.

222. K. Taki, Y. Kaneda, and S. Maekawa, "The Experimental Lisp Machine," *Proc. 6th Int'l Joint Conf. on Artificial Intelligence*, pp. 865–867, William Kaufmann, Los Altos, CA, August 1979.

223. K. Taki, M. Yokota, A. Yamamoto, H. Nishikawa, S. Uchida, H. Nakashima, and A. Mitsuishi, "Hardware Design and Implementation of the Personal Sequential Interface Machine (PSI)," *Proc. Int'l Conf. on Fifth Generation Computer Systems*, pp. 398–409, ICOT and North-Holland, New York, 1984.

224. W. Teitelman and L. Masinter, "The Interlisp Programming Environment," *Computer*, vol. 14, no. 4, pp. 25–33, IEEE, April 1981.

225. M. F. M. Tenorio and D. I. Moldovan, "Mapping Production Systems into Multiprocessors," *Proc. Int'l Conf. on Parallel Processing*, pp. 56–62, IEEE, 1985.

226. E. Tick and D. H. D. Warren, "Towards a Pipelined Prolog Processor," *New Generation Computing*, vol. 2, no. 4, pp. 323–345, OHMSHA Ltd. and Springer-Verlag, New York, 1984.

227. S. Torii et al., "A Database System Architecture Based on a Vector Processing Method," *Proc. 3rd Int'l Conf. on Data Engineering*, IEEE, February 1987.

228. P. C. Treleaven and I. G. Lima, "Japan's Fifth-Generation Computer Systems," *Computer*, vol. 15, no. 8, pp. 79–88, IEEE, August 1982.

229. A. B. Tucker, Jr., "A Perspective on Machine Translation: Theory and Practice," *Comm. of the ACM*, vol. 27, no. 4, pp. 322–329, April 1984.

230. S. Uchida, "Inference Machines in FGCS Project," *Proc. VLSI Int'l Conf.*, IFIP TC-10, WG 10.5, August 1985.

231. K. Ueda, *Guarded Horn Clauses*, Technical Report TR-103, ICOT, Tokyo, Japan, 1985.

232. L. M. Uhr, "Parallel-Serial Production Systems," *Proc. 6th Int'l Joint Conf. on Artificial Intelligence*, pp. 911–916, William Kaufmann, Los Altos, CA, August 1979.

233. D. Unger, "Generation Scavenging: A Non-Disruptive High Performance Storage Reclamation Algorithm, "*ACM. Sigsoft*, vol. 9, no. 3, pp. 157–167, May 1984.

234. D. Unger, R. Blau, P. Foley, D. Samples, and D. A. Patterson. "Architecture of SOAR: Smalltalk on RISC," *Proc. 11th Annual Int'l Symp. on Computer Architecture*, pp. 188–197, IEEE/ACM, 1984.

235. D. Ungar and D. Patterson, "What Price Smalltalk?" *Computer*, pp. 67–74, IEEE, January 1987.

236. S. R. Vegdahl, "A Survey of Proposed Architectures for the Execution of Functional Languages," *Trans. on Computers*, vol. C-33, no. 12, pp. 1050–1071, IEEE, December 1984.

237. J. S. Vitter and R. A. Simons, "New Classes for Parallel Complexity: A Study of Unification and Other Complete Problems," *Trans. on Computers*, vol. C-35, no. 5, pp. 403–418, IEEE, May 1986.

238. B. W. Wah, G. J. Li, and C. F. Yu, "Multiprocessing of Combinatorial Search Problems," *IEEE Computer*, vol. 18, no. 6, pp. 93–108, June 1985. Also in Tutorial: Computers for Artificial Intelligence Applications, ed. B. W. Wah, IEEE Computer Society, 1986, pp. 173–188.

239. B. W. Wah, "Guest Editor's Introduction: New Computers for Artificial Intelligence Processing," *Computer*, vol. 20, no. 1, pp. 10–15, IEEE, January 1987.

240. D. L. Waltz, "Applications of the Connection Machine," *Computer*, vol. 20, no. 1, IEEE, January 1987.

241. D. H. Warren, L. M. Pereira, and F. Pereira, "Prolog—The Language and Its Implementation Compared with Lisp," *Proc. Symp. on Artificial Intelligence and Programming Languages*, also *SIGART Newsletter*, vol. 64, pp. 109–115, ACM, August 1977.

242. P. Wegner and B. Shriver (eds.), "Special Issue on Object-Oriented Programming Workshop," *SIGPLAN Notices*, vol. 21, no. 10, ACM, October 1986.

243. M. Weiser, S. Kogge, M. McElvany, R. Pierson, R. Post, and A. Thareja, "Status and Performance of the ZMOB Parallel Processing System," *Proc. COMPCON Spring*, pp. 71–73, IEEE, February 1985.

244. L. C. West, "Picosecond Integrated Optical Logic," *Computer*, vol. 20, pp. 34–47, IEEE, December 1987.

245. T. Winograd, "Frame Representations and the Declarative/Procedural Controversy," *Representation and Understanding: Studies in Cognitive Science*, pp. 185–210, Academic Press, New York, 1975.

246. P. H. Winston and B. Horn, *Lisp*, Second Edition, Addison-Wesley, Reading, MA, 1984.

247. Y. Yamaguchi, K. Toda, and T. Yuba, "A Performance Evaluation of a Lisp-Based Data-Driven Machine (EM-3)," *Proc. 10th Annual Int'l Symp. on Computer Architecture*, pp. 363–369, IEEE/ACM, June 1983.

248. Y. Yamaguchi, K. Todo, J. Herath, and T. Yuba, "EM-3: A Lisp-Based Data-Driven Machine," *Proc. Int'l Conf. on Fifth Generation Computer Systems*, pp. 524–532, ICOT and North-Holland, New York, 1984.

249. C. F. Yu, *Efficient Combinatorial Search Algorithms*, Ph.D. Thesis, School of Electrical Engineering, Purdue University, West Lafayette, IN, December 1986.

250. L. A. Zadeh, "Fuzzy Sets," *Information and Control*, pp. 338–353, 1965.

251. L. A. Zadeh, "Approximate Reasoning Based on Fuzzy Logic,"*Proc. 6th Int'l Joint Conf. on Artificial Intelligence*, pp.1004–1010, William Kaufmann, Los Altos, CA, August 1979.

CHAPTER 2

ARCHITECTURAL FEATURES OF LISP COMPUTERS

Andrew R. Pleszkun and Matthew J. Thazhuthaveetil

Lisp has been a popular programming language for well over 20 years, and it has become the premier language of artificial intelligence research. Recent interest in fifth-generation computer systems has sparked renewed interest in systems for the efficient execution of Lisp and Lisplike languages. There are two main reasons for Lisp's popularity: extensibility and flexibility. Lisp allows the construction of powerful, friendly systems starting with just a few basic primitives and data types. In this connection, Lisp has been likened to a ball of mud [36]: you start with a small one, add features to it, and it is still a ball of mud. Lisp is dynamically typed and ideally suited for incremental program development, making it a good choice for the fast prototyping of software systems. Several current symbolic manipulation and algebraic systems [30], design and graphic description systems, expert systems, and other heavily used non-numeric systems are also based on an underlying Lisp system. Typically these systems do not run efficiently because of the large semantic gap between list-manipulating languages like Lisp and the conventional von Neumann machine.

In this survey, we discuss what takes place during typical Lisp execution. Based on this, we enumerate the run-time requirements of a Lisp system and identify potential obstacles to good machine performance. We classify the Lisp machines that we encounter during our survey, and we examine how these machines cater to Lisp's run-time requirements.

© 1987 IEEE. Reprinted, with permission, from *IEEE Computer*, vol. 20, no. 3, pp. 35–44, March 1987.

1 RUN-TIME REQUIREMENTS OF A LISP SYSTEM

Lisp originated in the late 1950s as a list-processing language [31]. This version of Lisp is known today as Lisp 1 or "pure Lisp" and was mathematically correct but awkward to program. Since the 1950s, Lisp has undergone a steady evolution, with Lisp 1.5 [52], Maclisp [33], Interlisp [50], Scheme [49], Franz Lisp [17], and T [42] being among the more prominent stages on the way. This proliferation of Lisps has prompted efforts to arrive at a Lisp standard, first in Standard Lisp [27] and, more recently, in Common Lisp [48].

A Lisp program is organized as a collection of functions that call each other. Lisp execution can therefore be thought of as a series of nested function evaluations. Extensive function calling is typical of Lisp. Another characteristic of Lisp is the data structures that these functions operate on—they are, by and large, lists. Understanding these Lisp hallmarks lays the foundation for discussing and contrasting different Lisp machine architectures.

1.1 Function Calling

In general, function calls and returns are expensive operations. Lisp function calls, however, are more complicated than those in a lexically scoped language like Pascal, since Lisp function evaluation takes place in a dynamically bound context; the *latest active value* bound to a variable name is used when that variable is referenced. At any instant in Lisp execution, there are a number of dynamically nested, uncompleted function calls. Only one of these, the most recent, is active. Each of these function calls has a *referencing context* associated with it. The referencing context is a set of name-value pairs that specifies the current bindings of the variable names used in the function. We will use the term *environment* to refer to the collection of referencing contexts corresponding to all the function calls that are uncompleted at a given time. When a function call completes and returns control to its caller, the referencing context of that caller must be restored to allow it to become active and continue execution. It is therefore necessary to update the environment upon every function call and function return. When a variable name is encountered during the evaluation of the body of a function, the environment is *interrogated* for the current binding of that name. We will also refer to environment interrogation as *name lookup*. Maintaining the environment is not conceptually difficult; a simple scheme using a name-value binding stack would simply have to add and delete items from the top of the stack on function calls and returns. In Lisp, however, where function calling is very frequent, it is essential that environment interrogation be fast; the simple name-value binding stack could result in slow lookup.

In a simple form of Lisp function evaluation, the definition of a function specifies exactly how many arguments the function expects. Calling a function with a different number of arguments is an error. When the function is called, the arguments in the call are evaluated and then bound to the formal arguments of the function. The environment must be modified so that these new bindings are present in the currently active referencing context. The body of the function is then evaluated, and a return value made ready for the calling function. Finally, the environment is again modified to mirror the referencing context of the calling function, to which control is now returned.

Several variations of this mode of function evaluation have arisen. In one variation, functions are allowed to accept a variable number of arguments. This is useful in program development and has made Lisp a popular prototyping language. Other variations do not evaluate arguments or lambda variables. The syntactic conventions used to specify these function calls vary from Lisp to Lisp [24]. Finally, there are variations in which a function is allowed to return more than one value to the function that called it. All of these variations make it difficult to provide generally applicable architectural support for function calling in Lisp.

Unfortunately, the woes of Lisp function calling do not end here. In Lisp, functions can be passed as arguments. Such a functional argument is bound to a formal argument, as with all arguments, but when it is executed, the evaluation must be conducted in the referencing context that was present when the functional argument was passed. This implies that function calling and returning are not strictly LIFO (last in first out) and complicates the implementation of environments. If function calling and returning were strictly LIFO, the environment could be implemented using a LIFO stack. With the advent of functional arguments, information about the referencing context of a function call might have to be retained even after the call has returned. The problem of maintaining the environment consistently under such conditions is called a *funarg problem*; a funarg is a function-environment pair [2]. A technique to deal with this problem is described in [5].

Thus, there are three problems that make efficient function calling critical to Lisp system performance—the high frequency with which function calls are made, the existence of a number of calling conventions, and the need to maintain the environment efficiently across function calls.

1.2 Dealing With Lists

The fundamental data structure in Lisp is the list. More specifically, Lisp data objects are called *s-expressions*, short for *symbolic expressions*. Two special cases of s-expressions are *atoms* and *lists*. Examples of atoms are numbers and names (character strings). A list is a collection of atoms and other lists; recursively defined data structures are thus possible. In Lisp notation, a list consists of a left parenthesis followed by zero or more atoms or lists separated by spaces and ending with a right parenthesis. Lists are typically represented as linked lists of *list cells*. A list cell is composed of a pair of pointers: the *car* pointer points to the contents of that list cell (which could be an atom or another list), and the *cdr* pointer points to the next list cell in the list. The special atom *nil* terminates lists and is also equivalent to the empty list. We will use the term *heap memory* to refer to the memory containing all the list

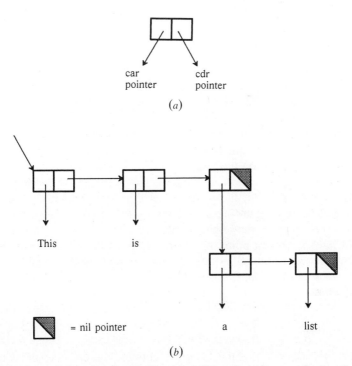

Figure 2.1 A list cell and a sample list: (*a*) A list cell, with car and cdr pointers; (*b*) Representation of (This is (a list)) using list cells.

cells. Figure 2.1 illustrates a list cell, with its car and cdr pointers, and shows how these list cells are used to represent lists.

Lisp provides a set of predefined primitive functions to manipulate lists. This set includes cons (used to create a new list cell), car (returns the first element of a list argument), cdr (returns everything following the first element of a list argument), rplaca (used to replace the car pointer of the list argument), and rplacd (used to replace the cdr pointer of the list argument). The data manipulated by Lisp programs is stored in lists, and accessing it often involves traveling down several pointers at run time using this set of primitive functions. A large fraction of Lisp execution time is spent in following these pointers [15].

Early Lisp system did not include type declarations, and even today they are not needed in a properly written program. (In more recent versions of Lisp, such as Common Lisp [48], type declarations can be included. This provision permits more efficient execution of compiled Lisp programs.) The lack of type declarations facilitates quick program development, but complicates run-time operation. Because of Lisp's dynamic nature, all data type checking must be done at run time, which is less efficient than conducting these checks at compile time. As a simple example of the need for type checking, consider the arithmetic function call (subtract X Y). The values of X and Y could be either integers or real numbers; a different action must be initiated in each case. This must be decided by run-time type-checking hardware. For example, the Symbolics 3600 provides hardware support for 34 data types, including symbols, list cells, infinite-precision integers, floating-point numbers, complex numbers, rational numbers, and arrays. Data words are tagged with a six-bit type tag (some six-bit sequences are unused), and hardware support is provided to enable type checking in parallel with the computation.

The management of heap space is also important in dealing with lists. In a language like Fortran, a compiler can predict the exact run-time memory requirements of any program. This is not possible in Lisp, where the number of list cells required during the evaluation of a program cannot be predicted by examining the program. Further, while the programmer explicitly causes new data objects to be created, the programmer does not explicitly cause them to be reclaimed when they are no longer referenced. Left to itself, then, a Lisp program would soon run out of heap space. To prevent this from happening, list cells that have no extant pointers pointing at them (which are called *garbage cells*), have to be recovered by the Lisp system and made available for reuse. Lisp did not attain widespread popularity until there were efficient solutions to this problem.

1.3 Summary

Based on the discussion of the last two sections, we consider the four major problem areas in efficient Lisp execution to be (1) efficient function calling, (2) environment maintenance, (3) list access and representation, and (4) heap maintenance. This assessment is confirmed by a survey of current Lisp machine designs; most of the special architectural features included in these machines address one or the other of the issues listed above. The benchmark programs offered by Gabriel [19] exercise one or more of these architectural features. The commentary given with each benchmark provides a guide for using the benchmark to evaluate a particular feature.

2 APPROACHES TO LISP MACHINE DESIGN

The first implementation of Lisp was done between 1958 and 1960 on an IBM 704. In fact, it is from the architecture of the 704 that the Lisp access primitives car and cdr get their names. The 36-bit data word of the 704 had two 15-bit fields, called the *address* and the *decrement*, which could be independently fetched to index registers using special instructions. The names *car* (contents of the address part of the operand) and *cdr* (contents of the decrement part of the operand) evolved from the representation of two-pointer list cells in these data words. Later on, it was with Lisp in mind that the designers of the DEC PDP-6 and PDP-10 computers included half-word instructions and stack instructions in these architectures. The Interlisp and Maclisp implementations on such machines, developed at Bolt, Bernark; and Newman (BBN) and MIT respectively, caused Lisp to gain popularity in the artificial intelligence community. The availability of these interactive Lisps led to a steady increase in the complexity of user programs and Lisp implementations in the years that followed.

The next landmark in Lisp implementation came in 1977, with the MIT Lisp Machine project [4]. Inspired by work done on personal Lisp machines at Xerox Palo Alto Research Complex (PARC) [13, 11], the MIT Lisp machine was a single-user computer, thereby assuring the user of a higher degree of service than that obtained on a time-sharing system. The MIT Lisp machine was implemented with a microprogrammed architecture, using a very unspecialized processor for reasons of speed, cost, and ease of microprogram debugging. Key Lisp primitive functions were implemented directly in microcode. Many of the commercial Lisp machines that have appeared on the market since then are based on the MIT Lisp Machine design. Examples include the Symbolics 3600 [35], the LMI Lambda, and the TI Explorer.* Another commercially available Lisp machine is the FACOM Alpha [23]. At the same time, there has been ongoing research on alternative Lisp machine architectures, mainly at the university level.

The Lisp machines surveyed here can be roughly divided into three classes, as illustrated in Figure 2.2. First, there are the *Class M* machines, which are unspecialized microcoded Lisp processors, like the MIT Lisp Machine or the Xerox PARC Lisp machine projects mentioned above. Second, there are the *Class S* machines, which are multiprocessor systems in which each processor serves a specialized function. Examples of this are MLS (an airborne multiprocessing Lisp system at the University of Illinois) [53], the Fairchild FAIM-1 multiprocessor AI machine [10], and a Lisp machine project undertaken at Keio University in Japan [55]. Finally, there are the *Class P* machines, which are multiprocessor systems composed of pools of identical processing elements, which concurrently evaluate different parts of a Lisp program. Examples of this class of Lisp machines are Guzman's AHR Lisp machine [20], the EM-3 machine [54], and the Evlis machine project at Osaka University [55]. We will discuss how the different machine classes address our four architectural issues.

2.1 Function Calling

In Section 1.1 we saw several Lisp function-calling conventions. In practice, it is not necessary for a Lisp machine to support all of these conventions. For example, suppose a function-calling convention expects a fixed number of arguments that are all to be evaluated. To implement a calling convention that does not evaluate the function arguments, the evaluation must be suppressed. In most Lisps this can be done using the quote function, which suppresses the evaluation of its arguments. To allow a variable number of arguments, the actual arguments can be passed as a single argument in the form of a list, which is then split up into its parts (using car and cdr) in the function. Similarly, to implement a calling convention that evaluates its arguments

---

* Explorer is a trademark of Texas Instruments, Inc.

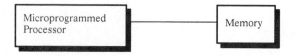

Class M Machine: Unspecialized Microprogrammed Processors

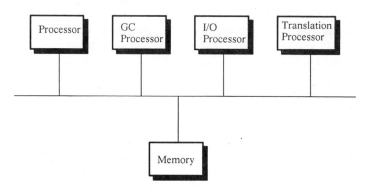

Class S Machine: Special-purpose Processors

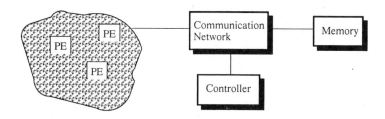

Class P Machine: Pool of Identical Processors

Figure 2.2 Alternative Lisp machine organizations.

using a calling convention that does not, we could cause the arguments to be explicitly evaluated in the function before the function body itself is evaluated.

Lisp functions are represented as lists. Low-level primitives like car and cdr are typically implemented as a few machine instructions. The interpretation of a user-defined function involves frequent accesses to the list representation of the function, interspersed with the execution of machine instructions corresponding to primitive operations. To speed up the evaluation process, Lisp compilers have been developed. A compiler reduces a function to a set of machine instructions, and this results in faster execution. On the negative side, compilation sometimes requires the programmer to declare the properties of functions and variables. This weakens the flexibility of Lisp programming. When comparing Lisp machines using a benchmark program, one must be careful to note whether performance results are for a compiled or interpreted version of the program.

Class M and Class S machines typically provide a range of function-calling support. In the MIT Lisp Machine, for example, a function can exist in any of three forms: as a slow interpreted code, as faster macro-compiled code, or as directly compiled microcode. The programmer chooses among the three based on how frequently a function is used. In the case of interpreted or macro-compiled code, there are two function call instructions, CALL0 for calls with no arguments, and the slower CALL instruction for other functions. CALL initiates the evaluation of the function's arguments and their binding to formal parameters. The Symbolics 3600 divides this functionality over several instructions—there are a set of function-calling instructions (including a special instruction for function calls that have no arguments) and a set of argument-binding instructions. Both the MIT Lisp Machine and the Symbolics 3600 have function return instructions that allow multiple values to be returned. An interesting variation on the CALL instruction was used in a MicroLisp project at Xerox [11]. Based on an observation that most functions have between 0 and 6 arguments, eight kinds of function calls are supported—for functions with 0, 1, 2, 3, 4, 5, 6, and more than 6 arguments. In the last case, the actual number of arguments is specified as a separate argument in the call instruction. A similar scheme is used in the Symbolics 3600.

Several Class P machines have attempted to add another dimension to function calling by evaluating the arguments of a function call in parallel. In Guzman's multimicroprocessor Lisp machine, AHR [20], and the Evlis multiprocessor Lisp machine at Osaka University [55], the evaluation of each argument is forked off onto a separate processor. As soon as all of a function's arguments have been evaluated, that function is scheduled on a free processor. The EM-3 machine [54] attempts even more. To bring about increased amounts of parallel evaluation, the EM-3 allows incomplete results to be sent forward. Thus, when all of the arguments of a function have become available, a pseudo-result is generated for that function and returned to its caller. Since some of those arguments could themselves be pseudo-results, the evaluation of a function proceeds in parallel with that of its arguments. Evaluation gets blocked when actual results, not pseudo-results, are needed. Such an evaluation scheme clearly involves a complicated control mechanism. To reduce the overhead of this control complexity, several researchers have suggested language extensions permitting the Lisp user to specify where parallel argument evaluation can be safely performed [21].

2.2 Maintaining the Environment

Recall that we refer to the collection of referencing contexts (made up of name-value binding pairs) in effect at a given time as the environment of Lisp execution. This environment must be modified on function call and on function return.

The most straightforward way to implement the environment would be as a linear linked list of name-value pairs. Items are appended to the head of this association list on function calls and deleted from the head of the list on function returns. Whenever a variable is referenced during the evaluation of a function, the association list is searched (from the head) for the first, and hence most recently active, instance of that variable name. In the worst case, this variable lookup might involve scanning the entire association list. This implementation of the environment is called *deep binding*. Figure 2.3 illustrates how the association list changes over a function call.

An alternative, called *shallow binding*, maintains a table with one entry for each variable name. Each entry contains the current value binding of that name. The table is often called the *oblist* or *global symbol table*. Each variable name thus has a value cell (in the table) associated with it. Figure 2.4 illustrates how a shallow-bound system maintains the environment over the same function call used in Figure 2.3. Shallow binding changes the name interrogation problem from a list search to a simple table lookup. To maintain consistency, the table of value cells has to be modified on each function call and return. Bindings that are in danger of being overwritten by a newer value have to be saved (typically on a binding stack), to be used in restoring the table to its original state on function return. This table modification procedure is more complicated than in the case of a simple deep-bound association list, where all deletions and additions to the list take place at the head.

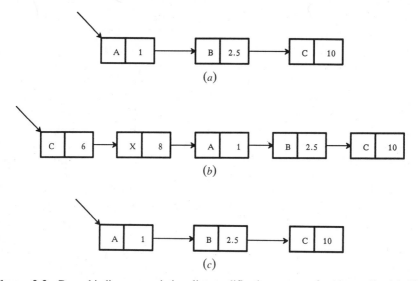

Figure 2.3 Deep binding: association list modification over a function call: (*a*) The association list is a linked list of name-value binding pairs. On variable lookup, the list is searched starting from its head. The environment shown has 'A' bound to value 1, 'B' bound to 2.5, 'C' bound to 10; (*b*) When a function call takes place new bindings get added to the head of the association list. In this example, two new bindings are added, one for 'C' and one for 'X'. The current environment still has 'A' bound to 1 and 'B' bound to 2.5. 'C' is now bound to 6 and 'X' to 8; (*c*) On function return the bindings that were added on that function's call are removed from the head of the association list, leaving the environment in the same state that it was in before the call.

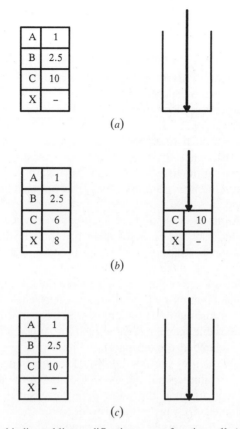

Figure 2.4 Shallow binding: oblist modification over a function call: (*a*) A shallow bound environment is maintained with a table of bindings (oblist) and a stack of old bindings for names that have been reused. This stack is initially empty; (*b*) On a function call some oblist entries get updated to reflect new bindings.The old values of these bindings are pushed onto the stack. Variable lookup is simplified to oblist table lookup; (*c*) On function return, the items are popped from the binding stack and used to update the oblist. The environment is thus modified to the state that it was in prior to the function call.

In choosing one implementation over the other, there is clearly a tradeoff between fast function calling and fast variable lookup. Further, it is possible to conceive of a continuum of implementation schemes between deep and shallow binding.

The MIT Lisp Machine supports a shallow-bound environment. On function call, some value cells get modified to reflect the new name-value bindings associated with the call. The old values are stored on a binding stack, from which they are recalled on function return, to restore the the modified value cells to their original values. A similar scheme is used in the Symbolics 3600. As an optimization, there is a bit in each frame on the control stack that indicates whether there are value cell modifications associated with that call. This saves simple function calls from having to pay the shallow binding overhead.

Most of the other machines surveyed here support a deep-bound impleme-ntation of environments. The penalties associated with variable lookup are reduced with the help of architectural support. For example, in [13] and [11] a caching scheme is used, so that repeated references to the same variable in the same function cause only one expensive lookup to be made, resulting in cost savings of as much as 80 percent.

Deutsch also describes a compiler optimization for references to variables that are bound at the top level and never re-bound. In a normal deep-bound implementation, each such reference would involve a search of almost the entire association list, since the bindings made at the top level would be at the end of the association list. A compiler can resort to a shallow binding technique for such references. Unfortunately, this optimization cannot be used for interpreted code.

The FACOM Alpha [23, 1] supports deep-bound environments with a *value cache*. The value cache is an associative memory device that is searched before the association list during the lookup process. The association list is maintained in the frames of the control stack; the name-value bindings added to the environment on a function call are stored in the control stack frame associated with that function call. Each value cache entry is composed of a validity bit, a stack frame number (to identify which function call it belongs to), and fields for the variable name and value binding. When a function is called, the value cache is searched for the names of its formal arguments and local variable. Such cache entries are invalidated. Then, for each variable lookup that is a miss in the value cache, the usual lookup (in the association list) is done, after which the corresponding value cache entry is updated, and the value cache entry is validated. On function return, the value cache is again searched, and all entries whose frame numbers are the same as that of the current function are invalidated. Figure 2.5 illustrates how a value cache improves lookup time in a system with deep-bound environments.

Class P machines have added problems in maintaining the environment. The environment of each ongoing evaluation must be available; instead of an association list, there is an association tree rooted at the global environ-ment of the top level. A new branch is added for each function evaluation. Since arguments are evaluated in parallel, these branches grow in parallel. To perform a variable lookup, a processor specifies the variable name and the head of the association tree branch corresponding to its environment. Class P machines typically use deep-bound implementations of environments. An effi-cient method of specifying to which branch of the association tree a reference belongs is described in [37].

Though we have described Lisp as being dynamically scoped, several mod-ern Lisps are lexically scoped; examples include Scheme, T, Common Lisp, and NIL. Under lexical scoping, a variable name is visible only in the lexical context of its binding, as in a language like Pascal. This modification simplifies the task of compiling Lisp functions, and removes the name lookup problem. The designers of Scheme argue against dynamic scoping since it violates *referential transparency*, that is, the requirement that "the meanings of the parts

Name	Value	Valid?	Frame no.
A	10	Yes	1
B	4	Yes	1
C	1	Yes	1

(*a*)

Name	Value	Valid?	Frame no.
A	10	No	1
B	4	Yes	1
C	1	No	1

(*b*)

Name	Value	Valid?	Frame no.
A	11	Yes	2
B	4	Yes	1
C	0	Yes	2

(*c*)

Name	Value	Valid?	Frame no.
A	11	No	2
B	4	Yes	1
C	0	No	2

(*d*)

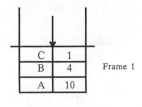

Figure 2.5 FACOM Alpha value cache operation: (*a*) The FACOM Alpha uses a value cache and an association list (maintained in the control stack) in its optimized deep-bound environment; (*b*) If a function call with 'A' as formal argument and 'C' as a local variable name is made, the value cache and stack are updated as shown; (*c*) When 'A' and 'C' are referenced in the function body, the a-list is searched for the latest binding. The value cache is then updated; (*d*) On function return a stack frame is popped and all value cache entries with that frame number are invalidated.

of a program be apparent and not change, so that such meaning can be reliably depended upon" [47].

In summary, there are two main schemes for maintaining run-time refer-encing environments: deep binding and shallow binding. Deep-bound environ-ments allow for fast function calls and returns at the expense of potentially slow name interrogation, while shallow-bound environments make name interrogation fast at the expense of slower function calls. Many machines use a combination of these techniques.

2.3 Efficient Lisp Representation

In early Lisps, lists were represented as linked lists of two-pointer list cells. Each of the two pointers was large enough to address all of memory. This representation proves to be efficient at list accessing. With two-pointer list cells, both car and cdr primitives can be implemented as simple memory read operations, rplaca and rplacd as simple memory write operations, and cons as a list cell allocation followed by two memory write operations. However, a problem with this representation can be seen when we try to *traverse* a list, that is, access all the elements of a list. During a traversal, the address of the list cell to be accessed next is contained in one of the pointers of the cell that has just been accessed. The address can be forwarded to the memory system only after the previous access has been completed. This addressing bottleneck slows down list traversal; therefore, the two-pointer list cell representation is time-inefficient. Further, the representation is highly space-inefficient; studies have shown that it is not necessary for both the car and cdr pointers to span the entire address space [6]. Lisp machines generally provide more compact representations of lists. We classify these compact list representation schemes as being either *vector coded* or *structure coded*.

2.3.1 *Vector-Coded Representations of Lists* The basic idea behind vector-coded list representation schemes is to represent *linear lists* as linear vectors of symbols. We call a list linear if none of its elements is itself a list. Under such a scheme, a list cell is represented by a vector element, with its car pointer assumed by default to be a pointer to a symbol and its cdr pointer assumed by default to be a pointer to the next element in the vector. Pointers that differ from these default types are dealt with by providing for exception conditions. In traversing a list, the address of the next list cell to be accessed is, by default, the next location in the vector. The address generation bottleneck appears only when an exception condition occurs. This kind of representation was first proposed in [22]. Two examples of this basic vector-coded representation are the *conc* representation [25] and the *linked-vector* representation [26].

The conc representation calls its vectors *tuples*. A tuple is a list of elements stored in contiguous memory locations. It is accessed through a descriptor that specifies the number of elements in the tuple and a pointer to the beginning of the tuple. There are special tuples called *conc cells* whose elements are pointers to other conc cells or to tuples. Conc cells are used to implement list concatenation without having to modify the list structure. The concatenation of lists L1 and L2 in the naive two-point list cell representation involves modifying a pointer at the end of L1 to point to L2; in the conc representation, the operation involves allocating a conc cell and setting its fields to L1 and L2.

The linked-vector representation is another basic vector-coded representation. To take care of exception conditions, each vector element is tagged as

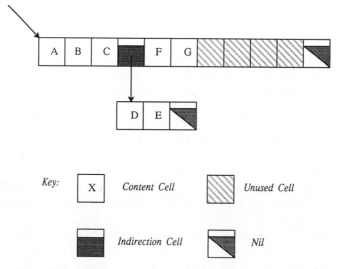

Figure 2.6 Linked-vector representation of (A B C (D E) F G).

either a default cell or an indirection cell. Default cells contain list elements, while indirection cells contain pointers to elements in other vectors (or to nil). The last cell in a vector is assumed to be an indirection cell. Further, as an optimization to make frequent vector compactions unnecessary, each vector element can be tagged as *unused*. All of the information about a vector element can be encoded into a two-bit tag, with the four possible tag sequences being used to differentiate between the cdr being nil, the cdr starting from the next cell, the current cell being an indirection cell, and the current cell being unused. Figure 2.6 illustrates the linked-vector representation.

Vector-coded representation schemes are both space-efficient and time-efficient. The main problem with the basic vector-coded representation scheme follows from the fact that the unit of memory allocation is the vector. If fixed-size vectors are used, either internal fragmentation (if the vectors are too large) or an excessive number of indirection cells (if the vectors are too small) results. If, on the other hand, variable-size vectors are used, memory management becomes difficult. A separate free list must be maintained for each vector size, and on every vector allocation, a decision must be made as to which vector size should be allocated. The cdr-coding schemes that we describe next overcome this problem by allocating memory in units of vector elements.

The cdr-coding representation scheme is used in the MIT Lisp Machine, the Symbolics 3600, and many other Lisp machines. With this scheme, lists are represented using cdr-coded list cells, which are made up of a large car pointer and a small cdr-code. For example, in the MIT Lisp Machine, the two fields are 29 bits and 2 bits wide respectively. The four possible cdr-code bit sequences are called cdr-normal, cdr-error, cdr-next, and cdr-nil. Cdr-next and cdr-nil approximate a vector-coded representation.

A vector is composed of a set of contiguous cells whose cdr-codes are cdr-next, except for the last cell in the vector, which has a cdr-code of cdr-nil. The cdr of each list cell is simply the cell next to it. The cdr-normal code is provided for cases where such a vector representation is not possible. A cell with a cdr-code of cdr-normal is assumed to have its cdr-code of cdr-error. Thus, if a cell has a cdr-code of cdr-normal, then that cell and its neighbor resemble a normal two-pointer list cell. When lists are destructively modified (by rplaca and rplacd operators) during Lisp execution, the compact vector-coded parts of a list might have to be modified into less space-efficient structures, possibly using indirect pointers. Dereferencing such indirect pointers during list access involves extra memory activity. *Invisible pointers* are used to reduce this indirection cost; an invisible pointer is automatically dereferenced by the hardware, thereby reducing the indirection cost and overhead associated with the data structure reference. Figure 2.7 illustrates the MIT Lisp Machine cdr-coded representation.

The compact list representation scheme employed in [13] is also called cdr-coding. This scheme uses a 24-bit car field and an 8-bit cdr-code. A cdr-code of 0 means that the cdr is nil, while cdr-codes between 1 and 127 are interpreted as offsets to be added to the current list cell address to obtain the cdr of that list cell. A cdr-code of 128 means that the cdr is located at the address specified in the car field of the list cell, and cdr-code values 129 to 255 represent the offset from the current list cell where the address of the cdr is located. This interpretation of cdr-codes was chosen largely from working set considerations, since the Lisp system operated in a paged virtual memory system with a page size of 256 words.

2.3.2   *Structure-Coded Representations of Lists*   The basic idea behind structure-coded list representation schemes is to attach to each list cell a tag that specifies the position of that cell in the list structure. In addition, list structures are stored on associative memory devices. The combination of detailed structural information at each node and associative search capabilities has the

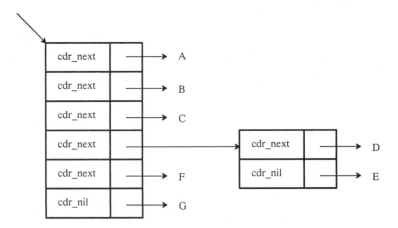

Figure 2.7   MIT Lisp machine cdr-coded representation of (A B C (D E) F G).

potential for fast list access and traversal. In list traversal, the address of a list element can be generated without having to look at the preceding list cell.

A scheme for representing binary trees on associative devices is described in [32]. Each node of the tree is tagged with a pair (l, k), where l is the depth of the node in the tree and k is the maximal number of nodes at level l that could precede the node in left-to-right order. Figure 2.8 illustrates this list representation. An extension of this node-tagging scheme is suggested for use in the BLAST (Better Lisp processing using ASsociative Tables) Lisp machine architecture [45], where the pair (l, k) is compressed into a single node number, $N = 2^{l-1} + k$.

A list can be mapped into such a tree with all the symbols in the list mapping

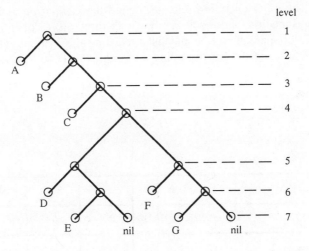

Node value	Minsky pair	BLAST number
A	(2,0)	2
B	(3,2)	6
C	(4,6)	14
D	(6,28)	60
E	(7,58)	122
F	(6,30)	62
G	(7,60)	126

Figure 2.8 Tree-coded representation of (A B C (D E) F G).

onto leaves in the tree; the structure of the list is described by the shape of the tree. Then the list can be represented compactly by remembering only its symbols and tagging each with its node number from the corresponding tree; a list is thus encoded as a set of (*node_number, symbol*) tuples. In BLAST, these tuples are stored in tables called *exception tables*. An associative searching capability on such tables is useful in implementing efficient list manipulation.

Two similar coding schemes are described in [39]. One of these, called *CDAR coding*, tags each symbol in a list with a string of 0s and 1s. This string specifies the series of car (represented by 0s) and cdr (represented by 1s) operations that, when applied to the list, yield that symbol as the result. This is equivalent to the node number used in BLAST; it specifies the position of the symbol in the list and can be used to make list access operations efficient. The other encoding scheme, called the *explicit parenthesis storage* (EPS) representation, tags each symbol with three pieces of information: the number of left parentheses in the list preceding the symbol, the number of right parentheses in the list preceding and immediately following the symbol,

(A B C (D E) F G)

Node Value	CDAR Code	EPS Representation		
		Left	Right	Position
A	000000	1	0	1
B	000001	1	0	2
C	000011	1	0	3
D	000111	2	0	4
E	010111	2	1	5
F	001111	2	1	6
G	011111	2	2	7

Key: Left   :   *Number of left parentheses in list to left of atom*

Right   :   *Number of right parentheses in list to left of and immediately following atom*

Position :   *Position of atom in list*

Figure 2.9   CDAR-coded and EPS representations of (A B C (D E) F G).

and the position of the symbol in the list. Figure 2.9 illustrates the CDAR-coded and EPS list representations.

The structure-coded schemes improve list access speed by eliminating repeated heap memory references to pointers when accessing a list. The heap address of a list item is basically generated by a base-plus-offset calculation. There are other ways to achieve the goal of improved list access speed. One approach includes storing structural information in a cache-like memory [38].

2.3.3 Summary The simplest way to represent lists is by using two-pointer list cells. The major drawback with this scheme is that the information needed to address a list cell is entirely contained in another list cell. Because of this, in performing a list access, the processor must wait for one memory access to complete before the next one can be initiated. We have seen two classes of more complicated list representation schemes: vector-coded and structure-coded. Simple linear lists can be represented compactly and accessed efficiently using vector-coded representation schemes. Unfortunately, not all lists are linear, and so vector-coded representation schemes include exception conditions to take care of more complex list structures. In such schemes, the information needed to address a list cell is still partially contained in another list cell. The Lisp machines that we surveyed used either the naive two-pointer list representation or a vector-coded list representation (typically cdr-coding.) Structure-coded representation schemes make it possible to address the elements of a list independently by attaching to each list cell a tag that describes its position in the list. There is no clear consensus on which type of list representation scheme is preferable. A major factor in this determination is how typical Lisp programs access lists [6, 7, 51]. For instance, if lists are usually accessed in a very well-structured manner (akin to the manner in which a Fortran DO loop accesses the elements of an array), then structure-coded representation schemes would be preferred.

2.4 Heap Maintenance

Finally, we turn our attention to heap maintenance strategies. By and large, Lisp machines have tagged architectures. Tags are used to specify data types; this greatly facilitates type checking in a dynamic language like Lisp, in which there are no type declarations. Tags also make it possible to distinguish between list pointers and atomic data, which turns out to be useful in managing the heap space.

Recall that in Lisp, it is the system's responsibility to manage the allocation and reclamation of list cells in the heap, since the user does not explicitly deallocate list objects. Since the reclamation of garbage cells is its most interesting facet, heap maintenance has come to be largely identified with *garbage collection*. Several basic garbage collection schemes are described in [81]. Clearly, the goal of garbage collectors is to reclaim garbage quickly and with minimal overhead.

Garbage collection is a two-stage process. We call the two stages *garbage detection* and *garbage reclamation*. In the garbage detection stage the system identifies which heap memory cells are no longer referred to, and then the garbage is treated and made ready for reuse in the garbage reclamation stage. Garbage reclamation often just involves adding the newly recognized garbage cell to a list of free list cells available for reallocation. In systems where lists are represented using compact representations, reclamation might also require compacting the parts of the heap that are in use. Garbage detection can be done in two ways, either by using *reference counts* or by *marking*.

In marking [44], all accessible list cells are marked starting with a set of root cells that are known to be nongarbage cells, following the pointers contained within them to other list cells, and so on. All list cells that are not marked at the end of this operation are not accessible and are reclaimable as garbage. Marking involves a space overhead of at least one bit per cell, for the mark. In marking schemes, garbage collecting need be initiated only when there is no more heap space available and a request for more space is made. So the overhead of heap management is felt only when there is no more free heap space.

In reference counting [9], a count is maintained for each heap cell of the number of extant pointers to it. A cell is known to be garbage when its reference count goes to zero. Reference counting has several drawbacks. One problem is the memory space required to maintain the reference counts. Each reference count could potentially become as large as the total number of list cells in the heap; accommodating this could require 100 percent space redundancy in the heap. Another problem is that heap users pay the price of garbage collection continuously in the form of this space overhead (a reference count field for every list cell) and a time overhead (the updating of these counts on heap accesses). Further, reference counting has the disadvantage of not, in general, being able to reclaim circular lists. Also, reference counting can lead to poor real-time performance. Consider what happens when the reference count of a list cell goes to zero. Before the list cell can be added to a free list of cells, the reference counts of the two list cells that it points at are decremented by one. This could cause the reference counts of these two cells to go to zero, in which case they would have to be reclaimed, with the reference counts of their descendants being decremented by one. Thus, the seemingly simple operation of reclaiming a list cell could initiate an arbitrarily large amount of reference count updating and list cell reclamation.

Several variations of these two basic schemes have been developed. Some variations are a combination of the two [12]. Others use clever marking schemes to permit marking in parallel with "useful" Lisp evaluation [14, 46, 41]. The two parallel processes are commonly referred to as the *collector* and the *mutator*. More than one bit is used to mark each list cell, implementing a form of mutual exclusion of access to the list cells by the collector and mutator. Another approach [3, 16] divides the heap space into two semispaces, the "oldspace" and the "newspace," only one of which will be undergoing garbage collection at a given time. The two semispaces are simultaneously active, but the garbage collector tries to relocate, or copy, all the accessible cells from the oldspace to the newspace. When that transfer has been completed, the two semispaces get flipped, with the oldspace becoming the new newspace and vice versa. These schemes are called *copying garbage collectors*. They can also be made to work in real time by performing a fixed number of relocations on every heap allocation request, resulting in copying incremental garbage collectors.

The MIT Lisp Machine supports Baker's real-time copying incremental garbage collection scheme [3]. Hardware features, such as tags, help in making the garbage collector efficient. Further, memory is organized as *areas*, each containing related list cells. The user can declare an area to be static and save the garbage collector the task of processing that area.

The Symbolics 3600 uses a similar garbage collection algorithm with additional hardware support [34]. Each memory page has an associated *page-tag*, which indicates whether or not that page contains pointers to memory areas that are of interest to the garbage collector. These tags help reduce the work of the garbage collector. Objects are identified as static (unlikely to become garbage) or ephemeral (assumed to contain cells of intermediate lifetimes), based on how much garbage collection activity they have required in the past. The virtual memory software maintains this information, as well as tables with other information useful to the garbage collector.

Despite its drawbacks, reference counting has also been incorporated into Lisp machine garbage collectors. The Machine for Lisp Like Languages (M3L) Project [43] uses a three-bit reference count field. This count does not include pointers that are on the run-time stack or in registers. Sansonnet et al. reported studies suggesting that this reference count suffices to reclaim about 98 percent of all inaccessible list cells. If stack and register pointers were included, reference counts would grow in proportion to the number of function calls, since arguments are passed on the stack. A separate one-bit reference flag is therefore maintained for each cell to indicate whether or not that cell is referred to by pointers on the stack and in registers. This has the unfortunate consequence that the reference flag might have to be updated on each stack operation. On a stack pop that involves a stack item containing a pointer to a list cell, the entire stack must be checked to see if there are any other references to that list cell, in order to determine whether or not to modify the list's reference flag.

The FACOM Alpha also uses a reference-counting garbage collection scheme [23]. The Alpha memory is organized as a number of subspaces. Reference counts are used to reclaim subspaces, not list cells. So, there is one reference count for each subspace. The reference count of a subspace counts only the pointers to cells in that subspace that originate from other subspaces. A marking scheme is used to detect garbage cells within subspaces.

In summary, mark-and-sweep algorithms are typically used to maintain the heap. (We include copying incremental garbage collectors in this class.) Different machines have variants of this basic technique that amortize time spent in garbage collection over the entire run of the program. This is important for responsive real-time performance. A few machines have incorporated reference-counting techniques into a basic mark-and-sweep strategy.

3 CONCLUSIONS

A survey of Lisp machines reveals that designers have, by and large, four issues in mind: fast function calls, environment maintenance, efficient list representation, and heap maintenance.

We do not see much variety in the architectural support for function calling and environment maintenance. Most machines provide a small set of call instructions to support the most frequent forms of function calls. They use either a shallow-bound implementation of environments or a deep-bound implementation. Some machines include a scheme for name-value bindings in order to make name lookups faster. The interesting research in these two areas involves concurrent execution and new evaluation paradigms.

How should lists be represented internally? The simple list representation scheme, with the two-pointer list cell, is both space-inefficient and time-inefficient in terms of the time required to traverse a list. We classified list representation schemes as either vector-coded or structure-coded; vector-coded representation schemes aim for space efficiency, while structure-coded schemes aim for time efficiency. Which representation scheme is optimal? The answer to this question will depend on the structure of typical lists, and how they are accessed. For instance, if most lists are simple linear lists that are not modified much, then vector-coded representation schemes will be preferable. If they are not, but happen to be accessed in well-structured ways, structure-coded list representation schemes can make efficient list access possible; list traversal can be performed more efficiently on structure-coded lists than on vector-coded lists.

All of the machines that we surveyed provide some degree of support for heap maintenance. Commercial machines generally provide support for real-time copying incremental garbage collection. Hybrid reference counting and mark-and-sweep schemes have also been used.

Three areas of Lisp machine design are ripe for further investigation. The first is the use of structure-coded list representations. The structure-coding schemes that we described in this chapter have not been used in actual Lisp machines. There have been few attempts at investigating the use of such techniques in improving Lisp machine performance [45]. The second area is the development of Multilisps. Lisp functions contain implicit parallelism; for example, the arguments to many functions could be evaluated concurrently [28, 54]. Multilisps extend Lisp with constructs that enable the programmer to explicitly indicate which parts of a function can be evaluated concurrently [40, 21, 18]. The third area of promising research is the development of multiprocessor Lisp systems that spread evaluation over several processors, taking advantage of both the implicit and explicit parallelism present in Lisp programs [21, 29]. Investigations in this area will require a re-evaluation of architectural features that support Lisp. In particular, we see the maintenance of the environment and the heap as the major areas of interest.

REFERENCES

1. H. Akimoto, S. Shimizu, A. Shinagawa, A. Hattori, and H. Hayashi, "Evaluation of the Dedicated Hardware in Facom Alpha," *Spring 1985 Compcon Digest of Papers*, pp. 366–369, 1985.

2. J. R. Allen, *Anatomy of Lisp*, McGraw-Hill Artificial Intelligence Series, 1978.

3. H. G. Baker, Jr., "Lisp Processing in Real Time on a Serial Computer," *Communications of the ACM*, vol. 21(4), pp. 280–294, April 1978.

4. A. Bawden, R. Greenblatt, J. Holloway, T. Knight, D. A. Moon, and D. Weinreb, "Lisp Machine Progress Report," MIT AI Laboratory Memo No. 444, August 1977.

5. D. G. Bobrow and B. Wegbreit, "A Model and Stack Implementation of Multiple Environments," *Communications of the ACM*, vol. 16(10), pp. 591–603, October 1973.

6. D. W. Clark and C. C. Green, "An Empirical Study of List Structure in Lisp," *Communications of the ACM*, vol. 20(2), pp. 78–87, February 1977.

7. D. W. Clark, "Measurements of Dynamic List Structure in Lisp," *IEEE Transactions on Software Engineering*, vol. 5(1), pp. 51–59, January 1979.

8. J. Cohen, "Garbage Collection of Linked Data Structures," *Computing Surveys*, vol. 3(13), pp. 341–367, September 1981.

9. G. E. Collins, "A Method for Overlapping and Erasure of Lists," *Communications of the ACM*, vol. 3(12), pp. 655–657, December 1960.

10. A. L. Davis and S. V. Robison, "The FAIM-1 Symbolic Multiprocessing System," *Spring 1985 Compcon Digest of Papers*, pp. 370–375, 1985.

11. L. P. Deutsch, "A Lisp Machine with Very Compact Programs," *Proc. 3rd Int'l Joint Conf. on Artificial Intelligence*, August 1973.

12. L. P. Deutsch and D. G. Bobrow, "An Efficient, Incremental, Automatic Garbage Collector," *Communications of the ACM*, vol. 19(9), pp. 522–526, September 1976.

13. L. P. Deutsch, "Experience With a Microprogrammed Interlisp System," *Proc. 11th Annual Microprogramming Workshop*, pp. 128–129, November 1978.

14. E. W. Dijkstra, L. Lamport, A. J. Martin, C. S. Scholten, and E. F. M. Steffens, "On-the-Fly Garbage Collection: An Exercise in Cooperation," *Communications of the ACM*, vol. 21(11), pp. 966–975, November 1978.

15. R. J. Fateman, "Is a Lisp Machine Different from a Fortran Machine?" *ACM SIGSAM Bulletin*, vol. 12(3), pp. 8–11, August 1978.

16. R. Fenichel and J. Yochelsen, "A Lisp Garbage-Collector for Virtual Memory Computer Systems," *Communications of the ACM*, vol. 12(11), pp. 611–612, November 1969.

17. J. K. Foderaro, *The Franz LISP Manual*, University of California, Berkeley, 1979.

18. R. P. Gabriel and J. McCarthy, "Queue-Based Multi-Processing Lisp," *Conf. Record of 1984 ACM Symposium on Lisp and Functional Programming*, pp. 25–43, August 1984.

19. R. P. Gabriel, *Performance and Evaluation of Lisp Systems*, The MIT Press, 1986.

20. A. Guzman, "A Heterarchical Multi-Microprocessor Lisp Machine," *Proc. IEEE Workshop on CAPAIDM*, pp. 309–317, November 11–13, 1981.

21. R. H. Halstead, Jr., "Implementation of MultiLisp: Lisp on a Multiprocessor," *Conference Record of 1984 ACM Symposium on Lisp and Functional Programming*, pp. 9–17, August 1984.

22. W. J. Hansen, "Compact List Representation: Definition, Garbage Collection, and System Implementation," *Communications of the ACM*, vol. 12(9), September 1969.

23. H. Hayashi, A. Hattori, and H. Akimoto, "Alpha: A High-Performance Lisp Machine Equipped with a New Stack Structure and Garbage Collection System," *Proc. 10th Int'l Annual Symposium on Computer Architecture*, pp. 342–348, June 1983.

24. M. A. Jones, "A Comparison of LISP Specifications of Function Definition and Argument Handling," *Sigplan Notices*, vol. 17(8), pp. 67–73, August 1982.

25. R. M. Keller, "Divide and CONCer: Data Structuring in Applicative Multiprocessing Systems," *Conference Record of 1980 ACM Lisp Conf.*, pp. 196–202, 1980.

26. K. Li and P. Hudak, "A New List Compaction Method," *Yale University Research Report*, February 1985.

27. J. B. Marti, A. C. Hearn, M. L. Griss, and C. Griss, "Standard LISP Report," *University of Utah Symbolic Computation Group Report No. 60*, 1978.

28. J. B. Marti, "Compilation Techniques for a Control-Flow Concurrent Lisp System," *Conference Record of 1980 ACM Lisp Conf.*, pp. 203–207, 1980.

29. J. B. Marti and J. P. Fitch, "The Bath Concurrent Lisp Machine," *Proc. EUROCAL 83, European Computer Algebra Conf.*, pp. 78–90, March 1983.

30. W. A. Martin and R. J. Fateman, "The MACSYMA System," *Proc. 2nd ACM Symposium on Symbolic and Algebraic Manipulation*, pp. 23–25, 1971.

31. J. McCarthy, "Recursive Functions of Symbolic Expressions and Their Computation by Machine, Part I," *Communications of the ACM*, vol. 3(4), pp. 184–195, April 1960.

32. M. Minsky, "Representation of Binary Trees on Associative Memories," *Information Processing Letters* (2), 1973.

33. D. A. Moon, *MACLISP Reference Manual*, Project MAC Technical Report, MIT, Cambridge, MA, 1974.

34. D. A. Moon, "Garbage Collection in a Large Lisp System," *Conference Record of 1984 Symposium on Lisp and Functional Programming*, pp. 235–246, 1984.

35. D. A. Moon, "Architecture of the Symbolics 3600," *Proc. 12th Annual Symposium on Computer Architecture*, pp. 76–83, July 1985.

36. J. Moses, "The Function of FUNCTION in Lisp," *Sigplan Notices*, June 1969.

37. J. A. Padget, "The Ecology of Lisp, or The Case for the Preservation of the Environment," *Proc. EUROCAL 83, European Computer Algebra Conf.*, pp. 91–100, March 1983.

38. A. R. Pleszkum and M. J. Thazhuthaveetil, "An Architecture for Efficient Lisp List Access," *Proc. 13th Annual Symposium on Computer Architecture*, June 1986.

39. J. L. Potter, "Alternative Data Structures for Lists in Associative Devices," *Proc. 1981 Int. Conf. on Parallel Processing*, pp. 486–491, 1983.

40. G. Prini, "Explicit Parallelism in Lisp-Like Languages," *Conference Record of 1980 Lisp Conf.*, pp. 13–18, August 1980.

41. A. Ram and J. H. Patel, "Parallel Garbage Collection without Synchronization Overhead," *Proc. 12th Annual Symposium on Computer Architecture*, pp. 84–90, July 1985.

42. J. A. Rees and N. I. Adams IV, "T: A Dialect of Lisp, or Lambda: The Ultimate Software Tool," *Proc. 1982 ACM Symposium on Lisp and Functional Programming*, August 1982.

43. J. P. Sansonnet, M. Castan, C. Percebois, D. Botella, and J. Perez, "Direct Execution of Lisp on List Directed Architectures," *Proc. Symposium on Architectural Support for Programming Languages and Operating Systems*, pp. 132–139, March 1982.

44. H. Schorr and W. Waite, "An Efficient Machine-Independent Procedure for Garbage Collection in Various List Structures," *Communications of the ACM*, vol. 10(8), pp. 501–506, August 1967.

45. G. S. Sohi, E. S. Davidson, and J. H. Patel, "An Efficient Lisp-Execution Architecture with a New Representation for List Structures," *Proc. 12th Annual Symposium on Computer Architecture*, 1985.

46. G. L. Steele, Jr., "Multiprocessing Compactifying Garbage Collection," *Communications of the ACM*, vol. 18(9), pp. 495–508, September 1975.

47. G. L. Steele Jr. and G. J. Sussman, "The Art of the Interpreter, or The Modularity Complex," *MIT AIL Memo No. 453*, May 1978.

48. G. L. Steele, Jr. et al., *Common Lisp Reference Manual*, Digital Press, 1984.

49. G. J. Sussman and G. L. Steele, Jr., "Scheme: An Interpreter for Extended Lambda Calculus," *MIT AIL Memo No. 349*, December 1975.

50. W. Teitelman, *INTERLISP: Interlisp Reference Manual*, Xerox PARC Technical Report, Palo Alto, CA, 1975.

51. M. J. Thazhuthaveetil, *SMALL: An Architecture for Efficient Lisp List Access*, Ph.D. Dissertation, University of Wisconsin, Madison, August 1986.

52. C. Weisman, *LISP 1.5 Primer*, Dickenson Press, 1967.

53. R. Williams, "A Multiprocessing System for the Direct Execution of Lisp," *Proc. 4th Workshop on Computer Architectures for Non-Numeric Processing*, pp. 35–41, 1978.

54. M. Yamamoto, "A Survey of High-Level Language Machines in Japan," IEEE *Computer*, pp. 68–78, July 1981.

55. Y. Yamaguchi, K. Toda, and T. Yuba, "A Performance Evaluation of a Lisp-Based Data-Driven Machine (EM-3)," *Proc. 10th Annual Symposium on Computer Architecture*, pp. 363–369, June 1983.

CHAPTER 3

SYMBOLICS ARCHITECTURE

David A. Moon

What is an architecture? In computer systems, an architecture is a specification of an interface. To be dignified by the name *architecture*, an interface should be designed for a long lifespan and should connect system components maintained by different organizations. Often an architecture is part of a product definition and defines characteristics on which purchasers of that product rely, but this is not true of everything that is called an architecture. An architecture is more formal than an internal interface between closely related system components and has farther-reaching effects on system characteristics and performance.

A computer system typically contains many levels and types of architecture. This article discusses three architectures defined in Symbolics computers:

1. System architecture—defines how the system appears to end users and application programmers, including the characteristics of languages, user interface, and operating system.

2. Instruction architecture—defines the instruction set of the machine, the types of data that can be manipulated by those instructions, and the environment in which the instructions operate, for example subroutine-calling discipline, virtual memory management, and interrupts and exception traps. This is an interface between the compilers and the hardware.

© 1987 IEEE. Reprinted, with permission, from *IEEE Computer,* vol. 20, no. 1, pp. 43–52, January 1987.

3. Processor architecture—defines the overall structure of the implementation of the instruction architecture. This is an interface between the firmware and the hardware, and also between the parts of the processor hardware.

1 SYSTEM ARCHITECTURE

System architecture defines how the system looks to the end user and to the programmer, including the characteristics of languages, user interface, and operating system. System architecture defines the product that people actually use; the other levels of architecture define the mechanism underneath that implements it. System architecture is implemented by software; hardware only sets bounds on what is possible. System architecture defines the motivation for most of the design choices at the other levels of architecture. This section is an overview of Symbolics system architecture.

The Symbolics system presents itself to the user through a high-resolution bitmap display. In addition to text and graphics, the display contains presentations of objects. The user operates on the objects by manipulating the presentations with a mouse. The display includes a continuously updated reminder of the mouse commands applicable to the current context. Behind the display is a powerful symbol processor with specialized hardware and software. The system is dedicated to one user at a time and shares such resources as files, printers, and electronic mail with other Symbolics and non-Symbolics computers through both local-area and long-distance networks of several types. The local-area network is integral to system operation.

The system is designed for high-productivity software development both in symbolic languages, such as Common Lisp [1] and Prolog, and in nonsymbolic languages, such as Ada and Fortran. It is also designed for efficient execution of large programs, particularly in symbolic languages, and delivery of such programs to end users. The system is intended to be especially suited to complex, ambitious applications that go beyond what has been done before; thus it provides facilities for exploratory programming, complexity management, incremental construction of programs, and so forth. The operating system is written in Lisp and the architectural concept originated at the MIT Artificial Intelligence Laboratory; however, applications are not limited to Lisp and AI. Many non-AI applications that are complex enough to be difficult on an ordinary computer have been successfully implemented.

Meeting these needs requires an extraordinary system architecture—just another PC or Unix clone won't do. The intended applications demand a lot of processor power, main and virtual memory size, and disk capacity. The system must provide as much performance as possible without exceeding practical limits on cost, and computing capacity must not be diluted by sharing it among multiple users. These purely hardware aspects are not sufficient, however. The system must also improve both the speed of software production and the quality of the resulting software by providing a more complete substrate on which to erect programs than has been customary. Programmers should not be handed just a language and an operating system and be forced to do everything else themselves.

At a high level, the Symbolics substrate provides many facilities that can be incorporated into user programs, such as user interface management, embedded languages, object-oriented programming, and networking. At a low level, the substrate provides full run-time checking of data types, of array subscript bounds, of the number of arguments passed to a function, and of undefined functions and variables. Programs can be incrementally modified, even while they are running, and information needed for debugging is not lost by compilation. Thus the edit-compile-test program development cycle can be repeated very rapidly. Storage management, including reclamation of space occupied by objects that are no longer in use, is automatic, so that the programmer does not have to worry about it; incremental, so that it interferes minimally with response to the user; and efficient, because it concentrates on ephemeral objects, which are the best candidates for reclamation. The system never compromises safety for the sake of speed. (A notorious exception, the dynamic rather than indefinite extent of &REST arguments, is recognized as a holdover from the past that is not consistent with the system architecture and will certainly be fixed in the future.)

In an ordinary architecture, such features would substantially diminish performance, requiring the introduction of switches to turn off the features and regain speed. Our architecture considers such compromises unacceptable, because complex, ambitious application programs are typically never finished to the point where it is safe to declare them bug-free and remove run-time error checking. It is essential for such applications to be robust when delivered to end users, so that when something unanticipated by the programmer happens, the application will fail in an obvious, comprehensible, and controlled way, rather than just supplying the wrong answer. To support such applications, a system must provide speed and safety at the same time.

Symbolics systems use a combination of approaches to break the traditional dilemma in which a programmer must choose either speed or safety and comfortable software development:

- The hardware performs low-level checking in parallel with computation and memory access, so that this checking takes no extra time.
- Machine instructions are generic. For example, the Add instruction is capable of adding any two numbers regardless of their data types. Programs need not know ahead of time what type of numbers they will be adding, and they need no declarations to achieve efficiency when using only the fastest types of numbers. Automatic conversion between data types occurs when the two operands of Add are not of the same type.
- Function calling is very fast, yet does not lose information needed for debugging and does not prevent functions from being redefined.
- Built-in substrate facilities are already optimized and available for programmers to incorporate into their programs.

- Application-specific control of virtual memory paging is possible. Prepaging, postpurging, multipage transfers, and reordering of objects to improve locality are supported [2].

These benefits are not without costs:

- Both the cost and the complexity of system hardware and software are increased by these additional facilities.
- Performance optimization is not always automatic. Programmers still must sometimes resort to metering tools. Declarations are available to optimize certain difficult cases, but their use is much less frequent than in conventional architectures.

Why Lisp Machines?   This is really three questions:

1. Why dedicate a computer to each user instead of time-sharing?
2. Why use a symbolic system architecture?
3. Why build a symbolic system architecture on unconventional lower-level architectures?

Why Dedicate a Computer to Each User Instead of Time-sharing?   This seemed like a big issue back in 1974 when Lisp machines were invented, but perhaps by now the battle has been won. A report from that era [3] states these reasons for abandoning time-sharing:

- Time-sharing systems degrade under heavy load, so work on large, ambitious programs could only be conducted in off-peak hours. In contrast, a single-user system would perform consistently at any time of day.
- Performance was limited by the speed of the disk when running programs too large to fit in main memory. Dedicating a disk to each user would give better performance.

The underlying argument was that increasing program size and advancing technology, making capable processors much less expensive, had eliminated the economy of scale of time-sharing systems. The original purpose of time-sharing was to share expensive hardware that was only lightly used by any individual user. The serendipitous feature of time-sharing was interuser communication. Both of these purposes are now served by local-area networking. Expensive hardware units are still shared, but the processor is no longer among them.

These arguments apply to all types of dedicated single-user computers, even PCs, not only to symbolic architectures.

Why Use a Symbolic System Architecture?   Many users who need a platform for efficient execution of large symbolic programs, a high-productivity software development environment, or a system for exploratory programming and rapid prototyping have found symbolic languages such as Lisp and symbolic architectures very beneficial. Programs can be built more quickly, and fit more smoothly into an integrated environment, by incorporating such built-in substrate facilities as automatic storage management and the flexible display with its presentation-based user interface. Full error checking saves time when developing new programs. The programmer can concentrate on the essential aspects of the program without fussing about minor mistakes, because the machine will catch them. The ability to change the program incrementally greatly speeds up development.

Once the initial exploration phase is over, it is possible to turn prototypes into products quickly. Good performance can be achieved without a lot of programmer effort and without sacrificing those development-oriented features that are also of value later in the program's life, during maintenance and enhancement.

Why Build a Symbolic System Architecture on Unconventional Lower-Level Architectures?   Conventional instruction architectures are optimized to implement system architectures very different from Symbolics'. For example, they have no notions of parallel error checking and generic instructions; they often obstruct the implementation of a fast function call, especially one that retains error checking, incremental compilation, and debugging information; and they usually pay great attention to complex indexing and memory-addressing modes, which have little utility for symbolic languages. Implementing Symbolics' system architecture on a conventional instruction architecture would force a choice between safety and performance: we could not have both. The type of software we are interested in either could not run at all or would require much faster hardware to achieve the same performance. Later I will discuss the special aspects of Symbolics' instruction and processor architectures that make them more suitable to support a symbolic system.

Comparing the performance of machines with equivalent cycle times and different architectures can sometimes be illuminating. The 3640, VAX 11/780, and 10-MHz 68020 all have cycle times of about 200 ns. (The 68020 takes two clock cycles to perform a basic operation, so its 100-ns nominal cycle time is equivalent to the other two machines' 200 ns.) On a Fortran benchmark (single-precision Whetstone), the VAX is 1.8 times the speed of the 3640 (750 versus 400). With floating-point accelerators on each machine, the ratio is 2.1. On the Lisp benchmark Boyer [4], the 3640 is 1.75 times the speed of the VAX running Portable Standard Lisp, 3.9 times the speed of the VAX running DEC Common Lisp, and 2.1 times the speed of the 68020 running Lucid Common Lisp. (The 68020 time at 10 MHz was estimated by multiplying its 16-MHz time by 1.6, no doubt an inaccurate procedure.) The VAX and 68020 programs were compiled with run-time error checking disabled and safety compromised, while the 3640 was doing full checking as always. Like any benchmark figures presented without a complete explanation of what was measured, how it was measured, what full range of cases was tested, and how it can be reproduced in another laboratory, these numbers should not be taken very seriously. However, they give some idea of the effect of optimizing the instruction architecture to fit the system architecture. One could say that the VAX is three times better at Fortran than at Lisp and that the 68020 and VAX are similar for Lisp. These figures also show the effect of different compiler strategies on identical hardware.

This comparison was scaled to remove the effect of cycle time and show only the effect of architecture. This is not completely fair to the conventional machines, because in general they can be expected to have faster cycle times than a symbolic machine. Running the 68020 at full speed and using a newer model of the VAX would have improved their times. Hardware technology of conventional machines will always be a couple of years ahead of symbolic hardware in cycle time and price per cycle, because of the driving force of their larger market. It's interesting to note that this hardware advantage applies only to the processor, which usually contributes less than 25 percent of system cost. Power supplies, sheet metal, and disk drives don't care whether the architecture is symbolic; they cost and perform the same for equivalent configurations of either type of machine.

This comparison is not completely fair to the symbolic machine, either. Software exploiting the full capabilities of the symbolic machine should have been compared, but this software won't run at all on the conventional machines. Software technology on symbolic machines will always be a couple of years ahead of conventional machines, because it is built on a more powerful substrate using more productive tools.

Performance. The best published analysis of performance of Lisp systems appears in Gabriel's work. [4] The various 3600 models perform quite capably on these benchmarks, as can be seen from a perusal of the book. Some of the reasons for such good performance will become apparent as we proceed.

However, one must always ask exactly what a benchmark measures. A problem with Gabriel's benchmarks is that they are written in a least-common-denominator dialect that represents Lisp as it was in 1970. This makes it easier to benchmark a broad spectrum of machines, but makes the benchmarks less valid predictors of the performance of real-world programs. Since 1970, there have been many advances in the understanding of symbolic processing and in the range of its applications. The basic operations measured by these benchmarks, such as function calling, small-integer arithmetic, and list processing, are still important today, but many other operations not measured are of equal importance. These benchmarks do not use the more modern features of Common Lisp (such as structures, sequences, and multiple values), do not use object-oriented programming, and are generally not affected by system-wide facilities such as paging and garbage collection. As predefined, portable programs, these benchmarks cannot benefit from the unusual aspects of Symbolics system architecture, such as large program support, full run-time safety, efficient storage management, substrate facilities, support for languages other than Lisp, and faster development of efficient programs.

2 INSTRUCTION ARCHITECTURE

Symbolics' philosophy is that different levels of architecture should be free to change independently, to satisfy different goals and constraints. Users see only the system architecture, leaving the lower levels free to change to utilize available technology, maximize performance, or minimize cost. Most other computer families allow users to depend on the instruction architecture and therefore are not free to change it. It tends to be optimized for only the first member of the family. Later implementations using newer technology, as well as implementations at the high or low extremes of the price/performance curve, are penalized by the need for compatibility with an unsuitable instruction architecture.

Symbolics system architecture has been implemented on three different instruction architectures. The LM-2 machine, based on the original MIT Lisp Machine[3], was the first; it was discontinued in 1983. The 3600 family of machines uses a second instruction architecture and three different processor architectures. A third instruction architecture, appropriate for VLSI implementation, is Ivory, the basis of Symbolics' current product line. Ivory is similar to the instruction architecture described here, but more streamlined.

The following sections summarize the instruction and processor architectures of the 3600 family, discuss some of the design tradeoffs involved, and show how these architectures are especially effective at supporting the desired system architecture. Further details can be found elsewhere [5, 6].

Data Are Object References.   The fundamental form of data manipulated by any Lisp system is an *object reference*, which designates a conceptual object. The values of variables, the arguments to functions, the results of functions, and the elements of lists are all object references. There can be more than one reference to a given object. Copying an object reference makes a new reference to the same object; it does not make a copy of the object.

Variables in Lisp and variables in conventional languages are fundamentally different. In Lisp, the value of a variable is an object reference, which can refer to an object of any type. Variables do not intrinsically have types; the type of the object is encoded in the object reference. In a conventional language, assigning the value of one variable to another copies the object, possibly converts its type, and loses its identity.

A typical object reference contains the address of the object's representation in storage. There can be several object references to a particular object, but it has only a single stored representation. Side-effects to an object, such as changing the contents of one element of an array, are implemented by modifying the stored representation. All object references address the same stored representation, so they all see the side-effect.

In addition to such object references by address, it is possible to have an immediate object reference, which directly contains the entire representation of the object. The advantage is that no memory needs to be allocated when creating such an object. The disadvantage is that copying an immediate object reference effectively copies the object. Thus, immediate object references can only be used for object types that are not subject to meaningful side effects, have a small representation, and need very efficient allocation of new objects. Small integers (traditionally called fixnums) and single-precision floating-point numbers are examples of such types.

In the 3600 architecture, an object reference is a 34-bit quantity consisting of a 32-bit data word and a 2-bit major data type tag. The tag determines the interpretation of the data word. Often the data word is broken down into a 4-bit minor data type tag and a 28-bit address (see Figure 3.1). This variable-length tagging scheme accommodates industry-standard 32-bit fixed and floating-point numbers with a minimum of overhead bits for tagging. Addresses are narrower than numbers to make additional tag bits available for the many types of objects that Lisp uses.

Addresses are 28 bits wide and designate 36-bit words in a virtual memory with 256-word pages. The address granularity is a word, rather than a byte as in many other machines, because the architecture is object-oriented and objects are always aligned on a word boundary. This results in one gigabyte of usable virtual memory. It is interesting to note that the 3600's 28-bit address can actually access the same number of usable words as the VAX's 32-bit address, because the VAX expends two bits on byte addressing and reserves three-fourths of the remaining address space for the operating system kernel and the stack (neither of which is large).

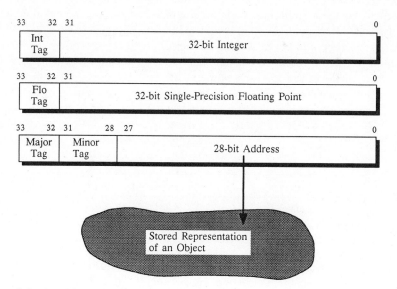

Figure 3.1 An object reference is a 34-bit quantity, consisting either of a 32-bit data word with a 2-bit data type tag, or of a 28-bit address with a 6-bit data type tag. (Reprinted from "Architecture of the Symbolics 3600," *12th Int'l Symp. Computer Architecture,* © 1985 IEEE.)

In addition to immediate and by-address object references, the 3600 also uses pointers, a special kind of object reference that does not designate an object as such. A pointer designates a particular location within an object or a particular instruction within a compiled function. Pointers are used primarily for system programming.[7]

Stored Representations of Objects.   The stored representation of an object is contained in some number of consecutive words of memory. Each word may contain an object reference, a header, a special marker, or a forwarding pointer. The data type tags distinguish these types of words. For example, an array is represented as a header word, containing such information as the length of the array, followed by one memory word for each element of the array, containing an object reference to the contents of that element (see Figure 3.2). An object reference to the array contains the address of the first memory word in the stored representation of the array.

A *header* is the first word in the stored representation of most objects. A header marks the boundary between the stored representations of two objects. It contains descriptive information about the object that it heads, which can be expressed as either immediate data or an address, as in an object reference.

A *special marker* at a memory location indicates that it does not currently contain an object reference. Any attempt to read that location signals an error. The address field of a special marker specifies what kind of error should be signaled. For example, the value cell of an uninitialized variable contains a special marker whose address field points to the name of the variable. An attempt to use the uninitialized variable provokes an error message that includes the variable's name.

A *forwarding pointer* at a location specifies that any reference to that location should be redirected to another memory location, just as in postal forwarding. These are used for a number of internal bookkeeping purposes by the storage management software, including the implementation of extensible arrays.

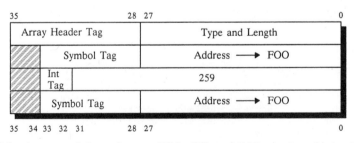

Figure 3.2   An array of three elements–FOO, 259, and BAR–consists of a header word defining the type and length of the array, followed by an object reference for each array element. (Reprinted from "Architecture of the Symbolics 3600," *12th Int'l Symp. Computer Architecture,* © 1985 IEEE.)

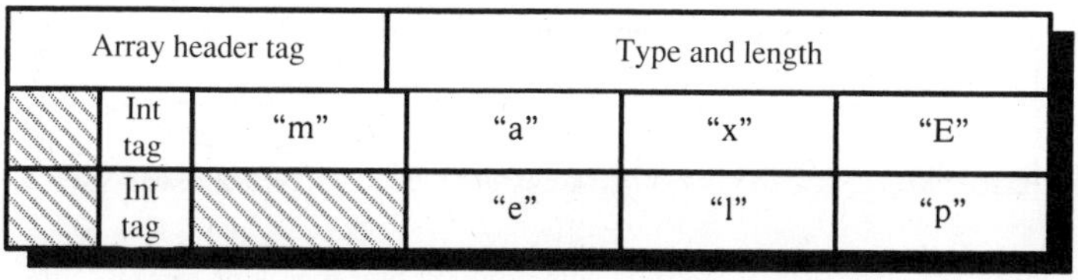

Array header tag		Type and length			
	Int tag	"m"	"a"	"x"	"E"
	Int tag		"e"	"l"	"p"

Figure 3.3 A string containing the seven characters "Example" stores each character in a single 8-bit byte. Bytes are packed into 32-bit integer objects. (Reprinted from "Architecture of the Symbolics 3600," *12th Int'l Symp. Computer Architecture,* © 1985 IEEE.)

Some objects include packed data in their stored representation. For example, character strings store each character in a single 8-bit byte (see Figure 3.3). For uniformity, the stored representation of an object containing packed data remains a sequence of object references. Each word is an immediate object reference to an integer, whose 32 bits are broken down into packed fields as required, such as four 8-bit bytes in the case of a character string.

A word in memory consists of 36 bits, 34 of which I have already explained. When a memory word contains a header or a machine instruction, the remaining two bits serve as an extension of the rest of the word. When a memory word contains an object reference, a special marker, or a forwarding pointer, the remaining two bits are called the cdr code. The representation of conses and lists (Ref. [1], p. 26) saves one word by using the cdr code instead of a separate header to delimit these small objects. In addition, lists are represented compactly by encoding common values of the cdr in the cdr code instead of using an object reference (see Figures 3.4 and 3.5).

Tagging every word in memory produces these benefits:

- All data are self-describing and the information needed for full run-time checking of data types, array subscript bounds, and undefined functions and variables is always available.

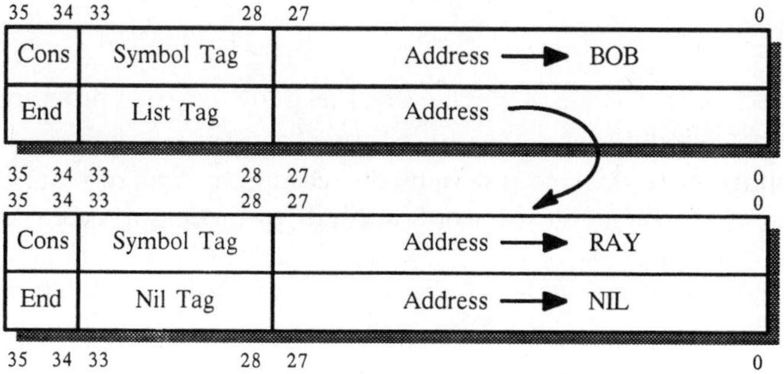

35 34 33	28 27	0
Cons	Symbol Tag	Address ⟶ BOB
End	List Tag	Address

35 34 33	28 27	0
Cons	Symbol Tag	Address ⟶ RAY
End	Nil Tag	Address ⟶ NIL

Figure 3.4 An ordinary list of two elements requires four words of storage. Unlike arrays, lists do not have headers. (Reprinted from "Architecture of the Symbolics 3600," *12th Int'l Symp. Computer Architecture,* © 1985 IEEE.)

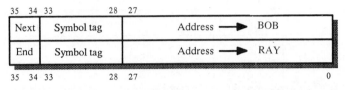

Figure 3.5 A compact list of two elements requires two words of storage. It uses the cdr to eliminate two object references. (Reprinted from "Architecture of the Symbolics 3600," *12th Int'l Symp. Computer Architecture,* © 1985 IEEE.)

- Hardware can process the tag in parallel with other hardware that processes the rest of a word. This makes it possible to optimize safety and speed simultaneously.
- Generic instructions alter their operation according to the tags of their operands.
- Automatic storage management is simple, efficient, and reliable. It can be assisted by hardware, since the data structures it deals with are simple and independent of context. The details appear elsewhere. [8, 5].
- Data use less storage because of compact representations. Programs use less storage due to generic instructions and because tag checking is done in hardware, not software.

The cost of tagging is that more main memory and disk space are required to store numerical information. Each main memory word includes seven bits for error detection and correction, so the four tag bits add 10 percent. Each 256-word disk sector includes about 128 bytes of formatting overhead, so the four tag bits per word add 11 percent. We feel that the benefits amply justify these costs.

Instruction Set. The 3600 architecture includes an instruction set produced by the compilers and executed by a combination of hardware and firmware. All instructions are 17 bits long, consisting of a 9-bit operation field and an 8-bit argument field. Instructions are packed two per word, which is important for performance in two ways:

1. Dense code decreases paging overhead by making programs occupy fewer pages and
2. simplifies the memory system by decreasing the ratio of required instruction fetch bandwidth (in words/second) to processor speed (in instructions/second).

A completed function consists of some fixed overhead, a table of constants, and a sequence of instructions (see Figure 3.6). The table of constants contains object references to objects used by the instructions, including pointers to definition cells of called functions. Indirection through the definition cell

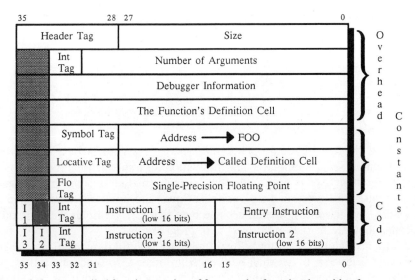

Figure 3.6 A compiled function consists of four words of overhead, a table of constants and external references, and a sequence of 17-bit instructions, packed two per word. (Reprinted from "Architecture of the Symbolics 3600," *12th Int'l Symp. Computer Architecture,* © 1985 IEEE.)

ensures that if a function is redefined its callers are automatically linked to the new definition.

Instructions operate within a stack machine model. Many instructions pop their operands off the stack and push their results onto the stack. In addition to these zero-address instructions, there are one-address instructions, which can address any location in the current stack frame. In this way the slots of the current stack frame serve the same purpose as registers. The one-address instructions include multioperand instructions, which pop all of their operands except the last off the stack and take their last operand from a location in the current stack frame.

An instruction can use its argument field in several ways. Table 3.1 lists the ways to develop the address of an operand in the stack or in memory by adding the argument to a base address. Table 3.2 lists nonaddress uses of the

TABLE 3.1   Ways to develop an operand address.

- Beginning of the current stack frame + offset
- End of the current stack frame − offset
- Lexical parent's environment captured in a closure + offset
- Current function's constants table + offset (possibly with indirection)
- An array used as a DEFSTRUCT structure + subscript
- A flavor instance + offset (possibly with mapping-table indirection)
- Address of the current instruction ± offset (for branching)

TABLE 3.2 Nonaddress uses of argument field.

- A byte field any number of bits wide, positioned anywhere within a 32-bit word. This specification takes ten bits and hence overflows into two bits of operation.
- An immediate integer between −128 and 127
- An immediate integer between 0 and 255
- Extended operation field, to allow more than 512 instructions

argument. Each individual opcode only uses the argument in a single way: there are no addressing modes. The motivation for implementing this particular set of argument uses is to provide for constants (including small integers as a special case), all types of Lisp variables (local and nonlocal lexical, special, structure slot, instance), branching, and byte fields. Byte fields were included because they are heavily used in system programming.

Many instructions are simply Lisp functions directly implemented by hardware and firmware, rather than built up from other Lisp functions and implemented as compiled instructions. These Lisp-function instructions are known as built-ins. They take a fixed number of arguments from the stack and from their argument field. They return a fixed number of values on the stack. Examples of built-ins are eq, symbolp, logand, car, cons, member, and aref [1], where logand and aret are called with two arguments. The criterion for implementing a Lisp function as a built-in instruction is that hardware is only used to optimize key performance areas. When a Lisp function is not critical to system performance, or hardware implementation of it cannot achieve a major speedup, it remains in software where it is easier to change, to debug, and to optimize.

Using an instruction set designed for Lisp rather than adapting one designed for Fortran or for a hand-crafted assembly language enhances safety and speed. 3600 instructions always check for errors and exceptions, so programs need not execute extra instructions to do that checking. Instructions operate on tagged data, so extra instructions to insert and remove tags are not needed. Instructions are generic, so declarations are not needed to tell the compiler how to select type-specific instructions and translate between data formats. In contrast, Lisp compilers for conventional machines [9] must generate extra shifting or masking instructions to manipulate tags, must use multi-instruction sequences for simple arithmetic operations unless there are declarations, and are always having to compromise between safety and speed.

Unlike many machines, the 3600 does not have indexed and indirect addressing modes. Instead it has instructions that perform structured, object-oriented operations such as subscripting an array or fetching the car of a list. This fits the instruction set more closely to the needs of Lisp and at the same time simplifies the hardware by reducing the number of instruction formats to be decoded.

Function Call. Storage whose lifetime is known to end when a function returns (or is exited abnormally) is allocated in three stacks, rather than in the main object storage heap, to increase efficiency. The control stack contains function-nesting information, arguments, local variables, function return values, and small stack-allocated temporary objects. The binding stack records dynamically bound variables [1]. The data stack contains stack-allocated temporary objects. This article concentrates on the control stack, which is the most critical to performance.

The protocol for calling a function is to push the arguments onto the stack, then execute a Call instruction that specifies the function to be called, the number of arguments, and what to do with the values returned by the function. When the function returns, the arguments have been popped off the stack and the values (if wanted) have been pushed on. Note the similarity in interface between functions and built-in instructions.

Every time a function is called, a new stack frame is built on the control stack. A stack frame consists of the caller's copy of the arguments, five header words, the callee's copy of the arguments, local variables, and temporary storage, including arguments being prepared for calling the next function (see Figure 3.7). The current stack frame is delimited by the frame-pointer (FP) and stack pointer (SP) registers, which are available as base registers in instructions that use their argument field to address locations in the current stack frame.

A compiled function starts with a sequence of one or more instructions known as the entry vector. The first instruction in the entry vector, the entry instruction, describes how many arguments the function accepts, the layout of

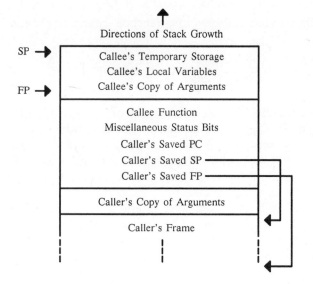

Figure 3.7 A stack frame consists of the caller's copy of the arguments, five header words, the callee's copy of the arguments, local variables, and temporary storage. The frame-pointer (FP) and stack-pointer (SP) registers address the current stack frame. (Reprinted from "Architecture of the Symbolics 3600," *12th Int'l Symp. Computer Architecture,* © 1985 IEEE.)

the entry vector, and the size of the function's constants table (see Figure 3.6), and tells the Call instruction where in the entry vector to transfer control. The Call instruction and the entry vector cooperate to copy the arguments to the top of the stack (creating the callee's copy), convert their arrangement in storage if required, supply default values for optional arguments that the caller does not pass, handle the &rest and Apply features of Common Lisp, and signal an error if too many or too few arguments were supplied.

Function Return. A function returns by executing a Return instruction whose operands are the values to be returned. The value disposition saved in the frame header by Call controls whether Return discards the values, returns one value on the stack, returns multiple values with a count on the stack, or returns all the values to the caller's caller.

Return removes the current frame from the stack and makes the caller's frame current, by restoring the saved FP, SP, and PC registers. If the cleanup bits in the frame header are nonzero, special action must be taken before the frame can be removed. Return takes this action, clears the bit, and tries again. Cleanup bits are used to pop corresponding frames from the binding and data stacks, for unwind-protect [1], for debugging and metering purposes, and for stack buffer housekeeping.

Motivations of the Function Call Discipline. The motivations for this particular function-calling discipline are

- to implement full Common Lisp function calling efficiently,
- to be fast, so that programmers will write clear programs,
- to retain complete information for the Debugger, and
- to be simple for the compiler.

To implement full Common Lisp function calling efficiently requires matching the arguments supplied by the caller–with normal function calling or with apply—to the normal, &optional, and &rest parameters of the callee, and generating default values for unsupplied optional arguments. The entry vector takes care of this. Common Lisp's &key parameters are implemented by accepting an &rest parameter containing the keywords and values, then searching that list for each &key parameter. Multiple values are passed back to the caller on the stack, with a count. The caller reconciles the number of values returned with the number of values desired.

Function calling historically has been a major bottleneck in Lisp implementations, both on stock hardware and on specially-designed Lisp machines. It is important for function calling to be as fast as possible. If it is not, efficiency-minded programmers will distort their programming styles to avoid function calling, producing code that is hard to maintain, and will waste a lot of time doing optimization by hand that should have been done by the Lisp implementation itself. The 3600's function call mechanism attains good speed (fewer than 20 clock cycles for a one-argument function call and return when no exceptions occur) by using a stack buffer to minimize the number of memory references required, by optimizing the stack frame layout to maximize speed rather than to minimize space, by arranging for the checks for slower exception cases to be fast (for example, Return simply checks whether the cleanup bits are nonzero), and by using the entry vector mechanism to simplify run-time decision-making.

The information that the debugger can extract from a stack frame includes the address of the previous frame (from the saved FP in the header), the function running in that frame (from the header), the current instruction in that function (from the PC saved in the next frame), the arguments (from the stack—the header specifies the argument count and arrangement), the local variables (from the stack), and the names of the arguments and local variables (from a table created by the compiler and attached to the function).

The compiler is simple because there is only a single calling sequence. Any call can call any function, and the argument patterns are matched up at run time. Everything is in the stack and no register-saving conventions are required, since there are no general-purpose registers.

The principal costs of this function-calling discipline are the five-word header in each frame and the copying of arguments to the top of the stack. The time to create the header is not a problem, because it is overlapped with necessary memory accesses, but the space occupied by the header and by the extra copy of the arguments is a substantial fraction of the typical frame size. This extra space is not a major problem because the stack buffer is large enough (1024 words) that it rarely overflows.

Argument copying is necessary because Common Lisp functions do not take a fixed number of arguments. In a function with &optional parameters, some of the arguments are supplied by the caller while the others are defaulted by the entry vector. The location in the stack frame of an argument must not depend on whether it was supplied or defaulted, since this varies from one call to the next, but the compiler must know the location in order to generate code to access the argument. The entry vector could not put default values in the standard location if the arguments were not at the top of the stack, because the frame header would be in the way. In a function with an &rest parameter, the caller can supply an arbitrary number of arguments. If these arguments were at the top of the stack, they would make it impossible for the compiler to know the locations of the local variables, which are pushed after the arguments.

Copying the arguments that are not part of an &rest parameter to the top of the stack solves both these problems. It gives the function complete control over the arrangement of its stack frame and makes the stack depth constant. Argument copying takes extra time, but typically only one clock cycle per argument, which is faster than the run-time decision making that would otherwise be necessary to access an optional argument or a local variable.

3 PROCESSOR ARCHITECTURE

Three processor architectures are used in three representative models of the 3600 family: 3640, 3675, and 3620. Since they all implement the same instruction architecture, there are substantial similarities among their processor architectures. They differ in technologies and cost/performance tradeoffs, but this overview largely glosses over the differences.

The main goal of each of these processor architectures is to implement the instruction architecture described earlier with the highest performance achievable within its particular cost budget. The costs are generally higher than most workstations' but lower than most minicomputers'. For high performance the number of clock cycles required to execute an instruction must be minimized; the goal is to execute a new instruction every cycle. Because the system architecture specifies that safety and convenience must not be compromised to increase performance, instructions typically make many checks for errors and exceptions. Minimizing the cycle count demands that these checks be performed in parallel, not each in a separate cycle.

Adequate bandwidth for access to operands is also required. In the 3600 instruction architecture, a simple instruction can read two stack locations and write one stack location. One of these is a location in the current stack frame specified by an address in the instruction, while the other two are at the top of the stack. Operands are supplied by the stack buffer, a 1K-word memory that holds up to four virtual-memory pages of the stack. The stack buffer contains all of the current frame plus as many older frames as happen to fit. When the stack buffer fills up (during Call), the oldest page spills into normal memory to make room for the new frame. When the stack buffer becomes empty (during Return), pages move from normal memory back into the stack buffer until the frame being returned to is entirely in the buffer. The maximum size of a stack frame is limited to what will fit in the stack buffer. A second stack buffer contains an auxiliary stack for servicing page faults and interrupts without disturbing the primary buffer.

Associated with the stack buffer are the FP and SP registers, which point to the current frame and to the top of the stack, and hardware for addressing locations in the current stack frame via the argument field of an instruction, which calculates a read address and a write address every clock cycle. The third operand access is provided by a duplicate copy of the top location in the stack, in a scratchpad memory, which can be read and written every clock cycle. The SP register is incremented or decremented by instructions that push or pop the stack.

The stack buffer provides the same operand bandwidth, two reads and one write every clock cycle, as in a typical register-oriented architecture. It has the advantage that register saving and restoring across subroutine calls is not required, since all registers already reside in the stack. As in a register-window design, overhead occurs only when the stack buffer overflows or underflows and requires a block transfer between stack buffer and main memory. Another advantage is that each instruction contains only one address instead of three, making the instructions smaller (so that they can be fetched from main memory more quickly and processed with less hardware) and allowing more registers to be addressed. A disadvantage of a stack architecture is that it requires address-calculation hardware, including a 10-bit (for a 1K-word buffer) adder. Since each instruction contains only one address instead of three, extra instructions are sometimes required to move data to the top of the stack so they can be addressed.

Instructions are processed by a four-stage pipeline (see Figure 3.8) under the control of horizontal microcode. Microcode is used as an engineering technique, not to create a general purpose emulator that could implement alternate instruction architectures. Knowledge of the instruction architecture is built into hardware wherever that achieves a substantial performance improvement.

For full performance, instructions must be supplied to the processor at an adequate rate. Each processor model has a different design, with different tradeoffs.

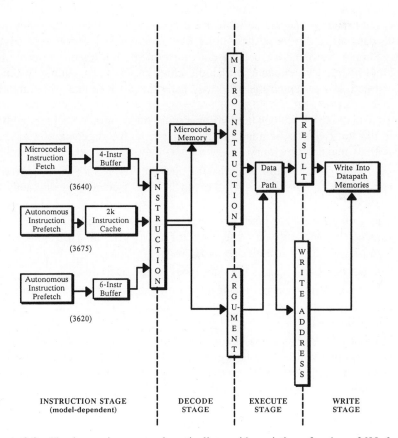

Figure 3.8 The instruction processing pipeline, with variations for three 3600 family models.

The 3640 uses a four-instruction buffer. When the buffer is exhausted, or a branch occurs, microcode reads two words from memory and refills the instruction buffer. This design uses much less hardware than the other two, but provides lower performance. Refilling the buffer takes five clock cycles, so in the worst case the performance penalty is about a factor of two. With a typical instruction mix, the observed slowdown is about 35 percent, because complex instructions such as function calls and memory references spend more than one cycle in the execute stage.

The 3675 uses a 2K-instruction cache. Program loops that fit in the cache execute at full speed, with no instruction fetching overhead. An autonomous instruction prefetch unit fills the cache with instructions before they are needed, in parallel with execution. At the cost of a substantial increase in hardware complexity over the 3640, this design ensures that the pipeline almost never has to wait for an instruction.

The 3620 uses a six-instruction buffer. An autonomous instruction prefetch unit fills the buffer in parallel with execution. The 3620 instruction stage is a compromise between the other two designs. Straight-line code executes at full speed, but branches execute at 3640 speed because they must refill the buffer.

The datapath contains several units that function in parallel (see Figure 3.9). Simple instructions such as data movement, arithmetic, logical, and byte-field instructions execute in a single clock cycle. For example, when an add instruction is executed the following activities all take place in parallel:

- The stack buffer fetches the two operands, one from a calculated address in the stack buffer memory and the other from the duplicate top of stack in the scratchpad memory.

- The fixed-point arithmetic unit computes the 32-bit sum of the operands and checks for overflow. This result is only used if both operands are fixnums.

- The optional floating-point accelerator, if present, starts computing the sum of the operands and checking for floating-point exceptions. This result is only used if both operands are single-floats.

- The tag processor checks the data types of the operands.

- The stack buffer accepts the result from the fixed-point arithmetic unit, adjusts the stack pointer, and in the write stage stores the result at the new top of the stack.

- The decode stage decodes the next instruction and produces the microinstruction that will control its execution. If the type-checking unit or either arithmetic unit detects an exception, control is diverted to a microcode exception handler.

When the operands of Add are not both fixnums, executing the instruction takes more than one machine cycle and more than one microinstruction. In the case of adding two single floats, the extra time is required only because the

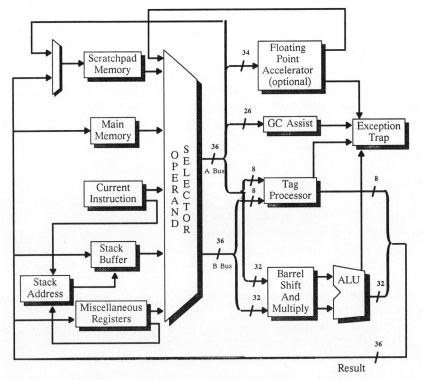

Figure 3.9 3640 data path, contained in the Execute and Write stages of the pipeline. Other 3600 family models have generally similar datapaths.

floating-point arithmetic unit is slower than the fixed-point arithmetic unit. In other cases, extra time is required to convert the operands to a common format, to perform double-precision floating-point operations, or to trap to a Lisp function to add numbers of less common types.

Memory-reference instructions such as the car and aref Lisp operations are limited mainly by the speed of the memory. Car, for example, takes four clock cycles. Complex instructions such as Call, Return, and the Common Lisp member function invoke microcode subroutines. A wide microinstruction word and fast microcode branching minimize the number of microinstructions that need to be executed. Simple and memory-reference instructions can be discovered to be complex at run time because of an exceptional condition, such as exceptional data types of the operands.

4 CONCLUSION

I have described here an unusual system architecture and presented an overview of the underlying architectures that implement it. When considering the type of applications that this system architecture targets, note how important to their success it is that we compromise neither safety nor speed. With this in mind, some of the unconventional design choices in these architectures were made based on rationales with varied benefits and costs. For example, a close fit between processor, instruction, and system architectures improves performance, but allowing users to depend on details of the instruction architecture can interfere with this. The lack of this close fit dissipates the hardware price/performance advantage of conventional architectures when measuring system-level performance on software suited to symbolic architectures.

REFERENCES

1. G. L. Steele, *Common Lisp*, Digital Press, Burlington, MA, 1984.

2. D. L. Andre, "Paging in Lisp Programs," Master's thesis, University of Maryland, 1986.

3. R. D. Greenblatt et al., "The LISP Machine," *Interactive Programming Environments*, eds. D. R. Barstow, H. E. Shrobe, and E. Sandewall, McGraw-Hill, Hightstown, NJ, 1984.

4. R. P. Gabriel, *Performance and Evaluation of Lisp Systems*, The MIT Press, Cambridge, MA, 1985.

5. D. A. Moon, "Architecture of the Symbolics 3600," *12th Int'l Symp. Computer Architecture*, pp. 76–83. IEEE Computer Society Press, Silver Spring, MD, 1985.

6. *Symbolics Technical Summary*, Symbolics Inc, Cambridge, MA, 1985.

7. *Symbolics Common Lisp: Language Dictionary*, Symbolics, Inc, Cambridge, MA, 1986.

8. D. A. Moon, "Garbage Collection in a Large Lisp System," Proc. 1984 ACM *Symp. Lisp and Functional Programming*, pp. 235–246.

9. R. A. Brooks et al., "Design of an Optimizing, Dynamically Retargetable Compiler for Common Lisp," *Proc. 1986 ACM Conf. Lisp and Functional Programming*, pp. 67–85.

ABOUT THE AUTHOR

David A. Moon is a Symbolics Fellow at Symbolics, Inc. Previously, he was a hardware designer, microprogrammer, and writer of manuals at Symbolics. His interests include advanced software development and architectures for symbolic processing. Moon received the B.S. degree in mathematics from MIT in 1975. Readers may write to the author at Symbolics, Inc., 8 New England Executive Park, East Burlington, MA 01803. His e-mail address is Moon@Symbolics.Com.

CHAPTER 4

MEMORY MANAGEMENT AND USAGE IN A LISP SYSTEM: A MEASUREMENT-BASED STUDY*

Rene L. Llames and Ravi K. Iyer

1 INTRODUCTION

Garbage collection in large virtual memory Lisp systems performs the dual function of postponing the exhaustion of address space by reclaiming unused memory and improving memory performance by increasing the locality of objects. Recent implementations of incremental and generational garbage collection perform these functions less intrusively and more efficiently than the original "stop the world" techniques, and have contributed to the popularity of Lisp workstations for developing and running large and complex applications. Nevertheless, with the ever-increasing size and performance requirements of such applications, garbage collection remains an important issue.

In this chapter, we present an analysis of memory system and garbage collection activity on a Symbolics 3620 Lisp machine. The data consists of regular samples of activity times, event counts, and memory space sizes maintained by the system. The measured programs include the Boyer benchmark and QPE, a large AI program, run with two different input data sets. Our analysis characterizes the dynamic behavior of the system and is organized into three parts.

1. The first part is an overall analysis of processor utilization, page faults, and other page management overhead. Distributions of utilization and of fraction-of-time and event rates are presented.

* This work was supported in part by the National Aeronautics and Space Administration under Grant NAG-1-613 and by the National Science Foundation under Grant (DCI) MIP-8604893.

119

2. The second part presents a detailed breakdown of time and page faults in the garbage collector. We use linear regression to derive empirical models that relate garbage collection time to collector work, page faults, and other overhead; the model parameters quantify the time cost of the modeled operations. We discuss the implications of the parameter estimates obtained by regression and test the accuracy of the models in predicting overall garbage collection time.

3. In the final part of our analysis we examine the time-varying characteristics of memory allocation, survival, and collector work and efficiency; collector efficiency is quantified as the amount of discovered garbage per unit time (or work) expended by the collector. This analysis demonstrates the utility of the cost-benefit metrics in guiding the tuning of the garbage collector.

Section 2 discusses related work. Section 3 contains background information on garbage collection and its implementation on the Symbolics Lisp machine. The data collection method and measured programs are described in Section 4. Sections 5, 6, and 7 correspond to the three main parts of the analysis outlined above, and the conclusions appear in Section 8.

2   RELATED WORK

Many studies have been made of the memory management of linked data structures, and in particular, of garbage collection algorithms and their analysis (for example, Cohen [6] presents a survey of work done before 1981). However, relatively few measurement-based studies have been published. In this section, we review some studies that have a significant experimental component and summarize their major objectives and results.

Clark and Green [5] investigated the potential for compact representation of Lisp lists, basing their conclusions on static measurements of the list structure existing after each of five large Interlisp programs was run on a DEC PDP-10. Their measurements showed substantial regularity among pointers to atoms and lists; for example, list pointers usually referred to a nearby memory location. Such predictability suggested the use of space-efficient, linearized encodings of list structure. The entropy or information content of pointers was calculated to measure the effects of linearizing lists.

In a subsequent study, Clark [4] ran three of the five Lisp programs on a PDP-10 simulator modified to produce a trace of dynamic list structure usage. The trace file was analyzed off line to yield measurements of the dynamic frequencies of Lisp primitives; the dynamic locality of list references; the tendency of linearized lists to become unlinearized during program execution; and the effect of linearization on paging behavior. Clark found that if all lists were linearized, the page fault probability decreased slightly; furthermore, because of the high cost of a fault, a small decrease may have a large effect on list referencing time.

Gabriel [8] compiled a set of Lisp benchmarks collected over a number of years, and tabulated timing results for many Lisp implementations. He presented the benchmarks together with a discussion of the various levels of Lisp system architecture that influence performance.

Several articles have reported instruction-level measurements of various Lisp implementations, including, recently, the simulations of Lisp programs compiled for the SPUR and the MIPS-X reduced-instruction-set processors [21, 17, 18]. These studies measured the dynamic frequencies of machine instructions and of primitive Lisp operations, for example, car, cdr, cons, type checking, and function call/return. The goal was to identify the areas of the processor architecture most suited for optimization. The studies did not consider the effects of system facilities such as virtual memory and garbage collection.

Shaw's analysis [16] of four Lisp programs on a 68020-based HP9000 workstation also measured the frequencies of instructions and low-level Lisp operations, but was intended to provide information for use in evaluating design options for new implementations of Lisp on conventional architectures. Shaw used "operations per Lisp function entry" as his primary metric rather than the time metric used in earlier studies; his goal was to provide a less system-dependent frequency measure. Data on the mortality of objects validated the lifetime heuristics on which current generational garbage collectors are based.

Andre [1] studied the effects of Lisp programming methods on paging and discussed optimizations in the system and their application to the minimization of page faults. He considered various methods for improving paging performance, including (1) proper selection of optimum data representations and algorithms to reduce working set size and assist the garbage collector; (2) system facilities for manual memory management; (3) system-level optimizations of object representations to reduce working set size for common operations; and (4) memory reordering schemes.

The measurements most closely related to ours are those of Moon [12]. Moon's measurements showed paging and garbage collector performance on a Symbolics 3600 for two programs, the Boyer benchmark and a compilation. The measurements were made primarily to evaluate the ephemeral garbage collector, an incremental, copying, generational, hardware-assisted collector designed for a large virtual memory. The algorithm seeks to minimize run time and paging by exploiting the lifetime heuristics—assumptions about Lisp object sizes, lifetimes, and connectivity. These assumptions were shown by measurement to be valid (or invalid in exceptional cases). To handle exceptional or performance-critical cases, the implementation provides facilities for user-directed memory management [20].

It is clear that more empirical studies are needed to improve our understanding of system behavior in general and of memory management in particular. We believe that, in such studies, it is important to describe dynamic variation in measured quantities if we are to build realistic models which relate system and program characteristics to performance. The rest of this chapter discusses measurements and analyses of this kind.

3 GARBAGE COLLECTION IN VIRTUAL MEMORY

This section addresses the problem of garbage collecting a large virtual memory and describes the Symbolics garbage collector; it is intended to provide background information helpful in understanding the rest of the paper. The interested reader is referred to [12, 13, 14, 19] for more detail.

What is required of a garbage collector? The function of garbage collection is to provide for automatic reuse of memory, a finite resource. Because Lisp systems are interactive, pauses for garbage collection must be kept short to minimize response time. Virtual memory imposes two specific performance criteria. Firstly, the collector must preserve the locality assumption; techniques which require traversal of the entire address space are incompatible with virtual memory because of run time and paging considerations. Secondly, the collector should improve the locality of the program's objects [23].

The Symbolics system fulfills these requirements by means of incremental copying garbage collection, approximately depth-first copying, generational garbage collection, and tagged architecture and special hardware. The rest of Section 3 describes these methods.

3.1 Incremental Copying Garbage Collection

Garbage collection in our measured system employs the incremental copying technique, based on modified versions of the Cheney [3] and Baker [2] algorithms. The Cheney algorithm performs breadth-first copying of linked structures without requiring an explicit stack. The Baker algorithm interleaves collection with normal processing, avoiding the long, unpredictable delays to the user that would result if garbage collection were to be performed without interruption. In the Baker algorithm, the heap is divided into two equal-sized spaces, *fromspace* and *tospace*, and a garbage collection copies all accessible objects in fromspace to tospace (an object is accessible if it can be reached starting from some set of root objects, called the *root set* or *base set*). After all accessible objects have been copied, fromspace can be reused. To begin another garbage collection, the labels of the two spaces are interchanged or flipped. The copying technique enhances locality by removing interspersed garbage.

In the Symbolics system, the heap is divided into static and dynamic areas. Only dynamic space (or some portion of it) is garbage collected; static space is assumed to contain objects that are unlikely to become garbage. During a collection, three kinds of dynamic space become meaningful:

- The portion of dynamic space to be garbage collected is turned into *oldspace*.
- Objects in oldspace discovered to be nongarbage—by a procedure to be described shortly—are copied to *copyspace*.
- New objects created during the collection are allocated in *newspace*.

After all accessible objects in oldspace have been copied, oldspace may be reclaimed. Another collection may then begin by flipping copyspace and newspace into oldspace and allocating a fresh copyspace and newspace. Hence, oldspace corresponds to fromspace in the Baker algorithm, and copyspace/newspace, to tospace. Unlike the Baker algorithm spaces, however, the three spaces are not fixed in size or location. The portion of dynamic space to be collected is turned into oldspace, and copyspace and newspace are allocated as necessary from free virtual address space.

The garbage collector consists of two threads of control, the *scavenger* and the *transporter*, which are interleaved with the user program and other system processes, collectively called the mutator. The scavenger's job is to scan through memory, searching for all possible references to oldspace by nongarbage objects outside oldspace. Initially, the only place where such references can exist is, by definition, the root set. When the scavenger encounters an oldspace reference, the transporter is called. The transporter

1. Copies the object from oldspace to copyspace, installing a forwarding pointer in the oldspace object that points to the version in copyspace; and
2. Changes the oldspace reference to point to the copyspace version.

If the transporter is called because of a reference to a previously copied object, it only has to perform the second step, using the forwarding pointer to redirect the reference from oldspace to copyspace. As nongarbage objects are transported, copyspace may come to contain references to oldspace. Thus, after scanning the root set, the scavenger needs to scan copyspace as well, to detect accessible structures remaining in oldspace. The root set and copyspace together constitute scavenge space; after both have been scavenged, references to oldspace no longer exist and oldspace can be reclaimed.

The mutator may also attempt to reference objects in oldspace, thereby triggering the transporter. Thus transporter calls can be either scavenger-induced or mutator-induced.

The scavenger is allowed to run if the system is idle. Otherwise, the rate of performing collection work (scanning and transporting) is constrained to be proportional to the rate of allocation; that is, the garbage collector is cons-driven, ensuring that consumption does not outpace production of free space.

3.2 Approximately Depth-first Copying

The garbage collector can copy objects in whatever order it chooses, and this degree of freedom can be exploited to improve the spatial locality of the surviving objects. The Symbolics garbage collector modifies the Cheney algorithm to realize an approximately depth-first order. Whenever it is likely to result in discovery of oldspace references, the scavenger temporarily suspends its normal linear scan of the root set and copyspace to scan instead the partially filled page at the growing end of copyspace. This *last-page* scavenging of copyspace tends to place objects on the same page as their parent. Courts [7] suggests another technique, in which objects evacuated by mutator-induced transporting are separated from those evacuated by scavenger-induced transporting; although possible, the technique is not implemented in our measured system.

3.3 Generational Garbage Collection

Our system provides two forms of garbage collection—the original dynamic collector and the more recently developed ephemeral collector. In dynamic collection, all dynamic space is garbage collected and the root set is taken to consist of all objects in static space. The policy for initiating collections is safety-based: A collection is begun when the system decides it has reached the latest time at which a collection, if begun, could safely complete without running out of free memory space.

A dynamic collection typically requires much run time and paging time due to the enormous size of static space and the large number of objects that have to be transported. Although collection is interleaved with the user program, response time increases considerably because of paging. Consequently, most users turn off the dynamic collector during interactive use.

The ephemeral garbage collector is an implementation of generational collection, which is based on two heuristics about objects:

- Younger objects are more likely to become garbage than older objects (infant mortality).
- There are many fewer references from older to younger objects than from younger to older objects.

The first heuristic suggests that we stratify dynamic space into several independently collectible generations or levels; place newly created objects in the first generation; advance surviving objects to the next higher generation; and garbage collect the younger generations more frequently. Collecting the younger generations will be more efficient since effort is expended on reclaiming areas with a high proportion of garbage and little transporting work is required. Furthermore, the root set need only contain references from older generations to the generations being collected. The second heuristic greatly reduces the size of the root set and suggests that it is not impractical to keep track of these intergenerational references from older generations to newer ones.

During ephemeral collection, ephemeral objects (those assumed to be short-lived) are created at the first level. The policy for initiating collections is capacity-based: a collection is begun when the first level exceeds its specified capacity. The first level is flipped simultaneously with higher levels that have also exceeded their capacities. Objects that survive a garbage collection graduate to the next level. Those surviving a collection of the highest level become normal, "tenured" dynamic objects and may be collected by dynamic collection. Two tables remember the pages into which ephemeral object references have been written. These tables determine the root set for garbage collecting a particular level. The tables are the Garbage Collector Page Tags (GCPT) for in-main-memory pages, and the Ephemeral Space Reference Table (ESRT) for on-disk pages. A greater effort is made to minimize the size of the ESRT to avoid the unnecessary fetching of on-disk pages during scavenging.*

3.4 Tagged Architecture and Special Hardware

The Symbolics 3600 provides a tagged architecture and special hardware that allow the techniques we have just discussed to be implemented with acceptable overhead. The processor allows hardware detection of the following references:

- Oldspace object references during memory reads (to indicate when to trap to transporter microcode)
- Ephemeral object references during memory writes (to indicate when to update the GCPT)

This hardware "barrier" between the processor and the memory includes memory for mapping a virtual address to a space type and ephemeral level. The GCPT is also implemented in hardware, preventing the performance degradation that would result from performing address checks in microcode or Lisp. Moon [12] shows that the chip overhead is small.

4 THE MEASUREMENT ENVIRONMENT

In this section, we describe the data collected and the workload used to obtain the results presented in succeeding sections. We also consider interference and the steps taken to minimize it.

---

* In other generational collection schemes, the entity performing the functions of the GCPT and ESRT has gone by such names as entry vector, remembered set, and indirection cells.

4.1   The Data

Our study is based on the periodic sampling of a large number (about 80) of software-accessible meters.* The meters track virtual memory management, garbage collection, and disk access activity. A meter may be classified as

- An activity timer—for instance, it may measure the cumulative number of milliseconds spent servicing page faults;
- A resource counter—it may, for example, count the number of pages of a specified kind; or
- An event counter—counting, for example, the cumulative number of page faults.

The meters are system wide (i.e., not process specific) and are maintained with little overhead by system software or microcode.

The data also includes samplings of meters monitoring certain interesting processes. Specifically, we sample the run time, disk wait time, and page fault count of the process running the workload under measurement, the garbage collector processes, and the process performing the sampling.

Finally, the data includes the sizes of various memory spaces relevant to garbage collection. The growth or shrinkage of these spaces over time may thus be observed.

4.2   The Workload

Two programs were used to obtain the results presented here:

- The Boyer benchmark [8], a theorem-proving kernel using a rewrite-rule–based simplifier and a dumb tautology checker; and
- QPE (Qualitative Process Engine), a qualitative simulator by K. Forbus which uses an assumption-based truth maintenance system as a substrate.

Boyer was measured on a 3620 running the Genera™ 7.0G1 system with 2 megawords of main memory and about 15 million words of paging area. QPE was measured on the same 3620 running the earlier Release 6.G2 system with about 25 million words of paging area. The earlier system release was used because no version of QPE for Genera 7 was available at the time the data was collected.

Data from six measurement runs is presented, each run being a particular combination of program, input data, and garbage collector mode. The runs range in length from about nine minutes to over 12 hours. The machine was cold-booted before measuring each program.

---

* The user can view a continuously updated display of these meters using the Peek facility

TABLE 4.1 Measured Programs, Number of Samples, and Total Real Time.

Program	GC	Nominal sampling period (seconds)	Samples	Real time (hh:mm:ss)
Boyer-24	off	1	449	8:47
Boyer-15	dynamic	1	1664	31:45
Boyer-24	ephemeral	1	820	14:58
QPE-short	off	1	903	16:12
QPE-short	ephemeral	1	1083	20:19
QPE-long (full execution)	ephemeral	15	2888	12:05:05
QPE-long (initial part)	ephemeral	5	3744	5:16:57

Table 4.1 shows some general information about the runs. Boyer-*n* refers to a workload consisting of *n* consecutive calls of the Boyer program. Note that Boyer is not dependent on any input data. QPE-short refers to the execution of QPE on input data representing a simple problem; QPE-long, a more complex one. Two measurements of QPE-long were made: one of the entire 12-hour execution, sampled every 15 seconds, and another of the initial 5 hours only of execution, sampled every 5 seconds. The choice of this set of runs allowed us to make informal observations of the following characteristics:

- Small benchmark versus large program behavior (Boyer, QPE-short versus QPE);
- The effect of input data (QPE-short versus QPE-long); and
- Reproducibility* and the effect of the sampling period (full execution of QPE-long versus initial part only of QPE-long).

4.3 Software Sampling and Interference Effects

A separate process—the sampler process—periodically saved the values of selected meters. As with any measurement experiment involving software sampling, interference effects must be carefully considered when designing the instrumentation software, and in analyzing and drawing conclusions from the results.

---

* The sensitivity of the results to the initial state of virtual memory is probably the strongest factor affecting reproducibility. Measurements were collected on a program starting from a freshly booted, minimally mutated world in order to provide a basis for comparison between programs. However, it is conceivable that even slight changes to the world—not to mention large changes occurring after the machine has been used for some time—could significantly alter the measurements. Short programs such as Boyer would probably be very sensitive, whereas longer ones might be able to establish a consistent working set and object population pattern after an initial time period and a few garbage collection cycles. This conjecture seems to be supported by the similarity of the results for the two measurements of QPE-long.

The real-time overhead attributable to the sampling is easy to measure. Since the sampler process is not interrupted while saving the meter values to memory, the time elapsed during this saving provides a good estimate of time overhead. Our data shows that saving the meter values typically takes on the order of one or two milliseconds. With sampling periods on the order of seconds, this overhead is relatively small.

However, there are other kinds of overhead: paging and memory usage overhead, for example. We used a number of manual storage management features to reduce or eliminate paging and garbage generation caused by the sampler process. For example, the buffer into which the sampled data was saved was pre-allocated and wired down in main memory, that is, declared non-pageable, and the memory area it used was declared to be static, so that the garbage collector did not consider it. Temporary structures were consed on the control stack, so that they were automatically reclaimed, and scheduling overhead was minimized by implementing the sampler itself as an especially efficient kind of process which did not need its own stack group [20].

5 PAGING AND STORAGE ALLOCATION ACTIVITY

In this section, we focus on low-level virtual memory management activity such as paging and page allocation. First we give the total time, total event count, mean fraction of real time, and mean event rate for each activity of interest. Variation in time is then shown in the form of distributions and time series of the fractions of time and event rates. A scatter plot is shown as a way of visualizing the distribution of the fraction of time taken by the major activities.

5.1 Preliminaries

When sampling the meters that measure activity times or event counts over the execution of a particular workload, several kinds of first-cut results can be obtained:

- A total, such as the total time spent in activity A_i or the total count of event E_k
- A ratio of totals; for example the fraction of real time spent in activity A_i, the ratio of time for activity A_i to time for activity A_j, the rate of event E_k measured over real time or over time spent in activity A_i, or the frequency ratio of two events E_k and E_l
- The time behavior of a ratio (of counts or times of interest) computed for each sampling interval, and presented in the form of a distribution or a time series.

Totals and ratios of totals are the usual results of benchmarking experiments. They express dynamic behavior to the extent that they quantify the cumulative (or aggregate) effect of operations over the measured period. No sampling is necessary to produce them (except, of course, at the beginning and end of a run).

Distributions and time series, however, quantify variability, which may be an important consideration. For example, variability in interactive response time, (a function of the variability of the underlying operations) could be as important a specification as the average value of response time; and its effect on the perceived quality of a user interface may be equally significant. Again, when tuning performance, it may be useful to know whether a program incurs system overhead, such as paging, uniformly throughout its execution or only during certain periods. Likewise, knowledge of how a Lisp application's memory usage varies over time would be useful when adjusting garbage collection parameters for maximum efficiency.

When sampling a system using a periodically activated software process, there will be slight variations in the length of the sampling period over the data collection session. In obtaining distributions of quantities computed for each sampling interval, we can account for nonuniform sampling periods by giving more weight to quantities computed for longer intervals. The weighting would be used in calculating statistics based on first and second moments (such as mean and variance), and in drawing frequency charts (histograms), but not in determining rank statistics (such as median and other quantiles). Real-time–based weighting is appropriate for such quantities as the fraction of real time spent in an activity, or an event rate measured over real time.

More formally, let t_i be the real time length of the ith sampling interval, T be the total real time for n intervals, and c_{1i} and c_{2i} be event counts or activity times for the ith interval, where the ratio c_{1i}/c_{2i} is of interest. Then $T = \sum_{i=1}^{n} t_i$, $C_1 = \sum_{i=1}^{n} c_{1i}$, and $C_2 = \sum_{i=1}^{n} c_{2i}$ are totals. C_1/C_2 is a ratio of totals. If the denominator c_{2i} is nonzero for all intervals, then the sequence of observations c_{1i}/c_{2i}, $i = 1, 2, \ldots, n$ has no missing values and its weighted mean and variance are given by

$$\overline{X} = \frac{1}{n} \sum_{i=1}^{n} w_i \frac{c_{1i}}{c_{2i}}$$

$$S^2 = \frac{1}{n-1} \sum_{i=1}^{n} w_i \left(\frac{c_{1i}}{c_{2i}} - \overline{X} \right)^2$$

where the weight is $w_i = n t_i / T$. For perfectly uniform sampling, $w_i = 1$.

For the distributions shown in this section, $c_{2i} = t_i$ and $t_i > 0$ for all i. That is, the ratio of interest is the fraction of real time spent in a particular activity, or is an event rate (measured over real time). In this case, the mean of the distribution equals the corresponding ratio of totals:

$$\overline{X} = \sum_{i=1}^{n} \frac{t_i \, c_{1i}}{T \, c_{2i}} = \frac{C_1}{T} = \frac{C_1}{C_2}$$

Sections 5.2 and 5.3 present totals, ratios, and distributions for time spent and work done in memory management at the page level.

5.2   Totals and Ratios

The upper halves of Tables 4.2 and 4.3 show the total real time spent measuring each of the programs and the average fraction of time spent in various page management activities. The lower halves show the total count and average event rate for various page management events.

TABLE 4.2   Total Real Time and Page Management Work Breakdown.

Totals and Ratios	Boyer-24 GC Off	Boyer-15 Dynamic	Boyer-24 Ephemeral
Real Time			
Total real time (hh:mm:ss)	8:47	31:45	14:58
Percentage of total time in			
runtime	52.71	21.22	86.32
page faults	14.26	64.16	2.29
page creation	26.51	5.80	5.80
sequence breaking	5.44	1.88	2.43
page prefetching	0.00	4.96	0.11
page destroying	0.00	0.17	2.74
(error/unknown)	1.08	1.81	0.31
Page Management Work			
Total number of			
page faults	441	5,231	148
page creations	42,482	32,907	53,152
page prefetches	190	41,117	40
discarded prefetched pages	469	1,643	439
forced modified page writes	2,186	759	160
Average rate (1/second) of			
page faults	0.8367	2.744	0.1648
page creations	80.60	17.27	59.20
page prefetches	0.3605	21.58	0.0446
discarding prefetched pages	0.8898	0.8621	0.4889
forced modified page writes	4.147	0.3983	0.1782

TABLE 4.3 Total Real Time and Page Management Work Breakdown.

Totals and Ratios	QPE-Short GC Off	QPE-Short Ephemeral	QPE-Long (Full Execution) Ephemeral	QPE-Long (Initial Part) Ephemeral
Real Time				
Total real time (hh:mm:ss)	16:12	20:19	12:05:05	5:16:57
Percentage of total time in				
runtime				
page faults	84.68	85.74	93.11	85.85
page creation	11.76	9.94	6.17	13.03
sequence breaking	3.22	4.05	0.55	0.79
page prefetching	–	–	–	–
page destroying	0.00	0.21	0.01	0.03
	–	–	–	–
(error/unknown)	0.34	0.06	0.16	0.30
Page Management Work				
Total number of				
page faults	1,768	1,697	48,677	49,561
page creations	10,336	21,920	197,885	93,308
page prefetches	3	482	6,785	6,769
discarded prefetched pages	0	0	0	0
forced modified page writes	369	346	403	553
Average rate (1/second) of				
page faults	1.818	1.392	1.119	2.606
page creations	10.63	17.98	4.549	4.906
page prefetches	0.0031	0.3953	0.1560	0.3559
discarding prefetched pages	0	0	0	0
forced modified page writes	0.3795	0.2838	0.0093	0.0291

Memory management time is broken down into time spent fixing page faults, time for page creation (the allocation of pages), and other page management activities, such as prefetching, destroying, wiring (making non-pageable), unwiring, and flushing (making pages replaceable). Sequence breaking time is time spent in the scheduler process, deciding which process to run next.* All other time is lumped together as run time. The run time expressed as a fraction of real time can be considered a measure of average "CPU utilization." Run time includes the time spent executing both user code and system code such as the garbage collector. Note that memory management time is that expended on behalf of both user and system code. The measures of time here reflect the effect of all active processes. However, system activity is expected to be heavily influenced by the process running the program being measured and by garbage collection.

Comparing the results for Boyer in Table 4.2, we observe heavy paging under dynamic garbage collection—64 percent of real time on average, 5,231 total faults, 2.7 faults/second—and consequent low CPU utilization. This is not surprising. Oldspace consists of all memory containing non-static objects, and the scavenging of such a large copyspace and the transportation of a large number of nongarbage objects requires a great deal of paging. A non-negligible amount of prefetching can also be observed under dynamic garbage collection (5.0 percent of real time, 41,117 total pages prefetched, 22 pages/second), because of the garbage collector's prefetching policy.

More interesting is the question of how the performance under ephemeral garbage collection compares with that for no collection. The comparison can be made for Boyer and QPE-short. For Boyer, paging under ephemeral collection is considerably less, and CPU utilization higher, than when the garbage collector is turned off. With no collection, paging took 14 percent of real time on average, utilization was 53 percent, 441 page faults occurred and the average fault rate was 0.84 faults/second. Under ephemeral collection, paging consumed only 2.3 percent of real time and utilization increased to 86 percent; there were only 148 faults and the rate was 0.16 faults/second.

The smaller number of faults is probably due to increased locality arising from the copying and compaction of nongarbage objects. However, even though there were fewer faults, the program ran 70 percent longer with ephemeral collection—898 seconds, compared with 527 seconds with ephemeral garbage collection—because of time spent in garbage collection. In Section 4.6, it will be seen that a total of 452 seconds, or about 50 percent of total real time, was spent in the garbage collector.

The trends observed in running Boyer can also be seen in QPE-short. Paging was slightly reduced—from 12 percent of real time, 1,768 total faults, and 1.8 faults/second with no collection, down to 10 percent of real time, 1,697 total faults, and 1.4 faults/second under ephemeral collection. Again,

---

* No sequence-breaking or page-destroying times are shown for QPE because the associated meters were not available under Release 6. Page-destroying, wiring and unwiring, and flushing times are not shown because they were negligible or zero.

total running time increased because of garbage collector run time—from 972 to 1219 seconds, a 25 percent increase. However, the reduction in paging and increase in total running time are less dramatic than for Boyer. Garbage collection consumed a total of 220 seconds, or 18 percent of total real time.

The reduction in paging indicates the effectiveness of the ephemeral collection algorithm in a virtual memory environment, especially for the Boyer program. The reduction is achieved by scavenging a smaller root set and copyspace, and transporting nongarbage objects from a smaller oldspace than that of the traditional dynamic garbage collector. The high percentage of time spent in garbage collection suggests that improving the garbage collector can still lead to increased performance.

Turning to page creation characteristics, note that more pages are created in running Boyer under ephemeral collection than with no garbage collection—53,152 versus 42,482 pages. Since Boyer is not a nondeterministic program, we expect it to allocate the same number and sizes of objects regardless of the collection type. We must assume, therefore, that the additional creation of about 11,000 pages is caused by the garbage collector. The bulk of additional creation probably occurs when copyspace pages have to be allocated to contain transported objects.

However, even though almost 11,000 more pages are created in running Boyer under ephemeral collection, less time is spent on page creation—5.8 percent of real time on average, compared with 27 percent with no collection. The reason for the high page creation time in Boyer with no collection is unclear, and this oddity is not observed in the case of QPE-short. For QPE-short, the change from no collection to ephemeral collection results in an increase in the total number of pages created—from 10,336 to 21,920 pages (caused by garbage collector allocation); but the page creation time also increases—from 3.2 percent to 4 percent of real time. In any case, the small amount of time consumed by page creation overhead for all runs with ephemeral garbage collection suggests that improving the page creation operations will result in, at best, only a small increase in performance.

In summary, running Boyer and QPE-short with ephemeral garbage collection reduced paging and increased CPU utilization, but also increased total execution time in comparison with runs with no collection. The longer execution time is due mostly to garbage collector run time. Although the ephemeral algorithm is much less intrusive than the original dynamic algorithm, its run-time overhead was significant for two of the runs—50 percent of real time for Boyer and 18 percent for QPE-short. The overhead of page creation time was small for all runs with ephemeral garbage collection.

5.3 Distributions

Distributions for CPU utilization and the fractions of time spent in the various memory management activities, as measured over each sampling interval, are contained in Tables 4.4 and 4.5. Distributions for some page management

TABLE 4.4 **Distributions of Fractions of Real Time Spent in Various Activities.**

	Boyer-24 GC Off	Boyer-15 Dynamic	Boyer-24 Ephemeral
Run Time			
	Mean: 0.5271	Mean: 0.2122	Mean: 0.8632
	Std: 0.2127	Std: 0.2056	Std: 0.1230
	Min: 0.0171	Min: 0.0124	Min: 0.1326
	P1: 0.0500	P1: 0.0252	P1: 0.3627
	Q1: 0.5596	Q1: 0.0888	Q1: 0.8588
	Med: 0.6021	Med: 0.1490	Med: 0.9089
	Q3: 0.6491	Q3: 0.2426	Q3: 0.9414
	P99: 0.8703	P99: 0.8692	P99: 0.9536
	Max: 0.8766	Max: 0.9412	Max: 0.959
Page Faults			
	Mean: 0.1426	Mean: 0.6416	Mean: 0.0229
	Std: 0.2955	Std: 0.2679	Std: 0.04065
	Min: 9.449e-4	Min: 0.0	Min: 0.0
	P1: 9.870e-4	P1: 0.002892	P1: 9.854e-4
	Q1: 0.004980	Q1: 0.5804	Q1: 0.005671
	Med: 0.006745	Med: 0.7204	Med: 0.01153
	Q3: 0.009440	Q3: 0.8078	Q3: 0.02755
	P99: 0.8982	P99: 0.9380	P99: 0.09289
	Max: 0.9595	Max: 0.9581	Max: 0.4381
Page Creation			
	Mean: 0.2651	Mean: 0.05800	Mean: 0.05796
	Std: 0.1493	Std: 0.1471	Std: 0.07506
	Min: 0.0	Min: 0.0	Min: 0.009796
	P1: 0.01134	P1: 0.0	P1: 0.01737
	Q1: 0.2380	Q1: 0.005889	Q1: 0.02596
	Med: 0.2800	Med: 0.01129	Med: 0.0414
	Q3: 0.337	Q3: 0.02322	Q3: 0.06551
	P99: 0.7879	P99: 0.7272	P99: 0.4015
	Max: 0.9189	Max: 0.8843	Max: 0.8433
Sequence Breaking			
	Mean: 0.05437	Mean: 0.01884	Mean: 0.02426
	Std: 0.02928	Std: 0.02136	Std: 0.007504
	Min: 8.336e-4	Min: 9.088e-4	Min: 0.003482
	P1: 0.005137	P1: 0.002453	P1: 0.01083
	Q1: 0.03828	Q1: 0.01013	Q1: 0.02074
	Med: 0.06865	Med: 0.01472	Med: 0.02233
	Q3: 0.07908	Q3: 0.02098	Q3: 0.02453
	P99: 0.1089	P99: 0.1148	P99: 0.04372
	Max: 0.1732	Max: 0.3617	Max: 0.06083

TABLE 4.4 (*Continued*)

Boyer-24 GC Off	Boyer-15 Dynamic	Boyer-24 Ephemeral

Page Prefetching

Boyer-24 GC Off	Boyer-15 Dynamic	Boyer-24 Ephemeral
Mean: 0.0	Mean: 0.04960	Mean: 0.001133
Std: 0.0	Std: 0.05462	Std: 0.003779
Min: 0.0	Min: 0.0	Min: 0.0
P1: 0.0	P1: 0.0	P1: 0.0
Q1: 0.0	Q1: 9.493e-4	Q1: 0.0
Med: 0.0	Med: 0.03944	Med: 0.0
Q3: 0.0	Q3: 0.07654	Q3: 0.0
P99: 0.0	P99: 0.2266	P99: 0.01573
Max: 0.0	Max: 0.3416	Max: 0.04186

event rates are in Tables 4.6 and 4.7. The means of the distributions are equal to the corresponding ratios of totals in Tables 4.2 and 4.3.

The distributions are useful in showing the extent to which the mean values are representative quantities. The mean value may be less meaningful if the distribution is skewed in the direction of high (or low) values, or has heavy tails, or exhibits multiple modes corresponding to certain, particularly frequent values.

Consider Boyer under dynamic garbage collection, for which heavy paging is clearly evident. The paging time distribution is skewed to the right (and the CPU utilization distribution, consequently, to the left). For 70 percent of all sampling intervals, the fraction of real time spent paging was greater than the overall mean of 0.64. The paging time distribution also shows a large spike at zero, but, interestingly, there is no corresponding spike at unity in the utilization distribution. That is, during most of the intervals in which absolutely no time was spent in paging, other kinds of overhead were present.

The shape of the page fault rate distribution roughly follows that of paging time but is more spread out over its range.* Although the distributions for page fault rate appear the same for all runs, a clue to the difference in the case of Boyer under dynamic garbage collection appears in the third quartile statistic (2.9 faults/second compared with zero for most other runs). A small, almost unnoticeable shift in the fault rate distribution, resulting in a small increase in the mean, has a large effect on the paging time distribution. Of course, this behavior reflects the high cost of a fault, arising from the large disparity between disk and main memory referencing time.

Earlier, we observed that page creation time was low for all runs (except Boyer with no collection) and therefore not an attractive target for optimization.

---

* It is somewhat difficult to see this from Table 4.6, but a greatly magnified view reveals the similarity.

136

TABLE 4.5   Distributions of Fractions of Real Time Spent in Various Activities.

	QPE-Short GC Off	QPE-Short Ephemeral	QPE-Long (Full Execution) Ephemeral	QPE-Long (Initial Part) Ephemeral
Run Time				
	Mean: 0.8468	Mean: 0.8574	Mean: 0.9311	Mean: 0.8585
	Std: 0.2735	Std: 0.2596	Std: 0.1344	Std: 0.2046
	Min: 0.0200	Min: 0.0143	Min: 0.04929	Min: 0.03422
	P1: 0.02649	P1: 0.03732	P1: 0.2843	P1: 0.08807
	Q1: 0.8802	Q1: 0.8781	Q1: 0.9605	Q1: 0.8423
	Med: 0.9816	Med: 0.9542	Med: 0.9809	Med: 0.9488
	Q3: 0.9895	Q3: 0.9854	Q3: 0.9838	Q3: 0.9795
	P99: 0.9990	P99: 1.197	P99: 0.9963	P99: 0.9987
	Max: 1.030	Max: 1.645	Max: 0.9987	Max: 1.101
Page Faults				
	Mean: 0.1176	Mean: 0.09935	Mean: 0.06174	Mean: 0.1303
	Std: 0.2661	Std: 0.2413	Std: 0.1298	Std: 0.1979
	Min: 0.0	Min: 0.0	Min: 0.001202	Min: 0.0
	P1: 0.0	P1: 0.001731	P1: 0.003530	P1: 0.001391
	Q1: 0.004745	Q1: 0.007361	Q1: 0.01112	Q1: 0.01288
	Med: 0.008414	Med: 0.01381	Med: 0.01344	Med: 0.04374
	Q3: 0.01776	Q3: 0.03021	Q3: 0.03344	Q3: 0.1473
	P99: 0.9415	P99: 0.9396	P99: 0.6740	P99: 0.8847
	Max: 0.9567	Max: 0.9631	Max: 0.9289	Max: 0.9436

Page Creation

Mean: 0.03223
Std: 0.07104
Min: 0.0
P1: 0.0
Q1: 0.003790
Med: 0.004761
Q3: 0.01994
P99: 0.3247
Max: 0.6583

Mean: 0.04045
Std: 0.07663
Min: 0.0
P1: 0.0
Q1: 0.003802
Med: 0.01969
Q3: 0.03734
P99: 0.3338
Max: 0.6852

Mean: 0.005543
Std: 0.009177
Min: 0.0
P1: 0.0
Q1: 0.003513
Med: 0.004345
Q3: 0.005582
P99: 0.03575
Max: 0.1797

Mean: 0.007939
Std: 0.02885
Min: 0.0
P1: 0.0
Q1: 9.943e-4
Med: 0.003203
Q3: 0.007985
P99: 0.06722
Max: 0.7513

Page Prefetching

Mean: 3.291e-5
Std: 0.001019
Min: 0.0
P1: 0.0
Q1: 0.0
Med: 0.0
Q3: 0.0
P99: 0.0
Max: 0.0315

Mean: 0.002143
Std: 0.01584
Min: 0.0
P1: 0.0
Q1: 0.0
Med: 0.0
Q3: 0.0
P99: 0.02610
Max: 0.2381

Mean: 1.277e-4
Std: 0.001079
Min: 0.0
P1: 0.0
Q1: 0.0
Med: 0.0
Q3: 0.0
P99: 0.002416
Max: 0.04278

Mean: 2.889e-4
Std: 0.002817
Min: 0.0
P1: 0.0
Q1: 0.0
Med: 0.0
Q3: 0.0
P99: 0.004670
Max: 0.1158

TABLE 4.6 Distributions of Rates of Various Page Management Events.

	Boyer-24 GC Off	Boyer-15 Dynamic	Boyer-24 Ephemeral
Page Faults			
	Mean: 0.8367	Mean: 2.743	Mean: 0.1648
	Std: 2.286	Std: 4.836	Std: 1.807
	Min: 0.0	Min: 0.0	Min: 0.0
	P1: 0.0	P1: 0.0	P1: 0.0
	Q1: 0.0	Q1: 0.0	Q1: 0.0
	Med: 0.0	Med: 0.0	Med: 0.0
	Q3: 0.0	Q3: 2.943	Q3: 0.0
	P99: 4.702	P99: 21.29	P99: 1.633
	Max: 10.79	Max: 32.89	Max: 31.18
Page Creations			
	Mean: 80.60	Mean: 17.27	Mean 39.20
	Std: 34.27	Std: 22.62	Std: 28.40
	Min: 0.0	Min: 0.0	Min: 9.280
	P1: 4.225	P1: 0.0	P1: 19.55
	Q1: 83.89	Q1: 5.885	Q1: 35.36
	Med: 92.59	Med: 11.23	Med: 44.03
	Q3: 101.26	Q3: 18.49	Q3: 86.95
	P99: 136.3	P99: 119.2	P99: 135.9
	Max: 140.4	Max: 140.5	Max: 140.1
Page Prefetches			
	Mean: 0.3605	Mean: 21.38	Mean: 0.04455
	Std: 1.249	Std: 20.83	Std: 0.4570
	Min: 0.0	Min: 0.0	Min: 0.0
	P1: 0.0	P1: 0.0	P1: 0.0
	Q1: 0.0	Q1: 3.047	Q1: 0.0
	Med: 0.0	Med: 16.07	Med: 0.0
	Q3: 0.0	Q3: 38.14	Q3: 0.0
	P99: 2.261	P99: 69.67	P99: 0.4176
	Max: 3.828	Max: 80.03	Max: 7.472
Discarding Prefetched Pages			
	Mean: 0.8898	Mean: 0.8021	Mean: 0.4889
	Std: 8.957	Std: 3.798	Std: 3.988
	Min: 0.0	Min: 0.0	Min: 0.0
	P1: 0.0	P1: 0.0	P1: 0.0
	Q1: 0.0	Q1: 0.0	Q1: 0.0
	Med: 0.0	Med: 0.0	Med 0.0
	Q3: 0.0	Q3: 0.0	Q3: 0.0
	P99: 17.45	P99: 24.31	P99: 4.859
	Max: 129.2	Max: 129.1	Max: 119.9

TABLE 4.6 (*Continued*)

	Boyer-24 GC Off	Boyer-15 Dynamic	Boyer-24 Ephemeral

Forced Modified Page Writes

Boyer-24 GC Off	Boyer-15 Dynamic	Boyer-24 Ephemeral
Mean: 4.147	Mean: 0.3983	Mean: 0.1782
Std: 3.131	Std: 1.448	Std: 1.116
Min: 0.0	Min: 0.0	Min: 0.0
P1: 0.0	P1: 0.0	P1: 0.0
Q1: 3.003	Q1: 0.0	Q1: 0.0
Med: 4.752	Med: 0.0	Med: 0.0
Q3: 5.870	Q3: 0.0	Q3: 0.0
P99: 13.16	P99: 6.245	P99: 4.313
Max: 29.82 0 20	Max: 27.83 0 20	Max: 20.45 0 1

Nevertheless, the distributions for page creation rate are interesting because they characterize the page allocation load placed on the memory system by all workload components acting jointly. In this study, the measured program and the garbage collector are the dominant components. However, in any multiprocess environment, there may be multiple mutators competing for processor and memory service. The characterization of multiprocess workload, especially its resource demands, could provide valuable information for system design and tuning, and for adaptive resource management.

The distribution of page creation rate for Boyer with garbage collection off has a number of distinct modes; one, at about 90 pages/second, stands out. Since page creation with garbage collection off directly reflects object creation, the multi-modal characteristic can be attributed to the program rather than to program/system interaction. Under dynamic garbage collection, the distribution is pushed to the left because of heavy paging activity. When running with ephemeral garbage collection, the 90 pages/second mode can still be seen, but there is now a dominant mode located at about 30 pages/second, most likely caused by garbage collector allocation.

5.4 Time Series

A time series plot is useful for showing when activities or events of interest occur; whether their intensity or frequency of occurrence is uniform or localized to certain portions of the measured period; and whether there is any pattern or regularity.

Figures 4.1–4.7 show the time series plots for utilization, paging, and page creation activity while running the measured programs.

The high paging overhead noted earlier in runs of Boyer under dynamic garbage collection is evident in Figure 4.2. The reduced paging for Boyer

TABLE 4.7 Distributions of Rates of Various Page Management Events.

	QPE-Short GC Off	QPE-Short Ephemeral	QPE-Long (full execution) Ephemeral	QPE-Long (initial part) Ephemeral
Page Faults				
	Mean: 1.818	Mean: 1.392	Mean: 1.119	Mean: 2.606
	Std: 4.658	Std: 4.330	Std: 3.454	Std: 5.390
	Min: 0.0	Min: 0.0	Min: 0.0	Min: 0.0
	P1: 0.0	P1: 0.0	P1: 0.0	P1: 0.0
	Q1: 0.0	Q1: 0.0	Q1: 0.0	Q1: 0.0
	Med: 0.0	Med: 0.0	Med: 0.0	Med: 0.0
	Q3: 0.0	Q3: 0.0	Q3: 0.1332	Q3: 2.185
	P99: 20.11	P99: 21.06	P99: 19.36	P99: 25.60
	Max: 28.24	Max: 26.68	Max: 35.35	Max: 35.13
Page Creations				
	Mean: 10.63	Mean: 17.96	Mean: 4.549	Mean: 4.907
	Std: 16.96	Std: 19.32	Std: 4.293	Std: 7.925
	Min: 0.0	Min: 0.0	Min: 0.0	Min: 0.0
	P1: 0.0	P1: 0.0	P1: 0.0	P1: 0.0
	Q1: 3.743	Q1: 3.798	Q1: 3.197	Q1: 0.7975
	Med: 3.798	Med: 15.43	Med: 4.512	Med: 3.169
	Q3: 11.22	Q3: 23.37	Q3: 4.780	Q3: 7.011
	P99: 56.90	P99: 88.95	P99: 21.86	P99: 33.72
	Max: 204.2	Max: 238.2	Max: 84.53	Max: 126.4

Page Prefetches

Mean: 0.003086
Std: 0.09553
Min: 0.0
P1: 0.0
Q1: 0.0
Med: 0.0
Q3: 0.0
P99: 0.0
Max: 2.958

Mean: 0.3953
Std: 4.654
Min: 0.0
P1: 0.0
Q1: 0.0
Med: 0.0
Q3: 0.0
P99: 0.0
Max: 64.70

Mean: 0.1560
Std: 1.334
Min: 0.0
P1: 0.0
Q1: 0.0
Med: 0.0
Q3: 0.0
P99: 4.609
Max: 35.61

Mean: 0.3559
Std: 2.115
Min: 0.0
P1: 0.0
Q1: 0.0
Med: 0.0
Q3: 0.0
P99: 7.241
Max: 58.15

Forced Modified Page Writes

Mean: 0.3795
Std: 1.504
Min: 0.0
P1: 0.0
Q1: 0.0
Med: 0.0
Q3: 0.0
P99: 4.610
Max: 26.22

Mean: 0.2838
Std: 1.070
Min: 0.0
P1: 0.0
Q1: 0.0
Med: 0.0
Q3: 0.0
P99: 4.456
Max: 21.26

Mean: 0.009263
Std: 0.09155
Min: 0.0
P1: 0.0
Q1: 0.0
Med: 0.0
Q3: 0.0
P99: 0.2664
Max: 3.534

Mean: 0.02908
Std: 0.3344
Min: 0.0
P1: 0.0
Q1: 0.0
Med: 0.0
Q3: 0.0
P99: 0.5966
Max: 13.47

141

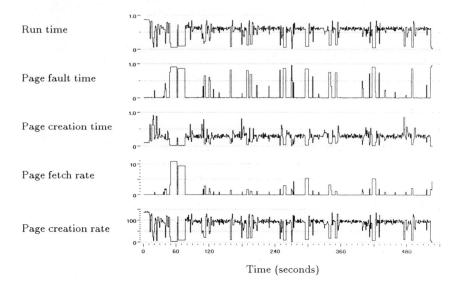

Figure 4.1 Run time, page fault, and page creation time expressed as a fraction of real time, and rate (1/second) of page fetches and creations for Boyer-24, gc off.

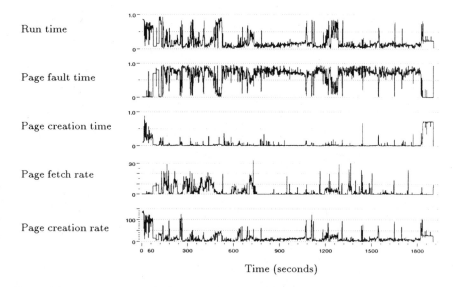

Figure 4.2 Run time, page fault, and page creation time expressed as a fraction of real time, and rate (1/second) of page fetches and creations for Boyer-15, dynamic gc.

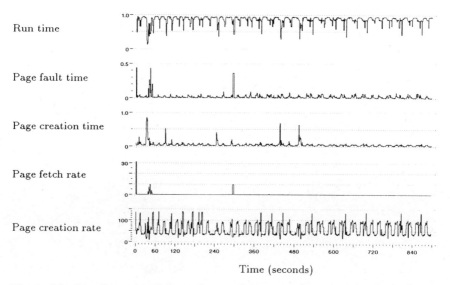

Figure 4.3 Run time, page fault, and page creation time expressed as a fraction of real time, and rate (1/second) of page fetches and creations for Boyer-24, ephemeral gc.

under ephemeral collection versus no collection can also be observed by comparing Figures 4.1 and 4.3. All plots for Boyer show that page management overhead is fairly uniform throughout the measured period. The uniformity is expected, since the workload consists of data-independent repetitions of the same program.

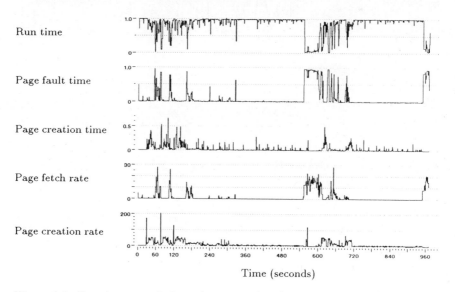

Figure 4.4 Run time, page fault, and page creation time expressed as a fraction of real time, and rate (1/second) of page fetches and creations for QPE-short, gc off.

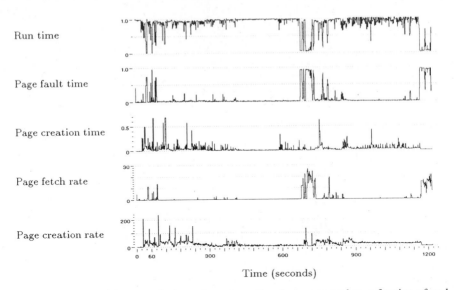

Figure 4.5   Run time, page fault, and page creation time expressed as a fraction of real time, and rate (1/second) of page fetches and creations for QPE-short, ephemeral gc.

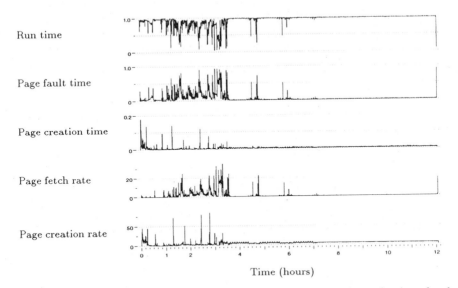

Figure 4.6   Run time, page fault, and page creation time expressed as a fraction of real time, and rate (1/second) of page fetches and creations for QPE-long (full execution), ephemeral gc.

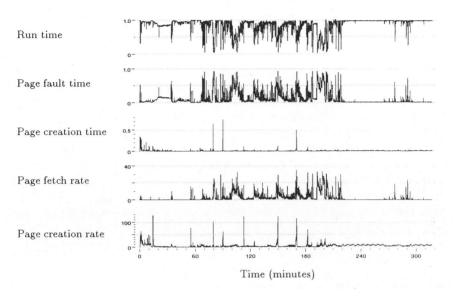

Run time

Page fault time

Page creation time

Page fetch rate

Page creation rate

Time (minutes)

Figure 4.7 Run time, page fault, and page creation time expressed as a fraction of real time, and rate (1/second) of page fetches and creations for QPE-long (initial part), ephemeral gc.

Since distributions summarize the spread in the data values but do not show time order information, it may be beneficial to consider a distribution alongside the corresponding time series plot to determine the pattern in the time behavior which produces the observed modes. Do particular modes in the distribution result from behavior during particular time segments?

For example, earlier we observed two dominant modes in the page creation rate distribution for Boyer with ephemeral garbage collection (Table 4.6). The time series (Figure 4.3) shows that these modes are due to a roughly periodic alternation between time intervals of high creation rate and intervals of low creation rate. The occurrence of this cyclic pattern is probably related to the occurrence of garbage collections. Section 4.7 shows that collections occur fairly regularly in the Boyer case.

While Boyer incurs system overhead uniformly, QPE-short and QPE-long do not. In particular, the time series for the full execution of QPE-long (Figure 4.6) reveal that most of the paging occurs during the first 3 hours and 40 minutes of execution. In optimizing the paging performance of this particular program, one would concentrate on this initial period, since 92 percent of the total 48,677 page faults occur during this time.

Tables 4.5 and 4.7 show that over the full 12 hours and 5 minutes of execution of QPE-long, paging takes 6.2 percent of real time on average (with 13 percent standard deviation) and the mean rate is 1.1 faults/second (with standard deviation of 3.5). When calculated over the first 3 hours and 40 minutes only, the figures increase to 17 percent of real time (19 percent standard deviation) and 3.4 faults/second (with standard deviation of 5.4).

Again, it is demonstrated that statistics measured over the entire execution lifetime may not be representative if there are large variations in the intensity of the activity over time.*

These results are a reminder that complex and long-running applications may exhibit quite different paging characteristics at different times, and that it is possible to identify periods of frequent paging. This observation is neither new nor specific to Lisp programs, and suggests the applicability and adaptation to Lisp systems of paging reduction techniques on conventional machines. For example, improvements in paging performance may be achieved by choosing more efficient representations, by restructuring control or data in the program, or by adjusting paging policies such as replacement and prefetching. Performance may also be improved by reordering objects or pages on the basis of a cluster analysis of references, thereby taking into account dynamic referencing patterns. Furthermore, it may be possible to perform some of these tasks automatically, by learning from traces or samples of past behavior.

5.5 Scatter Plots

Consider the three major activities: run time, paging, and page creation. Over the various measurement runs, they account on average for 91.2–99.9 percent of real time between samples. We can visualize the dynamic relationship between them by representing each sample as a point in three-dimensional space, with each axis representing one activity.

Figure 4.8 shows the resulting scatter plots, which depict the increase in run time caused by the two major types of low-level storage management overhead (paging and allocation). The lower the overhead, the higher the system may appear on the run time axis. As expected, most points lie on or slightly beneath the triangular surface whose vertices are at unity on each axis, and the dominance of paging in Boyer under dynamic garbage collection is evident.

When running Boyer without collection, the system spends more than a quarter of real time (27 percent) in page creation (see Table 4.2 or 4.4). This characteristic can be seen in Figure 4.8 as a concentration of points along the page-creation/run-time plane. In contrast, paging time is the more likely overhead when running QPE-long, as manifested by the concentration of points along the paging/run-time plane.

Less obviously, very few points are clustered near the bottom edge of the imaginary triangular surface, that is, along the paging/page-creation plane. This result indicates that, for the programs measured and sampling periods chosen (1, 5, and 15 seconds), paging overhead and page creation time overhead tend

---

* Note that while the standard deviation provides a useful and convenient measure of the degree of fluctuation in a quantity, it must be specified together with the sampling period chosen. If the sampling period were made very large (or, alternatively, if smoothing transformations are applied to the raw data), high frequency variation would be effectively "averaged out" resulting in a lower standard deviation.

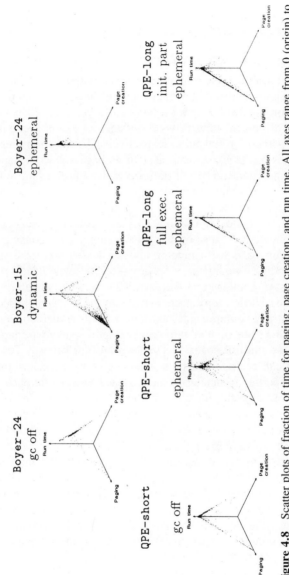
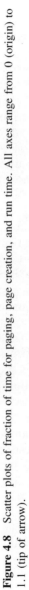

Figure 4.8 Scatter plots of fraction of time for paging, page creation, and run time. All axes range from 0 (origin) to 1.1 (tip of arrow).

147

not to appear simultaneously in large amounts; during a sampling period, there may be considerable faulting or considerable page allocation activity, but not both.

6   GARBAGE COLLECTOR ANALYSIS

Ephemeral garbage collection is much more efficient and less intrusive than dynamic collection. However, running a program with the ephemeral collector may still take significantly longer than running the same program without it (see Table 4.1 or Figure 4.11); for example, Boyer took 70 percent longer and QPE-short 25 percent longer, with garbage collection accounting for 50 percent and 18 percent of real time, respectively. This section presents a performance analysis of the constituent tasks of garbage collection, and develops and evaluates regression models of garbage collection time in terms of work done, page faults, and other overhead.

Figure 4.9 shows the decomposition of garbage collection tasks in the form of a state chart.* At the highest level of abstraction, storage reclamation comprises scavenging and transporting. Scavenging may be consing-induced — performed to keep pace with mutator allocation — or may take place while the machine is idle. Two memory spaces must be scavenged: the root set and copyspace. As explained in Section 3.2, the approximately depth-first copying technique allows copyspace scavenging to be decomposed into last-page and normal-page components; and ephemeral garbage collection allows root set scavenging to be divided into scavenging of in-memory pages (flagged by the GCPT) and on-disk pages (flagged by the ESRT).

Transporting can be triggered by scavenging or by normal computation; a call to the transporter occurs when an oldspace object reference is read by either the scavenger or the mutator. If the oldspace object has not already been copied, the transporter copies it and installs a forwarding pointer; in all cases the transporter replaces the oldspace reference with a reference to the copyspace version of the object.

Table 4.8 shows a breakdown of real time, page faults, and work done. The values are measured over the entire program execution. For all runs, scavenging was almost entirely consing-induced, since none of the program times include any idle-time caused by pauses for user input.

Total garbage collection time varied from 3.5 percent of total real time for QPE-long (full execution) to 74 percent for Boyer under dynamic collection. Although every program spent most of its garbage collection time (93–99 percent) in the scavenger, the programs differed in the way scavenging time

---

* A state chart is a particular kind of higraph [9] that combines a state transition diagram and a Venn diagram (for describing set inclusion relationships). The dashed line dividing the area representing the scavenger denotes orthogonality; that is, the scavenger state is specified by independent choices of items on either side of the line.

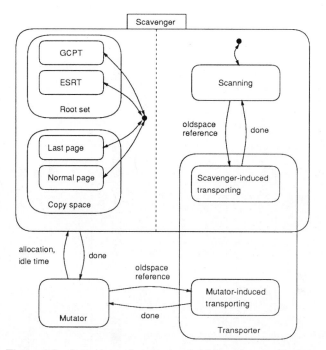

Figure 4.9 State chart of agents involved in garbage collection.

was distributed: both among the various types of pages to be scavenged, and between scanning memory and transporting objects.

When running Boyer and QPE-short under ephemeral collection, little time (0.3 percent of scavenging time for Boyer and 0.2 percent of scavenging time for QPE-short) was spent scavenging ESRT pages, suggesting that most of the pages into which ephemeral object references were stored were not purged from main memory. In contrast, QPE-long spent about one fourth of scavenging time in scavenging disk-resident pages.

As expected, transporting one word is much more time-consuming than scanning one word. For QPE-short, for example, the ratio of words scanned to words transported was 5.8 to 1, but the scavenger spent almost as much time in transporting as in scanning. Assume a simple linear relationship between scanning (transporting) time and words scanned (transported);* then for Boyer with dynamic collection, transporting one word could be considered time-equivalent to scanning 1.2 words. For the four ephemeral runs, the equivalent scanning work would be 11, 5.8, 8.9, and 4.3 words. This result may be useful in defining a more realistic single measure of collector work for use in efficiency metrics, such as that discussed in Section 4.7. When analyzing

---

* This assumption does not quite hold in general, as we will later show.

TABLE 4.8 Total Real Time, Page Faults, and Words.

Totals and Ratios	Boyer-15 Dynamic	Boyer-24 Ephemeral	QPE-Short Ephemeral	QPE-Long (Full Execution) Ephemeral	QPE-Long (Initial Part) Ephemeral
Real Time					
Total real time (hh:mm:ss)	31:45	14:58	20:19	12:05:05	5:16:57
Percentage of total time in garbage collector	74.41	50.38	18.08	3.48	6.95
Percentage of gc time in scavenger	99.91	98.96	93.51	93.38	93.16
mutator-induced transporting	0.09	1.04	6.49	6.62	6.84
Percentage of scavenger time spent in scavenging					
GCPT pages	—	16.14	19.02	29.98	26.23
ESRT pages	—	0.31	0.16	26.26	24.30
last page	8.26	60.36	47.26	22.63	25.86
normal page	89.87	10.37	22.79	16.19	18.66
Percentage of scavenger time in scanning memory	89.44	62.92	50.47	70.15	64.30
transporting	10.56	37.08	49.53	29.85	35.70

Page Faults

Total page faults	5,231	148	1,697	48,677	49,561
Percentage of total faults in garbage collector	84.32	6.76	1.47	24.90	27.59
Percentage of gc faults in scavenger	99.86	60.00	100.00	98.89	99.69
mutator-induced transporting	0.14	40.00	0.00	1.11	0.31
Percentage of scavenger faults in scavenging					
GCPT pages	—	33.33	64.00	28.94	36.07
ESRT pages	—	0.00	0.00	29.79	29.46
last page	24.06	16.67	12.00	22.62	19.21
normal page	75.89	50.00	16.00	18.60	15.24
Percentage of scavenger faults in scanning memory	81.88	0.00	84.00	63.40	66.63
transporting	18.12	100.00	16.00	36.60	33.37

Work

Total words scanned	15,293,327	50,723,692	11,297,323	142,813,693	51,871,841
Total words transported	1,552,663	2,701,806	1,929,255	6,818,632	6,759,927
Total oldspace reclaimed	3,526,906	13,123,890	2,759,212	39,428,316	16,358,752
Percentage of words scanned in scavenging					
GCPT pages	—	69.70	73.06	72.84	64.17
ESRT pages	—	0.46	0.24	20.38	17.44
last page	6.06	6.75	9.80	1.99	5.30
normal page	93.94	23.10	16.90	4.79	13.09

an algorithm that pegs the rate of collector work to the rate of consing, a definition of work that puts greater weight on transporting could increase the predictability of scavenging time variations.

The time spent in mutator-induced transporting indirectly reflects the number of nongarbage oldspace objects encountered by the mutator. Boyer spends 1.0 percent of total garbage collection time in mutator-induced transporting, and QPE, 6.5–6.8 percent. Courts [7] has proposed the heuristic that oldspace objects accessed by the mutator are much more likely to be active (i.e., part of the working set) than those accessed by the scavenger, and that such objects occupy only a small fraction of total memory. After flipping all of dynamic and static space into oldspace, and running an interactive workload with the scavenger inhibited, he found that only 13 percent of oldspace had been referenced. In our measured system, the number of oldspace objects touched by the mutator is only a lower bound on the total number of active objects — since the scavenger was not inhibited and could have discovered active objects before the mutator. Nevertheless, the relatively small amount of mutator-induced transporting in our CPU-bound programs lends support to the view that only a small amount of oldspace is being actively referenced.

6.1 Regression Models

How does garbage collection time relate to the amount of collector work performed, the number of page faults, and other system software overhead? It seems reasonable to assume a linear relationship. To evaluate this assumption, we consider various regression models for collection time.

A regression model is a kind of *linear model*. A linear model expresses a response (or dependent) variable as a linear combination of independent variables, plus an error term. The error term is assumed to be a normally distributed random variable with mean 0 and variance σ^2. For example, a model for a response Y in terms of two variables X_1 and X_2 could have the form:

$$Y = \beta_0 + X_1\beta_1 + X_2\beta_2 + \epsilon$$

Given a set of observations for Y, X_1, and X_2, a least-squares analysis is typically used to compute estimates of the unknown constant parameters β_i and error variance σ^2. The goodness of fit of the model is expressed by R^2, which is the fraction of the variation in the response variable explained or accounted for by the model without the error term. $R^2 = 0$ implies complete lack of fit and $R^2 = 1$ implies perfect fit.

Our model is developed as follows. The time spent in garbage collection within a given interval of time can be expressed as the sum of scanning and transporting time (see Figure 4.9):

$$Y_{GC} = \text{expected scanning time} + \text{expected transporting time} + \epsilon$$

Scanning time is the time spent in the scavenger, excluding the time spent in scavenger-induced transporting, and consists of run time, paging and page creation time, and other page management overhead, as discussed in Section 4.5. Run time can be modeled as the sum of the following times:

- Time in reading memory sequentially (proportional to the number of words scanned)
- Some scavenger overhead (proportional to the number of calls to the scavenger)
- Some overhead in calling the transporter from the scavenger (proportional to the number of scavenger-induced transporter calls).

Paging time can be assumed to be proportional to the number of scavenger page faults, excluding faults occurring during scavenger-induced transporting. Finally, it assumed that no page creation time is incurred when scanning memory, and other page management overhead, which was negligible, is ignored. The expected scanning time is therefore assumed to be

$$\text{expected scanning time} = f(X_{\text{ScWords}}, X_{\text{ScCalls}}, X_{\text{STcalls}}, X_{\text{ScPagef}})$$

where f denotes linear combination:

$$f(X_1, X_2, X_3, \cdots) = X_1\beta_1 + X_2\beta_2 + X_3\beta_3 + \cdots$$

A more detailed model for expected scanning time is made possible by the inclusion of separate terms for the words scanned during GCPT, ESRT, last-page, and normal page-scavenging, and for the number of calls to the corresponding scavenging routines:

$$\begin{aligned}\text{expected scanning time} = f(&X_{\text{GCPTwords}}, X_{\text{ESRTwords}}, X_{\text{LastWords}}, X_{\text{NormWords}}, \\ &X_{\text{GCPTcalls}}, X_{\text{ESRTcalls}}, X_{\text{LastCalls}}, X_{\text{NormCalls}}, \\ &X_{\text{STcalls}}, X_{\text{ScPagef}})\end{aligned}$$

Similar arguments lead to a model for expected transporting time:

$$\text{expected transporting time} = f(X_{\text{TrWords}}, X_{\text{TrCalls}}, X_{\text{TrPagef}})$$

Table 4.9 summarizes the independent variables appearing in the various equations. The regression parameters could be interpreted as

1. the average number of seconds for each word scanned by the scavenger, or copied by the transporter (e.g., β_{ScWords});
2. the average number of seconds per call to the scavenger, or to the transporter (e.g., β_{ScCalls}); and

TABLE 4.9   **Independent Variables.**

Model for Scanning Time (Simple)	
Words scanned	$X_{ScWords}$
Scavenger calls	$X_{ScCalls}$
Scavenger-induced transporter calls	$X_{STcalls}$
Page faults while scanning	$X_{ScPagef}$
Model for Scanning Time (Detailed)	
GCPT words scanned	$X_{GCPTwords}$
ESRT words scanned	$X_{ESRTwords}$
Last-page words scanned	$X_{LastWords}$
Normal-page words scanned	$X_{NormWords}$
GCPT calls	$X_{GCPTcalls}$
ESRT calls	$X_{ESRTcalls}$
Last-page calls	$X_{LastCalls}$
Normal-page calls	$X_{NormCalls}$
Scavenger-induced transporter calls	$X_{STcalls}$
Page faults while scanning	$X_{ScPagef}$
Model for Transporting Time	
Words copied	$X_{TrWords}$
Transporter calls	$X_{TrCalls}$
Page faults while transporting	$X_{TrPagef}$

3. the average number of seconds per page fault during scanning or transporting (e.g., $\beta_{ScPagef}$).

In the transporter model, note that the term involving the number of words copied accounts for time spent both in copying objects and in allocating copyspace. The regression parameter $\beta_{TrWords}$ expresses the average number of seconds per word copied due to the memory read and write, and to forwarding pointer processing and copyspace allocation amortized over each word transported.

6.2   Regression Results

For each program, we fit both the simple and the detailed model to data from each sampling interval, using a statistical analysis program [15]. The data was first normalized with respect to real time; the dependent variable Y_{GC} was defined to be the fraction of time in garbage collection; and the independent variables X_i were defined to be event rates.

TABLE 4.10 R^2 **Values and Estimates of Error Standard Deviation. All models are statistically significant at the level of $\alpha < 0.0001$.**

Regression Model		Boyer-24 Ephemeral	QPE-Short Ephemeral	QPE-Long (Full Execution) Ephemeral	QPE-Long (Initial Part) Ephemeral
Simple	R^2	0.9970	0.9713	0.9330	0.9027
	$\hat{\sigma}$	0.0334	0.0494	0.0238	0.0491
Detailed	R^2	0.9994	0.9736	0.9455	0.9141
	σ	0.0146	0.0474	0.0215	0.0462

Values of $R^2 > 0.9$ were obtained, as shown in Table 4.10, suggesting that the linear relationships are quite accurate. The detailed models have a slightly better fit than the corresponding simple models. We found that both simple and detailed models suffer from *multicollinearity*: certain subsets of variables were significantly correlated, thus contributing redundant information [11]. Although multicollinearity is not a problem if the goal is to make good predictions, it caused some parameter estimates to be negative, and therefore physically uninterpretable. The variables that tended to be correlated were the "words scanned" and "times called" variables; approximately, the same number of words are scanned on every call to the function to scavenge ESRT pages, for example.*

To reduce multicollinearity and determine whether the parameter estimates were reasonable, we discarded the "times called" variables, allowing the effect of per-call overhead to be absorbed by the "words" variables. Table 4.11 shows the parameter estimates for the reduced models.

The regression parameter estimates in the reduced models reveal differences in the modeled time costs of scanning a word that depend on word location. Since the simple model lumps all words scanned into one term ($X_{\text{ScWords}} = X_{\text{GCPTwords}} + X_{\text{ESRTwords}} + X_{\text{LastWords}} + X_{\text{NormWords}}$) it cannot represent these differences and consequently exhibits poorer fit. The cost is highest for words in the last page of copyspace; the lowest cost is for root set words in physical memory and for words scanned during the normal linear traversal of copyspace. Root set words on disk have an intermediate cost. Note that these costs include garbage collector overhead associated with scanning the various areas, amortized over the words scanned.

Transporting a word costs more than scanning a word by a factor of $\hat{\beta}_{\text{TrWords}}$ / $\hat{\beta}_{\text{ScWords}}$, a ratio that could be used to define a time-equivalent measure of collector work. If scanning one word defines one unit of work, then trans-

---

* Strong correlations between the number of words scanned and the number of calls to the appropriate function were noted for GCPT, ESRT, and last-page scavenging, but not for normal-page scavenging.

†An estimate for β_{ScPagef} for Boyer-24 is missing because no page faults occurred during scanning. All faults that were counted in the garbage collector occurred during transporting (See Table 4.8). All other missing parameters are statistically insignificant at the $a = 0.025$ level.

**TABLE 4.11   Parameter Estimates in Reduced Models for Garbage Collection Time.
All models are statistically significant at the level of $\alpha < 0.0001$.**

Regression Parameter Estimate	Units	Boyer-24 Ephemeral	QPE-Short Ephemeral	QPE-Long (Full Execution) Ephemeral	QPE-Long (Initial Part) Ephemeral
Simple Model					
$\hat{\beta}_{\text{ScWords}}$	μsec/word	1.833	5.250	4.626	7.116
$\hat{\beta}_{\text{TrWords}}$	μsec/word	128.5	56.34	46.76	39.52
$\hat{\beta}_{\text{ScPagef}}$	msec/fault	–	158.7	50.48	50.10
$\hat{\beta}_{\text{TrPagef}}$	msec/fault	–	–	22.35	17.07
R^2		0.9864	0.6905	0.8183	0.8122
$\hat{\sigma}$		0.0716	0.1618	0.0392	0.0682
Detailed Model					
$\hat{\beta}_{\text{GCPTwords}}$	μsec/word	1.729	3.117	2.589	3.465
$\hat{\beta}_{\text{ESRTwords}}$	μsec/word	11.47	18.76	9.690	13.76
$\hat{\beta}_{\text{LastWords}}$	μsec/word	45.67	108.4	118.6	101.4
$\hat{\beta}_{\text{NormWords}}$	μsec/word	2.008	2.864	–	2.818
$\hat{\beta}_{\text{TrWords}}$	μsec/word	73.64	23.27	26.82	26.24
$\hat{\beta}_{\text{ScPagef}}$	msec/fault	–	142.8	41.50	44.61
$\hat{\beta}_{\text{TrPagef}}$	msec/fault	–	–	24.38	19.08
R^2		0.9907	0.9223	0.8842	0.8623
$\hat{\sigma}$		0.0592	0.0812	0.0313	0.0584

porting one word could be said to perform $\hat{\beta}_{\text{TrWords}} / \hat{\beta}_{\text{ScWords}} = 70, 11, 10,$ and 5.6 units of work, respectively, for the four programs that were run under ephemeral collection. Note that this scheme ignores the large differences among the various spaces in the per-word scanning time, and excludes paging time. These ratios are to be compared with 11, 5.8, 8.9, and 4.3 ratios calculated earlier using the total scanning and transporting time, which include paging time.

The estimates for β_{ScPagef} and β_{TrPagef} show that the time cost of a fault is several orders of magnitude greater than the time cost of collector work. Interestingly, the average time per page fault is similar across programs (in the 19–50 ms range) except for QPE-short, which exhibits higher values (143–159 ms). It is not clear whether this unusual result is simply due to the small number of samples, or whether it indicates some undesirable interaction between program and system behavior: for example, high disk latencies occurring as a consequence of pathological placement and referencing of disk blocks.

To test the accuracy of the models, especially across programs, we applied them to the mean rates of scanning words, transporting words, page faults, and the like, and compared the predicted fraction of time spent in garbage

TABLE 4.12   Predicted Garbage Collection Times and 95% Confidence Intervals.

Run	Actual GC Time	GC Time Predicted By Detailed Model Developed From:			
		Boyer-24 Ephemeral	QPE-Short Ephemeral	QPE-Long (Full Execution) Ephemeral	QPE-Long (Initial Part) Ephemeral
Boyer-24	.5038	.5035 ± .029	.5690 ± .095	.6218 ± .066	.6529 ± .105
QPE-short	.1808		.1745 ± .093	.1901 ± .042	.1802 ± .090
QPE-long full execution	.0348		.0458 ± .094	.0355 ± .042	.0370 ± .090
QPE-long initial part	.0695		.0959 ± .094	.0679 ± .042	.0677 ± .090

collection with the actual mean fraction of time spent. It was anticipated, for example, that the model developed from QPE-short, would make a good prediction for the overall mean collection time in QPE-short, but how well would it predict Boyer's collection time?

In all but two cases, not only the "self" predictions but also the cross-program predictions turned out to be correct (a prediction was considered correct if the actual value fell within the 95 percent confidence interval about the predicted value). Table 4.12 shows results for the full (non-reduced) detailed model only. This model gives the tightest confidence intervals, but the correctness results are identical for the other models variations we have discussed.* The exceptions were the two QPE-long models' overestimates of Boyer's mean garbage collection time. The overestimates might be caused by differences in program characteristics and in the underlying system software. Measurements of other programs are needed to determine the effect of program characteristics on the model.

It has occasionally been assumed in the literature, usually implicitly, that collection time is linearly dependent on the work to be performed, that is, the number of words to be scanned and transported. To test this assumption, we evaluated the following "work only" models:

$$Y_{GC} = f(X_{\text{ScWords}}, X_{\text{TrWords}}) + \epsilon \qquad (1)$$

$$Y_{GC} = f(X_{\text{GCPTwords}}, X_{\text{ESRTwords}}, X_{\text{LastWords}}, X_{\text{NormWords}}, X_{\text{TrWords}}) + \epsilon \qquad (2)$$

* The Boyer-24 model was not tested on any of the QPE means, since the model does not account for page faults. The confidence intervals given in Table 4.12 are those for predicting an individual value of Y_{GC} given a set of X values rather than the expected value of Y_{GC} [11]. Confidence intervals for the expected values are much smaller, but inappropriate in the context of our test.

For the QPE runs, the simple model (1) yielded values of $R^2 < 0.52$, while the detailed model (2) had $R^2 < 0.68$. For runs of Boyer, which produced no faults while scanning memory, and very few faults while transporting, both models yielded a high $R^2 > 0.98$. These results suggest that the total numbers of words to be scanned and words to be copied are not very good linear predictors of scanning and transporting time, except when there is little paging, as in small programs.

To summarize, regression analysis showed that collection time is linearly related to collector work, page faults, and run time overhead in the collector routines. The linear models correctly predicted the overall mean collection times for all ephemeral runs except for Boyer, whose time was overestimated by the models derived from QPE-long data. Regression analysis also provided estimates of, and comparisons between, the average time costs of various operations in the model. The set of estimates arise from, and may be viewed as a description of, the joint behavior of program and system. Linear models which ignore page fault time and express garbage collection time in terms of collector work alone were found to be inaccurate when there is a significant amount of paging.

7 MEMORY USAGE

This section considers the time-varying characteristics of memory usage. The behavior under garbage collection is measured, particularly the survival of objects, the amount of garbage reclaimed, and the amount of work done by the collector. Cost-benefit measures for garbage collection are defined and used to evaluate the efficiency of collection cycles.

Figure 4.10 shows a comparison between memory usage under dynamic and under ephemeral garbage collection. Note that these are idealized, composite plots of the various memory spaces over time; in reality, the memory spaces are not allocated contiguously as these pictures might suggest.

Initially, the only spaces that exist are newspace and static space (memory containing static objects). A dynamic collection (Figure 4.10a) is initiated either by explicit user action, or by default at some "safe" time determined by the system, as explained earlier. Newspace is flipped, that is, turned into oldspace, and scavenging of the root set and copyspace begins. While scavenging is in progress, copyspace grows to accommodate oldspace objects that are found to be nongarbage. Newly allocated newspace grows to accommodate objects created by the mutator, and some static objects may also be created. When scavenging is complete, the collection ends, and oldspace can be reclaimed. Another cycle can now begin, immediately or at some later time.

During ephemeral garbage collection (Figure 4.10b) only a small portion of non-static space is garbage collected. Most notably, only those ephemeral levels are flipped that have exceeded their capacities. With no garbage collection, the idealized picture is simplified, since only newspace exists.

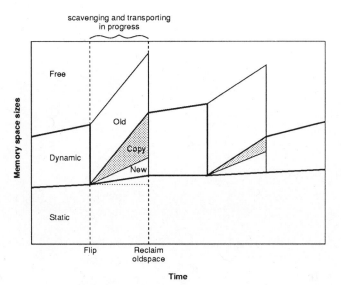

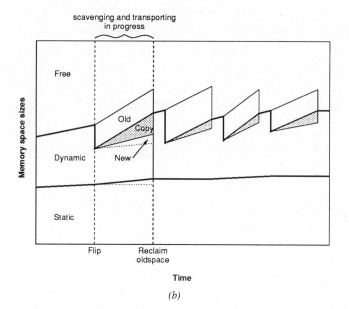

Figure 4.10 Idealized composite plot of memory space sizes under garbage collection. (*a*) Dynamic gc; (*b*) Ephemeral gc.

7.1 Dynamic versus Ephemeral versus No Collection

In Section 4.5 we compared Boyer's CPU utilization and page management overhead when running under dynamic collection, ephemeral collection, and no collection; Figure 4.11 compares its memory usage characteristics. Boyer

Memory space sizes
(millions of words)

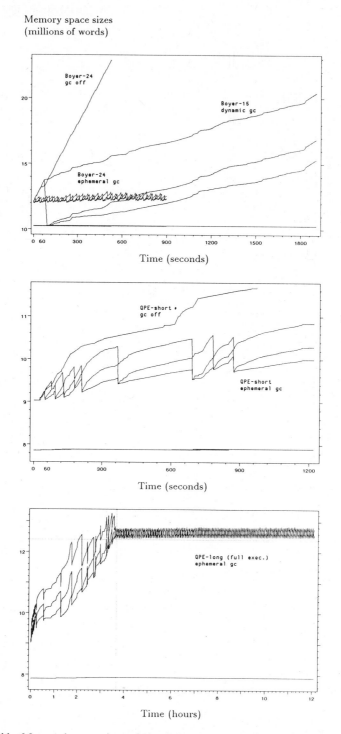

Figure 4.11 Measured composite variation in memory space sizes under garbage collection.

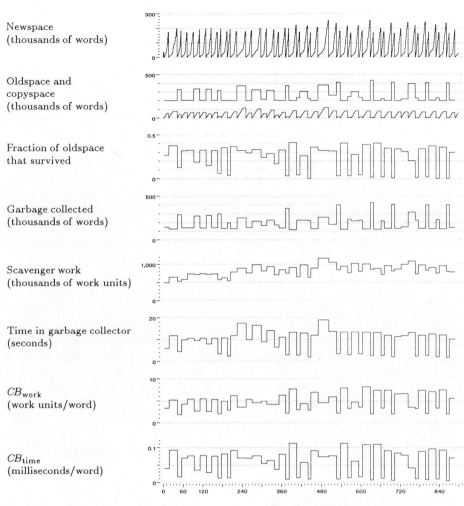

Newspace
(thousands of words)

Oldspace and
copyspace
(thousands of words)

Fraction of oldspace
that survived

Garbage collected
(thousands of words)

Scavenger work
(thousands of work units)

Time in garbage collector
(seconds)

CB_{work}
(work units/word)

CB_{time}
(milliseconds/word)

Time (seconds)

Figure 4.12 Consing, nongarbage discovery, scavenging work, collection time, and cost-benefit measures for each garbage collection in Boyer-24, ephemeral gc.

with no collection is the fastest, but fills up virtual memory. With ephemeral collection, the program takes longer, and many collection cycles occur. With dynamic collection, the program takes the longest, and there is only a single cycle.

The memory usage plots for Boyer are probably typical of workload that happens to fit the ephemeral assumptions: that newly created objects are likely to become garbage (so time is not wasted performing unnecessary collection of these newer objects); that they become garbage quickly (and hence will be caught before surviving all the ephemeral levels); and that there are relatively few ephemeral references (so the ESRT does not become very big).

7.2 Allocation Behavior

To observe the allocation behavior of a program without the influence of garbage collection, consider the runs of Boyer and QPE-short with no collection. Figure 4.11 shows that Boyer allocates at a much higher rate and more uniformly than QPE-short. For Boyer the mean rate of allocation is 22,800 words/second with a standard deviation of 6,560. For QPE-short, it is 2,690 words/second with a standard deviation of 4,120. A fairly constant allocation rate probably reflects the (lack of) complexity or diversity of the workload; this was certainly the case for Boyer, since the workload consisted of repeated executions of the same program.

Can future memory usage behavior, such as the rate of allocation of memory for new objects, be predicted from past behavior? The observation that the rate of allocation is relatively stable—both for Boyer and for QPE—points toward the possibility of predicting future allocation from past allocation. An interesting approach would be to investigate the predictive ability of time-series-based empirical models: models derived from memory usage measurements like the space sizes considered here, but including other variables for representational adequacy. Such a prediction could be used as part of an overall scheme, also based on empirical modeling, for adaptive tuning of garbage collection.

7.3 Collection Cycle Characteristics

Characteristics of each garbage collection cycle—and of the application—can be observed in the composite plots in Figure 4.11. They include oldspace size, the growth of copyspace as nongarbage is discovered, the fraction of nongarbage in oldspace, and the duration of a cycle.

These characteristics are shown explicitly in Figures 4.12–4.15, which are time series plots that depict several aspects of behavior. Consing (or object creation) activity is represented by the plot of the size of newspace.* The nongarbage discovery process is represented by the plots of oldspace and copyspace, and, for each collection, the surviving fraction of oldspace and the amount of garbage actually reclaimed. The cost of garbage collection is represented by the plots of scavenger work and garbage collection time; scavenger work is the sum of work done in scanning the root set, and in scanning and transporting objects found to be nongarbage. We adopt Moon's definition [12] that one scavenger work unit equals one word scanned or one

* New static objects are not included, but their exclusion makes little difference since no significant amount of static object creation was observed in any of the measured programs.

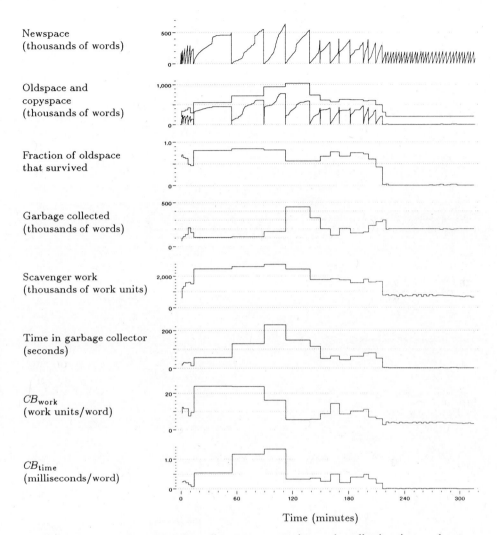

Newspace
(thousands of words)

Oldspace and
copyspace
(thousands of words)

Fraction of oldspace
that survived

Garbage collected
(thousands of words)

Scavenger work
(thousands of work units)

Time in garbage collector
(seconds)

CB_{work}
(work units/word)

CB_{time}
(milliseconds/word)

Time (minutes)

Figure 4.13 Consing, nongarbage discovery, scavenging work, collection time, and cost-benefit measures for each garbage collection cycle in QPE-short, ephemeral gc.

word transported, although this definition does not recognize that transporting one word is more work than scavenging one word (see Section 6.6). Finally, two plots of collector "inefficiency" are shown.

In explaining collector inefficiency, we must first note that performing garbage collection has two potential advantages. One is the possible increased locality of objects; the other, of course, is the ability to reuse storage and to continue running without exhausting it. We saw in Section 4.5 that ephemeral

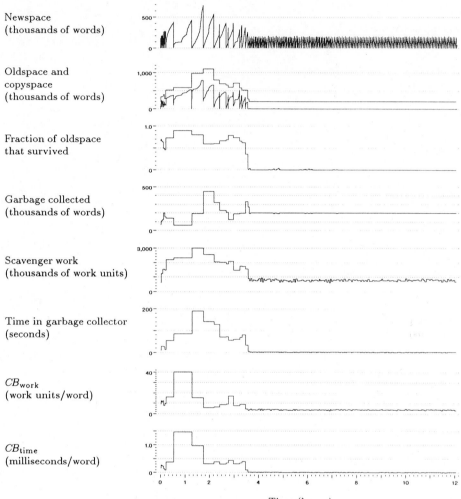

Figure 4.14 Consing, nongarbage discovery, scavenging work, collection time, and cost-benefit measures for each garbage collection cycle in QPE-long (full execution), ephemeral gc.

garbage collection for Boyer and QPE-short led to reduced paging, but longer running time, than with no collection. Thus there may be occasion to turn collection off, or inhibit it temporarily. However, when garbage collection is performed, we would like to know how efficient it is. To measure efficiency, the following cost-benefit metrics are defined and plotted for each collection cycle in Figures 4.12-4.15:

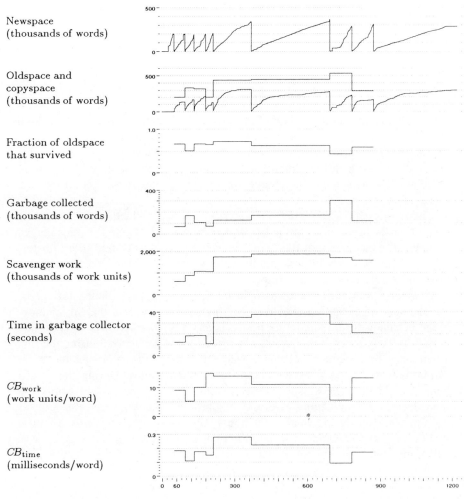

Newspace
(thousands of words)

Oldspace and
copyspace
(thousands of words)

Fraction of oldspace
that survived

Garbage collected
(thousands of words)

Scavenger work
(thousands of work units)

Time in garbage collector
(seconds)

CB_{work}
(work units/word)

CB_{time}
(milliseconds/word)

Time (seconds)

Figure 4.15   Consing, nongarbage discovery, scavenging work, collection time, and cost-benefit measures for each garbage collection cycle in QPE-long (initial part), ephemeral gc.

$$CB_{work} = \text{scavenger work done per word of garbage reclaimed}$$

$$CB_{time} = \text{gc time per word of garbage reclaimed}$$

Both metrics quantify benefit as the number of words of garbage discovered and reclaimed by the collection, but they differ in the cost measure. CB_{work} quantifies cost as the amount of scavenger work required, whereas CB_{time} quantifies cost as the time consumed by the garbage collector—a real, bottom line cost, since it includes the effect of system overhead such as page faults.

Note that only that portion of oldspace that is not transported is considered garbage.

For Boyer, collection cycles generally alternate between high and low cost-benefit values, and between collection of the first ephemeral level only and collection of both first and second levels. The cycles in which the second ephemeral level was flipped can be identified from the plot of oldspace size; Garbage collection used the default configuration of two ephemeral levels with capacities of 200,000 and 100,000 words, respectively. Most of the high efficiency cycles are those in which only a small fraction of oldspace survived and in which both ephemeral levels were flipped. These results suggest that it might be worthwhile to increase the capacity of the first level, in order to delay flips and allow more time for objects in the first level to become garbage.

Closer inspection of Figure 4.12 shows that two collections occur during each of the 24 invocations of the Boyer benchmark. Clearly, one of them is not very useful, since an invocation does not release its objects until it terminates. A single flip for every integral number of invocations seems more suitable for this (admittedly contrived) workload.

QPE-long was earlier shown to incur most of its page faults during the initial hours of execution (Figure 4.6). Figure 4.14 reveals that its memory usage behavior during this initial period is also remarkably different from that during the remainder of the program's execution. In particular, during the first 20 minutes of execution, oldspace sizes are relatively small—on the order of 300,000 words—and cycles complete in 1–3 minutes. During the next 3 hours and 20 minutes, oldspace sizes increase to over one million words and then decrease. Cycle times rise and fall proportionately, taking as long as 40 minutes. During the final eight hours, oldspace sizes remain constant at 200,000 words (indicating that only the first ephemeral level is being flipped); collections take about 2.3 minutes, with one minute between collections; all oldspace is garbage, and dynamic space does not grow.

The cost-benefit curves for QPE-long, like those for Boyer, are correlated to the survival of objects in oldspace. For example, the cycle of lowest efficiency had the highest fraction of nongarbage in oldspace—about 90 percent. The high cost-benefit values during the initial 3–4 hour period suggest flipping less frequently during this time, to reduce both collector work per word reclaimed and the number of objects that become garbage after promotion to dynamic status [22]. It remains to be seen whether the tradeoff in decreased locality is significant. One way to judge the tradeoff is to consider the effects of not performing a subset of the ephemeral collection cycles, taking care, of course, not to exceed available virtual memory. There would be two competing effects on total execution time. The elimination of time-multiplexing of garbage collection with mutator computation during the omitted cycles would tend to decrease execution time. On the other hand, the possible increase in scavenging and transporting during the retained cycles and the possible decreased locality of objects might increase execution time.

These results demonstrate the necessity of generation tuning to match an application's object usage characteristics with the ephemeral garbage collection algorithm. Generation tuning involves adjusting one or more of the following parameters: the number of ephemeral levels, the capacity of each level (in general, the policy for initiating collections), and the policy for promoting nongarbage objects to a higher level or to dynamic status. For some programs, it may be possible to achieve a significant speed-up by implementing alternatives to the usual criteria for initiating a garbage collection (reaching the "safe" flipping point in the dynamic case, or exceeding fixed ephemeral level capacities in the ephemeral case).

One approach to the discovery of such criteria is to perform a statistical clustering of the per-cycle data to quantify the patterns observed earlier by visual inspection, for example, high and low efficiency cycles observed during runs of Boyer. A state transition model [10] based on the clusters would then quantify time behavior; in the case of Boyer, alternation between states of high and low efficiency. This empirical model of past memory usage behavior could then be used in making adaptive and systematic decisions about what and when to garbage collect. This is an area for future research.

7.4 Static Space

An increase in static space size occurs when newly created objects are declared to be static. Static space size increased slightly over time during every measurement run; the largest increase observed was one of 50,000 words during QPE. The growth is not visible in Figure 4.11, however, unless the scale is greatly expanded.

7.5 Memory Usage Summary

In this section, we presented plots which provide a picture of memory usage variation, and cost-benefit metrics were defined to determine the efficiency of garbage collection. For Boyer, the variation in the values of these measures suggested synchronizing flips with invocations of the benchmark, whereas for QPE-long, the measures identified the initial phase of execution as the one over which generation tuning techniques, such as increasing ephemeral level capacities, could achieve significant gains.

We mentioned the possibility of developing empirical memory usage models: for example, time series analysis models for object creation, and state transition models based on cluster analysis of garbage collector performance. Such models may require more detailed memory usage data than that presented here: for example, information about the different generations, and the memory areas and regions [12, 20] into which related objects are stored. Measurements that reflect these additional dimensions, including object types and sizes, will serve as the basis for realistic models of memory usage.

8 CONCLUSIONS

This chapter has presented analyses of sampled memory system activity on a Symbolics 3620 Lisp machine. Some of the major results are as follows:

In general, when there is high variability in system or program behavior, single-number performance measures, such as mean values, become less representative.

Garbage collection accounted for up to 50 percent of real time, and up to 28 percent of all page faults, for the programs measured under ephemeral collection. More than 93 percent of garbage collection time was spent in scavenging, or scavenger-induced transporting; the remaining fraction was spent in mutator-induced transporting. That the amount of transporting caused by mutator references is small suggests that only a small amount of oldspace is being actively referenced.

Linear regression was used to derive empirical models relating garbage collection time to the amount of scanning and transporting work, page faults, and other overhead. The utility of the models for system or application optimization stems from two sources. Firstly, the model parameters quantify the time cost, and hence the relative influence, of the modeled operations. We observed the high time cost of a page fault, relative to that of scanning or transporting one word; moreover, transporting a word cost more than scanning a word by a factor of 5.6–70. The time cost of scanning one word depended on the location of the word; the highest cost was observed for words in the last page of copyspace, whereas the lowest cost was for root set words in physical memory and words scanned during the normal linear traversal of copyspace. Root set words on disk had intermediate cost. These differences are attributed to the differences in bookkeeping time overhead associated with scanning the various areas, amortized over the number of words scanned.

The second source of model utility is the ability of models to predict garbage collection time for a given amount of scanning and transporting work and a given number of page faults. Conversely, given a desired reduction in collection time for a constant amount of work, the model can specify the necessary reduction in page faults. The accuracy of the models was tested by using them to predict the overall mean collection time of the various runs. The actual mean times were within the 95 percent confidence intervals determined from the models.

We also examined the time-varying characteristics of memory allocation, survival, collector work, and efficiency. Collector efficiency was quantified as the time (or work) per word of garbage reclaimed. These cost-benefit metrics suggested synchronization of flips with benchmark calls in the case of the Boyer workload, and identified the portion of the program most suitable for optimization in the case of QPE-long.

It should be noted that the results presented were obtained from a particular Lisp system and measured programs. The usual care should be exercised in extrapolating the results to other systems or programs. Work is in progress to expand the scope of the analysis, and to develop other models for program and memory system behavior which would be useful in rapidly evaluating alternative collector configurations and policies, and in adaptive control.

REFERENCES

1. D. L. Andre, *Paging in Lisp Programs*, Master's thesis, Department of Computer Science, University of Maryland, 1986.

2. H. G. Baker, Jr., "List Processing in Real Time on a Serial Computer," *Communications of the ACM*, 21(4):280–294, April 1978.

3. C. J. Cheney, "A Nonrecursive List Compacting Algorithm," *Communications of the ACM*, 13(11):677–678, November 1970.

4. D. W. Clark, "Measurements of Dynamic List Structure Use in Lisp," *IEEE Transactions on Software Engineering*, SE-5(1):51–59, January 1979.

5. D. W. Clark and C. Cordell Green, "An Empirical Study of List Structure in Lisp," *Communications of the ACM*, 20(2):78–87, February 1977.

6. J. Cohen, "Garbage Collection of Linked Data Structures," *ACM Computing Surveys*, 13(3):341–367, September 1981.

7. R. Courts, "Improving Locality of Reference in a Garbage-Collecting Memory Management System," *Communications of the ACM*, 31(9):1128–1138 , September 1988.

8. R. P. Gabriel, *Performance and Evaluation of Lisp Systems*. MIT Press, Cambridge, 1985.

9. D. Harel, "On Visual Formalisms," *Communications of the ACM*, 31(5):514–530, May 1988.

10. M. C. Hsueh, R. K. Iyer, and K. S. Trivedi, "Performability Modeling Based on Real Data: A Case Study," *IEEE Transactions on Computers*, 37(4):478–484, April 1988.

11. W. Mendenhall and T. Sincich, *Statistics for the Engineering and Computer Sciences*. Dellen Publishing Company, Santa Clara, CA, 1988.

12. D. A. Moon, "Garbage Collection in a Large Lisp System," *Proc. 1984 ACM Symposium on Lisp and Functional Programming*, pp. 235–246, August 1984.

13. D. A. Moon, "Architecture of the Symbolics 3600," *Proc. 12th Annual International Symposium on Computer Architecture*, pp. 76–83, July 1985.

14. D. A. Moon, "Symbolics Architecture," *Computer*, 20(1):43–52, January 1987.

15. SAS Institute, Inc., Cary, NC. *Statistical Analysis System (SAS) User's Guide*, Version 5, 1985.

16. R. A. Shaw, "Empirical analysis of a Lisp system," Technical Report CSL-TR-88-351, Computer Systems Laboratory, Stanford University, CA, February 1988.

17. P. Steenkiste and J. Hennessy, "Lisp on a Reduced-Instruction-Set–Processor," *Proc. 1986 ACM Conference on Lisp and Functional Programming*, pp. 192–201, August 1986.

18. P. Steenkiste and J. Hennessy, "Lisp on a Reduced-Instruction-Set Processor: Characterization and Optimization," *Computer*, 21(7):34–45, July 1988.

19. Symbolics, Inc., Cambridge, MA. *Symbolics Technical Summary*, October 1985.

20. Symbolics, Inc., Cambridge, MA. *Symbolics Documentation Set*, June 1986.

21. G. S. Taylor, P. N. Hilfinger, J. R. Larus, D. A. Patterson, and B. G. Zorn, "Evaluation of the SPUR Lisp Architecture," *Proc. 13th Annual International Symposium on Computer Architecture*, pp. 444–452, June 1986.

22. D. Ungar and F. Jackson, "Tenuring Policies for Generation-Based Storage Reclamation,"*ACM SIGPLAN Notices*, 23(11):1–17, November 1988.

23. J. L. White, "Address/Memory Management for a Gigantic Lisp Environment or, GC Considered Harmful," *Conference Record of the 1980 Lisp Conference*, pp. 119–127, August 1980.

CHAPTER 5

MULTIPROCESSOR ARCHITECTURAL SUPPORT FOR BALANCED LISP PROCESSING

Raymond Chowkwanyun and Kai Hwang

1 INTRODUCTION

Concurrent Lisp processing requires computing architectures different from those used in numeric processing [1]. Two major features of Lisp processing are irregular memory access and a high demand for processing power [2], and much work has already been done on *dynamic memory management* to support Lisp processing with various garbage collection schemes [3,4]. In this chapter, we explore dynamic load balancing as a solution to the problem of allocating processes to processors. *Dynamic* and *run-time* are taken to have the same meaning when they appear in the text; likewise *compile-time* and *static*. We use the term *multiprocessor* throughout the chapter to mean a message-passing multiprocessor in which distributed local memory is attached to each processor—that is, the system does not have shared memory and its processors communicate by message passing.

Load balancing can be characterized as either *sender-initiated* [5] or *receiver-initiated* [6], depending on the way in which process migration is initialized. Processes always migrate from the busy processor to the lightly-loaded or idle processor; the busy processor is the *sender* while the idle processor assumes the role of *receiver*. While process migration always proceeds in a single direction, the flow of control may be in either direction; both the sender and the receiver can initiate the process migration. Thus the difference is defined between sender- and receiver-initiated types of load balancers.

In an earlier paper [7], we introduced a *hybrid* load balancing method that adapts to rapidly changing run-time conditions by switching between sender-initiated and receiver-initiated modes of operation. Mode selection depends on the system load; sender-initiated mode is used when system load is light, and receiver-initiated mode when it is heavy. System load is the processing load on the entire multiprocessor at a single time. It includes all currently executing processes and processes in the ready queue waiting to be run. Cybenko has proposed a dynamic load balancing scheme [8] restricting load migration to immediate neighbors, but our system is not restricted in this way.

In general, the sender-initiated mode is better suited to lightly loaded conditions because it eagerly pursues process migration and can quickly achieve a distribution of processes amongst the processors and consequently higher processor utilization. The receiver-initiated mode, on the other hand, is conservative about process migration and seeks to avoid process migration between busy processors, an activity that incurs unnecessary overhead. Thus the receiver-initiated mode is best suited to a heavy system load when most processors are busy.

In this chapter we propose hardware and software architectures for implementing the hybrid load balancer, with special emphasis on concurrent Lisp processing. Both architectures are based on *macro dataflow* execution with the insertion of operating system directives to interface with the applications code. Consequently, applications programs can be ported between the two architectures without modification. The software design needs less hardware and is suitable for the current generation of multiprocessors such as the Intel iPSC, but a run-time overhead is associated with the execution of the load balancer. To relieve the CPU of this overhead, we develop a design for specialized hardware units to execute the hybrid load balancer directly in parallel with the CPU—a technique which promises to yield performance better than that of the software approach.

A pure functional language has no side effects such as writing to global variables; thus all functions can be executed in parallel [9,10]. A pure functional subset of the Lisp language could be defined but the full Common Lisp standard has many side effect constructs that prevent us from exploiting full parallelism [11]. DeGroot has proposed a method of dealing with side effects that uses *synch-blocks* [12]. Although he addresses side effects in the context of logic programming, we believe the same concept could be used for Lisp. We have not, therefore, dealt with the issue of side effects in this chapter.

Closely related to the issue of side effects is the problem of how to handle large data structures in a hardware environment with no shared memory between processors. This problem has been addressed elsewhere, for example in FAIM-1 [13], Arvind's I-structure for dynamic dataflow machines [14] and Dally's object-oriented multiprocessor [15]. We concentrate here on the solution to the mapping problem for processes; the mapping problem for data structures is beyond the scope of this chapter.

The remainder of the chapter is organized as follows. Section 2 describes the multiprocessor architecture and the mapping problem; Section 3 addresses the issue of process representation and defines the O/S directives SUSPEND and RUN; Section 4 describes the hybrid load balancer and the macro dataflow execution model; and Sections 5 and 6 describe hardware and software designs for implementing the hybrid load balancer. Sections 7–9 discuss the sources of parallelism in Lisp programs, and parallelization techniques and their application to selected programs from the Gabriel suite of benchmark programs [16]. Section 10 presents the experimental results obtained on the Intel iPSC hypercube for the benchmark program *Tak* and analyzes performance relative to machine size, process granularity, threshold levels, and network topology. Finally, Section 11 summarizes the key results and discusses other open problems.

2 LOAD BALANCING IN MULTIPROCESSORS

Our methodology is directed towards message-passing multiprocessors, whose features are summarized below; these architectures have also been characterized as multicomputers or distributed-memory multiprocessors. We also discuss the mapping problem as it relates to Lisp processing on multiprocessors and conclude that a dynamic solution is desirable.

Multiple processors are connected by a network. Each processor consists of a CPU, local memory, and a switch that connects it to the network. In Figure 5.1, the circles represent processors and the lines represent communications links. The processors do not have shared memory and communicate by message-passing; the entire ensemble is controlled by a *distributed operating system* so all the processors work together on one problem. It is important to distinguish the distributed operating system from a *network operating system,* which links computers that typically work on different problems [17]. A distributed operating system implies *multiple-instruction multiple–data-stream*

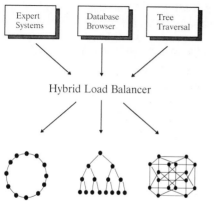

Figure 5.1 Load balancing for mapping Lisp programs on to message-passing multiprocessors.

(MIMD) operation, whereas a network operating system typically implies multiple *single-instruction single–data-stream* (SISD) operations, which are independent of each other [18].

Figure 5.1 shows interconnection networks consisting of rings, trees, and hypercubes, but a network may assume any topology. A processor can send messages to any other processor on the same network, although the message may have to pass through several *links* or *hops*. [(A link is a direct connection between two processors.) (Such processors are referred to as *neighbors*.)]

Available multiprocessors include the Intel iPSC™, the NCUBE™, the FPS T-series, and the Caltech/JPL Mark III, which have hypercube topologies, and the Ametek™, which is based on a mesh [19]. The DADO is a tree based multiprocessor [20], and Hypernets are a class of hierarchically structured multiprocessors [21]. Recently Texas Instruments announced a board-level product which allows multiple Lisp machine chips to be supported on computers with a *NuBus*, such as the Apple Macintosh.

We focus on exploiting *control-level parallelism* rather than the data-level parallelism found in the Connection Machine [22]. Figure 5.2 gives an example of control-level parallelism; the figure shows the *function invocation tree* of a Lisp program, representing functions by circles and function invocations by arrows. Processing begins at the root of the tree and spreads downwards as subprocesses are invoked. In control-level parallelism, subprocess calls are evaluated in parallel on different processors; generally, however, we expect that the number of processes in such a system would significantly exceed the number of processors, so a fundamental problem of such systems is the allocation of processes to processors—termed the *mapping problem*. A good solution to the mapping problem is required if the multiprocessor system is to realize its full potential. A poor allocation scheme results in idle processors, low utilization, and a system that is only a little faster than uniprocessor designs.

The problem of finding a good mapping becomes acute when Lisp processing is in question because Lisp programs have highly unpredictable run-time characteristics. The function invocation tree shown in Figure 5.2, for example, is dense and highly unbalanced. The leftmost branch springing from the root contains the largest number of function calls, the middle branch somewhat fewer, and the rightmost branch terminates after only one call. In general the shape and depth of these trees is highly data-dependent. For example, when the Gabriel benchmark Tak is invoked with the arguments (18, 16, 15), it results in nine function calls, whereas with the arguments (18, 16, 6), 648,305 function calls are made. A small change in a single argument has increased the computational load on the order of 10,000 times. Such variability limits the usefulness of static solutions [23, 24] to the mapping problem because they depend on reliable predictions of run-time characteristics. For the same reason we use a non-preemptive system; preemption presumes prior knowledge of run-time behavior. Thus Lisp processing demands a dynamic solution to the mapping problem just as it demands a dynamic solution to the memory management problem.

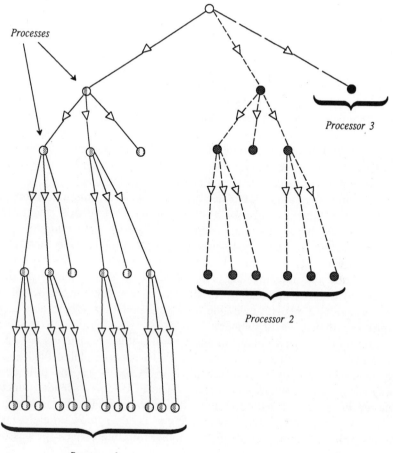

Processes

Processor 3

Processor 2

Processor 1

Figure 5.2 Imbalance in creating Lisp processes (function calls) among three processors in a system.

To illustrate the problems that can result from a bad mapping, we return to the example of Figure 5.2. Suppose each of the three branches springing from the root was mapped to one of three processors: the leftmost branch going to processor 1, the middle branch to processor 2, and the rightmost branch to processor 3. Assuming roughly equal processing times for each function invocation, processor 3 would finish executing first and remain idle while the other two processors did most of the work. A shorter processing time could be achieved if processes were migrated from the busy processors to the idle processor for execution (we use the terms *process* and *function* interchangeably). Such run-time process migration and architectural designs for its support are the basic concepts that we explore in this paper.

3 O/S FUNCTIONS FOR PROCESS MIGRATION

The use of run-time process migration requires that we choose between a policy of *complete copying* and one of *code migration*. Under a policy of complete copying, a copy of the application code is stored on every processor; moreover, the operating system code must also necessarily reside on every processor. Application functions are represented by *process control blocks* (PCBs) and it is PCBs that are migrated. Under a policy of code migration, the code itself is moved when a process is migrated for evaluation on a remote processor. The problem of selecting an appropriate policy is orthogonal to that of dynamic load balancing, because either policy may be used with a dynamic load balancer. Policy selection then becomes an engineering decision.

Our architecture uses complete copying; since communication takes more time than memory access it makes sense to trade off memory for communications. When this relationship is reversed we can switch to a policy of code migration. Fortunately Lisp code is very compact, and the price we pay for complete code copying is low. Table 5.1 shows the percentage of each iPSC processor total memory that is occupied by the code for the hybrid load balancer and the four applications programs. Each processor has 4.5 Mbytes of memory with 4.0 Mbytes available for Lisp. The largest application, Boyer, occupies less than 1.3% of memory.

A function is either executable or suspended. If it is executable it can be sent to the Lisp interpreter for execution. The PCBs for executable functions are maintained in a *ready queue* and are called *run* PCBs. A suspended function is waiting for its arguments to be evaluated. The PCBs for suspended functions are stored in a *suspend heap* and are called *suspend* PCBs. The ready queue and suspend heap are separate data structures. Only run PCBs are migrated in the hybrid architecture; suspend PCBs always remain on the processor on which they were created. The structures of run and suspend PCBs are shown in Figure 5.3. The *parent ID* field is common to both kinds of PCB, and is displayed separately.

The following example shows how suspend and run PCBs are used to represent the function invocation tree corresponding to the execution of a program. Figure 5.3 shows a function invocation tree for the function f, invoked

TABLE 5.1   Percentage of Processor Memory Occupied by Load Balancer and Benchmark Programs.

Program	Percentage of Memory Occupied
Hybrid	1.99
Tak	0.03
Boyer	1.27
Browse	0.16
Traverse	0.31

Run PCB

Executable Form	Parent ID

Suspend PCB

Process Address	Function Name	Parent ID	Argument Ports

Parent ID

Parent Processor	Process Address	Port ID

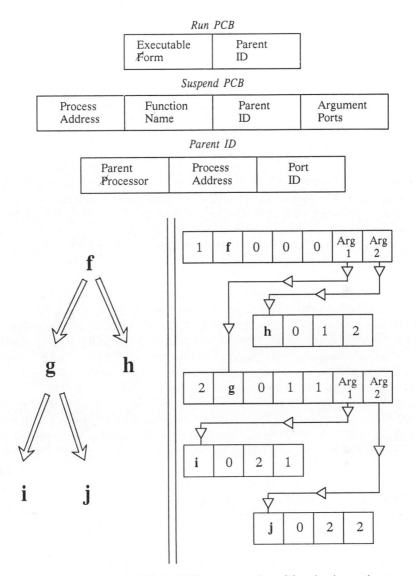

Figure 5.3 Process control block (PCB) representation of function invocation tree.

on processor 0. Function f's two arguments will be returned as the results of evaluating g and h. Function g also takes two arguments, the results of evaluating i and j. On the right of the figure is the corresponding tree of PCBs.

When a function such as f has arguments that need to be evaluated, a suspend PCB is created for it in the suspend heap, in this case at *process address* 1. The function's arguments are then spawned as run PCBs, parent ID fields of the spawned functions acting as pointers back to the parent. When the child process has completed evaluation, the system returns the result to the parent, locating the processor on which the parent resides by means of the parent ID. For example, if h is evaluated on processor 15, the result is sent back to processor 0 and placed in the suspend process at address 1 because the first two elements of h's parent ID are 0 and 1. (Recall that suspended processes do not migrate, so the parent processor field is always correct.) The *port ID* indicates the position of the result in the argument list of the parent; for example, g and h occupy the first and second positions (ports) in the argument list of f and so have port IDs of 1 and 2 respectively. When a function is suspended, the system sets the number in the *argument ports* field of the function's suspend PCB to the length of the function's argument list.

A suspend PCB remains in the suspend heap until all its arguments have been evaluated, at which time it is fired in dataflow style by adding it back to the ready queue. We emphasize, however, that our system differs from dataflow in two crucial aspects. First, dataflow exploits very fine grain parallelism at the instruction level (for example, a dataflow actor could perform addition on two integers) whereas we exploit parallelism at the level of Lisp programs, each of which is the equivalent of many dataflow actors. By doing this, we are able to amortize the overhead of returning results to parent processes over a much larger number of instructions than is possible in dataflow. Secondly, our system does not require tag matching when returning results, since a suspend PCB never moves and can therefore always be located by its address in the suspend heap. Thus, address decoding may be used to return results instead of the expensive tag matching operation used in dataflow.

We provide the user with two operating system directives, SUSPEND and RUN, which act together as a *fork* operation. When inserted into the applications program, they direct the hybrid load balancer to evaluate the application in parallel. This style of interfacing between applications and O/S is similar in concept to Halstead's *futures* [25]; we did not use futures, however, because the suspend/run system is simpler to implement, and we wanted to concentrate on load balancing rather than the application-O/S interface.

SUSPEND adds its argument to the suspend heap as a suspend PCB. The syntax of SUSPEND is:

```
(SUSPEND function numArgs groundedArgs)
```

function is the name of the function to be suspended, Which can be either a user-defined function or a built-in Common Lisp function, and numArgs is the total number of function's arguments. SUSPEND will establish an empty association list* of this length as the arguments field of the suspend PCB. groundedArgs is an optional list of grounded arguments. (A grounded argument is already bound to a constant which can be inserted into the appropriate port of the argument list.

---

*An association list is a list of tags and data pairs.

RUN converts its argument, a Lisp form, into run PCB form and adds it to the ready queue. The syntax of RUN is:

```
(RUN  form portID)
```

form is the Lisp form that is to be evaluated. RUN and SUSPEND are used in conjunction—a RUN always follows a SUSPEND. RUN creates a run PCB stamped with a parent ID that points to the function just suspended. SUSPEND writes the suspended function's process address into a global variable, which is then read by RUN into the parent ID of the run PCB which it creates. The parent processor field of the run PCB is assigned the ID of the processor on which the run is being executed.

4 THE HYBRID LOAD BALANCING SYSTEM

In what follows, we briefly describe the hybrid method by defining the receiver and sender-initiated modes. Our theoretical modeling of both the hybrid and other methods is described in [7]. In the absence of a central controller, the processors must make load balancing decisions on the basis of local information. Each processor has a static list of processors which lie within a *neighborhood diameter* of, typically, one or two hops (links) from the processor. The neighborhood diameter is defined thus in order to minimize the network traffic incurred during the exchange of load information. Ferrari has developed a theoretical approach to determining the load index [26], but it requires predictions of program behavior which are presumed to be unavailable, so instead we simply define the *internal load* of a processor to be the length of the ready queue. The *external load* of a processor is the load information broadcast by a processor to its neighbors.

4.1 Receiver-Initiated Mode

The receiver-initiated mode of load balancing is based on a *drafting protocol* proposed by Ni and his associates [6]. Idle processors initiate load balancing by requesting work from busy processors. The drafting protocol is a modification of the *bidding method* by Smith [27] but avoids the excessive message traffic caused by the broadcasting of requests. Each processor may send requests for work and exchange load information with its neighbors. In addition, a process may be migrated only once to avoid the possibility of circulating processes.

The external load information is obtained by dividing the range of the internal load into three bands: *light-load*, *normal-load*, and *heavy-load*. The bands are separated by threshold and saturation levels. When the internal load falls below the threshold level, the processor is in light-load state; when the load falls between the threshold and saturation levels, the processor is in normal-load state; and when the load exceeds saturation, the processor is in heavy-load state.

A heavy-load processor is a candidate for migrating processes, a normal-load processor does not participate in load balancing, and a light-load processor requests work from heavy-load processors in its neighborhood using a *drafting protocol*. A timing diagram for the drafting protocol is shown in Figure 5.4 with time increasing downwards along the vertical axis. The time line of a light-load processor is flanked by the time lines of two heavy-load processors; the dots indicate events. First, the light-load processor requests work from the heavy-load processors; this request is called a *draft*. The heavy-load processor rejects the draft if its load state has changed to normal-load or light-load, and broadcasts its new status to its neighborhood. All eligible heavy-load processors return their *draft age* to the drafting processor. (Draft age may be defined in several ways; we define it as the processor's internal load.) The drafting processor then requests a process from the processor with the highest draft age of those responding, and a process is migrated to the drafting processor.

4.2 Sender-Initiated Mode

The *sender-initiated mode* is based on a *gradient* method of load balancing by Lin and Keller [5]. In this mode the busy processors initiate load balancing. As in receiver-initiated mode, the internal load of a processor is the length of the ready queue, but the external load is now the number of hops to the nearest idle processor. A processor is considered idle, (and its external load set at zero) if its internal load drops below a static systemwide threshold, T;

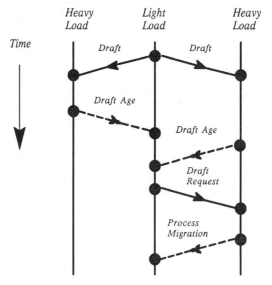

Figure 5.4 Timing diagram for the drafting protocol.

if on the other hand, its load exceeds *T*, its external load is calculated by incrementing the minimum external load of its neighbors by one. The external load of a processor is bounded by the diameter of the network to prevent circulating processes; its definition is summarized in the following equation.

$$External\ Load = \begin{cases} 0 \\ min(\Delta, 1 + min\{\ External\ Load\ of\ neighbors\}\ ) \end{cases}$$

$$if\ Internal\ Load < T$$
$$otherwise$$

An external load of zero signals to the rest of the system that a processor is lightly loaded and available to take on work from busier processors. The propagation of a processor's external load values provides an implicit routing to the nearest lightly loaded processor.

Figure 5.5 provides a snapshot of such an external load grid. Processor D is the only idle processor, with an external load of 0. All the other processors have internal loads greater than the threshold so they propagate the external load value, incrementing it by one at each hop. The external load indicates the number of hops to the nearest processor with load 0. A processor always picks the neighbor with the least external load when exporting a process. The exported process follows the gradients provided by the external load values until it reaches a 0 loaded processor. The gradients also ensure that the shortest path is used. For example, processor F would always route the process to processor E (since processors C and I have heavier external loads); processor E would then forward it to processor D (since processors B and H have heavier external loads). In this way the process is forwarded by the shortest path between F and D.

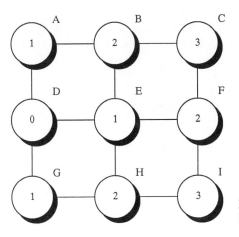

Figure 5.5 External loads in sender-initiated mode.

4.3 The Hybrid Load Balancing Method

In a very heavily loaded system, the receiver-initiated mode is preferable, since it allows processes to migrate only once, minimizing network traffic. Receiver-initiated mode reduces the possibility of all the heavily loaded processors migrating their work to a single lightly loaded processor and overwhelming its communications ports. However in a lightly loaded system the sender-initiated mode is preferable, since in this mode process migration begins as soon as a processor enters a heavily loaded state, with no necessity to wait for lightly loaded processors to send requests and to engage in a lengthy protocol before migration begins. We therefore propose a hybrid method, using the receiver-initiated mode when system load becomes excessive and the sender-initiated mode when system load is light.

The hybrid method requires a distributed mechanism to manage the switching between the sender-initiated and receiver-initiated modes of operation. Distributed control cannot guarantee that the entire system is in either mode exclusively; instead each processor operates in either sender-initiated or receiver-initiated mode according to the state of its local environment. All processors are initialized to sender-initiated mode. When the number of a processor's heavily loaded neighbors exceeds a threshold, the system is said to be *congested* and the processor switches to receiver-initiated mode. We refer to this threshold as the *hybrid threshold*, H, to distinguish it from T, the threshold value that defines light and heavy loading. If the number of heavily loaded neighbors falls below the hybrid threshold, the processor switches back to sender-initiated mode. Table 5.2 summarizes these changes. Processors in one mode can service messages and processes received from processors operating under the other mode, providing a smooth interface between the sender-initiated and receiver-initiated parts of the system.

4.4 The Macro Data-Flow Execution Model

Figure 5.6 shows the macro data-flow execution model which is the basis for implementing the hybrid load balancer. The figure shows the ready queue and suspend heap for a single processor. The components of the hybrid load balancer are shown as boxes with italic labels. The *decision maker* pops the

TABLE 5.2 Switching Modes of Each Processor for Hybrid Load Balancing.

Mode / Condition	Processor is in receiver-initiated mode	Processor is in sender-initiated mode
System becomes congested	No change	Switch to receiver-initiated mode
System becomes uncongested	Switch to sender-initiated mode	No change

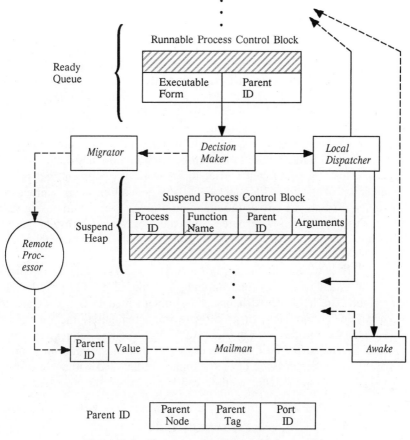

Figure 5.6 The macro dataflow execution model.

next process off the ready queue and decides whether to evaluate it remotely or locally, according to the processor mode; it also decides whether the processor should be in sender- or receiver-initiated mode (the making of these decisions was described in the previous section). PCBs destined for remote evaluation are sent to the *migrator*, which routes them to a remote processor. The returned values are picked up by the *mailman* and forwarded to the *awake* module, which matches the value with its parent in the suspend heap. When all arguments are present the parent is returned to the ready queue. If the decision-maker decides on local evaluation, the run PCB is sent to the *local dispatcher* and evaluated. Evaluation may cause additional functions to be suspended or spawned. Results are sent to the awake module on which the parent process resides (either local or remote). As discussed in Section 3, the parent ID field of the PCB acts as a pointer back to the parent, allowing the result to be returned.

5 HARDWARE SUPPORT FOR HYBRID LOAD BALANCING

Recent advances in VLSI technology have brought about the possibility of multiprocessor Lisp machines that are an order of magnitude faster than current architectures [28]. Lisp CPUs available from Texas Instruments, Xerox, and Symbolics provide faster Lisp processing than the general purpose CPUs used in present Lisp multiprocessors, and some fast switching chips are capable of handling 100 megabit/second interprocessor communications — 10 times faster than current rates. But although significant, these innovations increase component speed without resolving the difficult mapping problem, and the programmer is still left with the burden of deciding which process to execute on which processor. This is a significant barrier to software development and prevents the realization of the full potential of the new generation of multiprocessors. Dynamic load balancing solves the mapping problem but can impose a severe run-time overhead on CPU resources. We therefore propose an alternative — a hardware design for executing the hybrid load balancer which relieves the CPU of run-time overhead.

Figure 5.7 shows the architecture of a single processor. The fast switching chip is embedded in the postman module and the Lisp CPU [29] is shown on the right. The hybrid load balancer is implemented in the postman, load info, and decider modules. In a software implementation, the functions of the postman, the decider, and the Lisp CPU are executed sequentially by the CPU; the advantage of the hardware design is that these three units operate in parallel. Messages arriving at the postman are handled without waiting for CPU intervention, the decider module can begin exporting processes without

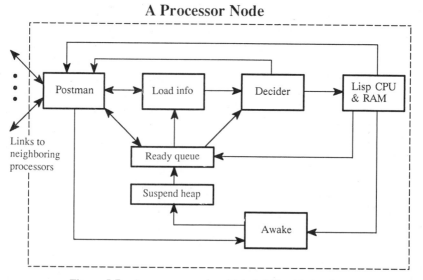

Figure 5.7 A processor node with hybrid load balancing.

waiting for the CPU to finish executing, and the CPU can continuously evaluate applications programs without waiting for the load balancer to complete execution.

The hardware design also dedicates memory to the ready queue and suspend heap, speeding access to these structures. In a software implementation, these structures are part of the general address space and access may be delayed while the CPU accesses another part of the same memory space.

The postman module, shown in Figure 5.8, handles all communications functions, and the switch handles all message routing functions. We are not concerned here with the design of the switch, simply assuming that it can handle all incoming and outgoing messages.

All messages are tagged, allowing the tag matcher unit to sort incoming messages and forward them to the appropriate modules. Imported run PCBs are sent to the ready queue where they are either executed locally or exported, depending on the actions of the decider module. Results of remote evaluation are sent to the awake module and returned to the corresponding parent in the suspend heap. Load information messages are sent to the load info module for use by the decider module and the draft protocol unit. (Recall that the drafting protocol is part of the receiver-initiated mode defined in Section 4.) A signal from the decider module activates a new drafting protocol, and both draft and response messages are handled by the dedicated protocol unit.

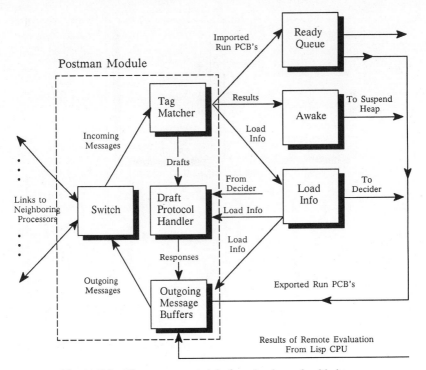

Figure 5.8 The postman module for a hardware load balancer.

The strategy of complete code copying results in the generation of a large number of short messages (on the order of 50 bytes). To accommodate this type of traffic, the outgoing message buffer unit is designed to handle a large number of short messages rather than longer messages. The continuing decline in the cost of memory makes it feasible to provide 10,000 to 100,000 buffers, and our experience in running programs on the iPSC indicates that they will be required.

The load information module shown in Figure 5.9 consists of *load information registers* and a *load calculator* unit. The registers hold the external load data broadcast by neighboring processors, while the load calculator monitors the ready queue and the registers to calculate the processor's internal and external loads.

The decider module appearing in the center of Figure 5.9 consists of the decision maker, migrator, and local dispatcher units. These hardware units perform the same functions as their counterparts in the macro dataflow model. The decision maker switches the processor between sender- and receiver-initiated modes of operation. In sender-initiated mode, the decision maker can signal the migrator to begin load migration, and in receiver-initiated mode it

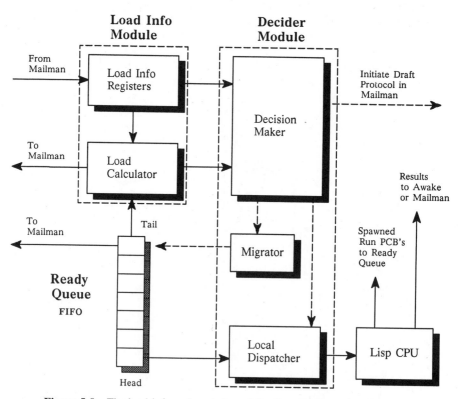

Figure 5.9   The load info and control modules for a hardware load balancer.

can initiate a draft protocol by signaling the draft protocol unit in the mailman. The local dispatcher is invoked if local processing is to be used.

The ready queue is a first-in first-out (FIFO) queue. Accessing the queue causes the length of the queue to be signaled to the load calculator. The execution of suspend and run directives in the Lisp CPU spawns run PCBs which are added to the ready queue. The results of evaluation are sent to the awake module when the parent process resides in the same processor; when the parent resides on a remote processor, the result is sent to the mailman for forwarding.

Figure 5.10 shows the awake module, which controls the interaction of the ready queue and the suspend heap. The ready queue accepts run PCBs from three sources: neighboring processors, the Lisp CPU, and processes activated by the awake module. Run PCBs are exported under the control of the decider and the draft protocol unit. The local dispatcher unit takes run PCBs from the head of the queue for evaluation by the Lisp CPU.

The awake module shown in Figure 5.10 consists of the heap controller and result inserter units. The heap controller accepts new suspended PCBs from

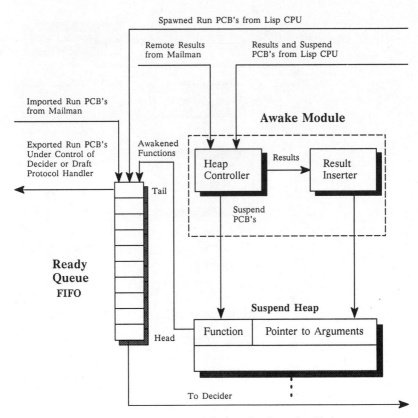

Figure 5.10 The awake module for a hardware load balancer.

the Lisp CPU and adds them to the suspend heap. When all the arguments of a suspend PCB have been returned, the heap controller reactivates the PCB as a run PCB and moves it to the ready queue. Results arriving at the heap controller are passed on to the result inserter where they are matched with the corresponding parent process.

6 SOFTWARE IMPLEMENTATION OF THE HYBRID LOAD BALANCER

In this section, we present a software implementation of the hybrid load balancer based on the macro dataflow execution model. The implementation is characterized by a high degree of modularity that enhances understandability, maintainability, portability, and reliability. Figure 5.11 shows the architecture of the software design. The architecture is duplicated on every processor, and there are no client/server relationships between processors. The hybrid load balancer is therefore a true distributed operating system with no central master, and has the desirable property of scalability—as the number of processors in the

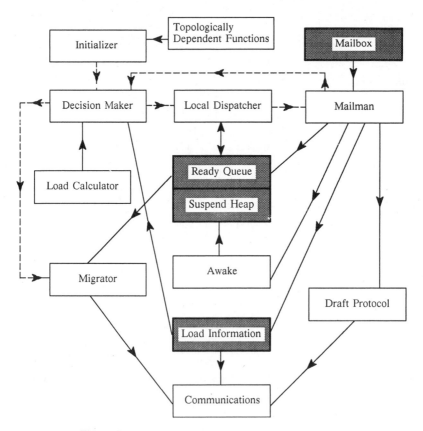

Figure 5.11 Components of a software load balancer.

system is increased, the hybrid load balancer will continue to perform, since there is no master component to become a bottleneck and eventually to choke the system.

In Figure 5.11, functions are shown in white boxes and data structures are shown in shaded boxes. The basic kernel is a tight loop consisting of the decision maker, the local dispatcher, and the mailman. The kernel is entered through an *initializer* which configures the hybrid load balancer according to information contained in a file of topologically dependent functions; in this way we are able to isolate the hybrid load balancer from the topology of the underlying hardware, enabling it to be ported to multiprocessors of any topology.

The architecture is centered around three data classes: the run PCB, the suspend PCB, and a class of messages of seven types. Each data class is contained in a separate *package* which defines that data class and all functions that access that class. A package is an information hiding mechanism, and only names declared as exported within the package can be accessed by objects outside the package. For example, run PCBs are stored in the first-in, first-out ready queue. All ready queue access functions have been concentrated in the local dispatcher package, and functions outside the local dispatcher package cannot access the queue directly. Table 5.3 shows the data classes together with their packages.

The run PCB and suspend PCB data classes were treated in Section 3, hence this discussion centers on the data types defined in the mailman package. All data types in the mailman are messages, and are passed back and forth between the processors. (In an object oriented system, we would have defined each message type as a subclass of a class *message*.)

Data structures of the type *result of evaluation* are used to return the results of function evaluation to the parent process. Instances of this data type are

TABLE 5.3   Three Packages and Their Corresponding Data Classes in the Hybrid Load Balancer.

Package	Data Class
Local dispatcher	Run PCB
Awake	Suspend PCB
Mailman	Message Result of evaluation Sender-initiated mode load information Receiver-initiated mode load information Receiver-initiated mode draft Receiver-initiated mode response-to-draft Receiver-initiated mode final-call Receiver-initiated mode request-process Kill-Kernel

produced by the local dispatcher and consumed by the awake module. The structure of a result of evaluation instance is shown in the diagram.

Result Type

Parent address	Port ID	Return value

Parent address and port ID allow the awake module to place the return value in the correct argument port of the correct suspended process.

Sender-initiated mode load information messages carry the external load information defined in Section 5.4. The structure of this data type is shown in the following diagram. The processor ID field identifies the processor whose load information is being broadcast to the neighboring processors.

Load Information Type

Processor ID	Load value

The three receiver-initiated mode types—draft, response-to-draft and request-process—are used to implement the drafting protocol defined in Section 5.4. (The stages of the drafting protocol and the corresponding message types were shown in Figure 5.4.) A new drafting protocol is initiated when the decision maker signals the draft protocol module, which generates and handles all protocol messages. The structures of the draft, response-to-draft, and request-process messages are shown in the following diagrams. When sending a draft or a request for a process it is necessary only to identify the sender of the message. The response-to-draft message requires two fields, one to identify the sender and another to hold the internal load information.

Draft and Request-Process Types

Processor ID

Response-to-Draft Type

Processor ID	Internal load

The *kill-kernel* message type is used to terminate the kernel when the application has finished executing. The completion of execution is detected by the following mechanism: The parent address 0 is reserved for use by the parent of the root process of the application; therefore when the awake module receives a result destined for a parent with address 0, processing must be complete. An instance of the kill-kernel type is immediately broadcast to all processors, and the kernel shuts down. The kill-kernel message consists simply of an empty dummy field.

7 PARALLELISM IN EXECUTING LISP PROGRAMS ON MULTIPROCESSORS

In this section we discuss how the hybrid load balancer exploits four classes of parallelism in Lisp programs: AND-parallelism, OR-parallelism, the parallel evaluation of list structures, and the parallel evaluation of arguments of functions.

AND-parallelism may occur in a conditional statement, when the arguments of the AND function can be evaluated in parallel. The following Lisp form contains an example; if the functions f and g evaluate true, then the function h will be executed.*

```
(cond ((AND (f)(g))
          (h)))
```

Care must be taken when employing AND-parallelism because the semantics of parallel AND can be different from those of sequential AND. In Common Lisp, evaluation stops at the first failure, so if f failed, g would not be evaluated. With AND-parallelism, however, all parts of the condition would be executed; hence if g gave rise to some side-effect, the semantics of parallel AND would be different from those of sequential AND. AND-parallelism may also incur heavy overhead. Continuing our example, suppose g gives rise to a great deal of computation, and f is false. In the case of sequential AND, g would not be executed and its computational overhead would be avoided; in the case of parallel AND, however, g would be evaluated and the overhead incurred. When we converted the application programs for parallel execution we avoided AND-parallelism in situations where it would give rise to different semantics or where there was a possibility of incurring heavy overhead.

OR-parallelism may also occur in a conditional statement, and the arguments to the OR can then be evaluated in parallel. The following Lisp form contains an example; if either f or g evaluates true, the function h will be executed.

```
(cond ((OR (f)(g))
          (h)))
```

---

* In Lisp syntax (f a) means a call to the function f with the argument a, and is equivalent to the Pascal function call f (a). Therefore (f) is equivalent to calling a function f which takes no arguments.

The remarks about the potentially different semantics of sequential and parallel AND apply equally to OR, since Common Lisp OR stops at the first success. Note that the sequential and parallel versions are not necessarily different, but only potentially so. Necessary and sufficient conditions for different semantics are that an interruption in execution should occur (for AND a failure, for OR a success) and that one of the forms that would not have been executed in the sequential version has a side effect. In the absence of side effects the semantics would be exactly the same.

List parallelism is the parallel application of a function to all the elements of a list. For example, the following code causes f to be applied to each element of longList, resulting in multiple calls to f which can be evaluated in parallel.*

```
(mapcar #'f longList)
```

Argument parallelism is a special case of list parallelism where the list happens to be the argument list of a function. In the following example, the function f has two arguments which are the result of the evaluation of g and h. In argument parallelism, g and h are evaluated in parallel.

```
(f (g) (h))
```

When exploiting parallelism in existing sequential programs, care must be taken if the sequential program uses a shared global variable to pass data between subprocesses (we use the more common term, *global variable*, instead of the Common Lisp terminology, *special variable*). In the example that follows, the function f writes its argument to the global variable channel using the Lisp assignment function setq. The function g reads channel, wraps up the result of the read and its argument into a list and prints out the result. Thus, channel serves as a channel for information from f to g. We shall call such global variables *channel variables*.

```
(defun f (y)
       (setq channel y))
(defun g (y)
       (format t "-%-a" (list channel y)))
```

To continue with this example, we declare a temporary function using lambda. The mapcar then causes the lambda body to be applied to each element of targetList. If the evaluation of f is true, then g will be evaluated.

---

* If longList were bound to (a b c), execution of the code would give rise to 3 calls to f: (f a) (f b) (f c).

```
(mapcar #'(lambda (z)
                (cond ((f z)(g z))))
        targetList)
```

If `targetList` consisted of the list (a b) then executing the code fragment above would result in two calls to the `lambda` body:

```
(cond ((f a)(g a)))
(cond ((f b)(g b)))
```

The `mapcar` cannot be evaluated using list parallelism because each call to f causes a value to be written to the channel variable `channel`, values which may not then be read by the correct call to g. In our example, the value of `channel` written by the call to (f a) could be read by (g b) instead of (g a), since a single global variable `channel` is being used as a channel by multiple copies of f and g and there is contention for the use of the channel. The solution is to localize `channel` so that each f and g pair has its own private channel.

```
(defun h (y)
        (let (channel)
             (cond ((setq channel y)
                      (format t "≠%≠a" (list channel y)))))))
```

```
(mapcar #'(lambda (z)(h z))
        (targetList)
```

We define a new function, h, that combines the code of f and g. The `let` causes `channel` to be treated as a local variable. The body of the `lambda` in the `mapcar` is now changed to a call to h. List parallelism may now be applied to the `mapcar` without danger of conflict in the use of the channel variable.

8 TRANSLATING SEQUENTIAL PROGRAMS FOR PARALLEL EXECUTION

We now show how the `run` and `suspend` directives are used to exploit the four parallelism classes defined in Sections 6–7.

We begin with an example of AND-parallelism. The sequential AND is converted to parallel form as follows: First the AND is suspended, and an argument of 2 given to SUSPEND to indicate that the AND is expecting 2 arguments. The arguments are supplied by evaluating the functions f and g; f is run with a port ID of 1 so that f's result will be returned to position 1 of the argument list of the suspended AND; and g is run with a port ID of 2. Recall that RUN does not cause its argument to be executed immediately; instead it is added to the ready queue for later evaluation. Thus f and g may be evaluated in parallel if one of them is migrated to another processor.

Sequential AND:

```
(and (f)(g))
```

Parallel AND:

```
(SUSPEND 'and 2)
(RUN '(f) 1)
(RUN '(g) 2)
```

OR-parallelism can be exploited in the same way as AND-parallelism.

Sequential OR:

```
(or (f) (g))
```

Parallel OR:

```
(SUSPEND 'OR 2)
(RUN '(f) 1)
(RUN '(g) 2)
```

List parallelism may be exploited by spawning a run PCB for each operation on a list element, as in the example that follows. We first suspend a list to accumulate the results of the spawned functions, duplicating the semantics of mapcar, which returns a list of the results of the calls to the function. Since one call to f will be made for each element of longList, the argument list of the suspended list is set to the length of longList. mapcar is used to execute RUN once for each element of longList; the macro '(f ,x) evaluates to an application of f to each list element.*

Sequential List Operation:

```
(mapcar #'f longList)
```

* If longList is bound to '(a b c), the macro evaluates to (f a) (f b) (f c).

Parallel List Operation:

```
(suspend 'list (length longList))
(mapcar #'(lambda (x)
          (let ((i 0))
               (RUN '(f ,x)  (setq i (1+ i)))))
     longList)
```

Argument parallelism can be exploited in a straightforward way; first the calling function f is suspended, then each of its arguments is run with the appropriate port ID.

Sequential Evaluation of Arguments:

```
(f (g) (h))
```

Parallel Evaluation of Arguments:

```
(SUSPEND  'f  2)
(RUN  '(g)  1)
(RUN  '(h)  2)
```

9 PARALLELIZATION OF LISP BENCHMARK PROGRAMS

We decided to use benchmarks from the Gabriel suite [16] rather than write our own benchmarks because the semantics and profile of Lisp operations for Gabriel's programs are well known and they have been used on a wide variety of computers. The Gabriel suite contains benchmarks oriented towards both numerical and AI processing. Since our interest is primarily in AI processing we selected four benchmarks from this area: Tak, Boyer, Browse, and Traverse. Because the semantics and profile of these programs are published elsewhere, we will concentrate on discussing the degree of parallelism found in these programs and on the *translation effort* required to prepare the sequential code for parallel execution. The following parallelization examples show that very little translation effort is required to obtain a high degree of parallelism.

9.1 The Tak Program

As a functionally oriented language, Lisp makes heavy demands on the function-calling capabilities of an implementation. Tak generates many function calls and is a good test of the amount of overhead an implementation incurs in setting up function calls. Moreover, Tak has a function invocation tree (shown in Figure 6.2) that is both bushy and unbalanced; thus it is an ideal test of the load balancing capabilities of the hybrid method. We reproduce here, for comparison, the sequential and parallel codes of Tak.

Sequential Tak:

```
(defun tak (x  y  z)
      (if (not (< y x))
              z
              (tak (tak (1-x) y z)
                   (tak (1-y) z x)
                   (tak (1-z) x y))))
```

Parallel Tak:

```
(defun tak (x y z)
      (if (not (< y x))
              z
              (progn
              (SUSPEND 'tak 3)
              (RUN '(tak ,(1-x) ,y ,z) 1)
              (RUN '(tak ,(1-y) ,z ,x) 2)
              (RUN '(tak ,(1-z) ,x ,y) 3))))
```

Sequential Tak consists of an *if* statement. If (not (< y x)) succeeds, that is to say, if y ≥ x, then Tak terminates, returning the value *z*. The *else* part of the statement consists of a recursive call on Tak whose arguments are three further calls on Tak. Parallelization is achieved by suspending the first call to Tak and running the three remaining calls; this is an example of *argument parallelism*, discussed in Section 7. The numbers 1, 2, and 3 shown as arguments to RUN indicate that the results of the respective calls to Tak should be returned to the first, second, and third places in the argument list of the suspended call on Tak.

9.2 The Boyer Program

Boyer is a theorem-proving program whose inner loop is a process of unification—also the inner loop process of the Prolog language. We identified three levels of parallelism of differing granularity and decided to exploit the two coarsest levels.

Level 1 This is the coarsest level of parallelism and exploits the list parallelism defined in Section 7. The Boyer program is given a logical term to prove correct. Since the term is a compound term constructed from simpler terms, we can use list parallelism to apply unification to each simple term. For example, the compound term (equal (implies x y) (equal z w)) may be broken up into the simpler terms (implies x y) and (equal z w), and unification may be applied to each (the Lisp prefix syntax is used here, so (implies x y) represents the logical term x ⊃ y. All logical terms are constants so we need not be concerned about consistency checking when applying parallel unification to ANDed terms.

Level 2 This level of parallelism has intermediate granularity. A term derived using level 1 parallelism can, potentially, be unified with multiple axioms; at level 2 we try to unify a single term with multiple axioms in parallel, a form of OR-parallelism. For example, we may attempt to unify the term (equal z w) with the following four axioms.

```
(equal (plus a b) (zero))
(equal (plus a b) (plus  a c))
(equal (zero)  (difference  x y))
(equal x  (difference  x y))
```

Level 3 This level has the finest grain of parallelism available in Boyer. Within a single level 2 unification, parallel unification of atomic terms such as z and w is possible. We conjectured, however, that this level of parallelism would be too fine-grained to justify the overhead of its exploitation; furthermore, it results in code of great complexity that is hard to understand, hard to maintain, and hard to test and debug.

9.3 The Browse Program

Browse performs pattern matching operations similar to the inner loop operations of expert systems. The two levels of parallelism were identified and used in Browse.

Level 1 This is the coarsest level of parallelism exploited. The Browse program runs the match subroutine 1200 times at the top level; since each run is independent of the others, we may run all 1200 in parallel. The semantics of the serial version of Browse requires that the result of the last match be returned. To duplicate this effect in the parallel version we wrote a new function last-elt that returns its last argument; last-elt is suspended and then serves as the parent to the 1200 runs of match.

Level 2 Within match we were able to make straightforward use of OR- and AND-parallelism arising in conditional statements. We give an example of OR-parallelism; the exploitation of AND-parallelism is similar and is discussed in Section 8.

Sequential OR:

```
(OR (match (cdr pat) dat alist)
    (match (cdr pat) (cdr dat) alist)
    (match pat (cdr dat) alist))
```

Parallel OR:

```
(SUSPEND 'OR 3)
(RUN '(match ,(cdr pat) ,dat ,alist) 1)
(RUN '(match ,(cdr pat) ,(cdr dat) ,alist) 2)
(RUN '(match ,pat ,(cdr dat) ,alist) 3))
```

In the parallel version we suspend the OR, which then serves as the parent to the three parallel invocations of match. Notice that the parallel code has almost the same syntax as the serial code, promoting ease of understanding and maintenance.

9.4   The Traverse Program

Traverse is a graph traversal program. A graph has 100 nodes; its edges are directed and reflexive edges, and multiple edges may exist between nodes. The graph also has cycles and a root. Traverse visits every node in the graph, marking visited nodes, and writing the marks to a global data structure. In all, 250 graphs are traversed. If all 250 traversals are to run in parallel, each must write its marks to a private copy of the data structure. A list of 250 graphs is stored on each processor; each graph in the list has a unique ID, and each copy of the traversal program works on its own graph. Information about graph nodes that have been visited is exchanged between processors by means of an *environment* field in the *run* PCB and *result* structures; thus processors working on the same graph can sometimes avoid marking the same node. We cannot, however, guarantee that the same node will not be marked twice by different processors working simultaneously on the same graph; hence the semantics of the serial and parallel Traverses are different. In the serial version each node is marked exactly once; in the parallel version at least once, but perhaps more. Whether visiting at least once is acceptable depends on the purpose of the traversal. If the purpose is simply to mark every node then the semantics is the same—only when we require that each node be marked exactly once does the parallel version fail.

We emphasize that the hybrid operating system knows nothing about the contents of the environment field; the use of this field is completely the responsibility of the applications programmer, and the operating system is completely insulated from the application. The operating system knows only that the application will provide a function called update-environment which must be run every time a *run PCB* or a *result* structure arrives at the *mailman*.

The two levels of parallelism exploited in Traverse are discussed below.

Level 1   At the coarsest level of parallelism all 250 graphs are traversed simultaneously, as discussed above.

Level 2 In the serial version of Traverse, an unmarked node is marked when encountered; then all the son's of that node are visited in turn. The parallel version of Traverse exploits list parallelism and visits all the children in parallel. The sequential and parallel codes follow.

Sequential Processing of Children List:

```
(do ((children (node-children node) (cdr children)))
    ((null children) ())
    (traverse (car children) mark))
```

Parallel Processing of Children List:

```
(SUSPEND 'AND (length (node-children node))
(do ((children (node-children node) (cdr children))
     (portID 1 (1+portID))
     ((null children) ())
     (RUN '(par-traverse,(car children),graphID) portID))
```

We review the sequential version first. The first argument to do is a list declaring loop variables (in this case ((children (node-children node) (cdr children))))) which initializes the loop variable children to the children list of node; on subsequent iterations, children will be set to its cdr, returning all the elements of children except the first. The main body of the do loop applies traverse to the head of the children list. In this way all the elements of the children list are processed until the list is empty, the exit condition ((null children) ()) is met, and the loop is exited.

In the parallel version, we suspend an AND to serve as the parent. The do loop is run in much the same fashion as the serial version, except that instead of executing par-traverse immediately we place it in the ready queue with RUN for later, possibly parallel, execution. (The traverse function of the parallel version was renamed par-traverse to avoid confusion.

9.5 Translation Effort

Table 5.4 shows the required number of operating system directives (SUSPEND and RUN combinations) as a percentage of the number of lines of code in the benchmark programs. This percentage suggests how much translation effort is required by indicating roughly how many lines of the sequential code must be changed. In the column headed *Number of SUSPEND/RUNs*, we treat each SUSPEND and its related RUNs as a single logical unit; for example, the parallel Tak program in Section 9 contains a SUSPEND followed by three RUNs that appear as one unit in the table. In the column headed *Number of Lines of Code* we list the number lines of code in the sequential version of each test program,

TABLE 5.4   O/S Directives as Percentage of Lines of Code.

Program	Number of SUSPEND/RUNs	Number of Lines of Code	Percentage
Tak	1	6	17
Boyer	5	501	1
Browse	3	96	3
Traverse	2	116	2

excluding comments; this gives a slightly more conservative estimate of the conversion effort than a listing of the number of lines of code in the parallel versions.

The percentage of directive insertions required for Tak is abnormally high because it is such a short program. The percentages for the other programs are more representative, ranging from 1 percent to 3 percent, low figures which suggest that the translation effort is not unduly onerous.

If we consider these percentages and the degree of parallelism we extracted from the programs, we see that we can elicit a high degree of parallelism for a small translation effort. In Browse and Traverse we extracted 1200 and 250 degrees of parallelism at the first level alone, adding more degrees of parallelism at the lower levels. Although our sample is too small to represent a scientific sampling, we believe it gives a reasonable picture of the ratio of parallelism to translation. The programs we used are standard benchmarks written originally for sequential machines, without any thought for parallelism; that we were able to extract parallelism from such programs suggests the same would be true of other sequential programs. Furthermore, in Section 7 we described four general sources of parallelism commonly found in Lisp programs; their existence suggests that the hybrid load balancer would be an efficient means of providing for the parallel execution of existing software libraries.

10   EXPERIMENTAL CONCURRENT LISP BENCHMARK RESULTS

Our experiments on the Intel iPSC multiprocessor demonstrate the feasibility of a software implementation of the hybrid load balancer on the current generation of multiprocessors. The experiments were run with the Gabriel [16] benchmark program Tak using the parameters (18 16 10), which result in 4,745 function calls, with speed-up as the performance measure. To calculate speed-up we timed the execution of the benchmark program first on a single processor, then on multiple processors. The experiments show that hybrid load balancing performs better than either sender-initiated or receiver-initiated methods alone. Results for ring and tree topologies demonstrate the portability of the hybrid load balancer. The experiments also indicate that the hybrid load balancer is insensitive to the setting of the threshold level.

Figure 5.12 compares dynamic load balancing with the hybrid method and static load balancing. For the static case processes were randomly allocated to processors, and did not migrate at run time; because run-time behavior cannot be predicted, we could not apply the more sophisticated static methods that assume some knowledge of run-time characteristics. The results indicate that faster processing times can be achieved with dynamic methods.

A multiprocessor is scalable if it provides increases in performance corresponding to increases in the number of processors. Figure 5.13 shows the scalability of the iPSC hypercube; the best performance is obtained under the hybrid method of load balancing. The low performance of the receiver-initiated method is attributed to the high overhead incurred during system initialization. Parallel processing begins at a single root processor and must propagate throughout the system. Under the receiver-initiated method, process migration cannot occur until load information has propagated, since receivers must have information about busy processors before they can begin drafting;

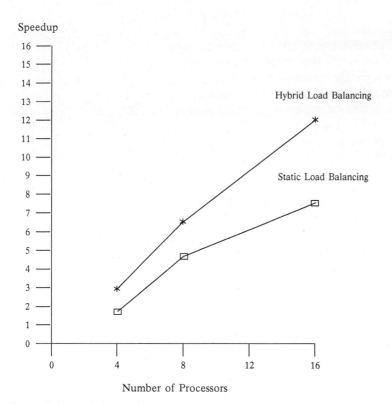

Figure 5.12 Relative performance of dynamic versus static load balancing.

thus the initialization period is extended. Because, during initialization, only the root processor is doing useful work, the receiver-initiated method should be reserved for the most heavily loaded stages of the computation, during which its conservative approach can best reduce overhead.

The sender-initiated method works well during the initial stages of computation when load is light and the primary need is to propagate processes. During the later stages of computation, however, this method produces excessive process migration since all the processors become heavily loaded and attempt to migrate processes to each other. The ability of the hybrid method to switch to the less expensive receiver-initiated method accounts for its better performance. The results in Figure 5.13 are conservative because they were obtained with an older, slower version of the hybrid load balancer.

Figure 5.14 compares speed-ups obtained for multiprocessors with 4, 8, and 16 processors. The results for the ring and tree topologies were obtained by simulations on the hypercube and indicate the applicability of the hybrid load balancer to multiprocessors with different topologies. Both applications and the hybrid load balancer itself are easily ported between topologies; applica-

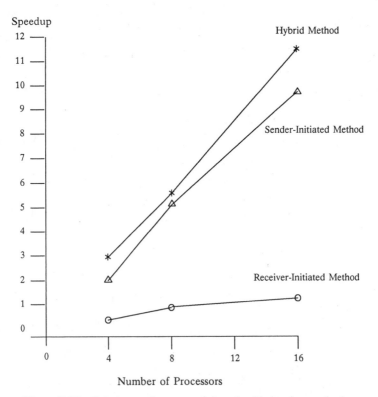

Figure 5.13 Relative performance of three load balancing methods.

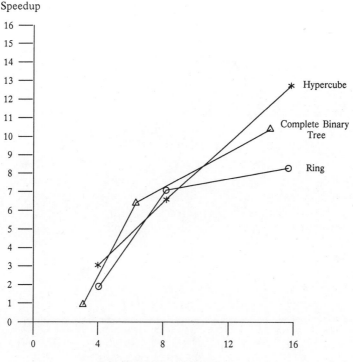

Speedup

Number of Processors

Figure 5.14   Relative performance of various multiprocessors.

tions can be moved without modification, and the hybrid load balancer needs only simple modifications to the set of topology-dependent files that list the neighbors of each processor. Different machine sizes can be handled in the same way. Such portability would have required expensive maintenance had the applications been written directly to the hardware.

With 16 processors, the hypercube performs best, followed by the tree and ring topologies. At this machine size, the hypercube has greater connectivity (each processor having four neighbors) than a tree, (in which a processor may have between one and three neighbors) or a ring (in which each processor has only two neighbors). At smaller machine sizes the hypercube does not have this advantage and so performance values are similar.

The hybrid load balancer has pairs of threshold/saturation parameters for both sender- and receiver-initiated modes. The parameters are set globally before run time and remain constant during the execution of the program. We interpret the results shown in Figures 5.15 and 5.16 as an indication that the performance of the hybrid load balancer is reasonably insensitive to the settings of these parameters. All experiments were performed with a hypercube topology.

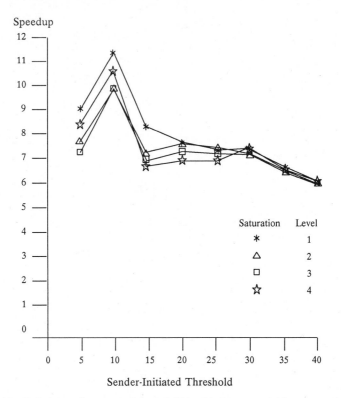

Figure 5.15 Relative performance of the hybrid load balancer on a 16-processor hypercube for different sender-initiated threshold values.

Recall that the threshold value in Figure 5.15 determines when processors will begin load migration in sender-initiated mode — when a processor's internal load exceeds the threshold, it begins load migration. The four graphs show results for four different values of the saturation parameter ranging from 1 to 4. Lower saturation levels increase the probability of switching to receiver-initiated mode.

At extremely high threshold settings load migration never occurs and the load balancer is effectively operating on one processor; therefore we would expect to see a smaller speed-up for higher threshold values, as indicated in Figure 5.15. The decline is gradual, indicating the desired insensitivity to the threshold setting for a wide range of values. We attribute the peak in the curves to the use of the hybrid load balancer to initialize the system. Consider that processing begins on the root processor and spreads to the other processors with load migration. A high value for the threshold delays this initialization process with consequent lower speed-up; on the other hand, if the threshold value is too low, excessive load migration occurs, and the resulting communications

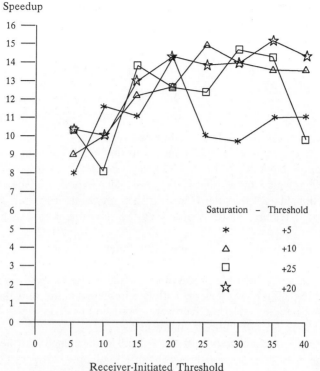

Figure 5.16 Relative performance of the hybrid load balancer on a 16-processor hypercube for different receiver-initiated threshold values.

overhead also reduces the speed-up. Best performance is achieved at a threshold value which balances the two effects. Since the peak in the curves is caused by using the dynamic load balancer to initialize the system, we could eliminate it by using a static initialization system. We do not explore this topic here because it has been investigated elsewhere [23].

The graphs in Figure 5.16 were obtained for different values of difference between threshold and saturation levels in receiver-initiated mode. Recall that in this mode, when a processor's internal load falls below the threshold, it begins to request load from its neighbors. The neighbors do not participate in load-balancing if their load falls between the threshold and saturation levels; only if their load exceeds the saturation level do they become candidates for load migration. The four graphs in Figure 5.16 were obtained by setting the difference to values from 5 to 20.

It is difficult to discern a pattern in these results. However, low threshold settings seem to affect performance adversely; if the request for more load is delayed until the internal load is low, there is a greater chance that the processor will run out of processes and become idle. In general, once the threshold reaches a reasonably high value, the speed-up displays the required insensitivity to the threshold setting.

11 CONCLUSIONS

The hybrid load balancer is a distributed operating system that allows all the processors in a multiprocessor to cooperate together on the solution of a single problem. The system's major features are its dual modes of operation that allow it to adapt to changing system loads, and a completely distributed mechanism for deciding which mode should be used. The hybrid load balancer allows properly translated sequential Common Lisp programs to be run in parallel — an important feature, since it preserves the investment in existing software libraries.

The hardware supports presented for direct execution of the hybrid load balancer are implementable with state-of-the-art VLSI components. The hardware design runs the Lisp-processing, communications, and load-balancing functions in parallel, whereas the software implementation runs them serially. The greater parallelism of the hardware approach promises higher throughput, but at the expense of additional hardware.

A software implementation of the hybrid load balancer on the Intel iPSC hypercube was used to obtain benchmark Lisp processing results confirming the superiority of dynamic methods for programs with unpredictable run-time characteristics. The hybrid method also produced better results than the sender- or receiver-initiated methods. We also demonstrated the portability of the hybrid load balancer, and the applications which it supports, between multiprocessors of different size and topology. The hybrid load balancer also appears relatively insensitive to the settings of internal parameters, performing well without the need for extensive tuning.

We have focused on Lisp applied to AI rather than numerical problems because AI programs generally exhibit more run-time unpredictability and therefore are in greater need of dynamic resource management. However, there is nothing in the design of the hybrid load balancer that restricts it to AI; it is equally suitable for numerical applications with similarly unpredictable run-time characteristics. Functional languages, for example, could be a candidate for execution under the hybrid load balancer. However, highly structured numerical algorithms with predictable run-time characteristics (for example, matrix operations or the fast Fourier transform) are best handled by compile-time resource management or special purpose architectures such as the BBN Butterfly. Since it has been conjectured that 80 percent of the numerical codes in use belong to the class of structured algorithms, it was natural that we should focus our attention on the AI area, where there are more programs suitable for dynamic load balancing.

Four classes of parallelism in Lisp programs are exploited: AND-parallelism, OR-parallelism, list parallelism, and argument parallelism. The hybrid load balancer constructs SUSPEND and RUN can be used to direct the hybrid load balancer to exploit such parallelism. Our experience with the Gabriel benchmarks leads us to believe that most sequential programs would respond well to such translation.

Future research is encouraged in the integration into a single coherent system of the various systems for load balancing and for handling side effects and data structures. Ongoing efforts to develop automatic serial-to-parallel code translators and distributed debuggers must be continued. All these systems are required if we are to succeed in making parallel hardware accessible to the user community.

12 ACKNOWLEDGMENTS

We would like to acknowledge the inputs of Dr. Les Gasser and Carl Braganza of the USC distributed AI group [30] and to thank them for the use of several communications primitives developed by them that we used in constructing our load balancer.

REFERENCES

1. R. Halstead, Jr., "Design Requirements of Concurrent Lisp Machines," *Parallel Processing for Supercomputing and Artificial Intelligence*, eds. Hwang and Degroot, McGraw-Hill, New York, 1988.

2. K. Hwang, J. Ghosh, and R. Chowkwanyun, "Computer Architectures for Artificial Intelligence Processing," *IEEE Computer*, pp. 19–27, January 1987.

3. H. Hayashi, A. Hattori, and H. Akimoto, "ALPHA: A High-Performance Lisp Machine Equipped with a New Stack Architecture and Garbage Collection System," *Proc. 10th Annual Int'l Symposium on Computer Architecture*, pp. 154–161, 1983.

4. D. Moon, "Garbage Collection in a Large Lisp System," *1984 ACM Symposium on Lisp and Functional Programming*, pp. 235–246, August 1984.

5. F. Lin and R. Keller, "Gradient Model: A Demand-Driven Load Balancing Scheme," *IEEE Conference on Distributed Systems*, pp. 329–336, 1986.

6. L. Ni, C. Xu, and T. Gendreau, "A Distributed Drafting Algorithm for Load Balancing," *IEEE Transactions on Software Engineering*, vol. SE-11, no. 10, pp. 1153–1161, October 1985.

7. J. Xu and K. Hwang, "A Simulated Annealing Method for Mapping Production Systems onto Multicomputers," *Proc. of Sixth Conf. on AI Applications*, March 7, 1990.

8. G. Cybenko, "Dynamic Load Balancing for Distributed Memory Multiprocessors," *Journal of Parallel and Distributed Computing*, to appear 1990.

9. S. Vegdahl, "A Survey of Proposed Architectures for the Execution of Functional Languages," *IEEE Transactions on Computers*, pp. 1050–1071, December 1984.

10. J. Darlington and M. Reeve, "ALICE: a Multiprocessor Reduction Machine for the Parallel Evaluation of Applicative Languages," *ACM/MIT Conf. on Functional Programming Languages and Computer Architectures*, 1981.

11. G. Steele Jr., *Common Lisp*, Digital Press, 1984.

12. D. DeGroot, "Restricted AND-Parallelism and Side-Effects in Logic Programming," in *Parallel Processing for Supercomputing and Artificial Intelligence*, eds. Hwang and DeGroot, McGraw-Hill, New York, 1988.

13. J. Anderson, W. Coates, A. Davis, R. Hon, I. Robinson, S. Robison, and K. Stevens, "The Architecture of the FAIM-1," *IEEE Computer*, pp. 55–65, January 1987.

14. S. Heller, *An I-Structure Memory Controller*, Master's Thesis, Department of Electrical Engineering and Computer Science, MIT, June 1983.

15. W. Dally, *A VLSI Architecture for Concurrent Data Structures*. PhD Thesis, Caltech, Technical Report No. 5209:TR:86, March 1986.

16. R. Gabriel, *Performance and Evaluation of Lisp Systems,* The MIT Press, 1985.

17. K. Hwang, W. Croft, G. Goble, B. Wah, F. Briggs, W. Simmons, and C. Coates, "A UNIX-Based Local Computer Network with Load Balancing," *IEEE Computer*, vol. 15, no. 4, pp. 55–66, April 1982.

18. J. Stankovic, "A Perspective on Distributed Computer Systems," *IEEE Transactions on Computers*, pp. 1102–1115, December 1984.

19. K. Hwang, "Advanced Parallel Processing with Supercomputer Architectures," *Proc. IEEE,* October 1987.

20. S. Stolfo, "Initial Performance of the DADO2 Prototype," *IEEE Computer*, vol. 20, no.1, pp. 75–83, January 1987.

21. K. Hwang and J. Ghosh, "Hypernet: a Communication-Efficient Architecture for Constructing Massively Parallel Computers," *IEEE Transactions on Computers*, December 1987.

22. G. L. Steele, Jr. and W. D. Hillis, "Connection Machine Lisp," *1986 ACM Conf. on Lisp and Functional Programming,* pp. 279–297, August 1986.

23. W. Chu and L. M. Lan, "Task Allocation and Precedence Relations for Distributed Real-Time Systems," *IEEE Transactions on Computers*, vol. VC-36, no. 6, pp. 667–679, June 1987.

24. G. Li and B. Wah, "Optimal Granularity of Parallel Evaluation of AND Trees," *Proc. Fall Joint Computer Conf.*, pp. 297–306, November 1986.

25. R. Halstead, Jr., "Multilisp: A Language for Concurrent Symbolic Computation," *ACM Transactions on Programming Languages and Systems*, vol. 7, no. 4, pp. 501–538, October 1985.

26. D. Ferrari and S. Zhou, "A Load Index for Dynamic Load Balancing," *Proc. Fall Joint Computer Conf.*, pp. 684–690, November 1986.

27. R. Smith, "The Contract Net Protocol: High-Level Communication and Control in a Distributed Problem Solver," *IEEE Transactions on Computers,* vol. c-29, no. 12, pp. 1104–1113, December 1980.

28. S. Sugimoto, K. Agusa, K. Tabata, and Y. Ohno, "A Multi-Microprocessor System for Concurrent Lisp," *Proc. Int'l Conf. on Parallel Processing,* pp. 135–143, 1983.

CHAPTER 6

DATA-FLOW COMPUTING MODELS, LOGIC AND FUNCTIONAL LANGUAGES, AND DATA-FLOW MACHINES FOR INTELLIGENCE COMPUTATIONS

Jayantha Herath, Yoshinori Yamaguchi, Susantha Herath, Nobuo Saito, and Toshitsugu Yuba

1 INTRODUCTION

Computer performance has increased by seven orders of magnitude with the implementation of new hardware while the sequential abstract computing model, sequential algorithms, languages, and architecture have remained the same. The von Neumann machine architecture consists of a processing element (PE) and a fixed size memory cell, as shown in Figure 6.1. The PE has a program counter naming the next instruction to be executed, the shared memory is used to communicate results, and control of execution is passed from instruction to instruction. Figure 6.2 shows the sequential computing model. Higher computing speeds in uniprocessor systems are achieved by using parallel control mechanisms such as interleaved memory, instruction fetch and execution overlap, extended instruction set, I/O processors, and multiple execution units. Circuit improvements that neglect the parallelism of a problem do not achieve the highest computing speed.

1.1 Parallelism in Computations

The multiplication of two *n*-by-*n* matrices requires computation of $n \times n$ inner products of pairs of *n*-element vectors. The sequential method for multiplying

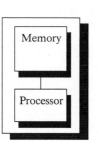

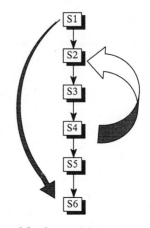

Figure 6.1 Von Neumann architecture.   **Figure 6.2** Sequential computing model.

two matrices computes the inner products as a sequence of multiplications and additions. Different approaches can be used for efficient parallel computations depending on the number of processors available. With n processors, the n multiplications of an inner product can be computed simultaneously. The terms of each inner product can be added by n processors in $O(\log n)$ time. With $n \times n$ processors, the $n \times n$ inner products can be computed simultaneously.

The discrete Fourier transform (DFT) is a computation analogous to the Fourier integral transform, which can be used to determine the frequency spectrum of a continuous, time-varying signal. The DFT of a time series is obtained as a weighted combination of the DFTs of two shorter time series. These DFTs in turn are computed from shorter series, and so on, until the trivial case of a single point is reached. DFT calculation on a sequential computer takes $O(n \times n)$ time, where n is the number of samples. Fast Fourier Transform (FFT) is an efficient computational technique to compute the DFT coefficients. FFT requires only $O(n \log n)$ time with a sequential computer and $O(\log n)$ time with n PEs.

The Monte Carlo integration method for evaluating multidimensional integrals over all possible degrees of freedom and the molecular dynamics integral technique for calculating the averages of physical quantities are powerful tools for problems in chemistry and physics. The simulations consist of a large number of parallel computations. The Monte Carlo simulation of the neutron transport problem generates several hundreds of neutrons and observes the system behavior. Neutrons move and collide with nuclei and are either scattered or absorbed. One or more neutrons are generated if the absorbing nucleus is fissionable. In the simulation, random numbers determine the event sequence of a neutron. Repetition of the calculation gives the system profile.

Partial differential equations can be solved using very high parallelism. The weather can be modeled by a set of partial differential equations with variables such as wind, air temperature, water content, and atmospheric pressure. Solutions to the system of differential equations give predictions of future weather on the basis of current and past observations.

Intelligence computations, consisting of large, parallel, non-deterministic numerical and non-numerical computations, model characteristics of human intelligence, such as understanding, learning, reasoning, and problem solving. Sequential and deterministic von Neumann machines are not oriented to intelligence computations. An ultra-high speed computing system is needed for such large, complex computations. The demand for such computing machines for analyzing physical processes is increasing every day. The major difficulty in satisfying this demand in uniprocessing is the physical constraints imposed by the hardware and the sequential and centralized control in the von Neumann model.

1.2 Ideal Parallelism

Parallelism in problems can be detected by users and compilers. The parallelism may be expressed differently by different algorithms. New languages map the algorithms to computing models by expressing all the possible parallelism of an algorithm and defining the parallel tasks. Parallel computing machines are based on vector processing, array processing, systolic processing, reduction, and data-flow computing.

The *ideal parallelism* of an algorithm is the number of parallel operations that can be executed in one time step in an ideal machine. The ideal machine consists of an unbounded number of processors and memories connected via an infinitely high-speed communication network. All operations have equal execution time and operations are executed as soon as operands are available. Figure 6.3 shows the ideal parallelism in computing Fibonacci(13), queen(4), and samefringe(32).

1.3 Overview

Data-flow computing [1], a radical departure from von Neumann computing, provides mechanisms for computing machines to achieve ideal computing machine behavior. Section 2 of this chapter discusses the acknowledgment static, strictly static, recursive dynamic, tagged-token dynamic, eduction, lazy-eager, pseudo-result, and Not(operation) data-flow computing models. The discussion is based on the representation of conditional computations, the root of iterative and recursive computations. Section 3 overviews the logic and functional languages used to represent data-flow computations and the process of transforming high-level languages to graphs. This process is applied to transform a Lisp program to a data-flow graph. Section 4 discusses the characteristics of representative data-flow computing machines for numerical and non-numerical computations and some common problems. Section 5 gives performance evaluation measurements made using EM-3.

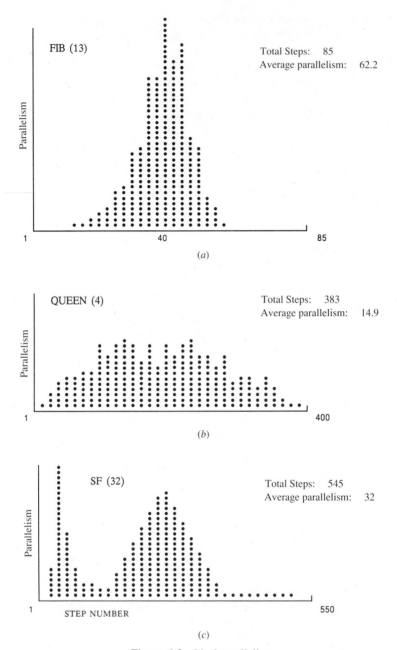

Figure 6.3 Ideal parallelism

2 DATA-FLOW COMPUTING MODELS

Data-flow computing provides multidimensional, multiple-pipelining instruction parallelism and hardware parallelism. Scheduling is based on availability of data. Processes are instruction-size. The basic elements of data-flow computing are *actors*, *arcs*, and *tokens*. Operators such as *add* and *multiply*, control actors such as *T gate, F gate, merge, whenever, ifthenelse,* and *switch*, and Boolean actors such as *or, and,* and *not* are data-flow actors. Data arcs and control arcs are data-flow arcs. Data tokens and control tokens are data-flow tokens. An actor is enabled as soon as its required operands arrive, and the partial results of the execution are passed directly as data tokens. The computations are free of side effects, and independent computations proceed in parallel. There is no concept of shared data storage. Every instruction is allocated by the computing element. A *data-flow graph* is a network of basic elements connected by directed arcs. These arcs represent the data dependencies. Execution of an operator consumes input tokens and produces or freezes a result on the output arc. All data-flow graphs shown in this chapter are drawn according to the following convention: Boxes represent operations; arrows represent arcs; arrows with black heads represent paths for data tokens; arrows with white heads represent paths for control tokens; black dots represent data tokens; and white dots represent control tokens.

In data-flow computing, data structures in storage are represented by pointer tokens. This reduces the parallelism of the computation, but provides safe execution. In static data flow, arrays are treated either as a set of scalars, which allows the elements of the array to be handled simultaneously by independent data-flow instructions, or as a sequence of values, which spreads the array out in time for pipeline execution. Heaps are functional directed acyclic graphs. They must be completely produced before consumption. The append, select, create, and delete actors are used to access these structures. The I-structures [9] allow selection of elements before complete production of the structure. The position of an element is defined by a tagged token. Presence, absence, and wait bits indicate the state of the element. Read of an unwritten storage cell is deferred by the controller until a write arrives. Pipelining between consumers and producers gives better performance; streams are sequentially allocated arrays.

2.1 Basic Model

In the basic model [1], the number of tokens per arc is restricted to one during the entire computation, which results in huge acyclic graphs. This makes the computations strictly iterative. Common data-flow evaluation techniques include strict and nonstrict evaluations. In strict data-flow computing, all the operands of an operation or arguments of a function must be presented to enable

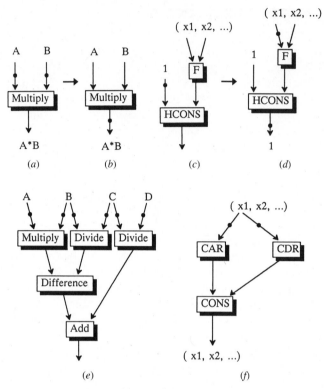

Figure 6.4 Data-flow computing.

the execution. Figure 6.4 shows snapshots of data-flow computing graphs. Figures 6.4*a* and 6.4*b* show the strict computation *multiply*. In nonstrict data-flow computing, a selected number of operands of an operation or arguments of a function is sufficient to enable the execution. This avoids unnecessary computations, eliminates nonterminating computations, and optimizes the computations to increase the parallelism and computing speed. Figures 6.4*c* and 6.4*d* show snapshots of the nonstrict computation HCONS. The HCONS operator has two arguments. The result of HCONS is simply the first argument, and it is generated as a result immediately after its arrival as an argument. Evaluation of the second argument, the result of function F, is not necessary. Figures 6.4*e* and 6.4*f* show the data-flow computing for the numerical computation (A*B) – (B/C) + (C/D) and the non-numerical computations CONS-(CAR(x1, x2, . . .), CDR(x1, x2, . . .)). Several data-flow computing models have been proposed for implementing the data-flow computing concept in practical machines. These models support the building of highly parallel and asynchronous computing machines, but differ in their approach as to how the computing should proceed.

2.2 Static Computing

Static data-flow computing was proposed by Dennis [2] for high-speed computing machines. The VIM, Texas DDP, LAU, Hughes, and NEDIPS systems are based on the static computing model. In static computing, concurrent re-entrance is inhibited. The model consists of operators, data and control arcs, and data and control tokens. Several tokens per arc are allowed, but there is a restriction of one token at a time. An actor fires when there are no tokens on any of the actor's output arcs. A token consists of a value and a tag identifying the target actor. No code copying or recursion is allowed; only iteration is supported. The *switch-t*, *switch-f*, and *merge* operations are introduced to support conditional computations. A *true* token at the input of switch-t copies the other token in the input to the output. A *false* token at the input of switch-t does not dispatch the other input token to the output. Similarly, the switch-f operator dispatches the input token to the output if and only if the boolean input token is false. Three-input merge operators are executed when the boolean input and appropriate data token are available. Figure 6.5 shows the implementation of

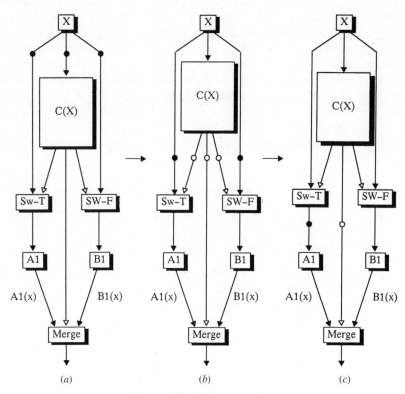

Figure 6.5 Static computing.

the conditional computation IF C(X) THEN A1(X) ELSE B1(X), and the firing sequence.

2.2.1 Strictly Static. In the strictly static model used in the Texas DDP system [3], it is prohibited to initiate a new iteration before the previous one is concluded. The branch node does not provide new tokens until the previous iteration is completed. This provides safe execution of the computation but limits the parallelism.

2.2.2 Acknowledgment Static. In the acknowledgment static model [2, 3, 4, 5], consumers send acknowledgment signals to producers, indicating the possibility of accepting a new set of tokens. This enables pipeline production of tokens and promotes parallelism by allowing initiation of a new iteration before the previous one has been concluded. This model provides safe execution of re-entrant graphs.

2.3 Recursive Dynamic Computing

In dynamic data-flow computing, proposed by Davis [16, 17] and Dennis et al. [4], several instances of a node can be fired at a time, and these nodes can be created at run time. Concurrent re-entrance is permitted using code copying to create a new instance of a subgraph. This enables recursive data-flow computations. FIFO queues are used in the recursive computations, which results in directed acyclic graphs. In each invocation, a maximum of one token is placed on an arc. An apply actor creates a new copy of the program graph. There are no merge actors because, in any instance of the graph, only one of the data inputs of the merge is used. A token can be represented by ⟨v ⟨u, s⟩ d⟩, for a data value v, an activation instance u, an actor s within the function, and an operand d of the target actor. The operand of the target actor is not necessary for single-operand operations. The DDM1 machine is based on this principle. Figure 6.6 shows the firing sequence for recursive dynamic data-flow computing.

2.4 Tagged-Token Dynamic

The tagged-token dynamic computing model, proposed separately by Arvind [8, 9, 10, 11] and Gurd-Watson [12, 13, 14] at Manchester, is more efficient in exploiting parallelism than the previous models. A tag assigned to each token distinguishes its identity. Identically tagged tokens enable the execution of an operation. Tagging allows many data values per arc at one time. Each node can be created at run time, and several instances of a node can be fired at a time. Recursion and iteration are represented directly. Successive cycles of an iteration are allowed to overlap by unfolding loops.

A token consists of a value and tag representing the target actor identity.

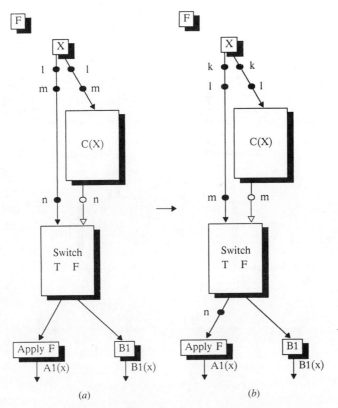

(a)   *(b)*

Figure 6.6   Recursive dynamic computing.

A token is represented by $< v, < u,c,s,i > d >$: a data value, v, an activation instance, u, a code block (loop body), c, an actor, s, within the function, an index representing the cycle of an iteration (data structures), i, and function activation, d. No merge actor is used. Identity actors are used, such as a D-operator for loop entry, which establishes a new context for iteration and sets the index of result tokens to one greater than the index of the input token, and a D-reset for loop exit, which restores the tag of the result token to that of the context surrounding the tag. Special mechanisms, such as loop throttling, are used to limit the parallelism exploited by tagged tokens.

The MIT TTDA, Manchester Dataflow machine, and all tagged-token dynamic data-flow machines are based on this model. Id, LAPSE, MAD, SASL, SISAL, and many other languages support these machines. The switch operation used to implement conditional computations has two input arcs: one for Boolean tokens and the other for data tokens to be switched. This operator also has two output arcs. The incoming Boolean token determines the output arc along which the incoming data token is sent. A TRUE (T) token copies the

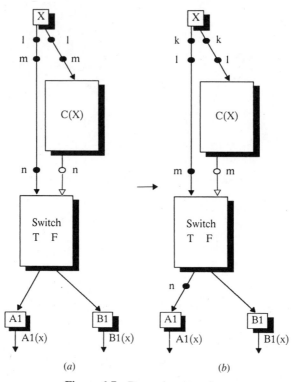

Figure 6.7 Dynamic computing.

other token in the input to the T output, and a FALSE (F) token copies to the F output. No merge operators are used. The BRR operation used by Gurd's group is similar to this switch operation. Figure 6.7 shows the firing sequence for tagged-token dynamic data-flow computing.

2.5 Eduction Computing

The eduction model, proposed by Ashcroft [18], is a hybrid of data-flow and demand-flow computing. Operator nets represent the eduction computations graphically. The demand for a result triggers its computation, which in turn triggers evaluation of its arguments. The demand propagation continues until constants are encountered, whereupon the values are returned to the demanding nodes and execution proceeds in the opposite direction. The arguments for branches of conditional computations are not evaluated in parallel; only necessary arguments are evaluated. The modal operators *where, first, next, followed by, as soon as, merge, whenever, upon,* and *is current* are used to express recursion and iteration in a purely functional way. The Eazy flow engine is proposed to execute operator nets described in LUCID language. Figure 6.8

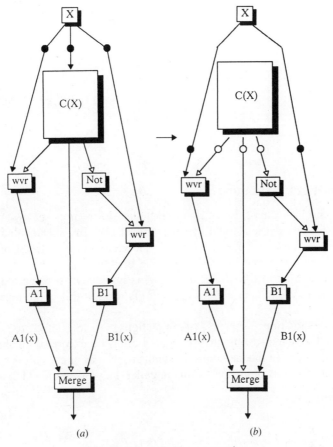

(a)   *(b)*

Figure 6.8 Eduction computing.

shows the firing sequence of operator nets. The *wvr* node is similar to the switch-t operation. The switch operation is occasionally used. Operators such as wvr and merge need extra memory to remember the last token arrived in dynamic data-flow computing.

2.6 Data-Flow–Control-Flow Computing

The data-flow–control-flow computing model was proposed by Treleavan et al. [20]. It uses two basic mechanisms. One instruction causes the execution of others using the control mechanism. Instructions receive and dispatch data using the data mechanism. Instruction execution is caused by the availability of data and control tokens. Data tokens carry partial result values while control tokens carry null values. Conditional computations are supported by the many-input, two-output switch operation.

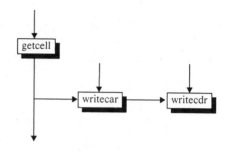

Figure 6.9 Eager-lazy computing.

2.7 Eager-Lazy Computing

The eager and lazy data-flow computing models were proposed by Amamiya et al. [21–26] for artificial intelligence applications. In eager evaluation, all possible computations are executed in parallel without optimizing. Conditional computations are executed in parallel to the branches. Car and cdr parts are evaluated in parallel to CONS. CONS(x, y) is implemented using the *getcell, writecar*, and *writecdr* operations shown in Figure 6.9; this is the lenient *cons* mechanism.

In lazy evaluation, selected computations are executed to optimize processing. The selected branch is executed after the execution of the conditional computation. In the lazy cons mechanism, the car or cdr part is evaluated only when its value is demanded. This is called lazy or optimized evaluation.

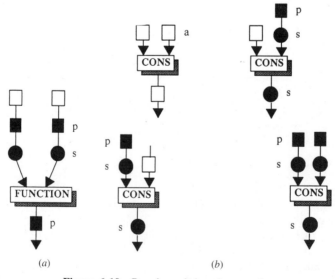

Figure 6.10 Pseudoresult-based computing.

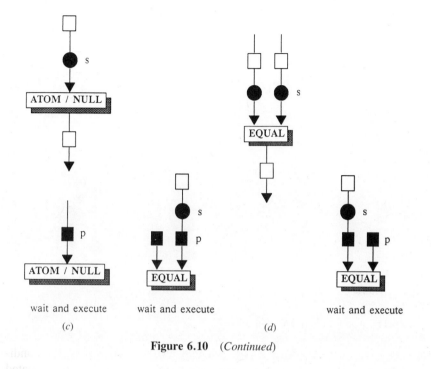

Figure 6.10 (*Continued*)

In the eager-lazy computing model, eager, lazy, nonstrict, and demand-driven computing mechanisms are selectively and efficiently implemented to obtain the maximum efficiency. The model is implemented in DFM using VALID language.

2.8 Pseudoresult Computing

Yamaguchi et al. [27, 30] proposed the pseudoresult data-flow computing model, particularly for AI applications. In Figure 6.10 black boxes represent pseudoresults, black dots represent semiresults, and white boxes represent actual results. Pseudoresults are generated immediately after the arrival of an argument, as a result of function execution; this is shown in Figure 6.10a. This pseudoresult enables successive computations relaxing the firing conditions. While the operations in the function are executed concurrently with evaluation of its successor. The identifiers of pseudoresults are realized by addresses in a result store and are eventually filled by actual results. A semiresult is a pseudoresult used in non-numerical computations, and a partial result is a pseudoresult used in numerical computations. When the input to an operation or function is actual, semi, or pseudo, the output is an actual or semiresult. The execution of a nonlist operation is deferred until the inputs become actual. Figure 6.10b shows four different instances of CONS execution, Figure 6.10c

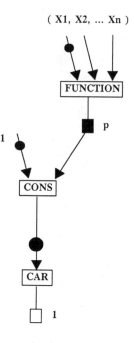

Figure 6.11 Application of pseudoresults.

shows two different instances of ATOM/NULL execution, and Figure 6.10*d* shows two different instances of EQUAL execution. Figure 6.11 shows an application example of pseudoresult generation. This model is implemented on the EM-3 using the languages EMLISP and EMIL.

2.9 Not(operation) Computing

In the not(operation) computing model [31–33], computations are represented by sequential, parallel, and decision-making computation segments. Ordered sequential computation segments ensure the logical correctness of the computation. Parallel computation segments are composed of independent computations. A conditional computation is represented using two parallel complementary computations. The transformation of traditional conditional computation to not(operation)-based computation is performed in two steps. First, the traditional conditional computation is disintegrated into two complementary basic operations, which must be executed for deadlock-free computation. The positive state is denoted by *operation* and the negative state is denoted by *not(operation)*. The not(operation) represents many other positive and negative states. A number *n* of sequential conditional computations are represented by *n* different independent parallel operations. Then the semantics of the execution are defined. One of the operations executed will give an output value if the operation is satisfied.

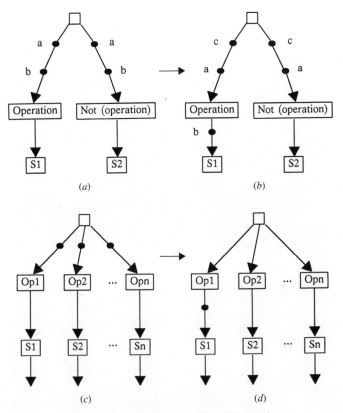

Figure 6.12   Not(operation)-based computing.

Figures 6.12*a* and 6.12*b* show the firing sequence of the conditional computation IF (operation) THEN S1 OR IF (not(operation)) THEN S2. Two input tokens that satisfy only one of the operations are required to enable the conditional computation. When the *operation* satisfies the input data, the data token is given as the result of execution, and the output of the *not(operation)* is frozen. Otherwise, the *not(operation)* gives the data value output, while the *operation* output is frozen. Figures 6.12*c* and 6.12*d* show two instances of *n* parallel conditional computations.

3   LOGIC AND FUNCTIONAL PROGRAMMING FOR DATA-FLOW COMPUTING

The language is very important in representing parallel algorithms and mapping them efficiently onto the computing environment. Functional and logic programming languages are two major declarative language paradigms to enhance computing productivity. In designing data-flow computing languages, it is possible to use an existing sequential language, functional language, parallel logic programming, or any other high-level language. The use of existing languages allows existing software to run on the new machine and gives the programmer a high degree of control over the run-time behavior. Conventional programs consisting of sequences of statements alter the data stored in the memory one piece at a time. Variables are used to represent storage cells, and a statement is necessary to alter data for each variable.

The use of sequential languages to represent parallel programs, and their compilation for a parallel execution environment complicate the execution process. Algorithms that form data-flow graphs from conventional languages are complex. The concurrency that can be detected by a compiler is also limited. The use of a language that reflects the parallel machine architecture exploits the machine parallelism but increases the programming complexity. Data-flow programming requires no knowledge of machine structure, and there is no need for explicit expression of parallelism. The compiler detects parallelism. Users do not need to consider the explicit control of memory allocations in using machines, but deal only with data values. Low-level languages for data-flow computing machines should describe data-flow computing efficiently.

3.1 Logic Programming

Drawing inferences at a very high speed is the future objective of expert systems. Logic programming, based on symbolic logic, is suitable for knowledge processing systems dealing with large databases. An implicit search strategy and parallelism support symbolic processing. Logic programming describes the facts and their relationships in a problem and controls the execution nondeterministically. Questioning gives the answer using declared facts and defined rules. A question is answerable if it is the head of any other clause and each of its goals is true. When answering a question, logic programming looks for matching facts in the database. Two facts match if their predicates and corresponding arguments are the same. The process of matching, unification, is the execution mode. Clauses in logic programming are transformed into data-flow graphs.

3.1.1 Unification. Robinson's resolution principle [34] applies only one powerful rule of inference to mechanical theorem proving. This enables the computer to make deductions from a set of logical formulas. Resolution establishes that a term is true for some binding of its free variables by showing that the particular term can be implied from a set of Horn clauses. The process consists of *unification* and *substitution*. Unification makes two terms identical by placing their free variables by a free variable. Unification of $U(X, a, Y)$ and $U(Y, Z, b)$ is $U(X, a, b)$. $U(a, b)$ and $U(X, X)$ cannot unify. A term is unified with an entire clause by unifying it with the term to the left of the implication. Every free variable replaced in the left-hand term is replaced in each occurrence on the right-hand side. Substitution replaces a term with the right-hand side of a clause that has unified with it. Unification of $U(b, Y, Z)$ with $U(X, a, X)$:− $G(X, Z)$, $H(Z)$ gives $U(b, a, b)$:− $G(b, Z)$, $H(Z)$. Substitution replaces $U(X, a, X)$ with the resolvent $G(b, Z)$ and $H(Z)$.

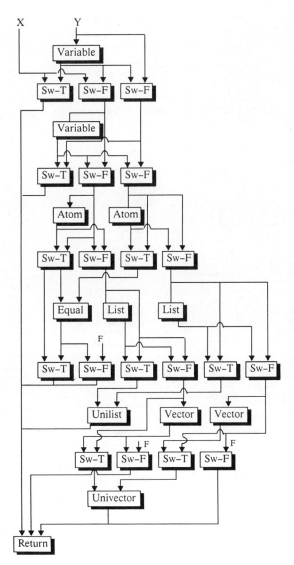

Figure 6.13 Unification—traditional.

Logic programs consist of terms such as symbols, integers, variables, lists, and vectors. To execute a goal, the system searches for the first clause whose head matches or unifies with a goal. The unification process finds the most general common instance of the two terms, in a goal and head literal of a selected clause, if it exists. If a match is found, each of the goals in the body is activated from left to right. If at any time the system fails to find a match for a goal, it rejects that clause execution and the output is "fail." The body of a nonunit clause head literal successfully unified with a goal becomes another unification process. The unification process terminates when a goal literal unifies with the head of a unit clause. Goal literals in a single statement that do not have any shared variables in a statement can be unified concurrently and easily. Alternative solutions for the goal and independent subgoals can be performed in parallel.

Figure 6.13 shows unification with the data-flow computing model. The data types are variable, atom, list, and vector, and the input arguments are

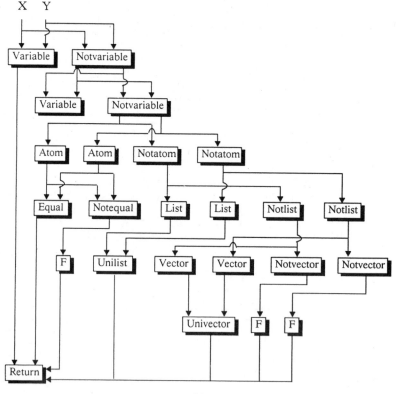

(a)

Figure 6.14 Unification—not(operation).

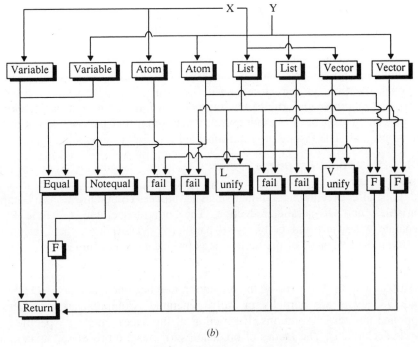

(b)

Figure 6.14 (*Continued*)

x and *y*. The variable checks whether *y* is a variable; if so, *x* is the output of unification. Otherwise, *x* is checked. If *x* is a variable, *y* is the output of unification; if not, the arguments are checked for atoms. If they are atoms, they are checked for equality. If they are equal, *x* or *y* will be the unification output. If the atoms are not equal, "fail" is the unification output. If *x* and *y* are not atoms, they are checked for lists. If they are lists, UNILIST gives the unified list as the unification output. If they are not lists, they are checked for vectors. If they are vectors, then UNIVECTOR gives the unified vector as the output of the unification. Otherwise, "fail" is the unification output.

Figure 6.14 shows unification with not(operation). VARIABLE checks whether the right operand of the two inputs is a variable, and if so gives the left operand as the output of unification. Otherwise, no output is given. NOTVARIABLE checks whether the right operand is a variable, and gives no output if it is. Otherwise, both input operands are given as the output. ATOM checks the input operand and gives it as the output if it is an atom; otherwise, there is no output. NOTATOM gives no output if the input operand is an atom; otherwise, the input is given as the output. Similarly, LIST, NOTLIST, VECTOR, and NOTVECTOR give output only for the appropriate type of input. FAIL gives a fail output.

3.1.2 Prolog. Colmerauer's Prolog language [36, 37] is based on language theory and mathematical logic with practical constraints. This sequential logic programming language draws inferences efficiently. Relationships are represented as predicates, and objects are represented as arguments. Facts declare the relationships between objects. An assertion, or fact, has no body. A conditional assertion, or rule, has a head and body. Rules are used to describe or define relationships. The execution mode is unification with backtracking. Prolog languages start the execution of a goal only after the completion of the previous goal. The Prolog program for quicksort [37] is shown in Figure 6.15.

3.1.3 Relational Language. Clark's relational language [38] focuses on parallel execution of logic programs. Relational language features include AND-parallel execution of conjuctive goals, process communications by shared variables, and OR-parallel reduction. The commit operator is introduced to separate the guard and body. AND-parallel processes are synchronized by defining the instances of the variables as producers or consumers.

3.1.4 Parlog. Clark's Parlog [39] augments the expressive power of the relational language. In Parlog the resolution tree has one chain at AND levels, and OR levels are partially or fully generated. Communicating processes combine the partial solutions. Restriction of the access mode is specified by mode declaration. The modes of predicate variables are predefined as input or output. The Parlog program [39] for quicksort is shown in Figure 6.16.

3.1.5 Concurrent Prolog. Many features of relational language are implemented in Shapiro's Concurrent Prolog [40]. In Concurrent Prolog, the search strategy is multiple, depth first. The resolution tree consists of one chain from top to bottom. Guards can bind variables. Read-only variables in a process are introduced to support process synchronization. Clause activation is suspended until the variable is assigned a value. The Concurrent Prolog program [40] for quicksort is shown in Figure 6.17.

```
quicksort ( Unsorted, Sorted ) :-
      qsort ( Unsorted, Sorted-[] ).
qsort ( [ X | Unsorted ], Sorted-Rest ) :-
      partition (Unsorted, X, Small, Large ),
      qsort ( Small, Sorted-[ X | Sorted1 ] ),
      qsort ( Large, Sorted1-Rest ).
qsort ( [], Rest-Rest ).
partition ( [ X | Xs ], A, Small, [ X | Large ) :-
      A < X, partition ( Xs, A, Small, Large ).
partition ( [ X | Xs ], A, [ X | Small ], Large ) :-
      A >= X, partition ( Xs, A, Small, Large ) :-
partition ( [], -, [], [] ).
```

Figure 6.15 Quicksort—Prolog.

```
mode quicksort ( ?, ^ ).
mode qsort ( ?, ^, ^ ).
mode partition ( ?, ?, ^, ^ ).
quicksort ( Unsorted, Sorted ) :-
        qsort (Unsorted, Sorted - [] ).
qsort ( [ X | Unsorted ], Sorted1-Rest ) :-
        partition ( Unsorted, X, Small, Large ),
        qsort ( Small, Sorted-[ X | Sorted1 ],
        qsort ( Large, Sorted1-Rest ).
qsort ( [], Rest-Rest ).
partition ( [ X | Xs ], A, Small, [ X | Large ] ) :-
        A < X : partition ( Xs, A, Small, Large ).
partition ( [ X | Xs ], A, [ X | Small ], Large ) :-
        A >= X : partition ( Xs, A, Small, Large ).
partition ( [], -, [], [] ).
```

Figure 6.16 Quicksort—Parlog.

3.2 Functional Programming Languages

In functional programming languages, programs are mathematical functions based on functional algebra. The major operation is function application, in which an object is mapped onto another object. There is no concept of storage, assignment, goto, or side effects. Programs are free building blocks for larger programs. Functional languages do not reflect von Neumann properties or the machine structure. They are zero or single assignment languages that provide specially controlled reassignment constructs for loops. Functional languages such as Pure Lisp and FP [41, 42] can be used effectively to execute computations in data-flow machines. In FP, programs are used to construct new programs using program-forming operations. This increases the expressiveness of algorithms; exploits the massive parallelism in scientific computations; per-

```
quicksort ( Unsorted, Sorted ) :-
        qsort ( Unsorted, Sorted - [] ).
qsort ( [ X | Unsorted ], Sorted-Rest ) :-
        partition ( Unsorted?, X, Small, Large ),
        qsort ( Small?, Sorted-[ X | Sorted1 ] ),
        qsort ( Large?, Sorted1-Rest ).
qsort ( [], Rest-Rest ).
partition ( [ X | Xs ], A, Small, [ X | Large ] ) :-
        A < X : partition ( Xs?, A, Small, Large ).
partition ( [ X | Xs ], A, [ X | Small ], Large ) :-
        A >= X : partition ( Xs?, A, Small, Large ).
partition ( [], -, [], [] ).
```

Figure 6.17 Quicksort—Concurrent Prolog.

mits abstract data structures, streams, and irregular data structures; and allows powerful programming constructs.

3.2.1 *VAL.*

VAL, the high-level language designed by Dennis's group [2–5], is value oriented, as opposed to traditional variable orientation. In a value-oriented system, new values are defined and used but can never be modified. Values may be bound to identifiers, but identifiers cannot be used as variables. The design principles of VAL provide implicit concurrency and synchronization by using completely functional language features. Expression-based features inhibit all forms of side effects. Once the values of all inputs are known, execution cannot influence the results of any other operation ready to be executed. Automatic detection of parallel computations by compilers (vectorization) has been used to exploit concurrency. Side effect features, memory update, and aliasing are banned. VAL helps simplify critical programming chores such as error handling, debugging, and speed analysis. A factorial function [4] written in VAL is shown in Figure 6.18.

3.2.2 *Id.*

Id [8–11] was proposed by Arvind and Gostelow. Id (for Irvine Dataflow) is a block-structured, expression-oriented, side-effect–free, single-assignment language. A program in Id is a list of expressions. The four basic expressions are blocks, conditionals, loops, and procedure applications. Id variables are not typed. SELECT and APPEND create new and logically distinct structures. The execution of operators is dynamic, as opposed to the static nature of Dennis's model. Id supports streams and nondeterministic programming. A quicksort program written in Id [9] is shown in Figure 6.19.

3.2.3 *LUCID.*

LUCID [19], proposed by Ashcroft and Wadge, incorporates iterations by regarding all values as histories. Everything, including constants, is an infinite history. Assignment statements are equations between histories. A program in LUCID is an unordered set of equations. Result generation is stopped when the required number of results is produced by stopping the machine. Conventional LUCID is implemented by employing demand-driven computing for infinite objects. A factorial function [19] written in LUCID is shown in Figure 6.20.

```
function Factorial ( n : integer returns integer )
   for i: integer := 0;
       p: integer := 1;
     do if i = n then p
       else iter i :- i + 1; p := p *i; enditer;
       endif
   endfor
endfun
```

Figure 6.18 Factorial — VAL.

```
procedure quicksort ( a, n )
m <-- a [i];
below, j, above, k <--
   ( initial below <-- A; j <-- 0;
        above <-- A; k <--
   for i from 2 to n do
            new below, new j, new above, new k <--
         ( if a [i] < middle then below + [ k + 1 ], a [ i ],
            j + 1, above, k
         else below, j, above + [k + 1 ], a [i ], k + 1 )
         return ( if j < 1 then quicksort ( below, j ) else
            below ), j
         if k < 1 then quicksort ( above, k ) else above ), k,
   return ( initial t <-- below + [ j + 1 ] middle
        for i from 1 to k do
            new t <-- t + [ i + j + 1 ] above [ i ]
        return t ))
```

Figure 6.19   Quicksort—Id.

3.2.4 Manchester Languages.

3.2.4 Manchester Languages. Languages used in the Manchester machine [12–15] are conventional languages, such as LAPSE, MAD, SASL, and SISAL. SASL, based on LUCID, treats functions as first-class objects. In particular, a function takes one argument; currying is used to obtain the effect of multiple-argument functions. LAPSE, a single-assignment language, has Pascal-like syntax. LAPSE stores arrays during iteration or for all loops that use them. MAD, based on Id, is typed using streams and has operators such as list-processing operations. MAD stores arrays for longer, and garbage collection is performed using reference counts.

SISAL [43, 44] (for stream and iteration in a single-assignment language) is a cooperative research effort of Colorado State University, DEC, Lawrence Livermore Laboratory, and Manchester University. This language is a value-oriented functional language for sequential, vector, multiprocessor, and data-flow computing machines. SISAL is implemented on the VAX, Cray, HEP, and Manchester data-flow machines. It is strongly typed. Recursion has been added, error values simplified, and some Id (/MAD) features added. Tokens are labeled to allow multiple use of arcs. Labels are used for data structures, loops, and functions. IF1, the intermediate language for SISAL, performs machine-

```
fac (n);
    where
        fac ( n ) = if n eq 0 then 1 else n*fac ( n - 1 ) fi;
    end
```

Figure 6.20   Factorial—LUCID.

```
define Quicksort
type Info = array [ integer ]
function Quicksort ( Data : Infor returns Info )
  function Split ( Data : Info returns Info, Info, Info )
    for E in Data
      returns array of E when E < Data [ 1 ]
             array of E when E = Data [ 1 ]
             array of E when E > Data [ 1 ]
      end for
  end function % Split
  if arraysize ( Data ) < 2 then Data
  else let L, Middle, R := Split ( Data )
    in Quicksort ( L ) Middle Quicksort ( R )
    end let
  end if
end function % Quicksort
```

Figure 6.21   Quicksort—SISAL.

independent optimizations and machine-dependent analysis. A quicksort function [43] written in SISAL is shown in Figure 6.21.

3.2.5 *VALID.* VALID, a value identification language [23] designed by Amamiya et al., is a functional language with implicit and explicit parallel constructs. Lenient cons computing is applied in function evaluation to achieve very high parallelism. List computations and higher-order functions are written using an Algol- and Lisplike syntax. A quicksort function [23] written in VALID is shown in Figure 6.22.

3.2.6 *EMLISP.* EMLISP is a single-assignment language [27–30]. To obtain side effect–free, pure functional list processing, the features added to conventional Lisp to increase efficiency in von Neumann computing, such as relatives of PROG, flow-controlling operations, list modifiers, relatives of array, and side effect operations such as RPLACA and RPLACD, are removed. The global and free variables and loops are inhibited. Special features such as parallel COND, parallel OR, parallel AND, and BLOCK are added. The low-level language used on the EM-3 to represent data-flow computing is EMIL.

3.3   DCBL Transformations for Data-Flow Computing Languages

The objectives of DCBL (pronounced "decibel") [32] design are to define operational semantics for data-flow computing languages, to develop high-level data-flow computing languages that free programmers from machine-specific considerations and make it easy for users to express concurrency, to facilitate the natural expression of parallelism, and to enable a compiler to generate optimized code that exploits the inherent parallelism without the application of sophisticated analysis techniques.

```
sort : function ( x ) return ( list )
= if x = nil then x
else clause y = list ( car ( x ));

   [ y1, y2, y3 ] = partition ( cdr ( x), y );

   return

   append ( sort ( y1 ), append ( y2, sort ( y3 )))

   end;

partition : function ( x, y ) return ( list, list, list )
= if x = nil then ( nil, y, nil )
else clause [ w1, w2, w3 ] = partition ( cdr (x ), y );

   x1 = car ( x ); y1 = car ( y );

   return

   case x1 = y1 --> ( w1, append ( list ( x1 ), w2 ), w3 );

       x1 < y1 ---> ( append ( list ( x1 ), w1 ), w2, w3 );
       x1 > y1 ---> ( w1, w2, append ( list ( x1 ), w3 ));

append : function ( x y ) return ( list )
= if x = nil then y else cons ( car ( x ), append ( cdr ( x ), y ))
```

Figure 6.22 Quicksort—VALID.

3.3.1 Specification of DCBL.

DCBL allows parallel algorithms to be expressed as a collection of expressions. The execution of a DCBL program consists of a sequence of parallel executions of expressions. An expression execution may generate zero, one, or two or more values. Tuple expressions, multivalue function expressions, conditional expressions, and parallel expressions generate two or more values. The syntax specification for iterative computations is as follows:

```
exp :: = function (exp)
         | exp, exp,...exp
         | IF exp THEN exp, IFNOT exp THEN exp
         | identifiers
         | constants
         | LET idlist = exp IN exp
         | IF exp THEN exp
         | FOR idlist = exp DO iteration

iteration :: = ITER exp NOTITER exp
             | LET idlist = exp IN iteration
```

```
    | IF exp THEN iteration
    | IF exp THEN iteration IFNOT exp THEN iteration

idlist :: = id
       | idlist id
```

The application of a function to an expression, `funct (exp)`, is used to represent sequential computations. The elementary functions are operators. The operations performed on expressions can be characterized by mathematical functions. The application of function F to the imports x, y, and z produces export $F(x, y, z)$. The expression tuple `|exp, exp, . . ., exp|` is used to represent parallel computations. *Identifiers* and *constants* are the most elementary expressions. Values can be bound to identifiers, which can be bound to simple types (integer and real), structured types, and function calls. The `LET ... IN ...` expression provides local binding to extend the execution environment.

Decision-making computations, conditional expressions, and `FOR ... DO ...` expressions sequence parallel computations to ensure logical correctness and to avoid initiating computations whose results can never be used. A conditional expression represents operational semantics for conditional, iterative, and recursive computations. General IF THEN ELSE expressions and case expressions are represented by `IF exp THEN exp`. All predicates in `IF exp THEN` are supported by expressions, and, depending on the imports, produce and seize exports, instead of producing Boolean values. `IF exp THEN exp` provides a single-branch conditional expression. `IF exp1 THEN exp2 IFNOT exp1 THEN exp3` is a two-branch conditional expression with two parallel complementary conditional computations. A set of n `IF exp THEN exp` expressions gives n parallel computations.

The `FOR idlist = exp DO iteration` expression implements iterative computations that depend on the previous iterative computation result. Loop initiation is performed by the `FOR idlist = exp` part, and the loop body appears in `DO iteration`. The expression is evaluated by binding the iterative identifiers, the elements of `idlist`, to the values of `exp`. The evaluation of the iteration body results in a `NOTITER` expression and an `ITER` expression, which are evaluated concurrently. If the `NOTITER` expression satisfies the condition, this terminates the iteration and gives the computation result. Otherwise, the output is given by the `ITER` expression. Here, the `ITER` expression is satisfied and continues iteration. The iteration is terminated when the evaluation of the `ITER` body results in an ordinary `NOTITER` expression. The value of this expression is the value of the `ITER` expression. Parallel expressions for computations of the type `For i := 1 to n do C[i] := A[i] * B[i]` represent iterations that do not depend on the previous computation result.

3.3.2 Data-Flow Graph Specification Language.
A data-flow graph representing data-flow computations can be defined by $N = [T, O, L]$ where T,

O, and L represent the set of tokens, the set of operations, and the set of links. For an element O_i in O, the set $\text{Im}(O_i)$ represents the import ports of O_i, and $\text{Ex}(O_i)$ represents the export ports of O_i. Firing an operator maps imports to exports. The semantics of firing define the minimum set of import ports, varying from one to the total number of imports, that must receive input to enable an operation or function. The output semantics depend on the execution of an operation. Firing an operator dispatches exports to zero or more export links. Expressions and compilers help identify concurrency in algorithms and their programs and map that concurrency into graphs. The graph of a computation, which connects subgraphs composed of operators, is an explicit representation of the concurrency available in evaluating expressions.

An element of a data-flow computation consists of import ports, imports, export ports, exports, import links, export links, and operators. Specifications of a data-flow graph include imports, exports, data links, and operators. Operators are defined recursively using local imports and exports. Imports to the operator embark at import ports. Exports of the operators disembark at export ports. The number of imports or exports in a link is unlimited. This gives the dynamic computing feature. The restriction of values to one gives the static computing feature. The operators communicate values through their import and export ports. The graph has an import port for each free variable of the expression and an export port for each value returned by the expression.

The exports produced are exported to defined destinations to enable successive computations. The destination of an export value is specified by the import port number of the destination operator. The export port of an operator is connected by a link to the import port of another operator. The export value of one operator is the import value to another operator.

The following notations are used to specify data-flow graphs. T(exp) represents the operators of the translated expression (exp). IM.T(exp) represents the set of imports to T(exp). EX.T(exp) represents the exports at the export ports of T(exp). Links are represented by EX.T(exp1)---> IM.T(exp2), which means that the exports of T(exp1) are linked as the imports to the defined import ports of T(exp2). The import ports of all parallel subgraphs are assigned the set of import values. The graph export ports are formed by concatenating the export ports of the component subgraphs. An extension of import-based language can be used to design demand-driven computing languages.

The complexity of a data-flow graph increases with the number of operators and arcs. This increases execution and communication times and creates many problems when resources are limited. The fundamental principle in managing the complexity is to reduce the size of the graph while preserving its original properties. In graph reduction the number of operators, arcs, and tokens generated are reduced without changing the final result of the computation.

3.3.3 DCBL Transformation.
The transfer function T maps DCBL expressions to data-flow graphs, and the DCBL operator F maps imports onto exports. The operational semantics are defined and derived by the application of $F(T(\text{exp}))$. The expression transformation to graphs gives the infor-

mal operational semantics of data-flow graphs. The transformation of funct(exp), T(funct(exp)), produces sequentially connected data-flow subgraphs. The transformation is made by connecting the export ports of T(exp) to the import ports of T(funct). The transformation of [T(exp1, exp2, . . . , exp-n)] consists of n subgraphs, [T(exp1)], [T(exp2)], . . . , and [T(expn)], that can be executed in parallel. Two subgraphs are connected sequentially in the implementation of the simplest conditional expression, IF exp1 THEN exp2. Predicate exp1 controls the evaluation of exp2. The import data value of T(exp1) is the export of T(exp1) if this data satisfies the condition expressed by exp1; if not, the data value is simply absorbed. This expression provides the facility to evaluate n parallel conditional expressions. The transformation of parallel complementary conditional expressions, IF exp1 THEN exp2 IFNOT exp1 THEN exp3, is illustrated in Figure 6.23. The transformation of an identifier, [T(id)], gives a graph with no operators. The transformation of a constant expression gives the const operator with import and export links. A trigger value import produces the value const as the export.

DCBL binds identifiers locally. In the evaluation of the iteration expression, FOR idlist = exp DO iteration, the elements of idlist are bound to the values of exp, and iteration is terminated when it results in an ordinary expression. ITER(exp) supports iteration if the imports satisfy the expression exp. NOTITER(exp) gives the result of the computation. The iteration body LET idlist = exp IN iteration is implemented in the same way as LET idlist = exp1 IN exp2. The data-flow graph implementation of the conditional iteration body IF exp THEN iteration is similar to that of the conditional expression. Both subgraphs, IF exp and IFNOT exp, provide a complete set of exports. T(exp) and T(notexp) are placed on the import paths of the iteration body subgraphs, T1(iteration1) and T1(iteration2). Exports of T(exp) or T(notexp) enable the evaluation of a selected iteration body.

3.3.4 Functionality.

The functionality of a data-flow graph encompasses the operational semantics of expressions and the formal simulation of data-flow graph execution. The operational semantics of a data-flow operator are given by its functionality, which maps its imports onto exports. The functionality of an operator is the usual arithmetic or boolean function associated with it. For example, F + $(x, y) = x + y$ and Fconst(x) = const. The arrival of the x token triggers the constant operator to give the defined export. The functionality can be extended for an ordered set of data-flow imports. The operation *plus* can be performed on the ordered sets $x.X$ and $y.Y$. Hence, Fplus$(x.X, y.Y)$ = Fplus(x,y).Fplus(X,Y) = $(x + y)$.Fplus(X, Y).

The characteristics of operators or functions can be distinguished by either imports or exports. According to imports, there are two types, strict and nonstrict, of operators or functions. In a strict operator or function, the availability of all imports enables the execution. Nonstrict operators or functions need the availability of specified operands or arguments to enable the execution. According to exports, there are two types of operators as follows: one produces

```
imports : (IM.T(exp1) U (IM.T(NOTexp1)=IM.T(exp1)) U (IM.T(exp
- (EX.T(exp1)) U (IM.T(exp3) - (EX.T(NOT(exp1))

exports : (EX.T(exp2)) or (EX.T(exp3))

links : (EX.T(exp1) -> IM.T(exp2)) U EX.T (NOTexp1) -> IM.T(ex
```

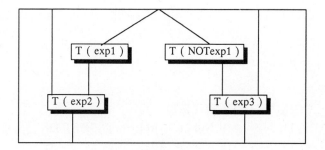

```
operators:

    T(exp1)
      imports: (IM.T(exp1) = EX.T(NOTexp1))
      exports: (EX.T(exp1))

    T(NOTexp1)
      imports: (IM.T(NOTexp1))
      exports: (EX.T(NOTexp1) = IM.T(exp1))

    T(exp2)
      imports: (IM.T(exp2))
      exports: (EX.T(exp2))

    T(exp3)
      imports: (IM.T(exp3))
      exports: (EX.T(ext2))
```

Figure 6.23 T(IF exp1 THEN exp2 IFNOT exp1 THEN exp3).

exports in the execution with the arrival of imports, and the other, used in the implementation of conditional, iterative, and recursive computations, produces and seizes or freezes the exports, depending on the arrival of imports.

In the execution of conditional operators, the data value imported is exported when it satisfies the conditional operator; if it does not satisfy the conditional operator, no export values are produced. The pair Fcond(x) and Fnotcond(x) is a complementary set of conditional operators that can execute concurrently. Here, N implies frozen or seized exports.

$$Fcond(x) = x \text{ if } x \text{ satisfies the condition}$$

$$Fcond(x) = N \text{ if } x \text{ does not satisfy condition}$$

Fnotcond(x) = x if x does not satisfy the condition

Fnotcond(x) = N if x does satisfy condition

3.3.5 DCBL Transformations in Lisp.

The DCBL transformation process can be used in any language to represent data-flow computations. Flow graph languages can be defined using imports or exports. Import-based languages can be used to represent demand-driven computations. Imports and exports show the relationship between data-flow and demand-flow computations. IL is the intermediate form for transformed Lisp languages. IL-1 represents the data-flow computations based on exports from an operator. The format of the codes is

(OPCODE CONSTANT DEST-LIST)
(CALL FUNCTION-NAME NO-OF-ARG NO-OF-RET DEST-LIST)
(PROC FUNCTION-NAME NO-OF-ARG DEST-LIST)

The opcode represents an operation or a function name. Constant type operands are placed in the constant datum field of the operation. DEST-LIST corresponds to the destinations of the result of the operation. A destination field consists of a label field and an attribute field. The label field gives the destination node and the attribute field gives the node attribute. The PROC line specifies a function name, total number of arguments, and destination list of each argument. Figure 6.24a shows a Lisp program and Figure 6.24b shows the IL code, based on exports, for Fibonacci computation, defined by

$$F(1) = 1$$
$$F(2) = 1$$
$$F(n) = F(n - 1) + F(n - 2)$$

Figure 6.24c shows the IL code based on imports to an operator. Instead of exports being defined, the origins of imports are defined. DEST-LIST in IL code is replaced by ORG-LIST, which represents the origins of the imports, to obtain the IL format for import-based computations.

```
Fib n = 1; if n = 1 or 2
   = Fim n-1 + Fib n-2

(defun fibonacci ( n )
( cond (( eq n 1 ) 1 )
(( eq n 2 ) 1 )
( t ( plus ( fibonacci ( difference n 1 ))
( fibonacci ( difference n 2 )))))))
```
(a)

Figure 6.24 Fibonacci—data-flow computing.

```
G0001 ( Procedure FIB 1. ( G0002 MONO-0 )
G0002 ( *DISTRIBUTE ( G0001 MONO-0 ))
G0003 ( *EQ ( C01 1.) ( G0012 ( RETURN 1.)) )
G0004 ( *EQ ( C-1 2.) ( G0006 MONO-0 )
G0005 ( *GT ( C-1 2.) ( G0007 MONO-0 ) ( G0008 MONO-0 )
G0006 ( *CONSTANT ( C-1 1.) ( G0012 ( RETURN 1.)) )
G0007 ( *DIFFERENCE ( C-1 1.) ( G0009 ( ARG 1. 1.)) )
G0008 ( *DIFFERENCE ( C-1 2.) ( G0010 ( ARG 1. 1.)) )
G0009 ( *CALL FIB 1.1. ( G0011 0.) )
G0010 ( *CALL FIB 1.1. ( G0011 1.) )
G0011 ( *PLUS ( G0014 ( RETURN 1.)) )
G0012 ( *RETURN 1.)
END
```
(b)

```
G0001 ( PROCEDURE FIB 1.)
G0002 ( *DISTRIBUTE ( G0001 MONO-0 )
G0003 ( *EQ ( C-1 1.) ( G0002 2.) )
G0004 ( *EQ ( C-1 2.) ( G0002 1.) )
G0005 ( *GT ( C-1 2.) ( G0002 1.) )
G0006 ( *CONSTANT     ( C-1 1.) (G0004 1.)
G0007 ( *DIFFERENCE   ( C-1 1.) ( G0005 1.) )
G0008 ( *DIFFERENCE   ( C-1 2.) (G0005 1.) )
G0009 ( *CALL FIB 1.1. ( G0008 0.) )
G0010 ( *CALL FIB 1.1  (G0008 1.) )
G0011 ( *PLUS ( G0009 1.) (G0010 2.) )
G0012 ( *RETURN ( G0003 1.) ( G0004 2.) G0011 3.)
END
```
(c)

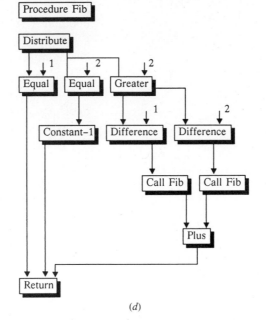

(d)

Figure 6.24 *(Continued)*

239

The non-numerical and numerical operations in IL are car, cdr, cons, add, multiply, subtract, and divide. The data-flow computing support codes are distribution, procedure, call, return, and constant. The first column of Table 6.1 gives the definitions of these operators. The second column gives the definitions of nonstrict operators used in IL. The third column gives the basic definitions of the conditional operations. CONSTANT is used to obtain constants, TRUE, FALSE, or any other required value. IL codes give the Figure 6.31d data-flow graph of the Fibonacci computation. Parallel EQUAL and NOT-EQUAL satisfy the conditional computation requirement. The functionality of DCBL operators for non-numerical, numerical, and conditional operations is as follows; E implies error value export.

List Operations

Fhead ((x1 x2 . . .))	= x1	Fhead (( )) = E
Ftail ((x1 x2 . . .))	= (x2 x3 . . .)	Ftail (( )) = E
Fcons (x1 (x2 . . .))	= (x1 x2 . . .)	

Numerical Operations

Fplus (x y)	= x + y	Fdifference (x y) = minus
Fquotient (x y)	= x/y	Fremainder (x y) = rem x/y
Ftimes (x y)	= x*y	

Conditional Operations

Fnull (( ))	= ( )	Fnotnull (x1 . . .) = (x1 . . .)
Fnull (x1 . . .)	= N	Fnotnull (( )) = N
Fatom ((x))	= (x)	Fatom (x1 . . .) = (x1 . . .)
Fatom ((x1 . . .))	= N	Fnotatom (x) = N
Fnumberp (1)	= 1	Fnotnumberp (1 . . .) = 1
Fnumberp (1 . . .)	= N	Fnotnumberp (1) = N
Fequal (x x)	= x	Fnotequal (x x) = N
Fequal (x y)	= N	Fnotequal (x y) = x

for integers x greater than y

Fgreater (x y)	= y	Fnotgreater (x y) = x
Fgreater (y x)	= N	Fnotgreater (x y) = N

Nonstrict Operators

Fhcons (x1(x2 . . .))	= x1	
Ftcons ((x1, x2 . . .) xn)	= xn	
Fand (. . F . .)	= F	Fand (T, T, T . . . T) = T
For (. . T . .)	= T	Fand (F, F, F . . . F) = T

TABLE 6.1   Basic Definitions—IL.

Common	Nonstrict	Conditional—IL	Conditional—EMIL
1. CAR: First element of an input list	1. HCONS: Availability of first argument gives that value as output	1. ATOM: Atom for atom input, freeze otherwise	1. ATOM: True for atom input, false otherwise
2. CDR: List of input list other than first element		2. NOTATOM: Freeze for atom input, input otherwise	2. NUMBERP: True for integer input, false otherwise
3. CONS: Combine list of two input lists		3. NUMBERP: Integer for integer input, freeze otherwise	3. EQUAL: True for equal inputs, false otherwise
4. PLUS: Addition of two inputs	2. TCONS: Availability of second argument gives that value as the output	4. NOTNUMBERP: Freeze for integer input, input otherwise	
5. DIFFERENCE: Difference of two inputs			4. NULL: True for null input, false otherwise
6. TIMES: Multiplication of two inputs		5. EQUAL: Right input if equal, freeze otherwise	
7. QUOTIENT: Division of one input by another		6. NOTEQUAL: Right input if not equal, freeze otherwise	5. GREATERTHAN: True if right input is greater than left, false otherwise
8. REMAINDER: Remainder of division of two inputs	3. AND: Availability of any argument with the value FALSE gives that value as the output. Otherwise TRUE is the output.	7. NULL: Null for null input, freeze otherwise	6. LESSTHAN: True if right input is less than left, false otherwise
9. DISTRIBUTE: Distributes input		8. NOTNULL: Input for not null input, freeze otherwise	
10. CONSTANT: Constant data when input is received		9. GREATER: Right input if greater than left, freeze otherwise	7. SWITCH-T: Freeze if not true, switch input otherwise
11. PROCEDURE: Defines procedure	4. OR: Availability of any argument with the value TRUE gives that value as the output. Otherwise FALSE is the output.	10. NOTGREATER: Right input if not greater than left, freeze otherwise	8. SWITCH-F: Freeze if not false, switch input otherwise
12. CALL: Calls procedure			
13. RETURN: Returns value of procedure			

4   DATA-FLOW COMPUTING MACHINES

Considerable progress has been made in building data-flow machines to support intelligence computations [52–56]. Data-flow machines have contributed to advances in building parallel systems. Recursive computations are implemented using tags or code copying. Software simulation is the most economical way to verify the effectiveness of the data-flow computing concept and to identify and solve some problems. Real hardware prototypes help to identify and solve hardware problems. Larger programs can be executed at a higher speed on large-scale prototypes with sufficient resources.

4.1 Static Machines

4.1.1 VIM. The Dennis group at MIT introduced the data-flow computing concept and laid the foundation for most other data-flow projects [1–5]. Research and development projects on data-flow computing started in 1968, and a 1 GFLOP VAL interpretive machine, VIM, is being developed. The group's contributions include basic and advanced data-flow computing models, design of data-flow graphs, data-flow computing languages, and computer architecture. Their main objective is to prove the feasibility of static data-flow computing with acknowledgment signals for large-scale numerical computations. Acknowledgment signals provide safe execution of the computation. In static computing, data tokens are stored in an instruction or a copy of the instruction. Each instruction has an operation code, operand values, and destination fields. The nodes of a program are loaded to memory before the computation begins and, at most, one instance of a node is enabled for firing at a time. To activate an instruction, operand fields must be filled and acknowledge signals must arrive. Enabled nodes are detected by associating a counter with each node. Resource allocation decisions are made by the programmer or compiler. Computations that do not contribute to the final result are avoided by demand-driven processing. The system has been implemented as an interpreter on a Lisp machine and an eight-PE multiprocessor prototype. Benchmark programs such as the weather model, the Navier-Stokes problem, and plasma simulation were executed. A larger prototype with 1031 cell blocks, 1031 functional units, and 32 array modules has been proposed.

1. Global configuration: The VIM machine consists of a routing network, cell blocks, functional units, and array memories, as shown in Figure 6.25. Interconnection networks tolerate the latency. Cell blocks store program graphs, operations, operands, and destination addresses of nodes, and recognize the instructions ready for execution. The functional unit performs operations on data values. Array memories store array structures.

2. Processing elements: Functions of PEs are performed by cell blocks and functional units. Simple instructions such as duplicating values and performing tests are executed within the cell block. A PE, with instruction enabling and execution mechanisms, consists of an update unit, an operation unit, a queue, a fetch unit, and an activity store. The activity store holds data-flow instructions. The fetch unit picks addresses of an enabled instruction from the queue, fetches that instruction with its operands from the activity store, and delivers

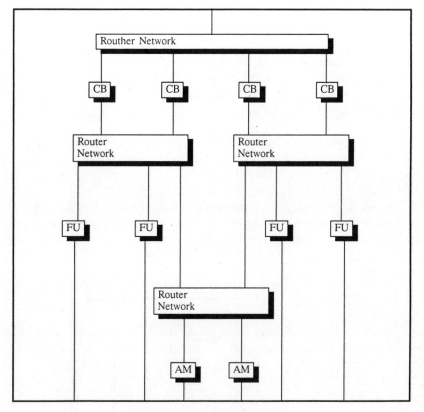

Figure 6.25 VIM architecture.

it to the operation unit. Instruction execution gives result packets, which are sent on to the update unit. The update unit enters the address of the enabled instruction in the FIFO queue. If the target instructions of a result packet reside in some other PE, the packet is sent off through the network. Program graph execution terminates when none of the nodes is enabled. Streams are handled by pipelining. There is no scheduler to assign nodes to a processor. Faults in the machine require restarting the computation from the beginning.

3. Packet formats and instructions: Result packets consist of a result value and a reference. Control packets contain boolean values and control values. Data packets contain integer or complex values. Floating point, fixed point, logical packet communication, and shift instructions are used in the processor instruction set.

4.1.2 *Texas Distributed Data Processor.* The Distributed Data Processor (DDP) was designed by Texas Instruments [6, 7]. The project started in 1976 and the DDP has been in operation since 1978. The main objective is to inves-

tigate the feasibility of static data-flow computing without acknowledgment signals for high-speed computing systems. The DDP uses a strict compound branch node to prevent the initiation of a new iteration before the completion of the previous iteration. The DDP is implemented in transistor-transistor logic (TTL). Each data-flow computer contains 32K words of MOS memory. Ada is used in the four-processor DDP at the computer science department of the University of Southwestern Lousiana.

1. Global configuration: The DDP, as shown in Figure 6.26, consists of four identical data-flow computers to execute programs and a TI 990/10 minicomputer acting as a front-end processor. These computing elements are connected by a DCLN ring.

2. Processing elements: Each data-flow computer consists of an arithmetic unit, which processes executable instructions; a program memory holding data-flow instructions; an update controller, which updates instructions with

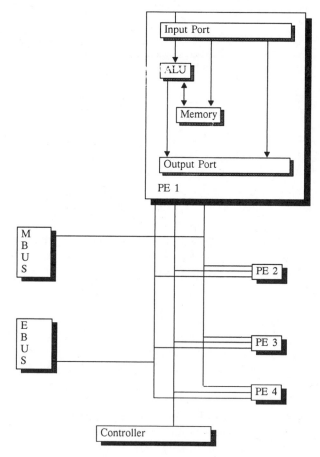

Figure 6.26 Texas distributed processor.

tokens; and a pending executable instruction queue. Each node is associated with a counter. When an instruction completes execution, a series of token packets is released to the update controller, which stores the token operand in the instruction and decrements the count by one. If this count is zero, the instruction is executable and is placed in the instruction queue. Two communication paths are used: one for transmitting instruction packets and result packets, and the other for maintenance and diagnostic purposes. A maintenance controller detects faulty processors. Computations can restart at the preceding checkpoint. A maintenance bus provides communication facilities to monitor the performance of each processor, to load and dump contents of the memory, and to diagnose the faults. The local memory of the processor has an instruction memory and a data memory. Result packets are stored in data memory. Recursive computations are not supported.

3. Packet formats and instructions: Instruction packets use up to fifteen 35-bit words. An instruction can have up to 13 input or 13 output arcs with a total of 14 input and output arcs. Result packets are two words long and contain routing information and data. Monitor call, semaphore instructions, and pipelining are used in implementing streams. Floating-point instructions, fixed-point instructions, logical and shift instructions, loop control, memory fetch, and communication with front-end processor-oriented instructions support FORTRAN IV programs. FETCH, STORE, and MC instructions handle instructions and data.

4.1.3 LAU System. The LAU project started in 1976 at the CERT Laboratory, Toulouse, France [45, 46]. The LAU machine has been in operation since 1979. The group designed the LAU high-level single-assignment language, programmed a large number of problems, and implemented a compiler and detailed simulator.

1. Global configuration: The machine consists of a memory unit, a control unit, and 32 processing units, as shown in Figure 6.27. The memory unit

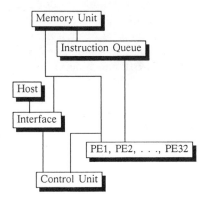

Figure 6.27 LAU system.

stores instructions and data. The control unit maintains the control memory. Six unidirectional buses are used for communication.

2. Processing elements: Each processing element has a 16-bit microprogrammed processor based on AMD 2900 bit slice microprocessors. Execution units read data from central memory. Enabled instructions are kept in a ready instruction queue until results come out of the processor. This helps ensure that instructions are reassigned to a healthy processor. Enabled nodes are detected by associating a counter with each node. The memory unit stores instructions and the data control unit maintains control memory. The von Neumann program counter is replaced by an instruction control memory, which handles instructions, and a data control memory, which handles data.

3. Packet formats and instructions: Each node can have a maximum of two input arcs and several output arcs. The length of instruction and data packets is 64 bits. The LAU system does not handle stream data structures. The instruction set includes fixed-point, logical, shift, and control instructions such as CASE, LOOP, CALL, RETURN, and EXPAND.

4.1.4 NEDIPS.
NEDIPS and IPP were the first commercially available data-flow processors [64]. They are special purpose data-flow processors with static architecture, well tuned to image processing applications developed by the Nippon Electric Co. NEDIPS is a 32-bit machine for scientific computation and uses high-speed logic. The IPP (Image Pipelined Processor) is a single-chip processor of similar architecture. This processor is a building block for highly parallel image processing systems. Special mechanisms are used to implement multiple tokens per arc. Special hardware operations are provided for generating, splitting, and merging streams of tokens.

4.2 Dynamic Machines

4.2.1 MIT Tagged-Token Data-Flow Machine.
The Irvine data-flow project started in 1975 at the University of California at Irvine and is being continued at MIT by Arvind's group [8–11]. The major contributions include the tagged-token dynamic computing model, I-structures, the Id language, and the computer architecture. The main objective is to exploit VLSI and provide highly concurrent program organization. A 32-PE machine using Symbolic Lisp machines is being constructed. A 256-board 1-BIP machine is under construction.

1. Global configuration: This asynchronous machine has 64 processing elements connected via an n-cube communication network. The organization minimizes communication overhead by matching at the processing element holding the storage instruction and bypassing the network to the processor itself.

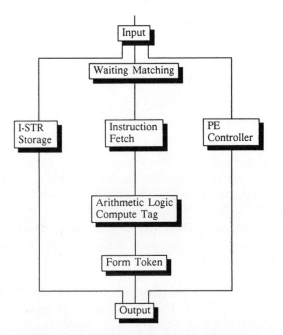

Figure 6.28   MIT tagged-token data-flow PE.

2. Processing elements: A PE consists of input, waiting matching, instruction fetch, service, and output sections, as shown in Figure 6.28. The input section accepts inputs from other processing elements, the waiting matching section forms data tokens into sets for one instruction, the instruction fetch section fetches executable instructions from local program memory, and the output section routes data tokens containing results to the destination PE. Enabled nodes are detected using tags carrying the information of the node. Tagged stream elements are processed in parallel using multiple instances, one for each element. Program memory stores instruction codes. The data memory, an I-structure memory, stores arrays. Waiting matching storage, an associative memory, matches or stores the incoming tokens. The allocation of memory and tags is controlled by a manager. Recursive computations are supported by tagged tokens. Processing units asynchronously evaluate the executable instruction packets. Faults require computations to be restored at the previous checkpoint.

3. Packet formats and instructions: There is a maximum of two input tokens and several output tokens per node. Thirty-two enabled nodes can wait for the ALU. The instruction set includes floating-point, fixed-point, and logical instructions. The instruction and data packet lengths are 33 and 71 bits.

4.2.2 Manchester Dataflow Computer. The Manchester project was started in 1975 by the Gurd-Watson group [12–15] at Manchester University. The group's main objective is to investigate the use of tagged-token data-flow computing for very high speed dynamic computing systems. They completed the construction of a 20-processor, strongly typed tagged data-flow machine in 1980 using Schottky bit slice microprocessors. Their contributions include the tagged-token dynamic computing model, several high-level data-flow computing languages, and the data-flow machine. The reported performance of the machine is approximately 1.6 MIPS.

1. Global configuration: The machine consists of a switch, a token queue, a matching unit, an instruction unit, and a processing unit (Figure 6.29). A switch provides input and output for the system. The token queue is the FIFO buffer providing temporary storage for tokens. The matching unit matches pairs of tokens, employing hardware hashing. The instruction store holds data-flow programs, and PEs execute instructions.

2. Processing elements: There are 15 functional units in each processor. One enabled node is assigned to each functional unit. The node store supplies enabled nodes to the processing unit. Enabled nodes are assigned to the functional units using any hardware distributor. Therefore, there are no multiple assignments to a functional unit. The matching unit can hold 16K units and employs dynamic hashing. The PE consists of distribution and arbitration systems and a group of microprogrammed microprocessors. Streams are processed in parallel using multiple instances, one for each element. Recursion computations are supported using tags.

3. Packet formats and instructions: The instruction set supports floating-point, fixed-point, data branch, token label, flow control, and token-relabeling instructions. A maximum of two input arcs and two output arcs is allocated to an operation. The lengths of instruction and data packets are 167 and 96 bits.

4.2.3 DDM1. The Data Driven Machine project [16, 17] was started in 1975 by Davis's group at Burroughs Interactive Research Center. Construction

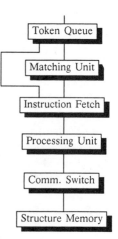

Figure 6.29 Manchester machine.

was completed in 1976, and the machine is now at the Utah University. This dynamic machine employs FIFO queues instead of tagged tokens to distinguish computations. Program execution and machine organization are based on recursion. The DDM1 is in operation and has been used to study basic issues in data flow. The DEC20/40 is used for software support. The graphs are generated from the high-level functional language GPL. Parenthesized strings in data-flow programs provide localized dynamic computing.

1. Global configuration: The machine is composed of an octary tree hierarchy of computing elements. This hierarchy exploits VLSI, utilizing the locality of reference to reduce the communication and control problems.

2. Processing elements: A PE consists of an atomic storage unit, an atomic processor, an agenda queue, an input queue, an output queue, and a switch (Figure 6.30). The atomic storage unit, a 4K–by–4-bit character store, is the program memory. The atomic processor is the execution unit. The agenda queue is the message store for the local atomic storage unit. The input queue is the buffer for the messages from the superior element. The output queue is the buffer for the messages to the superior element. The switch connects to eight computer elements. The tree structure inhibits immediate rerouting of the results before a fault.

3. Packet formats and instructions: Data tokens provide all communications. Each instruction is represented as a variable-length instruction packet. Each instruction has an enabling counter for input arcs. An instruction can have any number of input and output arcs. Streams are handled by pipelining the tokens. The atomic processor processes integer-oriented, logical, indexed read and write, and relational operator-oriented instructions.

4.2.4 SIGMA-1. The SIGMA-1 project [48–51] was started in 1982 by Yuba's group at the Electrotechnical Laboratory (ETL). Their main objective is to develop a large-scale tagged-token dynamic data-flow machine with 100 MFLOPS performance for scientific and technological computations. SIGMA-1 uses a C-like high-level data-flow computing language, DFC (data-flow C), and SAS intermediate language to describe the data-flow graphs. A preliminary version of the PE and structure element (SE) using advanced Schottky TTL

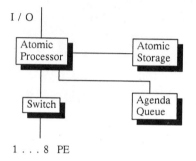

I / O

1 . . . 8 PE

Figure 6.30 DDM1.

logic and MOS memories has been in operation, with 1.3 MIPS, since November 1984. The final version of a single group that uses gate-array LSI chips is in operation now. The full hardware configuration with the total predicted performance of 100 MFLOPS is in operation now.

1. Global configuration: The SIGMA-1, as shown in Figure 6.31a, has 128 PEs and 128 SEs, which are divided into 32 groups connected by a two-level hierarchical network. This hierarchy corresponds to parallel execution of iterations and procedure calls that appear frequently in numerical computations. A single group consists of four PEs and four SEs connected by a 10-by-10 crossbar switch. The remaining two ports of the switch are used for the interfaces to the global network and the maintenance architecture. The global network is a two-stage omega network.

2. Processing elements: A PE, as shown in Figure 6.31b, consists of several functional units, each of which works synchronously and constitutes a two-stage pipeline. A chained hashing hardware with 64K cells is used as the matching memory unit. Each PE consists of about 81K logic gates, using nine types of 28 gate-array LSIs. An SE controls array structures, allowing single write and multiple read operations. It is implemented by memory of 256K cells, where each cell is attached with a waiting queue for asynchronous access control.

3. Packet formats and instructions: The data transfer between PEs and SEs is in fixed-length packet form. A packet consists of the PE or SE number (8 bits), the cancel bit, the destination identifier (28 bits), the tagged data (40 bits), and miscellaneous information (12 bits). The length of the instruction is 40 bits in a primitive format. The first 20 bits indicate the operation to be performed, and the next 20 bits indicate the destination address of the result

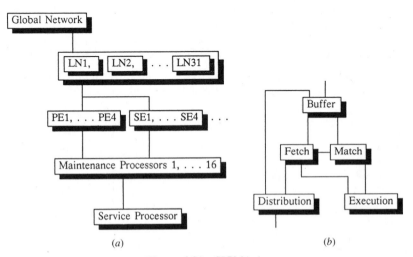

Figure 6.31   SIGMA-1.

data. It is possible to allocate a maximum of three different destinations to an instruction.

4.2.5 EM-3. The EM-3 project, started by Yuba's group at ETL [27–33] in 1982, is scheduled to continue until the end of 1987. The objectives of the project include evaluating the effectiveness of pseudoresult data-flow computing for symbolic manipulations, implementing new parallel architectures, and evaluating the performance of a hardware simulator by executing application programs. The 8-PE prototype started operation in 1984, and the 16-PE organization has been in operation since 1985. The EM-3 is used to implement new parallel control mechanisms. The maximum performance of the hardware is about 10 MIPS. An advanced version of the EM-3 that will be a more practical data-flow computer prototype is being developed.

1. Global configuration: Sixteen identical PEs are connected via a packet communication network. There is no locality in the network. The router network is adapted for communication; a special gate-array LSI chip has been developed for this purpose. The LSI chip is a 4-bit slice four-by-four router and the transfer rate of a packet through the network is 150 nanoseconds.

2. Processing elements: Each PE is constructed using an MC68000 microprocessor with special hardware; see Figure 6.32. Almost all the functions, including the function evaluation mechanism, are performed sequentially within the PE. The MC68000, the packet memory control unit used as the network interface, and the I/O interface to the host computer are connected by a common bus. Each PE comprises three boards excluding the interface to the host computer PDP-11/44 and the network

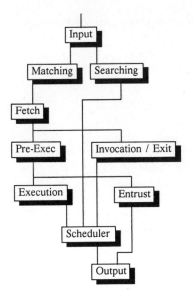

Figure 6.32 EM-3 PE.

boards. Packet memory is accessed from the microprocessor, and each packet is represented as a pointer to packet memory. Hence, there is no overhead in moving packets in a PE.

3. Packet formats and instructions: The 96-bit result packet carries output data of an operation. A result packet consists of the PE number (8 bits), the type-of-packet field (4 bits), the packet length (4 bits), the destination identifier (48 bits), and tagged data (32 bits). The packet is divided into six 16-bit segments in the network. The length of an instruction is 48 bits, comprising the 32-bit destination field and the 16-bit operation field. The immediate data (32 bits) can be contained, and the number of destination fields within an instruction is not fixed.

4.2.6 *EDDY.*

Amamiya et al. [22] at Nippon Telegraph and Telephone Corporation (NTT) started research and development on data-flow machines in 1980. A data-flow processor array system for scientific and technological computations, EDDY, was set up as a prototype in 1983. High speed was achieved by adapting the operational characteristics of scientific and technological computations to the machine architecture at a hardware level. Application programs were written in VALID. The machine exploited parallelism inherent in the application programs, and its performance was not sensitive to inter-PE communication delay or to load imbalance.

1. Global configuration: 16 PEs are connected in a four-by-four cellular array structure. Each PE connects directly to eight neighboring PEs. There are two broadcast control units for loading programs and data to each PE. These are located at the interfaces between the host computer, a PDP-11/60, and a set of PEs.

2. Processing elements: Each PE is constructed using two Z8000 microprocessors. One controls communication and the other controls the data flow and execution of instructions. The tagged-token concept is applied for function invocation and iteration handling. Each array element in a program is given a unique identifier, and all elements are processed in parallel. Each PE works logically as a circular pipeline, but practically, each functional unit within a PE operates sequentially.

3. Packet formats and instructions: A data packet consists of the identifier (color), the destination field, and the value field. An identifier comprises the array element name, the instantiation name, and the loop count. An instantiation name corresponds to a procedure instance name and is statically determined at compiling time by caller-callee analysis. The array elements are also statically allocated to the PEs according to a specific mapping strategy. The instruction contains almost the same information as the data packet except for an operation code.

4.2.7 *DFM.*

Amamiya et al. [21–26] at NTT started the DFM project in 1982. Their main objective is to develop a data-flow machine for symbolic manipulations [19–25] using lenient and lazy cons mechanisms. In 1985, the construction of the DFM-II was started using CMOS gate-array technology. Parallel processing on the DFM is realized by parallel evaluation of function arguments, partial execution of a function body, and pipeline processing of a delayed evaluation scheme. The two-PE version of the DFM has been in operation since the beginning of 1986. The basic cycle of a PE is 180 nanoseconds, and the maximum speed is about 1.4 MIPS per PE.

1. Global configuration: Several clusters were connected via a network, as shown in Figure 6.33. A cluster consists of eight PEs and eight structure memories connected by multiple buses. Structure memories are separate from PEs for efficient list processing. Each cluster is supervised by a cluster control unit. The cluster control unit controls the load balancing among PEs within the cluster and communicates with other clusters via the network or the host computer. The two-level network is based on clustering. The load distribution is within a cluster, and function distribution is among clusters. A blocked content-addressable memory scheme is applied to reduce the amount of hardware.

2. Processing elements: Each PE is composed of an instruction memory, an operand memory, and an execution unit. The matching unit contains content-addressable memory for each function activation. These units work as a circular pipeline. A hardware queue is placed at the entrance of the instruction memory to ease packet traffic in the circular pipeline. Each structure memory is constructed from multiple memory banks equipped with the list operation unit. Each cell of the structure memory is composed of the cell type field (one bit), the reference count field (9 bits), the CAR field, and the CDR field (23 bits each).

3. Packet formats and instructions: The size of a result packet is 56 bits and its contents are the destination identifier with the function name (31 bits) and data (32 bits). There are instructions to the cluster control unit and to the structure memory as well as to the execution unit. An instruction to the

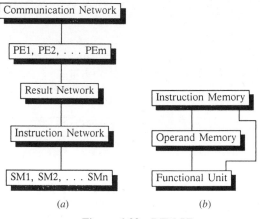

(a) (b)

Figure 6.33 DFM PE.

execution unit consists of the operation field (8 bits), two operand-fields (32 bits each), and the destination identifier. An instruction to the structure memory is 90 bits and is associated with the 3-bit PE number.

4.2.8 PIM-D. Itoh et al. [58] at ICOT began the research and development of a data-flow Prolog machine in the middle of 1982. The objective of ICOT is to develop all computer-related technology from the viewpoint of predicate logic. Data-flow architecture, logic programming, and natural language understanding are three research directions identified for this paradigm. Three different types of architecture—the PIM-D, a parallel reduction machine, and a parallel inference machine with an efficient task distribution mechanism—were studied to handle the highly parallel processing problems in logic programs. The PIM-D employs breadth-first search. To avoid the deadlock problem caused by the number of processes, each process is associated with execution priority. The eight-PE PIM-D is in operation now. An LSI implementation is being developed.

1. Global configuration: The machine consists of 16 PEs, 15 structure memories (SMs), and a three-level hierarchical network, implemented by a 113-bit bus, as shown in Figure 6.34. The PEs and SMs are divided into four clusters, each of which consists of four PEs and four SMs, except for one cluster. Each bus is connected by the network node with a 128-packet buffer. The minimum transmission time is 450 nanoseconds per packet.

2. Processing elements: A PE is composed of a packet queue, an instruction control unit, and two atomic processing units for execution, which are also connected via a bus. The instruction control unit serves as the matching function of data-flow control. Each hardware unit is constructed using bit-sliced microprogrammable processors and TTL ICs.

3. Data formats: A packet transferred between PEs via a bus consists of the PE/SM number (5 bits), the packet type (9 bits), the packet color (16 bits), the destination identifier (31 bits), and the operand data (32 bits). The length of the instruction is 59 bits. A cell of an SM is composed of data (32×2 bits), the type flag (2×2 bits), and the reference count area (10 bits).

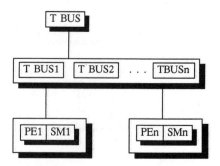

Figure 6.34 PIM-D architecture.

4.2.9 **TOPSTAR.** The TOPSTAR [59] is a macro data-flow machine, developed from 1978 to 1982 by Suzuki et al. at the University of Tokyo [59] to support the recognition of printed Chinese character patterns. The TOPSTAR-I, composed of three processing modules (PMs) and two communication and control modules (CMs), is the prototype of the more advanced TOPSTAR-II. The TOPSTAR-II was easily expandable by plugging in additional modules. Data buffers cause a pipeline effect. Using TOPSTAR-II as a testbed, some experimental studies, such as the data-flow Lisp compiler, logic simulation, and parallel Prolog implementation, were carried out. This led to the development of the parallel Prolog machine called the PIE.

1. Global configuration: Sixteen PMs and eight CMs are organized in a bipartite graph. Each PM is connected to a maximum of four CMs, while each CM is connected to a maximum of eight PMs. There are no direct paths between PMs or between CMs, but there are indirect paths through modules of the other type. A procedure-level data-flow graph is dynamically mapped into the PM-CM connection network. Each PM interrupts one of the connected CMs and requests a task. Each CM contains allocated procedures of an execution program, and if executable tasks exist, their instances as well as their argument data are sent to the PM that made the request.

2. Processing units: Each PM or CM is constructed using a Z-80 microprocessor and a direct memory access (DMA) controller. The communication between PM and CM is through the DMA system at high speed, because the data block is transferred when a new instance of a procedure is needed at an allocated PM. The CM's communication memory is shared with each PM and contains execution programs and their argument data.

3. Data formats: The data packet has a variable length and consists of the serial number, the field indicating the stack depth, the destination addresses, and the procedure instance. The serial number corresponds to a color, and the stack depth is used for recording the history of the data passed. The data format supports the implementation of the control mechanism of iteration and recursion.

4.3 Other Projects

In addition to the projects mentioned above, there are many other data-flow research projects. These include projects at the University of Southern California [47]; Hughes Aircraft Company; University of Adelaide [65]; University of New South Wales; Keio University, Japan [31, 57]; Osaka University [62]; Gunma University [57]; Tokyo University [61]; and Indian Institute of Technology.

4.4 Problems in Data-Flow Computing Machines

4.4.1 **Matching Bottleneck.** A data-flow processing element (Figure 6.35) basically consists of a matching unit, an execution unit, and a distribution

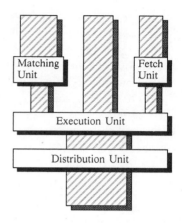

Figure 6.35 Data-flow PE.

unit. The execution unit performs the execution and structure handling. The matching unit sequentially matches and synchronizes the operands of double-operand operations for execution. No matching is necessary for single-operand operations. The larger the number of double-operand operations, the more matching to be performed in the matching unit. This narrows the pipeline between the matching unit and execution unit and reduces the computing speed. This is called the matching bottleneck.

4.4.2 Remaining Packet Garbage. The large number of unexecuted packets waiting in the matching unit after completing the execution of computation is called remaining packet garbage, or RPG. RPG is the result of vertical branches created by conditional computations and multi-argument functions. A conditional computation divides the data-flow computing into two vertical branches. Execution makes one branch *live* and the other *dead*. Data flow to the operations ignores the liveness of the branch. Loading only one operand of a double-operand operation in the dead branch results in RPG.

The multiple arguments in a function divide the data-flow computation into live vertical branches. RPG is created when a dead branch of a conditional computation in one vertical branch receives the data values from the same vertical branch and/or from some other vertical branch.

4.4.3 Control of Parallelism. Data-flow computations have huge parallelism, many times larger than the parallelism available in the hardware. Such computations tend to use excessive amounts of storage, since many partial results are created long before resources are available to process them. It is necessary to restrict excess program parallelism to match machine parallelism.

The tokens generated must be dispatched to the corresponding nodes. The more tokens generated, the more the communication delay. The parallelism can be used efficiently to hide latency. Balanced, light load distribution among PEs increases the performance of the system and utilization of the resources. Therefore, efficient techniques to reduce unnecessary token generation and efficient communication are necessary.

4.4.4   Sequential Computing Segments.   A sequential computing segment is a program segment in which the maximum parallelism is less than the number of processors and/or pipeline stages. Such computations are involved in conditional computations, recursive procedure calls, and iterative computations that use previously computed results to continue computation. The performance of the sequential computing segments in a program is important in parallel computing.

4.4.5   Parallel Execution of Conditional Computations.   Communication control systems and production systems consist of large numbers of parallel conditional computations. Non-determinate computations such as guarded expressions can be implemented in parallel with the use of the simple switch-t operation. A guarded condition evaluated as true enables the successive computations to return the value of the guarded expression. The switch operation with two outputs cannot be used without additional control mechanisms.

4.4.6   Optimization of Data-Flow Computations.   In functional languages, different occurrences of an expression always yield the same value. Loop-invariant expressions produce the same value on each pass of the loop. The value produced by such a common computation (CC) is evaluated once and can be used at all occurrences of the expression. Reusing CCs within conditional expressions saves computational effort. Detecting and arranging CCs in conditional expressions to optimize program execution in data-flow machines is a complicated process. Identifying nonstrict operators or functions and not evaluating the arguments that will not contribute to the final result of the computation optimizes data-flow computations. Such nonstrict operators are identified and added to enhance the efficiency of data-flow computing systems.

5   PERFORMANCE EVALUATION USING THE EM-3

5.1   EM-3 Operational Model

The functional configuration of the EM-3 data-flow computing element is shown in Figure 6.32. The result packets received at the input section are checked, and the packets corresponding to single-operand operations are sent to the operation fetch section. The result packets corresponding to double-operand operators or to functions of more than one argument are sent to the operand-matching section. This section matches and synchronizes the packets. The unmatched packets awaiting their partners are stored in the matching store and searched for when necessary. If the arrived result packet finds its partner in the matching store, both are removed from the store and sent to the instruction fetch section.

The program store is attached to the instruction fetch section and stores the program to be executed. The operation fetch section fetches operations from the program store according to the operation addresses and combines them with their operands to generate internal execution packets.

The invocation section is activated when a call operation is fetched at the operation fetch section. A call operation invokes a defined function. The result store, simulating the pseudoresult control mechanism, is handled by the search, invocation, exit, and execution sections. This store consists of a result table, a deferred buffer, and storage for list cells. The result table manages a set of pseudoresults. Each entry consists of tags and a result value. The deferred buffer is storage for entrust packets until the pseudoresults become actual. The list cell storage is for list cells created by the cons operation. The storage management of the result table and the deferred buffer is carried out by the reference count garbage collection scheme. A pseudoresult identifier is created for the newly invoked function and an invocation packet is generated. This packet is sent to the PE scheduler section.

The initiation section accepts invoke packets and extracts a function name. Its arguments are placed in the packets and generate result packets corresponding to each argument. The body of the function is activated by these result packets.

The search section is activated by an entrust packet that is associated with a pseudoresult identifier. The pseudo result table is searched using the pseudoresult identifier for the actual result. If found, the entrust packet is sent to the execution section. If not found, it is stored at the deferred buffer and waits for the completion of the predecessor operation assigned by the pseudoresult identifier.

The exit section stores the values of actual results or pointers to semiresults that correspond to each pseudoresult of the function and are stored in the pseudoresult table at the exit of each section. If the activated entrust packets corresponding to the exit operation and waiting for completion of the function are executable, they are sent to the execution section; otherwise they are sent to the entrust section.

The entrust section generates entrust packets and defers the execution of the operation when input packets include a pseudoresult. The generated entrust packets are sent to the PE assigned by the pseudoresult identifier. The scheduler section decides the destination PE by using a hashing function. The pre-execution section examines the operands of an execute packet. If there is a pseudoresult in an operand, the packet is sent to the entrust section, otherwise it is sent to the execution section. The output section sends external packets through the communication network to the corresponding destination PEs.

5.2 EMIL

EMLISP is the high-level language and EMIL is the low-level language used in the EM-3. The fourth column of Table 6.1 (see p. 241) shows the basic definitions of EMIL codes. NULL, NUMBERP, ATOM, LESSTHAN, EQUAL, and GREATERTHAN are conditional operations. SWITCH-T and SWITCH-F operations are executed with all the conditional operations. Distribute, procedure, constant, call, and return are data-flow computing support codes. The functionality of EMIL operations is as follows. Here, E, T, and F stand for error, true, and false, and integer x is greater than integer y.

List Operations

car ((x1 x2 . . .)) = x1	car(( )) = E
cdr ((x1 x2 . . .)) = (x2 x3 . . .)	cdr(( )) = E
cons (x1 (x2 . . .)) = (x1 x2 . . .)	

Numerical Operations

Fplus (x y) = x + y	Fdifference = x − y
Fquotient (x y) = x/y	Fremainder = rem x/y
Ftimes (x y) = x * y	

Attribute Checking

atom (a) = T	atom (x1 x2 . . .) = F
numberp (1) = T	numberp (1 . . .) = F
null (( )) = T	null (x1 . . .) = F
equal (x x) = T	null (x1 . . .) = F
greaterthan (x y) = T	greaterthan (y x) = F
lessthan (y x) = T	lessthan (x y) = F

5.3 Performance Evaluation Measurements

This section discusses the experimental results obtained by executing benchmark programs on the EM-3 [28–34]. The software simulator, written in SIMULA, describes the EM-3 data-flow computing environment, which interprets and executes the EMIL code that describes the data-flow computation. The Fibonacci function (F(13)), the Ackermann function (AK(2 9)), sequential and parallel versions of the n-queen problem (4QS and 4QP), a quicksort algorithm with maximum parallel data (QUI), same-fringe (SF), and copy (CP) are some of the non-numerical and numerical computations performed on the EM-3.

5.3.1 Effectiveness of Pseudoresult Model.

Figure 6.36 shows performance measurements for SF with and without pseudoresults. When not using pseudoresults, the execution of CONS is deferred until the operands become actual data values. In this case, pseudoresults are generated but never used in any instruction. The execution time difference in a single-PE configuration is due to the entrust packet overhead. The number of entrust packets used without pseudoresults is five times greater than with pseudoresults. Data-flow parallelism in SF is very small without pseudoresults, and the performance is not improved with the increase of PEs. The pseudoresult data-flow computing model revealed the hidden parallelism in Lisp languages and accelerated the program execution in the parallel computing environment.

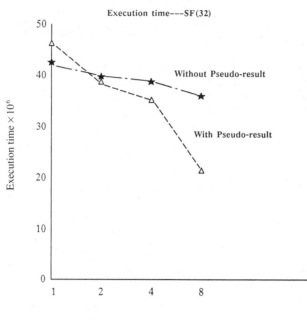

Figure 6.36 Execution time variation for SF(32).

5.3.2 Effectiveness of the Not(operation) Model.

The effectiveness of the not(operation) model is observed by comparing performance characteristics measurements. The timing parameters used for operation executions, shown in Table 6.2, are larger in the not(operation) model than the traditional model. Figures 6.37a and 6.37b show EMIL code and the corresponding data-flow diagram for computing Fibonacci numbers. Figure 6.31d represents not(operation)-based data-flow computing of Fibonacci numbers.

Figure 6.38a shows the ideal data-flow parallelism in F(13) with the traditional computing model. Here, the maximum parallelism, 495 concurrent operations, is observed at the 49th of 85 steps. Figure 6.38b shows the ideal data-flow parallelism in the not(operation) model. Here, the maximum parallelism, 261 concurrent operations, is observed at the 34th of 60 steps. This demonstrates the increase in real data-flow parallelism and the removal of a large number of unnecessary computations with pseudoparallelism, and hence the reduction in data-flow computing cost.

Table 6.2 shows the frequency of operations executed in each benchmark program. SWITCH operations account for a large percentage of all operations executed.

Figure 6.39 shows F(13) execution times in the traditional and not(operation) models for varying numbers of EM-3 processing elements. The shape of

TABLE 6.2   Execution Frequency of Operations.

Operation	Benchmark Program						
	A(29)	4QS	4QP	F(13)	QUI	SAF	CP
CAR	0	124	296	0	671	702	189
CDR	0	44	104	0	482	702	189
CONS	0	24	48	0	482	640	189
PLUS	110	125	308	232	0	0	0
DIFFERENCE	229	44	100	464	0	0	0
TIMES	0	0	0	0	0	0	0
DIVIDE	0	0	0	0	0	0	0
SWITCH-F	1060	958	2488	2434	2246	2330	763
SWITCH-T	820	383	900	841	2115	1052	381
ATOM	0	0	0	0	0	126	381
NULL	0	39	90	0	483	0	0
EQL	350	162	390	841	0	736	0
GREATERP	0	0	0	0	0	0	0
LESSP	0	0	0	0	189	0	0
CALL	229	101	241	464	522	798	382
DISTRIBUTE	460	366	875	465	1383	1469	381
PRINT	0	0	1	2	0	0	0
CONSTANT	0	0	3	3	0	0	0

```
( PROCEDURE FIB 1. ( G0002 MONO-0 ) ) G0002 ( *DISTRIBUTE ( G0003
MONO-0 ) ( G0006 DATA ) ( G0010 DATA ) ( G0014 DATA ) )
G0003 ( *EQ ( C-1 1. ) ( G0004 MONO-CONTROL ) ( G0006 CONTROL )
(G0007 MONO-CONTROL ) )
G0004 ( *SWITCH-T ( C-0 1. ) ( G0018 ( RETURN 1. )) )
G0006 ( *SWITCH-F ( G0005 0. ) )
G0007 ( *SWITCH-F ( C-0 2. ) ( G0005 0. ) )
G0005 ( *EQ ( G0008 MONO-CONTROL ) ( G0010 CONTROL ) ( G0011
MONO-CONTROL ) ( G0014 CONTROL ) ( G0015 MONO-CONTROL ) )
G0008 ( *SWITCH-T ( C-0 1. ) ( G0018 ( RETURN 1. )) )
G0010 ( *SWITCH-F ( G0009 0. ) )
G0011 ( *SWITCH-F ( C-0 1. ) ( G0009 1. )) )
G0009 ( *DIFFERENCE ( G0012 ( ARG 1. 1. )) )
G0012 ( *CALL FIB 1.1. ( G0017 0. ) )
G0014 ( *SWITCH-F ( G0013 0. ) )
G0015 ( *SWITCH-F ( C-0 2. ) ( G0013 1. ) )
G0013 ( *DIFFERENCE ( G0016 ( ARG 1. 1. )) )
G0016 ( *CALL FIB 1.1. ( G0017 1. ) )
G0017 ( *PLUS ( G0018 ( RETURN 1. )) )
G0018 ( *RETURN 1. )
END
```

(a)

Figure 6.37   Fibonacci—traditional.

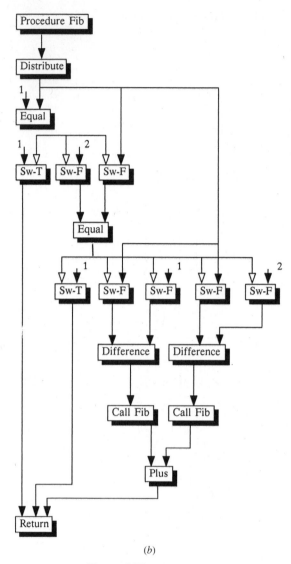

(b)

Figure 6.37 (*Continued*)

the graph does not change, but the computing speed approximately doubles in the not(operation) model. The reasons for the speed increase include reduction of double-operand operations, parallelization of sequential computing segments, balanced pipeline stages, and reduction of token generation and traffic. In traditional data-flow computing, each single-operand conditional operation must execute two or more additional double-operand operators and must cre-

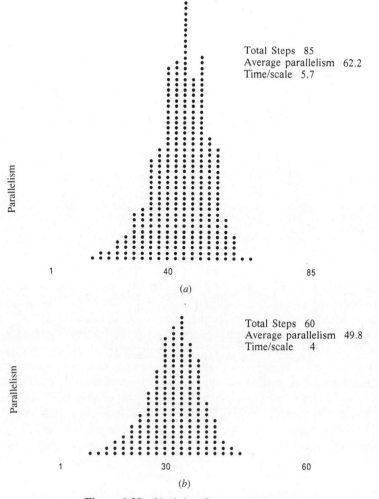

Total Steps 85
Average parallelism 62.2
Time/scale 5.7

Parallelism

1   40   85

(a)

Total Steps 60
Average parallelism 49.8
Time/scale   4

Parallelism

1   30   60

(b)

Figure 6.38   Ideal data-flow parallelism.

ate many packets to support intermediate executions. These packets contribute much to the congestion. Not(operation)-based single-operand conditional computation eliminates matching. Tokens are executed directly in the execution unit. In traditional data-flow computing, each double-operand conditional operation must execute two or more additional double-operand operators and must create many packets to support intermediate executions. Not(operation)-based double-operand conditional computation matches and executes directly to give the result immediately. Unnecessary packet creation is eliminated, thereby reducing congestion.

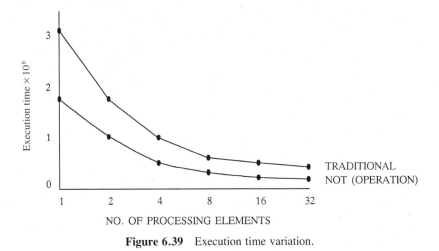

Figure 6.39 Execution time variation.

Figure 6.40 shows the average waiting time variation in the matching section in each PE of the 32-PE EM-3. The not(operation) model reduces the waiting time to one tenth that of the traditional model. This is due to the reduction of a large number of double-operand operations to be executed.

Table 6.3 compares the single-operand packets and double-operand packets generated in the execution of F(13). The number of single-operand packets generated decreased from 8673 to 4148. The number of result packets entering the matching section, double-operand operations, decreased from 4756 to 464. The not(operation) model significantly reduces the number of single-operand packets, double-operand packets, and operations executed. This results in a proportionate reduction of computing and communication costs and an increase in computing speed.

Table 6.4 shows the reduction in the number of operations executed. Table 6.5 shows the maximum number of packets waiting in the queue, and the congestion of each functional unit of the EM-3 at the busiest instance.

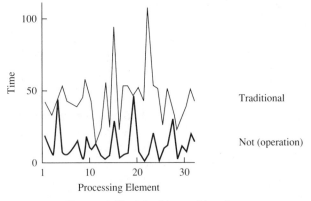

Figure 6.40 Matching waiting time.

TABLE 6.3 Packets Generated.

Packet type	Traditional	Not(op)
One-operand	8673	4148
Two-operand	4756	464

Figure 6.41 shows the maximum number of packets waiting in the queue to the matching section in each PE of the EM-3 for varying numbers of PEs. The maximum number of packets in a single-PE EM-3 with traditional computing is very high compared to the not(operation) model. The more tokens that are generated, the greater the communication delay and cost and waiting time in the queues. The not(operation) computing model provides an efficient way of reducing tokens generated and hence reduces the token traffic.

No remaining packet garbage is collected in the matching store when executing F(13) with the not(operation) model. The model completely stops the flow of data from the predecessor of the conditional computation to the dead branch. To remove RPG completely, an efficient garbage-collecting mechanism must be implemented, which may, for example, dispatch a special packet to execute all the operations in the dead branch, use a control operation to stop the flow of data into the dead branch, or set the life of double-operand packets.

The not(operation) computing model provides an efficient way of implementing parallel conditional computations. The optimization will reduce the number of arithmetic and conditional operations, the size of the dataflow graph, execution time, parallelism, total number of tokens, and token traffic.

TABLE 6.4 Operations Executed.

Operation	Traditional	Not(op)
Call/return	928	928
Distribute	465	465
Constant	—	144
Switch-t	841	—
Switch-f	2434	—
Notequal	—	841
Equal	841	841
Plus	232	232
Difference	464	464

TABLE 6.5   Packet Queue.

Section	Traditional	Not(OP)
input	1	1
match	355	1
ifetch	4	3
preexec	2	2
exec	10	218
invo	1	1
exit	1	1
init	1	1
search	1	1

6   CONCLUSIONS

The traditional computing model, languages, and architecture have not changed very much over the last 30 years, but applications of new computing models, languages, and machines to numerical and non-numerical computations show promise. The market is responding to the availability of such machines for intelligence computations. Intelligence computing research is highly dependent on high-performance low-cost parallel computing systems. Researchers

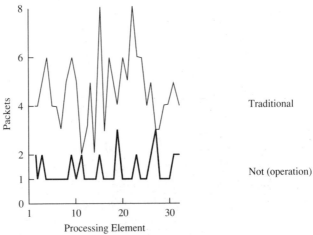

Figure 6.41   Maximum packets—matching queue.

are examining radically different approaches. The data-flow computing concept is the most effective, promising computing method to implement in machine architecture for high-speed computing. This chapter first analyzed the data-flow computing models, such as static computing with and without acknowledgment signals, recursive dynamic, tagged-token dynamic, eduction, data-flow–control flow, eager-lazy, pseudoresult, and not(operation).

Research on functional and logic programming languages and their applications to intelligence computations will revolutionize the computer paradigm. Logic programming allows high-level program specifications without explicit control directives. Functional languages free programmers from thinking in terms of storage. Functional programming transforms objects to other objects without naming. The functional languages designed to map algorithms into data-flow computing, such as VAL, Id, LUCID, VALID, DFC, and EMLISP were discussed. The DCBL transformation for data-flow computing and its application to Lisp were discussed.

The major difficulty in very high speed data-flow computing is the highly tuned, widely available, and familiar von Neumann machines. The data-flow machines VIM, DDP, LAU, NEDIPS, MIT TTDA, Manchester Dataflow Machine, DDM1, SIGMA-1, EM-3, EDDY, DFM, PIM-D, and TOPSTAR were discussed. The SIGMA-1, which is predicted to have the highest computing speeds, is a milestone in computing machines. Some general problems in data-flow computing and performance evaluation measurements made in the EM-3 data-flow computing environment were presented.

6.1 Further Research

Considerable progress has been made in building data-flow machines during the last few years. Further research is needed in the following areas:

1. Different data-flow computing models
2. High-level languages suited to data-flow programming
3. Low-level languages suited to data-flow architecture
4. Algorithms for specific applications and systems
5. Data-flow computing processors and system architectures
6. Operating systems with efficient resource allocation schemes
7. Optimum design of an instruction set processor
8. High-speed operand-matching mechanisms
9. Efficient structure memory implementations
10. Low-cost high-speed communication networks to interconnect PEs
11. Problems in data-flow computing
12. Fault-tolerant computing for 1000- to 10,000-processor data-flow computing machines

13. Impact of VLSI and device technology
14. Applications of the data-flow computing concept in other fields

7 ACKNOWLEDGMENTS

We wish to thank all the researchers in parallel computing for their efforts and contributions. We would also like to thank Professor Masahiro Ishii and Shuichi Itoh of University of Electrocommunications for their support in this research. We wish to acknowledge the invaluable support of and discussions with Mr. Toda, Dr. Shimada, Dr. Hiraki, Mr. Uchibori, and Mr. Nishida, the members of the computer architecture group at ETL. We would like to thank the members of the Saito lab at Keio University and appreciate the support extended by Dr. Rine and colleagues at George Mason University. We thank several unknown referees, Dr. Benjamin Wah, and Dr. C. V. Ramamoorthy for their helpful comments on an early version of this paper. Special thanks must go to Ajantha, Prasantha, and Ms. Rosemary Mattingley for helping in many ways in this work. The research was supported by the Ministry of Education and Ministry of Internal Trade and Industry, Japan, and the Ministry of Higher Education, Sri Lanka.

REFERENCES

1. J. E. Rodriguez, "A Graph Model for Parallel Computation," *Technical Report ESLR-398, MAC-TR-64, MIT Computer Science Lab.*, September 1969.

2. J. B. Dennis, "First Version of a Dataflow Procedure Language," *Proc. Colloque sur la Programmation*, vol. 19, pp. 362–376, Lecture Notes in Computer Science, Springer-Verlag, 1974.

3. J. B. Dennis, G. R. Gao, and K. Todd, "Modelling the Weather with a Dataflow Super Computer," *IEEE Transactions on Computers*, vol. C-33, no. 7, pp. 592–603, July 1984.

4. J. B. Dennis, "Data Flow Computation," NATO ASI Series, vol. F14, *Control Flow and Data Flow: Concepts of Distributed Programming*, ed. M. Broy, pp. 346–397, Springer-Verlag, 1985.

5. J. D. Brock, "Operational Semantics of a Data Flow Language," MIT/LCS/TM-120.

6. M. Cornish, "The TI Dataflow Architectures: The Power of Concurrency for Avionics," *IEEE Proc. 3rd Conf. on Digital Avionics Systems*, pp. 19–25, Fort Worth, TX, November 1979.

7. M. Cornish, D. W. Hogan and J. C. Jensen, "The Texas Instruments Distributed Processor," *Proc. Louisiana Computer Exposition*, pp. 189–193, Lafaytte, LA, March 1979.

8. Arvind and K. P. Gostelow, "Some Relationships between Asynchronous Interpreters of a Dataflow Language," *Proc. IFIP WG2.2 Conf. on Formal Description of Programming Languages*, St. Andrews, Canada, 1977.

9. Arvind and K. P. Gostelow, "The U-Interpreter," *Computer,* vol. 15, no. 2, pp. 42–49, February 1982.

10. Arvind, V. Kathail, and K. Pingaley, "A Dataflow Architecture with Tagged Tokens," TM-174, MIT Computer Science Lab. September 1980.

11. Arvind and D. E. Culler, "Dataflow Architectures," TM-294, MIT Computer Science Lab. 1986.

12. J. R. Gurd and I. Watson, "A Multilayered Dataflow Computer Architecture," *Proc. 7th Int'l Conf. on Parallel Processing*, August 1977.

13. I. Watson and J. R. Gurd, "A Prototype Dataflow Computer with Token Labelling," *Proc. 1979 AFIPS Computing Conf.*, vol. 48, pp. 623–628, June 1979.

14. J. R. Gurd, J. R. W. Glauert, and C. C. Kirkham, "Generation of Dataflow Graphical Object Code for the Lapse Programming Language," *Lecture Notes in Computer Science*, vol. 111, pp. 155–168, Springer-Verlag, June 1981.

15. J. Gurd and I. Watson, "Preliminary Evaluation of a Prototype Dataflow Computer," *IFIP*, pp. 545–551, 1983.

16. A. L. Davis, "The Architecture and System Method of DDM1: A Recursively Structured Data Driven Machine," *Proc. 5th Annual Symposium on Computer Architecture*, pp. 210–215, April 1978.

17. A. L. Davis, "A Dataflow Evaluation System Based on the Concept of Recursive Locality," *Proc. 1979 AFIPS Computer Conf.*, vol 48, pp. 1079–1086, 1979.

18. E. A. Ashcroft and R. Jagannathan,"Operator Nets," *Fifth Generation Computer Architecture*, IFIP, 1984.

19. E. A. Ashcroft and W. W. Wadge, "Lucid, a Nonprocedural Language with Iteration," *Comm. of ACM*, vol. 20, no. 7, pp. 519–526, July 1977.

20. P. C. Treleaven, D. R. Brownbridge, R. P. Hopkins, and P. W. Rantenbach, "Combining Dataflow and Control Flow Computing," *The Computer Journal*, vol. 25, no. 2, pp. 207–217, 1982.

21. M. Amamiya, M. Takesue, R. Hasegawa, and M. Mikami, "Implementation and Evaluation of a List Processing–Oriented Dataflow Machine," *Proc. 13th Int'l Symposium on Computer Architecture*, pp. 10–19, 1986.

22. N. Takahashi and M. Amamiya, "A Data Flow Processor Array System—Design and Analysis," *Proc. 10th Annual Int'l Symposium on Computer Architecture, IEEE*, pp. 313–250, 1983.

23. M. Amamiya, R. Hasegawa, and S. Ono, "VALID: A High-Level Functional Language for Dataflow Machine," *Review of Electrical Communication Laboratories*, vol. 32, no. 5, pp. 793–802, 1984.

24. M. Amamiya, R. Hasegawa, O. Nakamura, and H. Mikami, "A List Processing Oriented Data Flow Machine Architecture," *Proc. AFIPS Computer Conf.*, pp. 143–151, 1982.

25. M. Amamiya and R. Hasegawa, "Dataflow Computing and Eager and Lazy Evaluation," *Journal New Generation Computing* vol. 2, no. 8, pp. 105–129, 1984.

26. S. Ono, N. Takahashi, and M. Amamiya, "Optimized Demand-Driven Evaluation of Functional Programs on a Dataflow Machine," *Proc. Int'l Conf. on Parallel Processing '86*, pp. 421–428, 1986.

27. T. Yuba, Y. Yamaguchi, and T. Shimada, "A Control Mechanism of a Lisp Based Data-Driven Machine," *Information Processing Letters*, vol. 16, pp. 139–143, 1983.

28. Y. Yamaguchi, K. Toda, and T. Yuba, "A Performance Evaluation of a Lisp Based Data-Driven Machine (EM-3)," *Proc. 10th Annual Int'l Symposium on Computer Architecture*, pp. 363–369, 1983.

29. Y. Yamaguchi, K. Toda, J. Herath, and T. Yuba, "EM-3: A Lisp-based Data-Driven Machine," *Proc. Int'l Conf. on Fifth Generation Computing Systems, ICOT*, pp. 531–532, 1984.

30. K. Toda, Y. Yamaguchi, Y. Uchibori, and T. Yuba, "Preliminary Measurements of the ETL Lisp-Based Data-Driven Machine," *Proc. IFIP TC-10 Working Conf. on Fifth Generation Computer Architecture*, July 1985.

31. J. Herath, N. Saito, K. Toda, Y. Yamaguchi, and T. Yuba, "Not(operation) for High Speed Data-flow Computing Systems," *Proc. Int'l Conf. on Super Computing Systems*, pp. 531–532, December 1985.

32. J. Herath, N. Saito, K. Toda, Y. Yamaguchi, and T. Yuba, "Data-flow Computing Base Language with n-Value Logic," *Proc. Fall Joint Computer Conf.*, November 1986.

33. J. Herath, Y. Yamaguchi, T. Yuba, and N. Saito, "Extended Not(operation) Based Dataflow Computing for Intelligent Switching," *Proc. Int'l Conf. on Supercomputing Systems*, May 1987.

34. J. A. Robinson, " A Machine-Oriented Logic Based on the Resolution Principle," *Journal ACM*, pp. 23–41, January 1965.

35. R. A. Kowalski and M. H. van Emden, "The Semantics of Predicate Logic as a Programming Language," *Journal ACM*, pp. 733–742, October 1976.

36. A. Colmerauer, "Prolog and Infinite Trees," *Logic Programming*, Academic Press, New York, 1982.

37. W. F. Clocksin and C. S. Mellish, "Programming in Prolog," Springer-Verlag, New York, 1981.

38. K. L. Clark and S. Gregory, "A Relational Language for Parallel Programming," *Proc. 1981 ACM Conf. on Functional Programming and Computer Architecture*, pp. 171–178, October 1981.

39. K. L. Clark and S. Gregory, "Parlog: Parallel Programming in Logic," *ACM Trans. on Programming Languages and Systems*, pp. 1–49, January 1986.

40. E. Y. Shapiro, "A Subset of Concurrent Prolog and Its Interpreter," *TR-003, ICOT*, Tokyo, February 1983.

41. J. Backus, "Can Programming Be Liberated from the von Neumann Style? A Functional Style and Its Algebra of Programs," *Comm. ACM*, pp. 613–641, August 1978.

42. J. Backus, "Function Level Computing," *IEEE Spectrum*, pp. 22–27, August 1982.

43. J. McGraw, "SISAL: Streams and Iteration in a Single-Assignment Language Reference Manual," *University of California, Lawrence Livermore National Laboratory, Report M-146*, March 1985.

44. S. Skedzielewski and J. Glauert, "IF1—An Intermediate Form for Applicative Languages," *University of California, Lawrence Livermore National Laboratory, Report M-170*, July 1985.

45. O. Gelly et al., "LAU Software System: A High Level Data Driven Language for Parallel Programming," *Proc. 1976 Int'l Conf. on Parallel Processing*, August 1976.

46. D. Comte, N. Hifdi, and J. Syre, "The Data Driven LAU Multiprocessor System: Results and Perspectives," *Proc. IFIP Congress 80,* Tokyo, pp. 175–180, October 1980.

47. J. L. Gaudiot, M. Dubios, L. T. Lee, and N. Tohme, "The TX16: A Highly Programmable Multimicroprocessor Architecture," *IEEE Micro*, October 1985.

48. T. Shimada, K. Hiraki, and K. Nishida, "An Architecture of a Dataflow Machine and Its Evaluation," *Proc. COMPCON '84,* pp. 486–490, Spring 1984.

49. K. Hiraki, K. Nishida, S. Sekiguchi, and T. Shimada, "Maintenance Architecture and LSI Implementation of a Dataflow Computer with a Large Number of Processors," *Proc. 1986 Int'l Conf. on Parallel Processing.*

50. K. Hiraki, T. Shimada, and K. Nishida, "A Hardware Design of the SIGMA-1—A Dataflow Computer for Scientific Computations," *Proc. 1984 Int'l Conf. on Parallel Processing, IEEE*, pp. 531–531, 1984.

51. T. Shimada, K. Hiraki, K. Nishida, and S. Sekiguchi, "Evaluation of a Prototype Dataflow Processor of the SIGMA-1 for Scientific Computations," *Proc. 13th Annual Int. Symposium on Computer Architecture*, pp. 226–234, 1986.

52. T. Yuba, "Research and Development Efforts on Data-flow Computer Architecture in Japan," *Journal of Information Processing*, vol. 9, no. 2, 1986, pp. 51–62.

53. B. Wah and G. J. Li, "Computers for Artificial Intelligence Applications," *IEEE Tutorial*, 1986.

54. V. P. Srini, "An Architectural Comparison of Dataflow Systems," *IEEE Computer*, pp. 68–88, March 1986.

55. P. C. Treleaven, D. Brownbridge, and R. P. Hopkins, "Data-Driven and Demand-Driven Computer Architecture," *Journal ACM Computing Surveys*, vol. 14, no 1, pp. 93–142, March 1982.

56. J. Herath, T. Yuba, and N. Saito, "Dataflow Computing," *Proc. 1986 Int'l Workshop on Parallel Algorithms and Architectures, Lecture Notes in Computer Science*, Springer-Verlag, May 1987.

57. M. Tokoro, J. R. Jagannathan, and H. Sunahara, "On the Working Set Concept for Dataflow Machine," *Proc. 10th Annual Symposium on Computer Architecture*, pp. 90–97, June 1983.

58. N. Ito and M. Sato, "The Architecture and Preliminary Evaluation Results of the Experimental Parallel Inference Machine PIM-D," *Proc. 13th Int'l Symposium on Computer Architecture*, pp. 149–156, 1986.

59. T. Suzuki, K. Kurihara, H. Tanaka, and T. Moto-oka, "Procedure Level Dataflow Processing on Dynamic Structure Multimicroprocessors," *Journal of Information Processing*, vol. 5, no. 1, pp. 11–16, 1982.

60. M. Kishi, H. Yasuhara, and Y. Kawamura, "DDDP: A Distributed Data-Driven Processor," *Proc. 10th Annual Int'l Symposium on Computer Architecture, IEEE*, pp. 236–312, 1983.

61. K. Oyama, N. Nguyen, V. P. Shrestha, T. Saito, and H. Inose, "System Design of a Distributed Dataflow Computer and Its Experimental Evaluation," *Trans. Information Proccessing Society of Japan*, vol. 25, no. 1, 1984.

62. H. Nishikawa, K. Asada, and H. Terada, "A Decentralized Controlled Multi-Processor System Based on the Data-Driven Scheme," *Proc. 3rd Int'l Conf. on Distributed Computing Systems*, pp. 639–644, 1982.

63. M. Sowa and T. Murata, "A Dataflow Computer Architecture with Program and Token Memories," *IEEE Trans. Computer*, vol. C-31, no. 9, pp. 820–831, 1982.

64. T. Temma, S. Hasegawa, and S. Hanaki, "Dataflow Processor for Image Processing," *Proc. Mini and Microcomputers*, vol. 5, no. 3, pp. 52–56, 1980.

65. A. L. Wendelborn, "A Hybrid Data and Demand Driven Implementation of a Lucid-like Programming Language," *Proc. 9th Computer Science Conference*, Canberra, January 1986.

CHAPTER 7

DESIGN DECISIONS IN SPUR

Mark Hill, Susan Eggers, James Larus, George Taylor, Glenn Adams, B. K. Bose, Garth Gibson, Paul Hansen, Jon Keller, Shing Kong, Corinna Lee, Daebum Lee, Joan Pendleton, Scott Ritchie, David Wood, Ben Zorn, Paul Hilfinger, Dave Hodges, Randy Katz, John Ousterhout, and David Patterson

SPUR (Symbolic Processing Using RISCs) is a multiprocessor workstation being developed at the University of California at Berkeley to conduct parallel processing research. Its development is part of a multiyear effort to study hardware and software issues in multiprocessing, in general, and parallel processing in Lisp, in particular.* This chapter concentrates on the initial architectural research and development of SPUR.

Two key observations motivated the architecture of SPUR. First, although parallel processing hardware has existed for many years, these systems have been difficult to program. Often the architectural features of a parallel machine, particularly the interconnection network between the processors, had to be considered during programming [1]. The complexity of managing such details has kept parallel processing a novelty, rather than the norm. Consequently, we are designing SPUR to simplify parallel processing software by providing a single global memory that can be shared, with uniform access times. Implementing a high-performance shared memory system increases the system's hardware

---

* We distinguish between *multiprocessing* and *parallel processing*. Multiprocessing occurs whenever two or more processors in a computer are used at the same time. Parallel processing occurs when they are cooperating on the same job. All parallel processing is multiprocessing, but not vice versa.

complexity, but we believe the shared memory software model facilitates the rapid development of parallel processing software and permits implementation of other, more restricted, sharing paradigms (such as message passing).

Second, hardware is more difficult to design, construct, debug, and modify than most software. Consequently, most SPUR hardware features are simple, frequently used primitives. We migrate features from software into hardware only if doing so achieves a significant performance gain in return for reasonable design and implementation costs. The complex hardware features included in SPUR either facilitate parallel processing (for example, hardware-based cache consistency) or make large contributions to performance (for example, the instruction buffer and Lisp data-type tags).

The SPUR processor extends work on reduced instruction set computers (RISC) [2] and Smalltalk on a RISC (SOAR) [3] with some special support for two emerging standards: Common Lisp [4] and IEEE Standard 754-1985 for binary floating-point arithmetic [5]. We designed the Lisp and floating-point support so that software that does not use these extensions is not penalized by their existence. Thus, the SPUR processors are general purpose processors with some support for Lisp and floating-point, rather than special purpose Lisp or floating-point processors.

The SPUR project consists of SPUR workstation development and research efforts in integrated circuits, computer architecture, operating systems, and programming languages. Integrated circuit researchers are examining complementary metal oxide semiconductor (CMOS) design styles, the effects of scaling very large scale integration (VLSI) circuits, and control and clocking issues. Computer architecture researchers are studying multiprocessor address trace analysis, cache consistency, virtually tagged caches, in-cache address translation, multilevel cache design, coprocessor interfaces, instruction delivery, Lisp hardware support, and floating-point implementations. Operating systems researchers are investigating network file systems, network page servers, the effects of large physical memories on virtual memory implementations, and workload distribution. Programming languages researchers are examining parallel garbage collection algorithms, techniques for specifying parallel programs, and methods of compiling parallel Lisp programs.

1 SYSTEM OVERVIEW

SPUR contains 6 to 12 high-performance homogeneous processors (see Figure 7.1). The number of processors is large enough to permit parallel processing experiments, but small enough to allow packaging as a personal workstation.

The processors are connected to each other, to standard memory, and to input/output devices with a modified NuBus. Using a commercial bus reduces prototype design time by allowing the use of standard subsystems and memory.

SPUR supports sharing between cooperating processes with a global shared memory. System performance is improved by placing 128-Kbyte caches on each processor to reduce bus traffic, memory contention, and effective memory

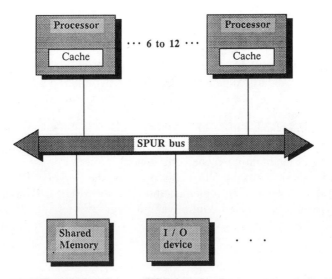

Figure 7.1 SPUR workstation system. SPUR is a shared-bus multiprocessor. The system supports several identical high-performance processors on a modified Texas Instruments' NuBus. Each of the custom processors contains a large cache to reduce the bandwidth required from the bus and shared memory.

access time. Each of these caches is accessed with virtual addresses, rather than physical addresses, so that address translation is not necessary on cache hits. On cache misses, virtual addresses are translated into physical addresses before accessing shared memory.

The caches are supplemented with hardware that guarantees that copies of the same memory location in different caches always contain the same data. This enables programmers to write software without considering the existence of cache memory.

The high-level architecture of SPUR (multiple processors communicating through shared memory over a common bus) is comparable to a few commercial multiprocessors, such as the Sequent Balance 8000 and the Elxsi 6400. However, since neither of these machines was intended to be a low-cost workstation, the differences in their more detailed architectures are fairly significant.

The Sequent is built from 10-MHz National Semiconductor 32032 central processor unit chips communicating over a pipelined, packet-switched bus. The high speed of the bus (relative to that of the CPUs) enables the Sequent to support between 2 and 12 processors with relatively small caches (8 Kbytes) and a write-through write policy.

The Elxsi 6400 processors and bus interface logic have been implemented with emitter-coupled logic (ECL) gate arrays. The bus is also packet switched, with a 25-ns cycle time, and easily supports up to eight CPUs. Cache consistency in the 16-Kbyte, two-way set-associative caches is maintained by a variety of software techniques [6].

2 PROCESSOR OVERVIEW

A SPUR processor is a general purpose RISC processor that provides support for Common Lisp and IEEE floating-point. It is implemented on a single board with about 200 standard chips and three custom two-micron CMOS chips: the cache controller (CC), the CPU, and the floating-point coprocessor (FPU). (See Figure 7.2.)

The CC chip manages the cache. This includes handling cache accesses by the CPU, performing address translation, accessing shared memory over the SPUR bus, and maintaining cache consistency.

The CPU chip is a custom VLSI chip based on the Berkeley RISC architecture [7, 2]. Like the RISC II implementation [8], the SPUR CPU uses a simple uniform pipeline, hard-wired control, and a large register file. It attempts to issue a new instruction every cycle. The SPUR CPU differs from RISC II

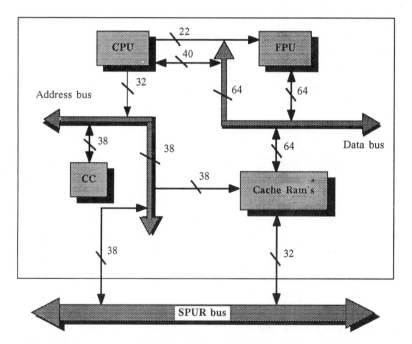

Figure 7.2 SPUR processor board. A SPUR processor is implemented on a single board that contains three custom VLSI chips and 200 standard chips. The three custom chips are the cache controller (CC), the CPU, and the floating-point coprocessor (FPU). Standard chips are used to hold the state, address tags, and data of the cache (cache RAMs), and to connect functional components together (not shown). Memory addresses and data are handled on separate buses. The address bus is 38 bits wide to accommodate global virtual addresses. The data bus is 64 bits wide to handle floatingpoint data. The FPU tracks instructions executed by the CPU with a special 22-bit connection. Some infrequently used data paths are not shown here.

because of the addition of a 512-byte instruction buffer, a fourth execution pipeline stage, a coprocessor interface, and support for Lisp tagged data.

The final custom chip is the floating-point coprocessor, which supports the full IEEE Standard 754 for binary floating-point arithmetic without microcode control. The FPU executes common operations under hardwired control. Infrequent operations cause traps and are handled by software.

Initial results with small Lisp benchmarks show that a single SPUR processor is comparable to the VAX 8600 CPU and the Symbolics 3600 CPU [9] (see Table 7.1).

3 THE MEMORY SYSTEM

The SPUR memory system appears to software as flat, global, shared memory, but is implemented with a hierarchy of levels. The fastest level, the *instruction buffer*, is an instruction cache on the CPU chip. The second level, the *cache*,

TABLE 7.1 Gabriel Benchmark Results.

Gabriel Benchmark	Execution Time (Seconds)			Execution Time Ratio	
	DEC 8600	Symbolics 3600	SPUR (Projected)	8600/ SPUR	3600/ SPUR
boyer	12.18	9.40	4.47	2.27	2.10
dderiv	6.58	3.89	1.13	5.82	3.44
deriv	4.27	3.79	0.99	4.31	3.83
destru	2.10	2.18	0.46	4.57	4.74
div2	1.65	1.51	2.77	**0.60**	**0.55**
fft (single)	9.08	3.87	9.47	**0.96**	**0.41**
frpoly (bignum)	1.40	2.10	7.17	**0.20**	**0.29**
frpoly (fixnum)	4.13	2.65	1.76	2.35	1.51
frpoly (flonum)	5.84	3.04	2.57	2.27	1.18
puzzle	15.53	11.04	7.47	2.08	1.48
stak	1.41	2.30	1.00	1.41	2.30
tak	0.45	0.43	0.13	3.46	3.31
takl	2.03	4.95	1.01	2.01	4.90
takr	0.81	0.43	0.23	3.52	1.87
traverse	46.77	41.71	18.12	2.58	2.30
triangle	99.73	116.99	66.55	1.50	1.76
geometric mean				1.97	1.73

This table presents execution times for Gabriel benchmarks on the DEC VAX 8600, Symbolics 3600 with instruction fetch unit, and a single SPUR processor. The times for the 8600 and the 3600 are from Gabriel's *Performance and Evaluation of Lisp Systems* [10]. The preliminary SPUR times are gathered with a functional-level simulator of a single-processor, assuming a 150-ns cycle time, single-cycle access to a 128K-byte cache, and 15-cycle cache miss time.

The last two columns compare SPUR with the 8600 and 3600. SPUR is slower for the ratios shown in bold. The geometric mean is used to combine the ratios in a manner that gives each benchmark equal weight. Garbage collection time is not included for any of the machines. The 8600 results were gathered with data-type declarations to reduce run-time type checking.

is a cache on the processor board for instructions and data. If information is not found in either of these local memories, then the virtual address is translated into a physical address and a global memory access is made via the SPUR bus. Both the virtual and physical addresses are transmitted on SPUR bus transactions. Off-the-shelf memory and I/O controllers use the physical address. Other cache controllers use the virtual address to preserve software's view of global, shared memory.

The Memory Model   SPUR presents software with a 256-Gbyte global virtual address space, divided into 256 1-Gbyte segments. Every process has direct access to four segments via a 32-bit process-specific virtual address. (The segments of a SPUR process resemble VAX-11 regions.) This address is mapped into a 38-bit global virtual address in parallel with the first part of a cache access (Figure 7.3).

A process's four segments will normally be used for system code and data, user code, a private stack, and a shared heap. Two or more processes that want to share information must share an entire segment. Support for sharing at the granularity of a segment is a compromise between using a single shared virtual address space and supporting sharing of arbitrary-size objects at this level. We rejected the former extreme because it does not permit hardware-guaranteed isolation of unrelated jobs. We rejected the latter because it is not clear that the benefits justify the hardware cost.

The memory system, except the instruction buffer, uses global virtual addresses instead of process-specific virtual addresses so that information can

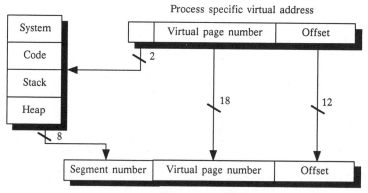

Figure 7.3   Virtual memory structure. All processes access virtual memory using a 32-bit process-specific virtual address. This address is converted into a 38-bit global virtual address during the cache lookup. The high-order two bits of the process-specific virtual address are used to select one of four segments from the 256 segments (eight bits) in the global virtual address space. The other 30 bits are used directly for the displacement within the selected 1-Gbyte segment.

be manipulated independently of processor and process identifiers. For this reason, cache flushes are not necessary on a context switch or when a process is migrated to a different processor.

We set the global virtual address space limit of 256 segments by balancing the projected needs of software against the cost of hardware implementation. The segment limit does not constrain the number of *lightweight* processes [10], which use the address space of their parent. This is important because we expect the parallel processing Lisp system to make extensive use of lightweight processes. This limit does, however, restrict the number of concurrently active heavyweight processes (such as Unix shell processes) depending on the number of segments shared. The limit ranges from 64 processes with no sharing to 253 processes with three segments shared.

The Instruction Buffer: An On-Chip Instruction Cache   The instruction buffer is a 512-byte instruction cache on the CPU chip. It reduces contention for the cache so that data references can use the single cache port without stalling the execution pipeline. By enabling instruction fetches and data references to be satisfied in parallel, the instruction buffer creates the illusion of a second cache port.

The instruction buffer also reduces effective instruction access time. This effect has little importance in SPUR because the cache can be accessed in approximately one cycle. Nevertheless, this effect will become increasingly important as technological improvements reduce cycle times faster than inter-chip communication times, thereby making it difficult to access off-chip caches in a single cycle.

The instruction buffer caches 128 thirty-two-bit instructions in 16 direct-mapped blocks. Each block contains eight instructions, divided into eight single-instruction *subblocks* [11,12] (see Figure 7.4). Preliminary estimates

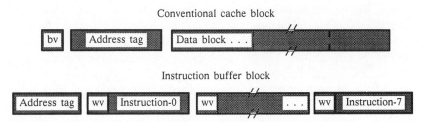

Figure 7.4   Cache block with sub-blocks. A conventional cache block (top) has three parts: the cache state bits (labeled bv for block valid bits), the address tag, and the cache data block. The state bits record whether the information in the block is valid (or dirty for a data cache); the address tag holds the block's memory address; and the data area holds the information cached. The instruction buffer's block (bottom) is divided into eight single-instruction sub-blocks (labeled instruction-0 to -7). Additional state bits, called word valid bits (labeled wv), are associated with each instruction so that only a single instruction, instead of the entire block, must be loaded on a miss.

show that the instruction buffer satisfies at least 75 percent of instruction fetches without cache accesses [13].

The instruction buffer handles misses differently than most caches. Instead of loading an entire block on each miss, it loads only the fetched instruction. The advantage of fetching single-instruction subblocks is that a miss can be completed more quickly. The disadvantages are that the state bits must be extended to include a valid bit for each sub-block in the block, and control must handle both block misses (the address tag does not match) and subblock misses (the tag matches, but the subblock is not valid). We believe the advantage of using subblocks justifies the small increase in chip area and design time needed to implement them [12].

The instruction buffer miss ratio is reduced by using prefetching from the cache to create the illusion that the entire 32-byte block is loaded on a miss. For example, near the end of a miss to the third instruction in a block, the prefetcher attempts to prefetch instructions 4, 5, 6, 7, 0, 1, and 2. Unless prefetches are blocked by data references, instructions 4 through 7 will be loaded into the instruction buffer before they will be accessed by the execution unit. If the execution unit fetches an instruction that misses in another block, the prefetcher begins prefetching in that block even if the old block is not completely loaded. The prefetcher never hurts performance because it never replaces instructions already in the instruction buffer or interferes with data references or instruction fetch misses.

The Cache   A 128-Kbyte cache for instructions and data on every processor board reduces SPUR bus traffic. For a fixed transfer size, bus traffic can be reduced by increasing a cache's size or complexity (degree of associativity) [14]. Memory chip technology makes it possible to build larger, unsophisticated caches with fewer packages than smaller, more complex ones. Consequently, the SPUR cache is simple, although larger than the caches in most mainframes. It is direct-mapped, does not permit unaligned accesses, and uses 32-byte blocks to transfer and cache information. It does not prefetch blocks from memory, because prefetching increases bus traffic. Instruction buffer prefetches that miss in the cache are terminated without a memory reference. Simulation results find cache miss ratios under two percent (see the column labeled "Ideal" in Table 7.2) [13].

The SPUR cache associates virtual address tags, rather than physical address tags, with blocks of data. A virtually tagged cache is accessed directly, without address translation. In contrast, physically tagged caches require that address translation be done before or in parallel with the first part of a cache lookup.

Unfortunately, schemes that permit parallel address translation limit the size of the cache, constrain the mapping of virtual pages to physical page frames, or require support for fast reverse mapping (translating physical addresses back to virtual addresses). For a physically tagged cache access to proceed in parallel with address translation, the cache must be indexed with bits that do not change in address translation. These bits are usually within the page offset, because systems designers are unwilling to constrain the mapping of virtual pages to

TABLE 7.2   In-Cache Address Translation versus Translation Buffers.

Aggregate Miss Ratios with Identical Caches
But Alternative Address Translation Mechanisms
Metric: (Cache Misses + Translation Misses) / References

		SPUR Cache Plus Translation Via:			
Trace	Ideal	SPUR In-Cache	VAX-11/780 TB	VAX 8600 TB	IBM 3033 TLB
Liszt	0.00584	0.00610	0.00775	0.00784	0.00614
		(1.000)	(1.270)	(1.285)	(1.006)
Vaxima	0.01844	0.01875	0.02432	0.02404	0.02001
		(1.000)	(1.297)	(1.282)	(1.067)
MVS	0.01677	0.01981	0.02208	0.02287	**0.01769**
		(1.000)	(1.115)	(1.154)	**(0.893)**

This table compares SPUR in-cache translation with translation using translation buffers. The metric used, the aggregate miss ratio, is the number of cache misses plus the number of translation misses divided by the number of processor references. Smaller values of this metric predict better performance if the cost of cache and translation misses are comparable (as they are in SPUR). Numbers in parentheses give the magnitude of the aggregate miss ratio relative to the SPUR in-cache aggregate miss ratio. Three comparisons are made. The first two are application programs running on a VAX-11 under Unix 4.2 BSD (Berkeley Software Distribution). Liszt is an address trace of the Franz Lisp compiler compiling a portion of itself. Vaxima is a trace of an algebraic system executing a representative repertoire of commands. The final trace, MVS, is a series of system calls executed by the MVS operating system on an Amdahl 470 machine.

This table assumes that each of the translation mechanisms are invoked only after a reference misses in the SPUR cache, which is 128 Kbytes large, has 32-byte blocks, and is direct-mapped. The first alternative, Ideal, sets the aggregate miss ratio to the cache miss ratio and assumes translation without cost. The second alternative, SPUR in-cache, uses the cache to hold page table entries for 4-Kbyte pages. The last three alternatives use translation buffers to do translation. The third uses half of the VAX-11/780 translation buffer (128 entries, 512-byte pages, and two-way set-associative) because the VAX-11 restricts process and system entries to different halves of the buffer. The fourth uses the VAX 8600 translation buffer (512 entries, 512-byte pages, and one-way set-associative) in the same manner. The last alternative uses the IBM 3033 translation look-aside buffer (128 entries, 4-Kbyte pages, two-way set-associative). In all but one case, shown bold, SPUR in-cache translation performs slightly better than systems that include translation buffer hardware.

physical page frames so that some bits of the virtual page number do not change. The bits used to index a cache will be within the page offset only if the cache size divided by its degree of associativity is equal to or smaller than the page size. Consequently, we believe that as cache sizes increase, virtually tagged caches will become more commonly used.

Another benefit of virtually tagged caches is that the address translation time does not affect cache hit time since address translation is necessary only for cache misses. Therefore, address translation can be done more slowly in a system with a virtually-tagged cache than in one with a physically tagged cache, where address translation must be less than the cache access time. SPUR exploits this freedom by eliminating the traditional translation buffer.

Most commercial computers use physically tagged caches rather than virtually tagged caches. This is because current commercial architectures include three features that make the implementation of virtually tagged caches difficult. The rest of this section explains how the use of a single, segmented virtual address space and a dual-address bus allows SPUR to avoid the problems commercial implementors have encountered.

Problems Implementing Virtually Tagged Caches The first problem is handling the virtual address space changes associated with most context switches. A virtual address space change means that virtual addresses refer to new locations. A virtually tagged cache must guarantee that references to the new virtual address space are not accidentally satisfied by data from the old locations.

Address space changes can be handled in a virtually tagged cache by flushing old data on every context switch or by attaching address space identifiers to cached data. The former method reduces performance for large caches. The latter increases the size of cache address tags and can increase cache complexity if synonyms, described in the next paragraph, are allowed. SPUR avoids the problems of virtual address space changes by using a global segmented virtual address space that does not change after a context switch. This results in an increase in tag size comparable to adding address space identifiers, but permits sharing without synonyms.

The second problem for implementors of virtually tagged caches is handling synonyms (aliases). Synonyms are multiple virtual addresses that map to the same physical address. They present a problem when the same physical location is read into a virtually tagged cache twice with two different virtual addresses, and then one of the copies is modified. To preserve the programmer's model of memory, the virtually tagged cache must guarantee that a read with the other virtual address gets the new value.

This problem is hard to solve in a single virtually tagged cache, and even harder to solve in a multiprocessor system with many such caches. SPUR avoids this problem by disallowing synonyms. Instead, two or more processes share information by putting it in a shared segment at the same displacement (the same global virtual address). Software resolves the location of static shared information at load time and uses operating system calls that allocate new storage to establish the location of dynamic shared information.

The third problem with virtually tagged caches is updating cache data that is being written by an I/O device using a physical address. In the long run, the problem of mapping physical I/O addresses back into virtual addresses can be avoided by having both I/O devices and memory use virtual addresses.

We rejected this approach in SPUR because we wanted to use off-the-shelf, physically addressed memory boards. Instead, the SPUR bus associates a virtual and a physical address with most bus transactions. The virtual address is used by other cache controllers for maintaining cache consistency. The physical address is used by memory and I/O controllers. The reverse mapping problem is solved by not permitting an I/O buffer to be cached while it is being written by an I/O device. The operating system can guarantee this by putting the buffer on a noncacheable page or by flushing the buffer from all caches before I/O begins. The latter does not imply a complete cache flush because the cache supports flushing of individual blocks.

Address Translation without a Translation Buffer   The mapping of virtual addresses to physical addresses is usually maintained in a structure called a *page table* [15]. The appropriate page table entry (PTE) is referenced during the address translation process.

Most computers use a special-purpose cache for PTEs, called the *translation buffer*, to reduce address translation time. Translation buffers are important in systems with physically tagged caches, which require address translation on every reference.

Fast address translation is less important with SPUR's virtually tagged cache, because address translation is necessary only on cache misses. Consequently, rather than using a translation buffer, the SPUR address translation mechanism always uses cache accesses to reference PTEs logically in shared memory [16].

The performance of SPUR *in-cache* translation compares to that with fixed-size translation buffers. Moreover, in-cache translation has two advantages. First, it avoids the design and implementation costs of a translation buffer. Second, it keeps PTEs consistent (translation buffer consistency) without special support.

Because traditional translation buffers are merely special-purpose cache memories, multiprocessors that use them suffer from a translation consistency problem (analogous to the data cache consistency problem). Solving this problem requires an increase in either hardware or operating system complexity. The SPUR in-cache translation mechanism avoids this problem by eliminating the translation buffer and storing the PTEs only in the cache, where they are kept consistent by the regular consistency mechanism.

In-cache address translation is invoked when referenced data is not in the cache. (In this discussion, *data* refers to instructions and data, in contrast to address translation information such as PTEs.) The cache controller performs address translation by the following steps. First, a page table base register and the virtual address of the data are used to construct the virtual address of the PTE. Second, the PTE is read from the cache. Third, the physical page address in the PTE and the page offset from the original virtual address are combined to form the physical address of the data. Fourth, a SPUR bus access for the data is made with both the virtual and physical addresses. Last, the data is loaded into the cache and passed on to the CPU.

On rare occasions, the PTE reference will also miss in the cache. Since SPUR places the first level of page tables in pageable virtual memory, a second translation effort is necessary to service the first-level PTE miss.

The second level of page tables is also in virtual memory and thus may be found in the cache. This level, however, is in nonpageable virtual memory at known locations. The physical addresses of second-level PTEs are computed by the cache controller to end the address translation process if the cache access for the second-level PTE misses. SPUR uses the two-level paging mechanism to reduce the physical memory dedicated to PTEs from 256 Mbytes to 256 Kbytes.

In-cache address translation works well for the traces shown in Table 7.2. Translation performance with in-cache translation is comparable to that achieved with translation buffers. In addition, other results show that the presence of PTEs in the cache does not significantly affect cache performance for data (non-PTE) references. The data miss ratio for Vaxima increased by only 0.00004, from 0.01844 to 0.01848. The increase for MVS was larger than for Vaxima, but still not significant (0.00142, from 0.01677 to 0.01819).

Cache Consistency Hardware   The problem of maintaining the shared memory model in multiprocessor systems with cache-shared, writable data is referred to as the *cache consistency* or *cache coherency* problem. Inconsistencies arise when two or more processors have copies of the same shared memory location in their private caches, and one processor modifies the location but fails to communicate the change to the other processors.

Cache consistency algorithms prevent the old data, called *stale data*, from being used. The two approaches traditionally used are (1) to update main memory and cause cache invalidations on each write, or (2) to use software assists. The first approach, called *write-through with invalidation*, generates bus traffic proportional to the number of writes. This seriously degrades performance in a system with several high-performance processors [17]. The second approach requires software to identify whether data is potentially shared and makes use of noncacheable pages or write-through with invalidation to keep that data consistent. This generates more bus traffic than our approach for unrestricted sharing, because bus transactions are generated on many references to shared pages even if most of the data is not in simultaneous use.

Other researchers are currently investigating how to improve the effectiveness of the software approach by using synchronization primitives to delay the invalidation of stale data [18,19]. The principal weakness of the software approach is that it may require extra effort from the programmer, possibly discouraging the development of parallel processing software.

The cache consistency algorithm used in SPUR, called *Berkeley Ownership*, is based on the concept of ownership of cache blocks [20]. The responsibility for maintaining consistency is distributed among the caches. If a cache owns a block, then no copies of the block occur in any other caches. The owner may update the cached entry locally without broadcasting its actions. If a cache does not own a block, it must first obtain ownership before it can update the block. Ownership is obtained by a broadcast to other caches, causing them to invalidate their copies of the block. In addition to the local update privilege, ownership carries the obligation to update main memory on block replacement (copy-back) and the responsibility of overriding main memory if another cache requests the block.

SPUR implements Berkeley Ownership with standard memory, a dual-address bus, and snooping caches. The bus broadcasts ownership requests and transfers cache blocks. Most bus transactions begin with a type field (such as read or read-for-ownership) and a block address (both virtual and physical), and end with a data transfer [21].

Each processor cache controller is supplemented with hardware, called the *snoop*, that monitors the bus for transactions involving blocks that it has cached. The snoop compares the virtual addresses of all bus transactions with a second copy of the cache's address tags. If a match occurs, the snoop may have to invalidate its copy of the block or override main memory and provide the data to complete the bus transaction. The latter action only occurs for blocks that have been modified and are simultaneously shared by processes on more than one processor. While we have little data on sharing, we expect this to occur on a small fraction of all transactions.

Berkeley Ownership, implemented in hardware with snooping caches, serves the goals of SPUR in several ways. First, it preserves the shared memory model. This model facilitates parallel processing experiments by providing a simple, flexible mechanism for sharing data among processes. Second, it simplifies parallel processing software by relieving programmers of the responsibility of understanding shared caches. Third, the Berkeley Ownership protocol has good multiprocessor performance, because it can be restricted to generate extra bus transactions only when two or more processors are simultaneously accessing writable shared data. For example, our protocol allows semaphores to be cached. No bus traffic is needed to modify a semaphore if only one process happens to be using the semaphore for some period of time, or if all the processes using the semaphore are on the same processor. Other methods generate bus transactions after shared data has been modified even if no processes on other processors are trying to access the same data. Fourth, our protocol yields good uniprocessor performance. When no interprocessor sharing can occur, no consistency-preserving bus transactions will be made. Fifth, the algorithm is not difficult to implement. It requires an additional state machine in the cache controller, two additional state bits for each 32-byte cache block, a second copy of all cache state bits and address tags, and a change to the system bus to permit snooping. It does *not* require centralized control or any memory board modifications.

4 THE CPU AND FLOATING-POINT COPROCESSOR

The SPUR CPU design evolved from the RISC II design [8]. Like RISC II, SPUR has a streamlined instruction set and a large register file with multiple, overlapping register windows to speed up procedure calls. For several reasons, the instruction set is well suited for a high-performance VLSI implementation without microcode. First, the instructions are easy to decode because of their fixed size and few formats. Second, computational instructions operate exclusively on registers, while memory can be accessed only with load and store instructions. Register-to-register instructions execute quickly and deterministically because they cannot generate cache misses or page faults once they begin execution. Third, the instructions perform simple operations implemented in a short, uniform pipeline. Every instruction uses a particular resource in the same pipeline stage. For example, all SPUR instructions use the arithmetic and logic unit (ALU) to combine operands or calculate an effective address in the second stage of the pipeline. This simplifies the hardware by predetermining the scheduling of resources.

The differences between SPUR and RISC II are products of technological improvements and the new goals of supporting Lisp and floating-point arithmetic. Technological improvements in the past few years have increased the number and speed of transistors possible on a VLSI chip. In SPUR, the additional transistors are used in an on-chip instruction cache, for tagging Lisp data, and in a low-overhead interface to a floating-point coprocessor.

4.1 General Purpose Features

The Register Set The SPUR register set, shown in Figure 7.5, includes 32 general-purpose registers. Like the RISC II chip, the SPUR CPU contains several copies of the general purpose register set (not shown in Figure 7.5) so that these registers do not have to be saved in memory and restored on most procedure calls and returns. In addition, the register windows for a caller and a callee overlap by six registers so that most arguments and returned values can be passed in place in registers instead of in memory. For both of these reasons, overlapped register windows reduce the time required for procedure calls and returns. The primary cost of the multiple register sets is a significant amount of chip area and, to a lesser extent, slower register access time and increased process switching overhead.

The Execution Pipeline The SPUR execution pipeline is one stage longer than the three-stage RISC II pipeline (see Figure 7.6). RISC II could issue a register-to-register instruction every cycle. It used resources efficiently: in every cycle two registers were read, one was written, the ALU was utilized, and the path to memory was used to fetch an instruction. Unfortunately, this arrangement left no memory bandwidth for data references. Consequently, loads and stores had to stall the pipeline one cycle to use the path to memory. Thus, RISC II did a memory reference per cycle rather than completing an instruction per cycle.

SPUR uses an instruction buffer and a four-stage pipeline to attempt to issue and complete an instruction every cycle. The instruction buffer satisfies most instruction fetches without cache accesses. The new pipeline stage allows memory-referencing instructions to make cache accesses without stalling the pipeline, but forces register-to-register instructions to delay their register write

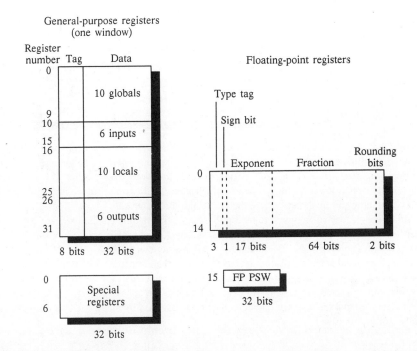

Figure 7.5 SPUR registers. SPUR's registers divide into three groups: general-purpose, special, and floating-point. The general-purpose registers are organized into fixed-size overlapping windows so that the output registers of one window become the input registers of the next window after a procedure call. Only one 32-register window is visible at a time. The entire general-purpose register file contains eight windows (not shown) for a total of 138 registers. The general-purpose registers are 40 bits wide, consisting of an 8-bit tag and 32 bits of data. The special registers include the user and kernel processor status words, register window pointers, and several program counters. The floating-point registers are 87 bits wide to accommodate SPUR's representation of IEEE extended-precision numbers. The representation includes a three-bit type tag to simplify detection of infrequent floating-point types (such as Not-a-Number). The 15 floating-point registers and the floating-point processor status word (a special register) are implemented on the FPU chip rather than on the CPU chip to improve operand access time for floating-point instructions. Multiple windows of these registers were not implemented because of insufficient FPU chip area.

for one stage. All instructions modify the general purpose register file in the fourth pipeline stage, thereby avoiding write conflicts. Internal forwarding is done by the hardware so that the result of a register-to-register instruction can be used by the next instruction even though that result has not yet been written into the register file.

In practice, SPUR will not be able to execute one instruction every cycle, principally because of instruction buffer and cache misses. On simulations with the Gabriel benchmarks (see Table 7.1), SPUR executed an instruction every 1.59 cycles with instruction buffer and cache miss ratios of 14 and 1

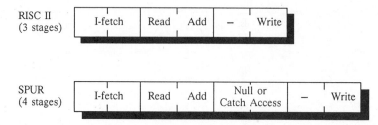

Figure 7.6 RISC II and SPUR pipelines. RISC II used a three-stage pipeline (top) that required the pipeline to stall one cycle on every data memory reference so that precisely one memory reference (instruction fetch or data access) was done every cycle. The first stage fetched the next instruction from memory; the second read two registers and performed an ALU operation; the final stage wrote the result into a register.

SPUR uses a four-stage pipeline (bottom) so that an instruction can be issued every cycle. Memory-accessing instructions use the additional stage to do a cache access. Register-to-register instructions do nothing in the additional stage. All instructions write the register file in the fourth state to guarantee that no two instructions try to write at the same time. Pipeline hardware guarantees that the result of a register-to-register instruction can be used by the subsequent instruction even though the result has not yet been written into the register file. However, software must guarantee that the result of a load instruction is not used by the instruction that (dynamically) follows the load.

percent. Performance varied across the benchmarks from 1.03 to 1.99 cycles per instruction. Larger programs are likely to have more cycles per instruction because of poorer locality of reference. Even if the instruction buffer and cache miss ratios double, however, SPUR still executes an instruction every 2.06 cycles.

The Instruction Set This section focuses on a few important decisions in the SPUR instruction set. (See Taylor's "SPUR Instruction Set Architecture" [22] for the complete design.) To simplify decoding, all instructions are four bytes long and use fixed positions for the opcode and register specifiers. Most instructions use either two source registers and one destination register or a source register, an immediate constant, and a destination register. Table 7.3 lists the basic instruction set, not including instructions for Lisp and floating-point operations.

Memory accesses are made with loads and stores. The effective address for a load is either the sum of two registers or the sum of an immediate displacement and a register. SPUR uses a *delayed* load, which requires software to guarantee that the result of a load instruction is not used by the instruction that (dynamically) follows the load. Cache misses on data references stall the entire pipeline and thus are not visible to software. The effective address for a store is always a register plus immediate displacement, so that a two-port register file suffices (one register for the address and one for the data). A store stalls the execution pipeline for one cycle, because less-common cache writes take longer than cache reads.

Cache reads access cache data in parallel with examining cache address

TABLE 7.3 **Basic SPUR Instructions.**

Instruction	Operands	Action	Cycles
Load/Store			
load — 32	dest, src1, ri	dest ← M[src1 + ri]	1
load — external	dest, src1, ri	dest ← external (cache) state	1
test — and — set	dest, src1, ri	dest ← M[src1 + ri]; M[src1 + ri] ‹00› ← 1	2
store — 32	src2, src1, imm	src2 → M[src1 + imm]	2
store — external	src2, src1, imm	src2 → external (cache) state	1
Compute			
add, subtract	dest, src1, ri	dest ← src1 op ri	1
add (no traps)	dest, src1, ri	dest ← src1 + ri	1
and, or, xor	dest, src1, ri	dest ← src1 op ri	1
sll, srl, sra	dest, src1, ri	dest ← src1 op ri ‹01.00›	1
extract	dest, src1, ri	dest ‹07:00› ← one byte from src1 selected by ri	1
insert	dest, src1, ri	dest ← ri ‹07:00› inserted into one byte of src1	1
Branch/Jump			
cmp — branch — delay	cond, src1, rci, offset	if (src1 cond rci) pc ← pc + signed word offset	1
cmp — branch — likely	cond, src1, rci, offset	if (src1 cond rci) pc ← pc + signed word offset else change next instruction into no-op	1
jump	address	pc ← word address (in same segment)	1
jump — register	src1, ri	pc ← src1 + ri	1
Call/Return			
call, call — kernel	address	increment current window pointer; save pc; pc ← word address	1
return, return-trap	src1, ri	pc ← src1 + ri; decrement current window pointer	1
Access Specials			
read — special	dest, src1	dest ← special register src1	1
write — special	dest, src1, ri	special register dest ← src1 + ri	1
read — kernel — psw	dest	dest ← kernel psw	1
write — kernel — psw	src1, ri	kernel psw ← src1 + ri	1

This table lists the basic SPUR instruction set. The column Cycles shows the minimum number of cycles consumed by an instruction. Many instructions operate on two sources (src1 and ri) and write a result into a destination (dest). Src1 and dest are five-bit register specifiers. Ri is either a 5-bit register specifier or a 14-bit signed immediate constant. Rci stands for a five-bit register specifier or a five-bit unsigned immediate constant. Pc stands for the program counter. The Action column describes what happens in the data portion of the destination and source registers. Exceptional conditions and Lisp tag manipulations are described in the SPUR instruction set architecture [22].

tags. Cache writes begin in a similar fashion, but cannot write into a cache block until after the address tag has been examined. In our initial design, stores did not stall the pipeline, because we set the cycle time to the cache write time. We were able to improve performance by reducing the cycle time to the cache read time, thus forcing the less frequent cache writes to take two cycles.

SPUR supports synchronization with a test-and-set instruction implemented in the cache. Under the best of conditions it does not require any bus transactions. To simplify the cache interface, SPUR does not have load or store instructions that manipulate individual bytes. A load-byte instruction would increase the cache access time, and a store-byte instruction would increase cache complexity. Instead, byte insert and extract instructions assist in loading and storing individual bytes.

SPUR adopted the delayed branch from RISC II. The execution of a branch instruction on most pipelined processors requires that the branch target be fetched and the execution pipeline flushed before the target instruction is executed. A branch instruction on SPUR allows—in fact, requires—the next sequential instruction to be executed while the branch target is fetched. A delayed branch saves program execution time if a useful instruction can be scheduled in this *delay slot*. Gross found this could be done on 63 percent of delayed branches dynamically encountered in the traces studied [23]. Gross also found that delayed branches did not significantly increase code size, since 87 percent of the statically examined delay slots contained useful instructions.

SPUR also includes a *canceling* compare and branch instruction, which dynamically turns the instruction in the delay slot into a no-op if the branch is not taken. The technique is also being used in the Lawrence Livermore S-1 AAP. This variant of the delayed branch makes it easier to schedule a useful instruction in the delay slot. The natural use of this instruction is at the bottom of a loop, with the branch target set to the loop's second instruction and the delay slot filled with a copy of the loop's first instruction.

An arbitrary shift instruction was not included, because most shifting done in high-level language programs is for effective address computation in arrays and records [8]. SPUR provides shift instructions only to shift one bit right and one, two, and three bits left. Shift operations are not needed for integer multiplication or division since these operations are done with the FPU.

Supporting Lisp   The Lisp programming language has some features difficult to implement efficiently on conventional computers. These include frequent function calls and returns, polymorphic operations, and automatic garbage collection. Most machines designed to run Lisp use a stack-based architecture with extensive microcode support (such as the Symbolics 3600 [24], LMI Lambda [25], and the Xerox D-Machines [26]. Our approach emphasizes a simple, regular instruction set, overlapping register windows, and tagged data [9]. Table 7.4 lists the instructions tailored for Lisp.

Fast function calls and returns are particularly important for Lisp, because Lisp programs are constructed from many small functions. SPUR provides fast function calls and returns through the overlapping register window mechanism. Studies have shown that this mechanism, developed for C, effectively speeds up Lisp calls and returns [27]. The complicated argument options allowed by Common Lisp (such as default and keyword parameters) are handled by software rather than by special purpose instructions or microcode. This approach increases the size of functions that use these options, but ensures that simple function calls execute rapidly.

TABLE 7.4 SPUR Lisp Instructions.

Instruction	Operands	Action	Cycles
load __ 40	dest, src1, ri	dest ← M [src1 + ri]	1
car/cdr	dest, src1, ri	dest ← M [src1 + ri]	1
store __ 40	src2, src1, imm	src2 → M [src1 + imm]	2
read __ tag	dest, src1	dest ‹7:00› ← src1 tag	1
write __ tag	dest, ri	dest tag ← ri ‹7:00›	1
tag __ cmp __ branch __ delayed	cond, src1, tab__ imm, offset	if src1 ‹tag› cond tab __ imm pc ← pc + signed word offset	1
tag __ cmp __ branch __ likely	cond, src1, tab__ imm, offset	if src1 ‹tag› cond tab __ imm pc ← pc + signed word offset else change nest instruction into no-op	1
compare __ and __ trap	cond, src1, rci	if (src1 cond rci) trap	1
tag__compare __ and __ trap	cond, src1, tag __ imm	if (src1 cond rci) trap	1

This table lists Lisp instructions. The column Cycles shows the minimum number of cycles consumed by an instruction. Load __ 40 and store __ 40 move tagged words into and out of registers. Car and cdr are special forms of load __ 40 that check for a proper list element. Read __ tag and write __ tag move a tag to and from the data part of a register. Compare __ and __ branch __ delayed and compare __ and __ branch __ likely, presented in Table 7.3, compare the tags and values of two Lisp data items. In addition, tag __ compare __ and __ branch __ delayed and tag __ compare __ and __ branch __ likely are available to determine the value of a tag (by comparing it with an immediate constant). Compare __ and __ trap and tag __ compare __ and __ trap are used to test for error conditions.

Tagged Architecture   Lisp uses *polymorphic* functions with operands whose type is not known until run time. A polymorphic function operates on arguments of more than one data type [28]. For example, the addition operator (+) is a polymorphic operator in most high-level languages, because it is defined to operate on both integers and floating-point numbers. Lisp complicates the implementation of polymorphic operations because it associates the type of data with the data values instead of the program variables. For example, a variable is not an integer variable, known at compile time, but rather a variable that may contain an integer at run time. When a Lisp function is evaluated, the types of operands must be determined before the appropriate routine is executed.

SPUR handles polymorphic operations by manipulating the 6-bit data-type tags of operands in parallel with operating on the 32-bit data values (see Figure 7.7). Type checking in SPUR, like in several other machines [29,3], assumes that most arithmetic operands are integers. For example, a polymorphic add operation in SPUR is implemented with an add instruction that begins by adding the 32-bit operands as if they were integers and, in parallel, checking the data type tags to verify that they are integers. If both operands are integers, the instruction finishes by writing the sum into the result register. Otherwise, the register write is suppressed and the instruction traps to software that determines the types of the operands and performs the appropriate form of addition.

The power of SPUR to manipulate data type tags is increased by several instructions that allow conditional traps and branches based on tag values (see Table 7.4). The conditional traps allow efficient checking of error conditions. Explicit tag comparison instructions are used to implement polymorphic operations in the more complicated cases not handled by the hardware.

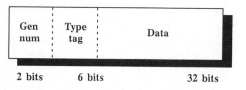

Figure 7.7 Lisp tagged data. SPUR augments Lisp data words with an eight-bit tag that includes a six-bit data-type tag and two-bit generation number. Lisp integers and characters are represented as immediate data. All other types of Lisp objects are referenced by typed pointers. Some of the tag values are used by the hardware to do tag checks in parallel with data operations. Other tag values are interpreted only by software. The generation number is used to implement a generation-scavenging garbage-collection algorithm.

Data type tags also assist list manipulation, which is fundamental in Lisp. A list is a sequence of elements (such as (*a b c*)). The Lisp functions that manipulate lists are called **car** and **cdr. Car** returns the first element of a list (*a*) and **cdr** returns the rest of the list ((*b c*)).

Car and **cdr** can be implemented with load instructions because lists are stored as linked lists in main memory. However, the semantics of Common Lisp strongly encourage generation of an exception if the argument of **car** or **cdr** is not a list. Conventional architectures must execute one or more instructions to check this condition even if the arguments of all **car**'s and **cdr**'s in a correct program are lists.

SPUR provides a car/cdr instruction that checks the data type tag in parallel with the load. A trap is generated if the type of the operand is inappropriate. This is an ideal use of parallel tag checking because it allows SPUR to execute **car** and **cdr** at the same speed as a load and still generate exceptions on errors.

SPUR also uses part of the tag field to assist in garbage collection. Lisp encourages programmers to dynamically create and use data structures in memory. Automatic garbage collection reclaims structures that are no longer in use. This feature relieves the Lisp programmer of the responsibility of explicitly discarding obsolete structures, a task that leads to subtle bugs and complicated programming. SPUR stores a two-bit *generation number* in the tag to assist a *generation scavenging* garbage collection algorithm (see Figure 7.7) [3]. The algorithm exploits a property of dynamic data: new data structures are likely to become garbage soon and old data structures are likely to stay in use. Therefore, most garbage collection activity focuses on the new data. The generation number records the number of garbage collections that an item has survived and hence its age.

Poor Data Density We designed the SPUR architecture with more emphasis on speed and simplicity than concern for code or data density. The prototype implementation has particularly poor Lisp data density because we decided not to build a complete 40-bit system.

The CPU manipulates 40-bit data (an 8-bit tag and 32-bit data). That data must often be loaded from and stored to the cache and the rest of the memory system. Three approaches exist for doing this:

1. Build the whole system with 40-bit words,
2. Allow unaligned cache accesses, or
3. Place 40-bit words in aligned 64-bit words.

We rejected a 40-bit-word memory system because it would preclude the use of many off-the-shelf subsystems, which would substantially delay completion of the prototype. It would also have complicated non-Lisp software in such areas as string manipulation and file transfer with non-SPUR machines. We rejected unaligned cache accesses because of the complexity they would add to the cache. An unaligned access can cross a cache block boundary, possibly forcing the cache to handle two cache misses and the associated address translation. Consequently, we chose to store 40-bit Lisp words in aligned 64-bit words. The other 24 bits are wasted for tagged Lisp data, but not for instructions, data for other languages, and some Lisp data structures. At worst, this storage strategy uses 60 percent more Lisp data memory than the first two schemes, but it allows us to explore ideas more quickly by simplifying the design of the prototype.

Floating-Point Support SPUR implements the IEEE 754 binary floating-point standard [5] with a mixture of hardware and software. Floating-point instructions are executed on the floating-point coprocessor chip (FPU). The FPU hardware is optimized to execute common floating-point operations quickly. Effective use of the FPU depends on a low-overhead floating-point interface and support for concurrent execution of floating-point and CPU instructions.

The SPUR FPU is one of the first implementations of IEEE floating-point that does not use any microcoded control. The Fairchild Clipper CPU also has a hard-wired IEEE floating-point unit.

The Floating-Point Coprocessor Floating-point instructions are either register-to-register instructions or loads and stores (see Table 7.5). The register-to-register instructions include add, subtract, multiply, and divide. Except for multiply (7 cycles) and divide (19 cycles), a new floating-point instruction can be issued every four cycles.

Data are transferred between the FPU and the cache with floating-point load and store instructions. Floating-point load instructions convert all single- (32 bits) and double- (64 bits) precision numbers to extended precision to simplify the computational instructions. A convert instruction must be executed before a store to perform the inverse operation. For example, an extended_to_single convert instruction must be executed before a store_single instruction.

TABLE 7.5   SPUR Floating-Point Instructions.

Instruction	Operands	Action	Cycles
load — single	dest, src1, ri	FPU dest ← (convert to extended) M[src1 + ri]	1
load — double	dest, src1, ri	FPU dest ← (convert to extended) M[src1 + ri]	1
load — extended1	dest, src1, ri	FPU dest ← M[src1 + ri]	1
load — extended2	dest, src1, ri	FPU dest ← M[src1 + ri]	1
load — integer	dest, src1, ri	FPU dest ‹63:32› ← M[src1 + ri]	1
store — single	src2, src1, i	FPU src2 → M[src1 + i]	2
store — double	src2, src1, i	FPU src2 → M[src1 + i]	2
store — extended1	src2, src1, i	FPU src2 → M[src1 + i]	2
store — extended2	src2, src1, i	FPU src2 → M[src1 + i]	2
store — integer	src2, src1, i	FPU src2 → M[src1 + i]	2
from — fpu	dest, src2	CPU dest ← FPU src2 ‹63:32›	1
to — fpu	dest, src2	FPU src2 ‹63:32› ← CPU src2	1
fadd, fsub	dest, src1, src2	FPU dest ← FPU src1 op FPU src1 op FPU src2	4
fmul	dest, src1, src2	FPU dest, FPU src1*FPU src2	7
fdiv	dest, src1, src2	FPU dest ← FPU src1/FPU src2	19
fcmp	src1, src2	FP — PSW ‹branch — bit› ← (FPU src1 cond FPU src2)	4
fnegate	dest, src1	FPU dest ← FPU src1 with opposite sign	4
fabs	dest, src1	FPU dest ← FPU src1 with positive sign	4
fmov	dest, src1	FPU dest ← FPU src1	4
int — to — extended	dest, scr1	FPU dest ← (convert to extended) FPU src1 ‹63:32›	4
extended — to — int	dest, scr1	FPU dest ‹63:32› ← (convert to integer) FPU src1	4
extended — to — single	dest, scr1	FPU dest ← (convert to single) FPU src1	4
extended — to — double	dest, scr1	FPU dest ← (convert to double) FPU src1	4
sync		CPU waits until FPU is not busy	1

This table lists SPUR floating-point instructions. The column Cycles shows the minimum number of cycles consumed by an instruction in normal operating mode. If the FPU and CPU are operated concurrently, a CPU instruction can begin one cycle after an FPU instruction has started (see the section labeled "The Floating-Point Coprocessor Interface"). There are floating-point load and store instructions for each floating-point format and for integers. Extended-precision numbers require two different loads and two different stores to move the first 64 bits and the last 64 bits. Loads do implicit conversion to extended precision, but stores merely copy bits. Store — single, store — double, and store — integer must be preceded by the corresponding convert instruction (such as extended — to — single). The to — cpu and from — cpu instructions transfer integers directly between the integer and floating-point register sets so that the FPU can be used effectively for integer multiply and divide. A conditional branch based on floating-point data is done in two steps. First, the CPU executes an fcmp instruction to set a bit in the floating-point procedure status word (FP PSW). Second, the CPU executes a conditional branch instruction that tests this bit.

Most floating-point operations execute in four cycles using the add/subtract hardware. Multiply and divide use additional special purpose hardware. A sync instruction is used when the CPU and FPU are executing instructions in parallel. It forces the CPU to stop executing new instructions until the FPU completes its current instruction (if any). This can be used to guarantee that the store of a floating-point result does not begin before the result has been computed.

The FPU contains fifteen 87-bit floating-point registers organized as a single register set (see Figure 7.5). There is no analog to the overlapping windows used for the general purpose registers because there is not enough FPU chip area to implement more registers. Furthermore, more research is needed to determine how to use overlapping windows for floating-point registers.

The floating-point register set is independent of the general purpose register set for four reasons:

1. To reduce access time for floating-point operands,
2. To allow more freedom in setting the width of floating-point registers,
3. To permit concurrent execution of integer and floating-point operations, and
4. To permit implementation of a separate FPU chip.

SPUR divides the floating-point standard into two parts. One part is implemented by a set of instructions (see Table 7.5) with hard-wired control, and the other is implemented by software trap handlers. The standard defines six types of floating-point numbers: zeros, normalized numbers, denormalized numbers, infinities, and two types of Not-a-Number symbols. The FPU manipulates normalized numbers and zeros entirely in hardware. The other four less common types require software assistance.

The FPU manipulates single- (32 bits), double- (64 bits), and extended-precision numbers (at least 79 bits) in a common 87-bit format to reduce hardware complexity. SPUR enlarges the minimum extended-precision format in four ways.

First, a three-bit tag identifies the type of a number. This tag reduces the time needed by a load instruction to convert numbers to extended precision by allowing the load to handle exponents for all types of numbers in a uniform fashion. In addition, the hardware for computational instructions can determine whether software assistance is necessary by examining the three-bit tag rather than the entire number.

Second, SPUR expands the exponent by two bits so that trap handlers can adjust denormalized operands. This enables SPUR to multiply and divide denormalized numbers using hardware designed for normalized operands.

Third, two rounding bits are added so that SPUR can mimic rounding from an infinitely precise result to a precision shorter than extended precision. This feature is necessary to correctly handle a denormalized number produced by an underflow exception.

Fourth, one bit is used to hold the most significant fraction bit in explicit form.

The Floating-Point Coprocessor Interface The FPU is sufficiently fast that the performance of floating-point operations is sensitive to the overhead associated with starting floating-point operations and the overhead of transferring floating-point operands to and from the FPU. Consequently, 28 CPU pins are used to implement a low-overhead interface between the CPU and the FPU. Unfortunately, the close coupling of the two chips may make it difficult to use the SPUR FPU without the SPUR CPU.

To reduce the overhead of starting floating-point operations, the FPU tracks all CPU instructions using 22 pins dedicated to carrying opcode, register specifiers, and other control information to the FPU (and possibly other coprocessors). Some commercial floating-point coprocessors track instructions by monitoring CPU instruction fetches to memory (such as the Intel 8087). However, this will not work in SPUR because the CPU fetches most instructions from the on-chip instruction buffer.

The SPUR floating-point interface reduces operand overhead in three ways. First, the floating-point registers reside on the FPU. Since all floating-point computation instructions operate with operands in these registers, intermediate results can be efficiently used.

Second, floating-point load and store instructions transfer data directly between the FPU and the cache. In contrast, many commercially available interfaces require floating-point data to be transferred through the CPU. The following sequence occurs when a floating-point load instruction is issued by the CPU: the FPU recognizes the floating-point load instruction and saves the destination register specifier; the CPU calculates the effective memory address and sends the address to the cache; the cache sends the data to both the FPU and the CPU; the FPU reads the data and loads it into the appropriate floating-point register; the CPU ignores the data, but recognizes that the load is complete.

Third, the data path between the FPU and the memory system is 64 bits wide, in contrast to more commonly used 8-, 16-, and 32-bit interfaces. This allows load and store instructions to move single- and double-precision numbers with a single transfer and extended-precision numbers with two transfers. SPUR's wide FPU interface reduces the probability that operand movement will limit floating-point throughput, which can easily occur for double-precision computations.

The coprocessor interface also allows concurrent CPU and FPU operation. Subject to some software constraints, the CPU can continue executing general purpose instructions, Lisp instructions, and floating-point loads and stores while the FPU is busy. Overlapping operand movement/index calculations with floating-point operations can halve the execution time of many inner loops of floating-point-intensive programs [30]. However, software must restrict the interaction between concurrently executed instructions by reordering instructions or by inserting *sync* instructions. For example, a sync instruction must be inserted between a floating-point operation and an instruction that stores the result of the operation in memory if the store could read the result register before it is written.

5   STATUS

The implementation of SPUR is in progress. As of September 1986 the custom components and processor board have been described at the register-transfer level with a variant of the ISP language and simulated with a software package called N.2. The layouts of the custom chips are near completion. They all use four-phase nonoverlapping clocks with a projected cycle time of 100 to 150 nanoseconds. The processor board has been designed and simulated with N.2, and is nearly ready for physical implementation. We expect to have working components by early 1987 and a working system later in the year.

SPUR is a multiprocessor research vehicle, but we have not as yet been able to run multiprocessing experiments. Nevertheless, we have some preliminary results.

First, selected architectural changes can significantly ease implementation and, at the same time, improve performance. For example, disallowing synonyms enabled us to build virtually tagged caches without complex reverse-translation mechanisms. Virtually tagged caches improved performance by reducing cache access time and permitting slow address translation.

Second, in-cache address translation keeps PTEs consistent and offers performance comparable to a translation buffer at less cost.

Third, cache consistency can be maintained in hardware at reasonable cost and without any modifications to main memory boards.

Fourth, Lisp can be supported without a stack-based architecture and without a microcoded implementation. However, data type tags or some other direct support of Lisp's dynamically typed data is advantageous.

Fifth, IEEE standard floating-point can be implemented without microcoded control if software handles the less common cases.

Sixth, floating-point coprocessor interfaces can be designed to significantly reduce operand movement overhead by putting the floating-point registers on the floating-point coprocessor and loading these registers directly from a cache using a 64-bit data path.

The goal of the first phase of the SPUR project is to design and implement several working prototypes. If the prototypes meet our expectations, we hope to find partners to help us transfer SPUR from academia to industry.

6 ACKNOWLEDGMENTS

The SPUR project is a cooperative project that benefits from the contributions of many people within the Berkeley community besides the authors of this chapter. The implementation of the CPU, FPU, and CC was begun by class members of CS 292I, taught in Spring 1985 by Randy Katz. Members of this class who assisted included Chien Chen, Lifan Pei, Rick Rudell, Trudy Stetzler, Sinohe Villalpando, Albert Wang, Don Webber, and Tom Wisdom. The implementation of three VLSI chips would not have been possible without computer-aided design software developed by Gordon Hamachi, Bob Mayo, John Ousterhout, Walter Scott, and George Taylor. The architecture of SPUR has been strongly affected by interactions with the SPUR operating systems group, consisting of Andrew Cherenson, Fred Douglis, John Ousterhout, Mike Nelson, and Brent Welch.

We would also like to thank Sue Dentinger, Dave Ditzel, Gregg Foster, Jim Goodman, Robert Henry, Louis Monier, Prabhakar Ragde, Dick Sites, Alan Smith, Jim Smith, Chuck Thacker, and John Wakerly for their suggestions, which improved the quality of this chapter.

SPUR was first presented at the 1985 Asilomar Microcomputer Workshop.

Principal funding for the SPUR project is provided by the Defense Advanced Research Projects Agency under contract N00039-85-C-0269. Additional support for this research was provided by the State of California MICRO program, by a Digital Equipment Corporation CAD/CAM grant, by the National Science Foundation under grant DCR-8202591, by equipment donations from Texas Instruments, Inc., and by computer resources provided under DARPA contract N00039-84-C-0089.

REFERENCES

1. J. Deminet, "Experience with Multiprocessor Algorithms," *IEEE Trans. on Computers,* vol. C-31, no. 4, April 1982.

2. D. A. Patterson and C. H. Sequin, "A VLSI RISC," *Computer,* vol. 15, no. 9, pp. 8–21, September 1982.

3. D. Ungar et al., "Architecture of SOAR: Smalltalk on a RISC," *Proc. 11th Int'l Symposium on Computer Architecture,* p. 1887, June 1984.

4. G.L. Steel, "Common Lisp: The Language," Digital Press, Burlington, MA, 1984.

5. IEEE Standard 754-1985 for Binary Floating-Point Arithmetic, 1985, order number CN953.

6. R. Olson, "Parallel Processing in a Message-Based Operating System," *Software,* pp. 39–49, July 1985.

7. D. A. Patterson, "Reduced Instruction Set Computers," *CACM,* pp. 8–21, January 1985.

8. M. G. H. Katevenis, "Reduced Instruction Set Computer Architectures for VLSI," Ph.D. thesis, University of California, Berkeley, October 1983.

9. G. S. Taylor et al., "Evaluation of the SPUR Lisp Architecture," *Proc. 13th Int'l Symposium on Computer Architecture,* Tokyo, Japan, June 1986.

10. R. P. Gabriel, *Performance and Evaluation of Lisp Systems,* MIT Press, 1985.

11. B. W. Lampson and D. D. Redell, "Experience with Processes and Monitors in Mesa," *CACM,* February 1980.

12. J. R. Goodman, "Using Cache Memory to Reduce Processor Memory Traffic," *Proc. 10th Int'l Symposium on Computer Architecture,* pp. 124–131, Stockholm, Sweden, June 1983.

13. M. D. Hill and A. J. Smith, "Experimental Evaluation of On-Chip Microprocessor Cache Memories," *Proc. 11th Int'l Symposium on Computer Architecture,* Ann Arbor, MI, June 1984.

14. R. H. Katz et al., "Memory Hierarchy Aspects of a Multiprocessor RISC: Cache and Bus Analyses," Computer Science Division Technical Report UCB/CSD 85/221, University of California, Berkeley, January 1985.

15. A. J. Smith, "Cache Memories," *Computing Surveys,* vol. 14, No. 3, September 1982, pp. 473–530.

16. P. J. Denning, "Virtual Memory," *Computing Surveys,* vol. 2, no. 3, September 1970.

17. D. A. Wood et al., "An In-Cache Address Translation Mechanism," *Proc. 13th Int'l Symposium on Computer Architecture,* Tokyo, Japan, June 1986.

18. L. M. Censier and P. Feautrier, "A New Solution to Coherence Problems in Multicache Systems," *Trans. on Computers,* vol. C-27, no. 12, pp. 1112–1118, December 1978.

19. A. J. Smith, "CPU Cache Consistency with Software Support Using One-Time Identifiers," *Proc. Pacific Computer Communication Symposium,* Seoul, Republic of Korea, October 1985.

20. R. H. Katz et al., "Implementing a Cache Consistency Protocol," *Proc. 12th Int'l Symposium on Computer Architecture,* Boston, MA, June 1985.

21. G. Gibson, "SPURBUS Specification," *Proc. CS292i: Implementation of VLSI Systems,* ed. R. H. Katz, University of California, Berkeley, September 1985. Also Computer Science Division Technical Report UCB/CSD 86/259.

22. G. Taylor, "SPUR Instruction Set Architecture," *Proc. CS292i: Implementation of VLSI Systems,* ed. R. H. Katz, University of California, Berkeley, September 1985. Also Computer Science Division Technical Report UCB/CSD 86/259.

23. T. Gross, "Code Optimization of Pipeline Constraints," Ph.D. thesis, Stanford University, August 1983.

24. D. A. Moon, "Architecture of the Symbolics 3600," *Proc. 12th Symposium on Computer Architecture,* Boston, MA, June 1985.

25. "The Lambda System: Technical Summary LMI," Lisp Machine, Inc., 1983.

26. R. R. Burton et al., *Papers on Interlisp-D,* Xerox PARC tech. report SSL-80-4, September 1980.

27. C. Ponder, "But Will RISC Run Lisp? (A Feasibility Study)," unpublished master's report, University of California, Berkeley, April 1983.

28. J. Donahue and A. Demers, "Data Types Are Values," *ACM Trans. on Programming Languages and Systems,* July 1985.

29. C. B. Roads, "3600 Technical Summary," Symbolics, Inc., Cambridge, MA, 1983.

30. P. M. Hansen, "Coprocessor Architectures for VLSI," unpublished thesis research proposal, University of California, Berkeley, May 1985.

CHAPTER 8

WHAT PRICE SMALLTALK?

David Ungar and David Patterson

1 INTRODUCTION

In anticipation of tremendous hardware advances, software researchers have fashioned expansive programming environments to improve programmer productivity. Even with the march of technology, exploratory programming environments such as Smalltalk-80 [1, 2] require such expensive computers that few programmers can afford them. With the hope of increasing that community, a research project at the University of California created a reduced instruction set computer [3] for Smalltalk called SOAR, which stands for Smalltalk on a RISC [4, 5]. Figure 8.1 shows SOAR, a 32-bit NMOS microprocessor.

We are now able to estimate the performance implications of the Smalltalk-80 programming environment. In the next section we list the demands that Smalltalk-80 places on traditional computer systems, then present software and hardware ideas that meet those demands.

We have learned a number of lessons from the SOAR microprocessor, including the hard lesson that we now call "The Architect's Trap."

2 THE DEMANDS OF SMALLTALK-80

To improve programmer productivity, Smalltalk-80 removes four restrictions found in conventional programming systems.

©1987 IEEE. Reprinted, with permission, from *IEEE Computer*, vol. 20, no. 1, pp. 67–74, January 1987.

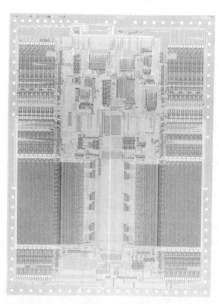

Figure 8.1 Microphotograph of the SOAR microprocessor. Using 4-micron line widths, Joan Pendleton led the implementation of this 32-bit microprocessor [6,7]. It is 10.7 mm by 8.0 mm, uses 35,700 transistors, dissipates about 3W, and runs at about 400 ns per instruction. These chips were fabricated by MOSIS, and have performed better than expected. Xerox also volunteered to make some SOAR chips. Although they run at 330 ns per instruction, the Xerox chips do not correctly perform all tests. We are considering a fabrication attempt with 3-micron line widths to further reduce the cycle time.

1. *Run-time Data Typing Instead of Compile-time Type Declarations.* Smalltalk-80 programmers do not have to declare the types of variables at compile-time [2]. The Fortran statement $I = J + K$ denotes integer addition and can be performed with a single Add instruction; but Smalltalk-80 has no type declarations, so J and K may hold values of any type, from Booleans to B-trees. Thus, every time a Smalltalk-80 system evaluates $J + K$ it must first check the types and then perform the appropriate operation.

   Measurements of conventional Smalltalk-80 systems show that over 90 percent of the + operations do the simplest possible operation, integer addition [8]. Since a type check takes at least as long as an Add instruction, most Smalltalk-80 systems waste a lot of time checking types for integer arithmetic.

2. *Dynamically Bound Messages Instead of Statically Bound Procedures.* Smalltalk was probably the first system to be called *object oriented.* The object-oriented approach to programming associates routines with data structures, encouraging programmers to think of programming as sending messages to data objects rather than calling pro-

cedures to update shared data. The routine to be invoked depends on the type of data, and since the type of data can change at run time, the equivalent of a simple procedure call in most systems is, in Smalltalk-80, a comparison and table lookup.

3. *Fast Compilation to Interpreter Instead of Optimizing Compilation to Native Hardware.* One way to improve programmer productivity is to reduce the time it takes to change and test a program. To shorten the edit-compile-debug cycle, the Smalltalk programmer recompiles a routine using a simple and fast compiler that produces code for a Smalltalk interpreter. The Smalltalk interpreter provides the foundation for an excellent debugger, which further improves programmer productivity.

    Like many other systems, Smalltalk supports separate compilation so that the programmer waits for only one routine to be recompiled, instead of the whole system; unlike other systems, Smalltalk further shortens this cycle by avoiding the conventional linking step: routines are linked on the fly, as needed. To shorten program development, Smalltalk sacrifices potential compile-time optimizations and static linking that lead to faster program execution.

4. *Automatic Dynamic Storage Allocation and Reclamation Instead of Programmer-Controlled Dynamic Structures.* The designers of Smalltalk believed that programmers should not be encumbered with managing dynamically varying data, so storage management was left to the system. Exploratory programming environments are also required to avoid distracting pauses while collecting garbage—a fast and predictable interaction with the programming environment is a major contributor to programmer productivity [9]. Smalltalk is very challenging for garbage collection, since Smalltalk programs generate garbage about 10 times faster than most Lisp programs.

3 THE COSTS ON TRADITIONAL SYSTEMS

Given the foregoing demands, how well does Smalltalk-80 run on conventional computer systems? The first step in answering this question is the determination of a fair way to measure performance. It is always dangerous to rely on architects to supply the tests of their creations, and fortunately we did not have to do so. The creators of Smalltalk have a standard set of benchmarks that all Smalltalk implementors use to compare systems [10]. Xerox rates performance by taking the mean of 13 macrobenchmarks, plus the text-scanning and BitBlt microbenchmarks. Compilation, decompilation, and searching the Smalltalk program hierarchy are examples of macrobenchmarks.

Table 8.1 compares the performance of several Smalltalk-80 systems using five of the macrobenchmarks. Both Tek 4404 and Berkeley Smalltalk (BS) use

TABLE 8.1   Speed of Smalltalk-80 Systems.

System	Maker	Host Processor	Instruction Time in ns	Execution Model	Speed
BS	UCB	68010	400 (18%)	interpreter	11%
Tek 4404	Tektronix	68010	400 (18%)	interpreter	25%
Dorado	Xerox	Dorado	67 (100%)	microcode	100%
SOAR	UCB	SOAR	400 (18%)	compiler	107%

software interpreters; BS was the experimental vehicle that we used to study Smalltalk. Dorado is the machine used at Xerox to run Smalltalk. The SOAR chip has been placed on a board, the board has been placed in a workstation, and this package is running diagnostic programs, but because we have not brought up the complete Smalltalk system, we must use simulations to estimate performance (see Figure 8.2). The SOAR system was described at a recent conference [11].

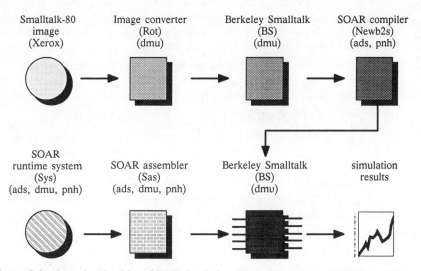

Figure 8.2   Steps involved in a SOAR simulation. First, Rot removes the object table from the Xerox Smalltalk-80 image. We then use BS to make any modifications necessary in the image (to eliminate some *becomes*). Newb2s produces a Smalltalk image for SOAR by converting the BS objects to SOAR format, and Hilfinger's Slapdash compiler translates the bytecoded programs into SOAR instructions. We have also coded the Smalltalk primitive operations and storage management software in SOAR assembly language. After this is assembled, it is fed to Daedalus, our SOAR simulator, with the Smalltalk image. The initials below each system identify its author: ads is Dain Samples, phn is Paul Hilfinger, and dmu is David Ungar.

TABLE 8.2 Summary of Innovations.

ST-80 Demands	Category	Name	% slower if omitted
Interpretation		Compiler	100%
Dynamic messages		In-line cache	33%
	Software	Direct pointers	20%
Automatic storage		Generation scavenging	12%
Slow message sends		Register windows	46%
	Architecture	Tagged integers	26%
Run-time data types		Two-tone instructions	16%

The Dorado, a $120,000 ECL* personal computer that supports a single programmer at a time is generally agreed to be fast enough for Smalltalk. (Think of it as a personal computer for people who consider the Mercedez-Benz a commuter car.) The Dorado is four to eight times faster than the Smalltalk interpreters on Motorola 68010 microprocessor-based systems; this might be expected since the instruction time of the Dorado is about six times faster than the instruction time of the 68010. SOAR, however is about the same speed as the Dorado even though SOAR's instructions are six times slower; SOAR is faster than expected because it uses several innovations in software and hardware (see Table 8.2) explained in the next two sections.

4 REDUCING COSTS THROUGH SOFTWARE INNOVATION

The first place to look for performance improvements is in software; it will be a very long time before it is as easy to build and debug hardware as it is to write and debug an equivalent program. Architects should consider hardware implementations only when there are tremendous gains to be had in total system performance, and when they are sure that system function will never change. As we shall see, architects don't always follow this sound advice.

4.1 Interpretation

The Smalltalk-80 system is defined by a stack-oriented virtual machine based on the Dorado Smalltalk-80 implementation [12]. Each instruction consists of one to three bytes; instructions generally correspond to tokens in the source program. Instructions are usually called *bytecodes*. Bytecodes have the following advantages:

* Emitter-coupled logic is an expenive technology used in mainframes and supercomputers.

- The simple correspondence between source and object code simplifies the compiler and debugger.
- Smalltalk can be transported to a new machine by writing only the virtual machine emulator. This approach has drawbacks too:
- Decoding such dense instructions takes either substantial hardware or substantial time. For example, the instruction fetch unit consumes 20 percent of the Dorado CPU [13], and in Berkeley Smalltalk decoding a simple bytecode takes twice as long as executing it.
- Some of the high-level instructions require many microcycles to execute. These multicycle instructions must be sequenced by a dedicated control unit. The alternative is simply to compile to the native instruction set of the machine, hoping to gain performance with the traditional advantage of compilation over interpretation. We estimate compilation improves the performance of SOAR by 100 percent. A negative consequence of our decision to abandon bytecodes is that it forces us to rewrite the Smalltalk-80 debugger. Lee has designed a debugger for SOAR and has built a prototype in Berkeley Smalltalk [14]. He has exploited the hardware organization of SOAR in the design of the debugger to add a conditional breakpoint facility and increase execution speed during debugging.

SOAR may also have played a role in inspiring a compiler project at Xerox by Deutsch and Schiffman [15]. This implementation, called PS, provided the speed of compilation while maintaining consistency with the interpreter so that the debugger would not have to be modified. Their system runs almost twice as fast as interpreters using the same microprocessor.

4.2 Caching Call Targets in Line

Another way to improve Smalltalk performance is by decreasing the time taken to find the target of a call. Once computed, the target's address can be cached in the instruction stream for later use, as suggested by Schiffman and Deutsch [15]—Figures 8.3 and 8.4 illustrate this idea. In-line caching exacts a price for its time savings: the processor must support non-reentrant code.

Although complicated, however, in-line caching pays handsome rewards. Smalltalk systems conventionally save call targets in a hash table, but the overhead of probing into a hash table would slow SOAR by 33 percent. The penalty for in-line caching is a software trap mechanism to synchronize process switches and cache probes. If we were forced to omit this, we could use a table containing the last addresses with an indirect in-line cache, slowing SOAR down by only 7 percent. Even with in-line caching, SOAR still spends 11 percent of its time in call-target cache probes and another 12 percent handling misses. Further research into computing the call's target might therefore yield even more substantial savings.

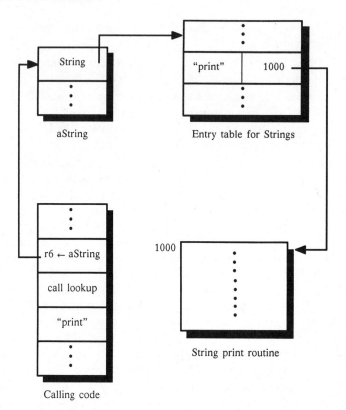

Figure 8.3 Caching the target address in the instruction stream. In this example, the print routine is called with an argument that is a string. (The argument is passed in r6.) The first time the call instruction is executed, the call contains the address of a lookup routine and the word after the call contains a pointer to the name "print." The lookup routine follows the pointers to the entry table for strings, and finds the entry for "print." It then overwrites the call instruction with a call to that routine and replaces the word after the call with the type of argument (string). (Reprinted from "Architecture of SOAR: Smalltalk on a RISC," 11th An. Int'l Symp. Computer Architecture, ©IEEE.)

4.3 Object-Oriented Storage Management

Software is also the architect's best option for managing dynamically vary-ing data structures, called objects in Smalltalk-80. Smalltalk-80 objects are smaller and more volatile than data structures in most other exploratory pro-gramming environments, averaging 14 words in length and living for about 500 instructions. Smalltalk-80 systems face three challenges in managing storage for objects:

- *Automatic storage reclamation.* On average, 12 words of data are freed and must be reclaimed per 100 Smalltalk-80 virtual machine bytecodes executed.

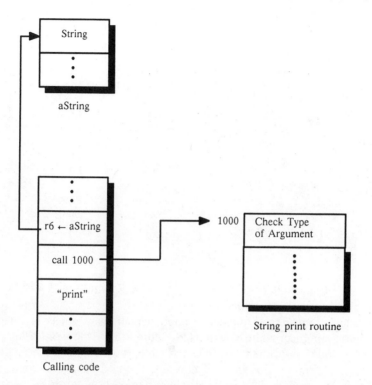

Figure 8.4 Caching the target address in the instruction stream. The next time the call is executed, control goes directly to the string print routine. A prolog checks that the current argument's type matches the contents of the word following the call instruction. This word contains the type that the argument had the previous time the call was executed. If the types match, control falls through to the string print routine; otherwise another table lookup is needed. (Reprinted from "Architecture of SOAR: Smalltalk on a RISC," *11th An. Int'l Symp. Computer Architecture,* © IEEE.)

- *Virtual memory.* All objects must be in the same address space.
- *Object-relative addressing.* Although offsets into objects are known at compile time, base addresses are not. Code must be compiled to address fields relative to dynamically determined base addresses.

4.4 Automatic Storage Reclamation

SOAR supports generation scavenging [16] to reclaim storage efficiently without requiring costly indirection or reference counting. This algorithm is based on the observation that most objects either die young or live forever; hence objects are placed into two generations, and only new objects are reclaimed.

Storage reclamation has a strong impact on performance; most other algorithms would squander 10 percent to 15 percent of SOAR's time on automatic storage reclamation, whereas generation scavenging uses only 3 percent (see

TABLE 8.3 Summary of Generation Scavenging's Performance.

	Berkeley Smalltalk	SOAR
Execution model	Interpreted	Compiled
Source of data	Measurements	Simulations
Processor		SOAR
Cycle time	400 ns	400 ns
	MC68010	
CPU time overhead		
mean	1.5%	0.9%
worst case	N.A	3.3%
Pause time		
mean	160 ms	19 ms
worst case	330 ms	28 ms
Peak main memory usage	200 KB	200 KB

Table 8.3). Without generation scavenging, SOAR would take 4–15 percent more cycles to run the benchmarks.

Traditional software and microcode implementations of object-oriented systems rely on an object address table (see Figure 8.5). Each word of an object contains an index into this table, and the table entry contains the address of each object; the indirection through the table primarily supports compaction. Generation scavenging, however, provides compaction for free, permitting SOAR to function without an object table (see Figure 8.6); without this algorithm, the extra work to follow indirect pointers through the object table would slow SOAR down by 20 percent. These results seem to be confirmed by a recent Smalltalk-80 interpreter that used generation scavenging with direct

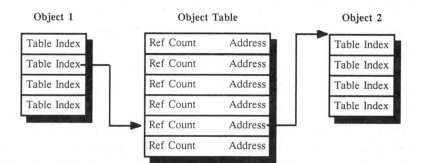

Figure 8.5   Indirect addressing. In traditional Smalltalk-80 systems, each pointer is really a table index. The table entry contains the target's reference count and memory address. This indirection required previous Smalltalk-80 systems to dedicate base registers to frequently accessed objects. The overhead to update these registers slowed each procedure, call and return. (Reprinted from "Architecture of SOAR: Smalltalk on a RISC," *11th An. Symp. Computer Architecture*, © IEEE.)

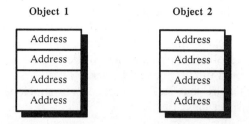

Figure 8.6 Direct addressing. A SOAR pointer contains the virtual address of the target object. This is the fastest way to follow pointers. (Reprinted from "Architecture of SOAR: Smalltalk on a RISC," *11th An. Symp. Computer Architecture, © IEEE.*)

pointers [17]; although we would expect it to be a factor of two slower than PS because it is interpreted rather than compiled, this implementation is close to 90 percent of the speed of PS. It may be that direct pointers and generation scavenging, not used in PS, explain the surprisingly close performance.

5 REDUCING COSTS THROUGH ARCHITECTURAL INNOVATION

The basic theme of the SOAR architectural additions is to allow the normal case to run fast in hardware and to trap to software in the infrequent complicated case.

5.1 Tags Trap Bad Guesses

SOAR follows the Smalltalk-80 virtual machine in supporting only two data types with tags: integers and pointers [1]. However, SOAR departs from the Smalltalk-80 virtual machine by starting arithmetic and comparison operations immediately and simultaneously checking the tags (Figure 8.7 shows the SOAR

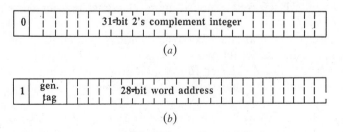

Figure 8.7 SOAR tagged data types. SOAR supports two data types, 31-bit signed integers and 28-bit pointers. Pointers include a generation tag. SOAR words could have contained 32 bits of data plus one bit of tag for a total of 33 bits. The scarcity of 33-bit tape drives, disk drives, and memory boards led us to shorten our words to a total of 32 bits including the tag (31 bits of data.) (Reprinted from "Architecture of SOAR: Smalltalk on a RISC," *11th An. Symp. Computer Architecture,© IEEE.*)

tag layout). Most often both operands are integers and the correct result is available after one cycle, but if it is not, SOAR aborts the operation and traps to routines that carry out the appropriate computation for the data types. SOAR is the only Smalltalk-80 system that overlaps these operations. Without tagging, SOAR would run 26 percent slower.

A tagged architecture that lacks microcode must include instructions that manipulate and inspect tags. Because the Smalltalk system already relies on the compiler to ensure system integrity, we can allow the compiler to mix instructions that manipulate tags with instructions constrained by tags.

Each SOAR instruction contains a bit that either enables or disables tag checking. The untagged mode turns off all tag checking and operates on raw 32-bit data. In untagged mode the tag bits are treated as data, and the complete instruction set can be used to manipulate this data. Untagged instructions also allow programs written in conventional languages such as C and Pascal to run on SOAR. Instead of providing two versions of each instruction, we could have defined a mode bit in the program status word (PSW), but this would have been expensive, increasing execution time by 16 percent.

5.2 Multiple Overlapping On-Chip Register Windows

We used hardware to improve the performance of arithmetic operations, and also to improve the performance of the frequent calls and returns. SOAR, like RISC I, optimizes them by providing a large, on-chip register file. The registers are divided up into overlapping windows, and instead of saving or restoring registers, calls or returns merely switch windows (Figure 8.8).

When we compared Smalltalk subroutines with C language subroutines of RISC I, we found that the shorter Smalltalk subroutines passed fewer operands and used fewer local variables, and so needed fewer registers. Hence, each SOAR register window has 8 registers, compared with RISC I's 12. Figures 8.8 and 8.9 show the register organization of SOAR. The inclusion of register windows requires the addition of 56 more registers; it also requires the addition of a register to select the current window, a register to detect overflows by recording the last saved window, more elaborate register decoders, and trapping logic [7].

Despite the cost of all the added hardware, Smalltalk-80's predilection for procedure calls makes the register windows very important. Saving and restoring a conventional register file would slow the machine down by 46 percent, even with load- and store-multiple instructions, which we discuss next.

Register windows allow the use of many registers in each procedure without making procedure calls and returns expensive [3]. Many RISC machines use optimizing compilers to allocate variables to registers and thus avoid register windows, but while this may be the right way to go with conventional programming systems like C and Pascal, the extra compile time used in register allocation may be inappropriate for exploratory programming environments such as Smalltalk.

Physical Registers   Local Registers

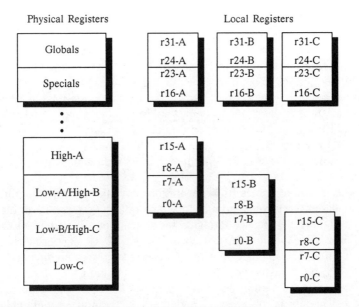

Figure 8.8 SOAR's register windows. Like RISC I, SOAR has many physical sets of registers that map to the logical registers seen by each subroutine. (Reprinted from "Architecture of SOAR: Smalltalk on a RISC," *11th An. Symp. Computer Architecture*, © IEEE.)

6 THE ARCHITECT'S TRAP

After the careful and clean design of the first Berkeley RISC projects, SOAR went on to try many architectural ideas. Although fully aware of the dangers, we fell into what we now call the "architect's trap" several times:

- Each idea was a clever idea;
- Each idea made a particular operation much faster;
- Each idea increased design and simulation time; and
- Not one idea significantly improved overall performance!

Put another way, the primary impact of these clever ideas was to increase the difficulty of building SOAR and thus to lengthen the development cycle.

	R31
Global	R24
Special	R23
	R16
High	R15
	R8
Low	R7
	R0

Figure 8.9 Logical view of the register file. The Highs hold incoming parameters and local variables. The Lows are for outgoing arguments. The Specials include the PSW and a register that always contains zero. The Globals are for system software such as trap handlers. (Reprinted from "Architecture of SOAR: Smalltalk on a RISC," *11th An. Symp. Computer Architecture*, © IEEE.)

One example is the load- and store-multiple instructions mentioned in the previous section. When the number of activations on the stack exceeds the on-chip register capacity, SOAR traps to a software routine that saves the contents of a set of registers in memory. Unlike RISC II, SOAR has load- and store-multiple instructions to speed register saving and restoring. These instructions can transfer eight registers in nine cycles (one instruction fetch and eight data accesses). Without them, the system would need eight individual instructions that would consume sixteen cycles (eight instruction fetches plus eight data accesses). In retrospect, these multicycle instructions added some complexity to the design, and the benefit of a three percent reduction of execution time is not worth the cost.

Another example comes from generation scavenging, which requires that a list be updated whenever a pointer to a new object is stored in an old object. While designing SOAR, we thought that stores would be frequent enough to warrant hardware support for this check; hence SOAR tags each pointer with the generation of the object that it points to. While computing the memory address, the store instruction compares the generation tag of the data being stored with the generation tag of the memory address. For 96 percent of the stores, list update is unnecessary and the store completes without trapping. Once again we rely on tags to confirm the normal case and trap in the unusual case. Surprisingly, however, tagged stores are so infrequent that hardware support saves only 1 percent of the total time. This feature does not seem worth the effort.

Alas, we have more than two examples of being caught in the architect's trap. We have put these observations to use by calculating the performance of some variations of SOAR and comparing them to some real systems (see Table 8.4). Our predictions of SOAR's performance are based on simulated macrobenchmark times and do not include virtual memory, operating system, and I/O overhead; however, all of the Smalltalk-80 systems we know about tend to be limited by the raw CPU speed rather than the overhead time for virtual memory, the operating system, and I/O. For a fair comparison, we assume a 400-ns cycle time for SOAR, RISC II, and MC68010. By comparing the speeds of different systems, we can gain some insight into the reasons for SOAR's good performance.

If we were to build a second-generation SOAR, we would implement Pendleton's reorganization of the pipeline, which would shorten SOAR's cycle time by 25 percent [7]. When combined with the winning features from the previous table the system would run faster than SOAR, even though it lacks a half dozen of SOAR's architectural innovations. Although the overall measurements confirm that reduced instruction sets can run Smalltalk-80 efficiently, this calculation goes further; it suggests that SOAR is not RISCy enough.

The software techniques developed for SOAR may eventually be used to construct fast, compiled Smalltalk-80 systems on general purpose processors. The RISC II row of Table 8.4 reflects the impact of removing the Smalltalk-specific architectural features of SOAR. The next row, RISC II with-

TABLE 8.4 **Trimming the Fat from SOAR. (Assumes 400-ns Cycle Time for SOAR, RISC II, and 68010. Dorado and MC68010 Figures are Measurements. Same Benchmarks as Table 8.1.)**

Configuration	Speed
Important architectural features + rearranged pipeline	103%
Full SOAR	100%
Dorado	93%
Important architectural features only	81%
Deutsch-Schiffman 68020	72%
RISC II	62%
RISC II without register windows	50%
Full SOAR without software ideas	41%
Deutsch-Schiffman 68010	36%
Full SOAR without hardware ideas	34%
Tek 68010 interpreter	24%
Stripped SOAR	22%

out register windows, is representative of the commercial reduced-instruction-set microprocessors. These machines would run Smalltalk-80 twice as slowly as SOAR, but this may not be too high a price to pay for a simpler and more general off-the-shelf design.

Falling into architects' traps will delay your project, making it less attractive as competing machines are announced. For example, near the end of the SOAR project, the Deutsch-Schiffman Smalltalk system PS doubled performance when Motorola announced their 32-bit follow-on to the 68010, called the 68020. The 68020 includes a small on-board instruction cache, executes instructions in 180 ns, and uses about five times as many transistors as SOAR. Thus, delays in your project inevitably reduce the relative attractiveness of your machine, since it is unlikely that every other project will face the same delays.

Table 8.2 summarized the innovations used in SOAR to improve the performance of the exploratory programming environment Smalltalk-80. Compared with the Motorola 68010, SOAR more than doubles performance while halving the number of transistors. Its features reduce the performance penalty for Smalltalk from a factor of 3.6 to a factor of 1.3, increasing the chances that programmers will use exploratory programming environments (see Table 8.5).

We recommend that anyone faced with the task of building a computer for an exploratory programming environment to consider compilation to a reduced instruction set with register windows and one-bit tags. There are four places to look for further performance gains: compiler technology, implementation technology [7], optimization of the run-time support primitives (which consume about two thirds of SOAR's time), and better hardware or software algorithms to cache call-target lookups (which consume 23 percent of SOAR's time). Of these, implementation technology—circuit design and VLSI processing technology—have the most dramatic impact. Since we started this project, the

TABLE 8.5   Cost of Smalltalk-80 Features: The Standard Approach Versus SOAR.

ST-80 Feature	Cost (% Slower)	Optimization	Cost (% Slower)
Interpretation	100%	Compiler	0%
Dynamic messages	56%	In-line cache	23%
Automatic storage	15%	Generation scavenging	3%
Object table	20%	Direct pointers	0%
Run-time data types	26%	Tagged integers	0%
Many subroutines	46%	Register windows	0%
Total cost of ST-80 (standard architecture)	3.6:1	(SOAR)	1.3:1

standard VLSI technology available to universities has improved from four-micron line widths to three-micron. This change alone could reduce our cycle time from 400 ns to 290 ns, a contribution as important as that of register windows.

Finally, we warn all architects to watch their step. Be sure you evaluate your proposal in terms of benefits to the whole system and not just to one part, and remember that a software solution is better than a hardware solution in every facet except speed. The bait of the architect's trap is your own ingenuity, but the trap ensnares a whole project, not just the fool who springs it. Beware!

7   ACKNOWLEDGMENTS

The SOAR project was sponsored by DARPA, and many students and faculty participated. We thank all of them for their hard work. We also wish to thank Walter Beach and Garth Gibson for their comments on an early draft of this article.

REFERENCES

1. A. J. Goldberg and D. Robson, *Smalltalk-80: The Language and Its Implementation*, Addison-Wesley, Reading, MA, 1983.

2. T. Kaehler and D. Patterson, *A Taste of Smalltalk*, W. W. Norton and Company, New York, 1986.

3. D. A. Patterson, "Reduced Instruction Set Computers," *Comm. ACM*, vol. 28, no. 1, pp. 8–21, January 1985. (Special issue on computer architecture.)

4. D. Ungar et al., "Architecture of SOAR: Smalltalk on a RISC," *11th Annual Int'l Symposium Computer Architecture*, pp. 188–197, Ann Arbor, MI, June 1984.

5. D. M. Ungar, "The Design and Evaluation of a High Performance Smalltalk," Computer Science Technical Report No. UCB/CSD 86/287 (Ph.D. dissertation), Computer Science

Division (EECS), University of California, Berkeley, March 1986. Also to be published by MIT Press as part of the ACM Doctoral Dissertation Award series.

6. J. M. Pendleton et al., "A 32b Microprocessor for Smalltalk," *IEEE Journal Solid State Circuits*, October 1986.

7. J. Pendleton, "A Design Methodology for VLSI Processors," Ph.D. dissertation, Department of E.E.C.S., University of California, Berkeley, September 1985.

8. R. Blau, "Tags and Traps for the SOAR Architecture," *Smalltalk on a RISC: Architectural Investigation*, ed. David A. Patterson, Computer Science Division, University of California, Berkeley, pp. 24–41, April 1983.

9. A. J. Thadhani, "Interactive User Productivity," *IBM Systems Journal*, vol. 20, no. 4, pp. 407–421, 1981.

10. K. McCall, "The Smalltalk-80 Benchmarks," *Smalltalk: Bits of History, Words of Advice*, ed. Glenn Krasner, pp. 151–173, Addison-Wesley, Reading, MA, 1983.

11. A. D. Samples, D. M. Ungar, and P. N. Hilfinger, "SOAR: Smalltalk without Byte-codes," *Proc. ACM Conf. on Object-Oriented Programming Systems, Languages, and Applications*, Portland, OR, September 1986.

12. L. P. Deutsch, *The Dorado Smalltalk-80 Implementation: Hardware Architecture's Impact on Software Architecture*, pp. 113–126, Addison-Wesley, Reading, MA, September 1983.

13. K. A. Pier, "A Retrospective on the Dorado, a High-Performance Personal Computer," *Proc. 10th Annual Symposium Computer Architecture*, pp. 252–269, Stockholm, Sweden, June 1983.

14. P. K. Lee, "The Design of a Debugger for SOAR," Master's thesis, Computer Science Division, Department of E.E.C.S., University of California, Berkeley, September 1984.

15. L. P. Deutsch and Allan M. Schiffman, "Efficient Implementation of the Smalltalk-80 System," *Proc. 11th Annual ACM SIGACT-SIGPLAN Symposium Principles of Programming Languages*, Salt Lake City, Utah, January 1984.

16. D. Ungar, "Generation Scavenging: A Non-Disruptive High Performance Storage Reclamation Algorithm," *Proc. ACM SIGSOFT/SIGPLAN SE Symposium Practical Software Development Environments*, Pittsburgh, PA, pp. 157–167, April 1984.

17. P. J. Caudill and A. Wirfs-Brock, "A Third Generation Smalltalk-80 Implementation," *Proc. ACM Conf. on Object-Oriented Programming Systems, Languages, and Applications*, Portland, OR, September 1986.

CHAPTER 9

SPECIAL PURPOSE CHIP FOR PRODUCTION SYSTEMS

G. T. Alley, W. L. Bryan, R. O. Eason, D. F. Newport, and D. W. Bouldin

1 INTRODUCTION

This paper presents an architecture that uses massive parallelism to improve the execution time for matching in rule-based systems by at least an order of magnitude. We describe implementation of the architecture in a special purpose VLSI chip which, when attached to a rule processor, looks like a write-only memory. The chip is first programmed with tests for a specific set of rules; then, during operation, the chip performs, in parallel, the symbol-matching tasks on all the data. Finally, the integration of a collection of these chips into a production system is described.

Expert systems based on rule processing (i.e., OPS-type languages) imitate human expertise in problem solving by employing a set of *rules* (the expert knowledge); a *global database* (observations about the problem); and an *inference engine* that performs an iterative *recognize-act cycle* of match, conflict resolution, and rule firing (conclusion or action) [2, 3, 11]. All rule-processing production systems use rules of the form *if-then* or *then-if*. The *recognize* part of the recognize-act cycle performs multiple matching of a global database with a set of **if** conditions for each rule or production. The size of the global database gives an initial limit on the effort required to complete one recognize-act cycle. The cycle is repeated as intermediate firings modify global data until either a rule fires and solves the problem or the production system halts without a solution.

Of the three phases of the recognize-act cycle, the rule firing phase requires the shortest time, followed by conflict resolution; matching generally requires

316

many (ten to twenty) times the effort of the other two phases [7]. This is especially true if the matching is implemented as a brute force sequential computing process. Several techniques may decrease the problem solution time: the use of faster or high data-bandwidth processors, optimized matching algorithms that reduce repetitious matching, a parallel approach to the matching task, or some combination of all these [1, 2, 4, 5, 6, 8, 9, 10].

2 THE ARCHITECTURE

The architecture described here was designed to address the matching task with a combination of the repetition-reducing Rete algorithm and parallel processing. In an expert system employing this architecture, all rules would be programmed (similar to rule compilation in OPS) and made simultaneously available to all processors in a single broadcast of the complete global database (also called working memory elements or WMEs) during the initial recognize-act cycle. Following the initial broadcast, only WMEs added, removed, or modified by rule firing must be rebroadcast during later matching cycles. Full implementations of expert systems based on this architecture should run ten to twenty times faster than brute force sequential implementations.

The architecture invokes parallelism to perform the matching task associated with production system rules. During matching each rule generates a number of structures of the form

$$x:, y, > 4, \leq z, \neq 7$$

where x, called a binding variable, is compared to or *matched* with variables and constants using the associated conditional operators $<, =, >, \leq, \geq, \neq$. The working memory contains a list of the possible values for subsets of the variables. The problem is to find a combination of WMEs such that every associated structure *simultaneously* holds for the current values of the variables and constants. If such a combination is found, the rule is said to have fired.

As the process of finding a combination of WMEs begins, all variables are unbound. At each step in the process a WME is selected from a class, the variables are bound to the given values, and then checks are made to see if all the given conditions still hold. Note that tests involving a variable are not made while that variable is unbound. If the conditions hold, the process is repeated by choosing a WME from the next class. If any of the conditions do not hold, the next WME representing the first class is tried. Backtracking becomes necessary when all the WMEs from a class have been tried without success; all variables which were bound while testing WMEs from this class must be unbound. The process then resumes, trying a new WME from the previous class. In this manner all possible valid combinations of WMEs may be found. The rule processor decides which valid combination to use.

3 THE VLSI IMPLEMENTATION

The architecture just described has been implemented in VLSI as OOPS-MOP (Our OPS Matching Only Processor), and prototypes using 3-micron NMOS technology have been fabricated and tested. In a typical system the rule processor selects the WMEs to be tried and the OOPS-MOP chip performs the required tests on all bound variables. The OOPS-MOP chip is divided into eight *chunks*, each of which tests a single binding variable against five constants or other variables. This is accomplished by each chunk being in turn divided into six *slivers*, one containing the binding variable and the other five each being responsible for a single comparison. By means of this regular structure the tests are performed in parallel. If more than five tests are required for a binding variable, more than one chunk may be used.

For any given chunk to pass, all five slivers must pass. Similarly, for the chip to report a pass condition, all eight chunks must pass. Beyond the chip level a high degree of parallelism may be achieved by ANDing together the outputs of multiple chips. Note that any unused (possibly defective?) sliver or chunk may be disabled in such a way that it always passes. The OOPS-MOP architecture thus allows multiple rule structures to be tested in parallel.

4 OPERATIONAL MODES

To the rule processor, the OOPS-MOP chip looks like a write-only memory. The data path is sixteen bits wide, and the address path is fourteen bits wide with two of the lines used for control. One of the control lines selects the operational mode: program or run. Each mode (shown in Figure 9.1) contains two instruction types. The second control line selects the instruction type.

Program mode is used during compilation to load the tests which are to be performed on the binding variables. In this mode the sliver address is included as part of the address field. The comparison to be performed by the indicated sliver is extracted from the opcode portion of the address field. Of the two types of instructions in program mode, the first indicates that the sliver is to test a constant (which does not change during run mode); the value of the constant is extracted from the data field. The second type of instruction indicates that the sliver is to test a variable (which may be bound and unbound during run mode); the variable label or *tag* value is extracted from the lower twelve bits of the data field. The process of loading a sliver with a variable label unbinds that variable by default. The value of the variable itself is defined during run mode.

After compilation, run mode is used by the rule processor to perform the matching tasks. In run mode all slivers representing variables compare their tag with the one represented in the address field, and any and all slivers with a match perform the operation indicated by the instruction type. Of the two instruction types in run mode, the first indicates that the variable is to be

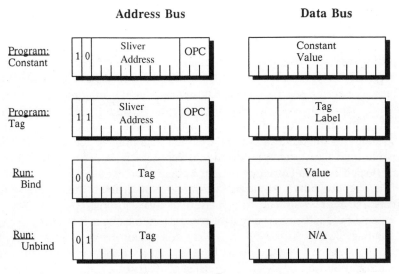

Figure 9.1 Operational modes.

bound to the value represented in the data field, and the second indicates that the variable is to be unbound and the data field ignored. At the end of each cycle, each sliver will report the result of its test.

5 FUNCTIONAL ELEMENTS

Figure 9.2 shows the major functional elements of a chunk of the OOPS-MOP chip. A chunk contains two types of slivers, which are identical except that the binding sliver (the top sliver of each chunk of six slivers) does not have a comparator and handles the result differently. Each sliver is composed of four

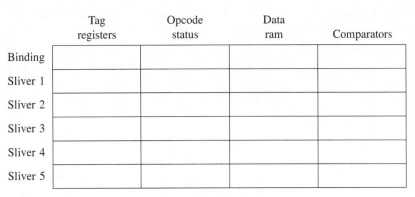

Figure 9.2 Chunk diagram.

major elements: the tag register, the opcode/status register, the data RAM, and the comparator.

The 12-bit tag register consists of a static memory with a built-in matching capability that allows it to act as a content-addressable memory. The tag register is used only in slivers that represent variables; the variable label is loaded into the register during program mode. During run mode the tag register will signal when the tag value is identical to the one presented on the address bus. In the event of a match the next occurrence depends on the type of instruction: either the variable is unbound or the value on the data bus replaces the value of the variable (and the variable is bound). Slivers which do not show a match and those which contain constants remain in their current state.

The opcode/status register performs a slightly different function in binding slivers and regular slivers. The register includes a three-bit field; in regular slivers the field indicates the type of comparison the sliver is to perform against its binding variable (i.e., $<, =, >, \leq, \geq, ! =$, or always pass), whereas in binding slivers only one bit of the field is used to enable or disable the chunk. A fourth bit of the opcode/status register is used to indicate whether the sliver contains a variable or a constant. These four bits are defined in program mode and do not change while in run mode. A fifth bit indicates if a variable is bound or unbound. Unbinding a variable in a regular sliver temporarily disables that sliver (it passes), whereas unbinding a variable in a binding sliver temporarily disables the chunk of which it is a part. The fifth bit will automatically be set to the unbound state as the sliver is loaded with a variable label in program mode. In run mode it is set or cleared according to the instruction type. Constants always act as bound variables.

The data RAM in each sliver consists of a 16-bit dynamic memory. If the sliver is to represent a constant the data RAM is loaded with the constant value during program mode and the contents do not change during run mode. If a variable is represented the data RAM holds its current value when bound and is ignored when it is unbound.

Comparators appear only in the regular slivers. Each comparator is composed of two bit-serial comparators, one for the low byte and one for the high byte, that compare the value in the data RAM with that in the binding sliver. For each constant or bound variable, the result of the high and low byte comparisons and the opcode field determine whether the sliver reports a pass or fail. The bit-serial action of the comparator is also used to refresh the dynamic RAM or to load it with new values.

A photomicrograph of the OOPS-MOP chip is shown in Figure 9.3.

6 ADDITIONAL FUNCTIONS

Aside from the chip select, address, and data lines, several pin functions are worthy of mention. The results of all tests made on a chip are reported to the outside world on the *pass* line. The validity of the pass line is indicated by

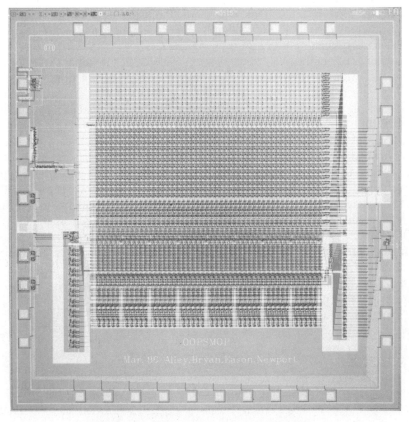

Figure 9.3 Photomicrograph of OOPS-MOP.

the *completion* line. Another line allows the system to select between two's complement and unsigned magnitude comparisons. A *reset* line is provided to set all opcode registers to their *always pass* state, preventing unprogrammed slivers and chunks from interfering with the test results generated by the programmed slivers. Finally, the clocking is provided by an external source.

7 CONCLUSIONS

The OOPS-MOP chip allows a system to perform a virtually unlimited number of comparisons in parallel. First the system compiles the desired operations and loads them into the OOPS-MOP chip; then the chip may be accessed as content-addressable memory, allowing a variable to be altered in numerous locations and all tests pertaining to that variable to be reevaluated simultaneously. Because each sliver or chunk may be disabled, chip defects may be bypassed, OOPS-MOP is extendable to wafer-scale integration. OOPS-MOP may also be used in applications other than OPS language processors including Prolog inference engines, fuzzy logic inference engines, and query searches for database systems.

A number of additional features were considered at the initial stages of the design but were not implemented because of area or design time shortages. Among these features were the use of an internal clock, read access to internal data, variable-size chunks, and a FIFO for burst transmission. The internal clock would allow selection of the internal or external clock, by means of an additional input pin. Although currently the rule processor can write only to slivers, allowances could be made to read the sliver data as well as to query which tests it had passed. The use of variable-sized chunks would make better use of the area since all undefective slivers could be used. A FIFO could be attached to the input to allow high-speed rule processors to transmit several variable changes without having to wait for test results from the OOPS-MOP chip.

In addition to these features, future versions might enhance the processing capability of the OOPS-MOP chip by means of word width expansion (32-, 64-bits) and the inclusion of arithmetic operations in the testing process. With more processing power on the chip, a rule-processing system would be faster and more powerful.

REFERENCES

1. P. L. Butler, J. D. Allen, and D. W. Bouldin, "Parallel Architecture for OPS5," *Proc. 15th Int'l Symposium on Computer Architecture,* pp. 452–457, Honolulu, May 1988.

2. L. Brownstron, R. Farrell, E. Kant, and N. Martin, *Programming Expert Systems in OPS 6—An Introduction to Rule-Based Programming,* Addison-Wesley, Reading, MA, 1985.

3. B. G. Buchanan and E. H. Shortliffe, eds., *Rule-Based Expert Systems,* Addison-Wesley, Reading, MA, 1983.

4. T. W. Curry and A. Mukhopadhyay, "Realization of Efficient Non-numeric Operations through VLSI," *Proc. IFIP TC WG 10.5 Int'l Conf. on Very Large Scale Integration,* Trondheim, Holland, August 16–19, 1983, North-Holland, Amsterdam, Netherlands.

5. C. L. Forgy, "RETE: A Fast Algorithm for the Many Pattern/Many Object Pattern Match Problem," *Artificial Intelligence,* vol. 19, pp. 17–37, September 1982.

6. C. L. Forgy, A. Gupta, A. Newell, and R. Wedig, "Initial Assessment of Architectures for Production Systems," *Proc. American Association for Artificial Intelligence,* pp. 116–120, August 1984.

7. A. Gupta and C. L. Forgy, *Measurements on Production Systems,* Technical Report CMU-CS-83-162, Carnegie-Mellon University, December 1983.

8. A. Gupta, *Implementing OPS5 Production Systems on DADO,* Technical Report CMU-CS-84-115, Carnegie-Mellon University, March 1984.

9. H. Tanaka, "A Parallel Inference Machine," *Computer,* IEEE Computer Society, pp. 48–54, New York, May 1986.

10. M. Togai and H. Watanabe, "A VLSI Implementation of Fuzzy Inference Engine: Toward an Expert System on a Chip," *Proc. 2nd Conf. on Artificial Intelligence Applications,* Miami Beach, FL, December 1985.

11. P. H. Winston, *Artificial Intelligence,* 2nd Edition, Addison-Wesley, Reading, MA, 1984.

CHAPTER 10

APPLICATIONS OF THE CONNECTION MACHINE

David L. Waltz

1 INTRODUCTION TO THE CONNECTION MACHINE SYSTEM

The Connection Machine (CM) development effort was initiated [17] in the belief that parallel processing and artificial intelligence could together accelerate the rate of progress toward truly intelligent machines. In the brief time since the first Connection Machine was switched on in April 1985, we have begun to realize this dream at a pace and with an ease that has far exceeded our expectations. This progress is the result of the ease with which the machine can be programmed and the dramatic increase in computing power which the machine can bring to bear on applications. We have been able to run multiple experiments in the time previously required for a single problem, enabling us to explore a great many more hypotheses and work on much larger problems than had been possible on artificial intelligence workstations of the previous generation. The ease of programming is in part the result of a decision to use existing serial machines (e.g., Symbolics, Sun 4, or Digital Equipment Corporation VAX) as front ends. Operating systems, editors, file systems, debuggers, network communication systems, and the like are left unchanged and provide familiar programming environments. The Connection Machine is programmed in Fortran or conservative extensions of Common Lisp and C. Users familiar with these languages and front-end computer systems have been able to produce results from the first day.

2 DATA LEVEL PARALLELISM

Programming the CM has proved to be easy. Because there are so many processors, it is practically impossible to program each one separately; instead, we have generally used a programming method—called *data-level parallelism*—that is similar, from the user's point of view, to that used on ordinary serial machines. In data-level parallelism the data elements for the problem are stored in the processors, one element per processor. The front-end computer then executes a serial program, each step of which can involve computation by all the Connection Machine's processors; for example, rather than using loops for performing a single operation on each element of an array or set of data elements, a program can carry out the operation simultaneously in all of the processors.

Each processor can, under software control, appear to the user as a number of *virtual processors*. To accomplish this, the memory of each processor is divided up, giving each virtual processor a fraction of the memory. We have routinely programmed applications involving millions of virtual processors. The speed penalty for computation is approximately linear; thus, a program with n virtual processors operates on n times as many data elements at about $1/n$ of the speed of a program without virtual processors.

As mentioned above, the CM can be programmed via the front end using the Fortran 8X proposed standard, C*, an extension of C, or *Lisp, an extension of Common Lisp. The extensions consist of a few additional instructions for creating and accessing parallel data structures. There are also parallel analogs of most common instructions. For example, the expression:

```
(+!! a b)
```

simultaneously carries out the addition of variables stored in field a in each of 64K physical processors to the contents of each of the 64K instances of the variable b, and stores the result on an internal stack within each processor. If n virtual processors are in use, this instruction will be carried out n times, adding 64K operands each time. Other *Lisp constructs are natural extensions of well-known operations. For instance,

```
(*if test operations)
```

executes the test and then executes the operations only in those processors in which the test succeeded (the *currently selected set* of processors which may, of course, be further restricted by including another *if among the operations). Examples of *Lisp and C* code for the Connection Machine are given in Section 5.

Such solutions seem far more natural and easy than those typically proposed or implemented for coarse-grained parallel machines. Coarse-grained parallel computers (those with two to a hundred or so processors) must generally be programmed using a method that might aptly be called "control-level parallelism." Control-level parallelism requires a programmer to divide a program into fragments, one for each of the processors in the coarse-grained machine. Synchronization is a problem in coarse-grained systems; moreover, it is often not easy to find parallelism in programs, let alone exactly the right amount of parallelism to fit the number of available processors.

In a wide variety of cases we have demonstrated that the Connection Machine can exploit data-level parallelism to achieve high computation rates. Data-level parallelism is particularly appropriate for database, natural language, and image-processing operations, as well as for finite element modeling, graphics, automata-based simulations of physical processes, and many others.

2.1 Hardware and Software: System-Level Specifications

The Connection Machine system is at the radically fine-grained end of the spectrum of parallel machines; it comes in units of 8K (2^{13}) processor-memory units up to a maximum of 64K (2^{16}) The processor–memory units are interconnected by a high speed communications network that allows all processors to communicate in parallel with any or all other processors. A 64K Connection Machine system consists of the processor–memory–communications system, plus a front-end machine.

The Connection Machine system supports up to eight front-end computers. Currently each front end is either a Digital Equipment VAX, Sun 4, or a Symbolics Lisp Machine. Physically, the Connection Machine consists of 4096 chips of proprietary design, each containing 16 processors and a hardware router. The CM-2's, on which most of the benchmarks in the chapter were run, contain up to 8 Gbytes of memory (128K bytes per processor). Processors operate in a bit-serial fashion; each operation can combine two bits from memory with one bit from a register, producing one bit to memory and one bit to a register. Each of these basic operation cycles (load-load-store) requires about one microsecond.

The processors are organized as a 12-dimensional hypercube with 16 processors at each vertex of the hypercube. A packet-switching network handles communication between the processors. Each 16-processor chip also contains a network router that can perform the following operations: it can accept a message from one of the processors on its chip (or from a router on a different chip) and send it either to a processor on its chip or to another router on a different chip; it can buffer messages if there are no channels available over which to send them; it can combine messages destined for the same processor, provided that the user has specified an (associative) combining operation, for example, SUM, OR, AND, MAX, or MIN; and it can run a combining pattern in reverse, resulting in a distribution of messages via a fan-out tree. A message header consists of 16 bits; each bit corresponds to a dimension along which the message must move to be delivered. The header can be computed by comparing the 16-bit address of the source processor with the 16-bit address of the destination processor and putting ones in each dimension in which the addresses differ. Each time the message is routed along a dimension of the machine, the bit corresponding to that dimension is set to zero; when all the bits are zero, the message has arrived at its destination processor. The CM-2 also contains proprietary indirect addressing chips that allow rapid delivery of messages to the appropriate virtual processors. Typical message delivery times, with all processors sending, are on the order of 1170 microseconds for 32-bit messages, with times ranging from 440 microseconds for very regular communications patterns to a worst case of 1400 microseconds when every processor is sending to some other randomly chosen processor.

Parallel channels support aggregate I/O rates up to 500 Mbytes per second. Each 16-processor chip has a single I/O port, which is connected via multiplexers to eight I/O ports, each 256 bits wide. The channels support high speed devices such as the DataVault™ —Thinking Machines' parallel, error-correcting disk unit—and Thinking Machines' graphic display system.

In its standard configuration, the DataVault contains 42 $5\frac{1}{4}$ in. Winchester drives, which are written and read in parallel. Thirty-two of the drives deliver 32-bits of data, seven drives provide single-bit error correction, and the final three drives are spaces. If a drive fails, DataVault's controller (a MicroVax™) can switch out the failed drive and switch in a spare, notifying the operator, who can replace it without interrupting the DataVault's operation. The spare disk can be "healed" in about 10 minutes, using the full error-correction capabilities of the system. The expected mean time between observed I/O failures is on the order of five years. Total storage capacity, using 740-Mbyte drives, is 40-Gbytes. The aggregate transfer rate for each DataVault is 25 Mbytes per second. Forty-two additional disks can be installed in a DataVault cabinet, and although only 39 disks can be read or written at a time, data latencies can be reduced by alternately using the two halves of the DataVault.

Thinking Machines also offers a framebuffer, for fast (potentially interactive) graphics and envisionment of the results of CM calculations. The framebuffer supports a 1024 x 1280 display with 8- or 24-bit color, and can be refreshed up to 50 times per second.

3 DOCUMENT RETRIEVAL

We have built an easy-to-use document retrieval system that mixes AI ideas with methods from information science research. Its basis is *relevance feedback,* a weighted associative memory algorithm. Document retrieval has traditionally been implemented as Boolean search on inverted files. The main difficulties of Boolean search are that (1) users require considerable training in the use of a query language; and (2) the search results in too many documents when users use a search pattern that is too general, and too few documents when the search pattern is more restrictive. Boolean search systems also exhibit poor precision and recall [4]. In contrast, a naive user can be trained on our system in a few minutes, and the system operates very rapidly, with high precision and recall.

3.1 Document Retrieval by Relevance Feedback

From the user's point of view, the search process on the Connection Machine document retrieval system has two distinct phases. In the first phase, the user types a list of a few keywords or an English question; for example, in Figure 10.1 the user has typed "Marcos Wealth". The front-end machine broadcasts to all processors coded forms of these two words and two associated numerical weights. Each processor contains stored representations of one or more articles, and each notes whether either word occurs in its article; if so, it increments the total score of its article by the appropriate numerical word weight. (Word weights are inversely proportional to the log of their frequencies in the database—for example, "platinum" appears in the database

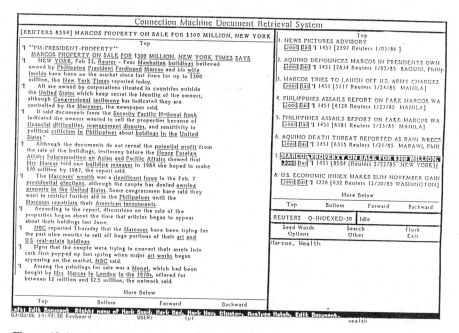

Figure 10.1 The search process on the Connection Machine document-retrieval system. Here, the user had typed "Marcos" and "Wealth". The first eight documents containing "Marcos", "Wealth", or both are shown in the upper right window. The user has marked article number seven as good by clicking on its mouse-selectable "Good" box.

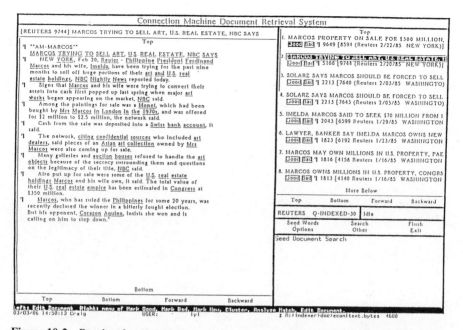

Figure 10.2 Results of seed-document search. Shown at right are the top eight documents—those most similar to the document selected in Figure 10.1. (Note that since only one document was used to form the search pattern, it appears first because it is perfectly similar to itself.) Scores for the documents are shown to the right of the "Good" and "Bad" boxes. Any number of "Good" and "Bad" selections can be batched together. Terms in "Bad" documents that do not appear in "Good" ones are given negative weights.

less frequently than "gold", and therefore has a higher associated weight. This weighting mechanism ensures that uncommon words have more of an influence than common words during the document lookup process. Assuming that the terms in a search profile are all from closely related articles, the greater the number of search terms, the more precise the search.)

When all the words have been broadcast, the articles with the largest scores are retrieved, sorted so that the best matches appear first, and presented to the user.

In the second phase, the user can browse through the returned documents and find one or more documents that bear on the topic of interest, using a mouse to mark words, paragraphs, or entire documents as "good" or "bad". Once one or more documents have been marked, the user can repeat the search. This time, all the marked words in the selected document or documents are broadcast and documents are sought that share a large number of words with the selected or *seed* document. Figure 10.2 shows the results. This method, termed *relevance search*, has been known since the 1960s to have excellent properties of precision and recall [32] but has not been used in practical systems

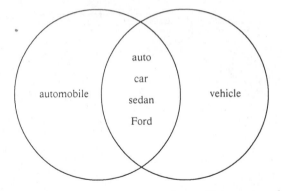

Figure 10.3 Documents on the same subject have a high overlap of vocabulary.

because its computational requirements are too high to allow fast turnaround interactive search. The algorithm takes advantage of the innate characteristics of documents. Every document is, in effect, a thesaurus for its subject matter; synonyms appear for many topics because writers usually try to avoid word repetition. Moreover, word variants (such as plural, singular, and possessive forms) and semantically related terms often appear among the words in a particular article. The use of many of those terms in forming a search pattern makes the search highly selective; for instance, although the word "chip" may appear in an article about cookies, "VLSI", "integrated", and "circuit" simply will not. Thus, in the overall scoring, truly useful documents are reliably separated from random matches (see Figures 10.3 and 10.4.)

Document search requires only about 170 milliseconds on the Connection Machine, given a 100-term search pattern and assuming that compressed representations of about 2K bytes of free text are stored in each CM virtual processor. A more detailed discussion of performance appears in Section 4.3.

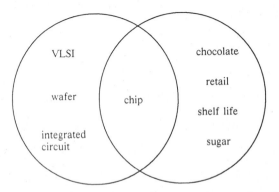

Figure 10.4 Documents on different subjects have a low overlap of vocabulary.

4 BUILDING A DATABASE ON THE CONNECTION MACHINE SYSTEM

The document database can be constructed using text sources such as a wire service, an electronic mail system, documentation, memoranda, encyclopedias, journal or magazine articles, or any other electronic text database. The system we have used for benchmarking contains about 1,000,000 Dow Jones newswire articles. 16 articles are stored on each processor; each article averages on the order of 2000 bytes of ASCII data. The text of the articles is compressed (using techniques that are described later) so that the 2000 bytes are encoded in about 500 Bytes.

To understand the text representation we use, it is important to draw a distinction between *source documents* and *content kernels*. A *source document* contains the full text of a particular article, book, letter or report, and is stored on the file system of the front-end machine. A *content kernel* is a compressed source document, encoding only the important words and phrases and omitting the commonplace words. Content kernels are stored in the memory of the Connection Machine system.

The content kernel is produced automatically from the source document. First, the source document is processed by a Thinking Machines Corporation document indexing program that selects the most significant terms (phrases) in the text [24]. The resulting list of terms contains on the average about 130 words per article, a data compression of about 2:1 over the original article. Next, these words are encoded into a bit-vector data structure, using a method called *surrogate coding* [1,30,31] that maps each word in the content kernel into n different bits in a 4096-bit vector using n hash functions. If more than one word hashes into a particular bit of the bit string, the result is simply ORed. For our database, $n = 10$. Given 130 terms, the process of surrogate coding marks about a third of the bits as ones, adding another factor of 1.6:1 to the data compression for a total compression of about 3.3:1.

Surrogate coding allows the content kernel to be stored more compactly and speeds up the search process. To test for the presence of a given word within a document, the same 10 hash functions are applied to the word, generating 10 bit positions. The bit positions are then checked for each of the bit vectors representing each article, and if all 10 positions match, the score of the overall article is incremented by the word "weight". This scheme is probabilistic; it is possible that all 10 bits for a word might happen to be set on account of other words, even though that word doesn't really appear in the source document. Such an accident will result in a *false hit* on that word. However, this will not seriously affect the results of the lookup, for two reasons. First, the probability of a false hit is small: about $(1/3)^{10}$, or less than one in 50,000. Second, a false hit will be only one of many terms that contribute to the score, and so will have only a small effect even when it does occur.

The source document in its original form is available for retrieval and presentation to the user when needed. The location of the original document (on the front-end system's disk) is stored with the content kernel.

5 PROGRAMMING THE DOCUMENT RETRIEVAL SYSTEM

The Connection Machine system enjoys what is probably its greatest advantage over its competition in the area of programmability. This section illustrates this point by showing part of the program code for the document retrieval system described above. Both *Lisp and C* versions of the program are included. Programs for several of the other applications in this paper can be found in [8].

The content kernel data structure is made up of the following fields:

score is used by the document lookup program to accumulate the ranking of each content kernel in the database according to how closely the content kernel matches the user's search pattern. Each time a match is found, *score* is updated.

document-id contains a reference to the original source document from which the associated content kernel was derived. When a content kernel is selected from the database lookup, the user is shown the source document referred to by this index.

kernel is a table of surrogate-coded bit-vector encoding.

The necessary declarations for these fields are as follows. (All of the code for this example is presented twice, first in the *Lisp language and then in the C* language, to make it easy to compare the two languages. Because the characters * and ? may not appear in C* identifiers, *Lisp names such as *score* and word_appears? are rendered in C* simply as score and word_appears.)

```
;;; Declarations for the *Lisp version:

(defconstant table-size 1024)
(defconstant hash-size 10)
(*defvar *score*)
(*defvar *document-id*)
(*defvar *kernel*)
```

```
/* Declarations for the C* version. */

#define TABLE_SIZE 1024
#define HASH_SIZE 10
poly unsigned score, document_id;

poly bit kernel [TABLE_SIZE];
```

The modifier poly declares variables present in all processors. There is also a modifier mono that declares variables resident in the front end-machine.

5.1 Document Lookup on the Connection Machine System

During the first stage of document lookup, the user lists a set of terms to be used in searching the database, and receives back a list of documents that contain all or some of those terms. This part of the search process is similar to Boolean search, except that (1) no special syntax is needed, and (2) the documents are retrieved in order, starting with the closest match. The user then points to the one or more documents (or portions of documents) that are most relevant. From these documents an overall *search profile* of content-bearing words is assembled: a list of these words, with weights which, as Section 3 explained, are proportional to the negative log of each word's frequency in the database.

Next, the search profile is broadcast to all processors in the Connection Machine system. The same mechanism that is used to encode each word in the content kernel as a series of bits is applied to the words in the search profile. For each word in the search profile, a set of ten bit indices is computed and broadcast. All content kernels that have these same ten bits set will have the weight of that word added to their *score* field.

The following code is used to broadcast one search pattern word to all the processors in the system. Upon receipt, each processor checks its content kernel and adds the value of word-weight to its *score* if the kernel contains the word. Each word is represented by a list of 10 bit locations (bit-locs).

```lisp
;;; *Lisp code for testing for the presence of a single word.
(*defun increment-score-if-word-appears (bit-locs
word-weight)
    (*let ((word-appears? t!!))
        (dolist (bit bit-locs)
            (*set word-appears?
                (not!! (zerop!! (load-byte!! *kernel* (!! bit)
(!! 1)))))))
                (*if word-appears?
                  (*set *score*(+!!*score*(!!word-weight)))))))
```

```c
/* C* code for testing the presence of a single word. */

poly void increment_score_if_all_bits_set
    (mono unsigned word_bit_position [HASH_SIZE], mono
int weight)

{
   mono j;
   poly bit word_appears = 1;
   for (j = 0; j < HASH_SIZE; j+ +)
      word_appears & = kernel [word_bit_position[j]];
   if (word_appears)
      score + = weight;
}
```

The main search program simply calls this routine once for each keyword in the keyword list. Since word_appears is a parallel variable, it is handled in the following way when it follows an if: each processor whose word_appears value satisfies the test is marked as belonging to the currently selected set of processors by pushing a stack bit in each processor. Operations broadcast by the front-end machine are carried out only by processors in the currently selected set. When one exits the scope of the conditional, the stack is popped, and the previous currently selected set is restored.

5.2 Retrieving the Highest-Scoring Documents

The code that follows is used to retrieve the *document-id* for each of the highest-scoring content kernels in the database, returning a list containing the *document-id* of each high-scoring kernel. The program retrieves the *document-id* for the highest score, then that for the next highest score, and continues until a list of length document-count is retrieved. The already-retrieved? flag is set when a processor's *document-id* has been retrieved so it will not be retrieved again.

```
;;; *Lisp code for retrieving documents in order, highest score first.

(*defun retrieve-best-documents
  (let ((top-documents-list nil))
    (*let ((already-retrieved? nil))
      (dotimes (i document-count)
        (*when (not!! already-retrieved?)
        (*when (=!! *score* (*max *score*))
      (*let ((next-highest-document (*min (self-address!!))))
        (setq top_documents-list
          (append top-documents-list
            (list(pref *document-id* next-highest-document))))
            (setf (pref already-retrieved? next-highest
document) t))))))
    top-documents-list))

/* C* code for retrieving documents in order, highest score first. */

poly void retrieve_best_documents
    (mono document_count, mono unsigned *document_id_array)
{
    poly bit already_retrieved = 0;
    mono i;
    for (i = 0; i < document_count; i+ +) {
        if (!already_retrieved) {
            if (score == (><= score)) {
                processor *next_highest_documents = (><=this);
                document_id_array[i] = next_highest_document->
document_id;
                next_highest_document->already_retrieved = 1;
                }
            }
        }
}
```

5.3 Timing and Performance

Benchmark tests have been run for document search using Reuters news-wire articles. The tests were performed on a 16K processor Connection Machine. Each document is typically represented by approximately 130 content words (selected from the roughly 400 words in an average article). Terms were selected in the following manner: terms rarer than a certain frequency are always kept; terms more common than another frequency are always dropped; terms in between these frequencies are compared with their neighboring words, and kept as a phrase if the joint frequency of a pair is sufficiently greater than would be predicted by the frequencies of the individual words. Thus phrases such as "White House" or "New Mexico" are kept as units, since the individual words occur much more frequently together than one would expect given the separate occurrence frequencies of the individual words. The entire phrase is hash-coded and stored as though it were a single term, and, to be safe, the individual words are also stored separately.

A relevance feedback search is based on the most important words taken from the union of the sets of content words from several articles, and typically involves 100 words. Only 170 milliseconds are required to search 200,000 articles (representing about 800 Mbytes of free text) against a 100-term pattern and assign scores to each; in another 100 milliseconds the system can select the 20 articles that most nearly match the set of 100 terms. Exactly the same amount of time is required to search four times as many documents (500,000—equivalent to 2 Gbytes of free text) on a 64K processor Connection Machine.

Larger databases can run on a Connection Machine, using different algorithms. We plan to use DataVault's high speed disk system to implement inverted index search, which assigns a large portion of a Connection Machine's memory to 16-bit "mailboxes", each corresponding to a document in the system. Thus, mailboxes for 0.5 billion documents (equivalent to 1.5 Terabytes, or 1.5×10^{12} bytes) can be stored on a single 64K Connection Machine. Inverted lists of term occurrences and term weights are stored on the disk system. Mailbox "addresses" (each consisting of a processor number and memory location) are loaded into each processor, sent via the router to the right mailboxes, and added to the mailboxes' scores.

Another promising approach is the use of cluster analysis. If the database has many documents on the same subject, the system need not store all their content kernels individually. It can store one kernel for the whole cluster, then retrieve the full set of related documents when needed. (A single document might, of course, participate in more than one cluster.) As the total database size grows, the size of the average cluster grows with it, making this a particularly appropriate technique for large databases. The addition of paging and clustering should allow one to extend the algorithm described above to the 10 Gbyte range and beyond.

6 MEMORY-BASED REASONING SYSTEMS

One of the unrealized goals of artificial intelligence is the creation of programs which base their reasoning on experience. AI researchers have assumed that most of the relevant experience of experts is encoded in the form of rules. While there has been some progress in automatically generating rules of inference [27], the construction of intelligent programs has for the most part proved to be a difficult task, one that depends on the ability of highly skilled individuals to generate sets of rules that capture their expertise.

We believe that, for humans, memories of specific or general events, not rules, are the keys to reasoning from experience. For example, a program or person can diagnose diseases by comparing a patient's symptoms and characteristics with those of all other patients it knows about; if it can find similar patients with similar symptoms, these patients can be used as a basis for reasoning about the current patient. Memories of instances have not been used in the past as a basis for reasoning because sequential computers are too slow to execute some of the necessary basic operations, for example, finding the item most similar to an input item when the amount of data is non-trivial. (This ability is sometimes called "associative memory)". It is important that such systems work with large amounts of data, so that it will in general be possible to find instances in memory that are similar to any given instance. With new parallel architectures, specifically the Connection Machine system, such operations suddenly become sufficiently fast, even with large amounts of data, to use as the basis of reasoning.

A further major reason that this solution has been overlooked in the past is that von Neumann machines do not support it well. There have been systems, such as Samuel's Checkers player [33], which worked from memory, but they have typically required an exact match between an item in memory and the current situation, and non-general search algorithms (e.g. hashing or indexing) to retrieve items from memory.

In the real world, there may be no exact matches, so a best-match is called for. Unfortunately, there is no general way of searching memory for the "best match" without examining every element of memory. On a von Neumann machine, this makes the use of large memories impractical. On a fine-grained parallel machine, such as the Connection Machine system, problems do not arise until the size of the database exceeds the size of the machine's memory; this allows us to extend the size of tractable problems solvable using the current workstations by three or four orders of magnitude.

Although it is not yet possible to encode experience in the sense familiar to humans, it is possible to build practical systems that use databases as memories. Like the document search system described in the previous section, a memory-based reasoning system depends on the ability of the CM to find within a few seconds the global nearest match, among 512 Mbytes of data, for patterns consisting of hundreds of attributes and values. Such a system must be able to compute statistics very rapidly. The CM is able to count the number of items with particular properties in time proportional to the log of the total number of data items, as described in Section 7; this sort of ability makes statistical computation tractable, even for large amounts of data, and the system that has it can bypass problems of knowledge acquisition and brittleness that have plagued expert system researchers.

6.1 The Operation of Memory-Based Reasoning

The key operation in memory-based reasoning is weighted associative search for items in memory that are similar to a present case that we hope to understand. Once some matches (or near matches) have been found, they may suggest hypotheses. Other operations, based on statistics, test these hypotheses, and separate important features from unimportant ones.

The operation of memory-based reasoning systems can be made clearer through an example. Suppose that we have a large relational database of medical patient records, with fields for age, sex, symptoms, features of patient history (pack–years smoked, previous heart attack(s), etc.), diagnosis, test results, treatment, and the like. We refer to these fields as *features*. A program can find diagnostic hypotheses for a new patient by (1) broadcasting each feature of the new patient; (2) having each processor compute a numerical measure of closeness to each feature, and add this number to a total similarity score, and (3) selecting the patients whose total scores are closest to that of the new patient. There are several possible outcomes: (1) all n similar patients received the same diagnosis; (2) no patients are very similar to the new patient; (3) only a small number are similar, and (4) the general case—there are several diagnoses among the nearest n patients.

In all these cases, we run a hypothesis-testing phase. If its results suggest that only one diagnosis is plausible, we are done; if more than one diagnosis remains, a second hypothesis-testing method is invoked that either completes the diagnosis, or proposes tests that would differentiate the remaining possibilities. These processing phases are described in more detail in Section 5.2. However, even at this point, we can make some observations: If (1) is true (i.e., if all n patients have the same diagnosis), it is very likely that the current patient should have that diagnosis; it is also likely that we could extract a rule for diagnosing this disease that could be added even to a microcomputer-based expert system. If case (2) holds, the system knows that it has never seen this set of symptoms before; thus it "knows that it does not know" how to diagnose this case. (Alternatively, the data on the new patient may be in error, and the system can propose this possibility.) In case (3), even if only one patient is similar (as in the second reported case of legionnaires' disease), the system may be able to make a diagnosis.

6.2 A System for Medical Reasoning

Memory-based reasoning has been successfully demonstrated in a preliminary test of feasibility. A medical reasoning system called QUACK was constructed [36]. Lacking quick access to a real database of medical cases, a database was created by stochastic modeling, drawing on commonsense knowledge of common medical conditions (hence the name). What follows is not offered as compelling evidence of the correctness of our theories, but only as an illustration and as a limited demonstration of their feasibility.

QUACK simulates the diagnosis of a patient. It starts with a database of past cases, including symptoms, personal data, results of tests, and the ultimate diagnosis. It is then given the data for a series of patients, starting with personal data and symptoms. QUACK forms and evaluates a number of hypotheses, eventually narrowing them down to between one and three possibilities. It then selects some test (e.g., taking the patient's temperature) and uses the results to make a final determination. QUACK does all this with no rules or information other than the database of past cases.

The database is created by a multiphase stochastic process. Firstly, personal attributes such as sex, hair, and eye color are chosen at random. (Sex is a significant feature; hair and eye color are added in an effort to distract the system.) Secondly, a medical condition is generated; the possible conditions are pregnancy, influenza, colds, over-exertion, health, and a variety of other conditions. Thirdly, directly observable symptoms are generated; they include generalized aches and pains, backaches, chills, nausea, and sneezing. The symptoms of some conditions overlap; in other cases symptoms are absent; symptoms unrelated to the primary condition sometimes appear (e.g., tiredness and aches). Finally, the process predicts the results of several tests which a physician might order, including the patient's temperature and a throat culture for strep throat. Figure 10.5 shows four cases from this synthetic database.

The basis of QUACK is retrieval from a database of cases similar to the current patient. The most important aspect of this retrieval problem is the metric used in computing similarity. This metric must give low weight to irrelevant features, such as hair and eye color, high weight to definitive features such as the results of pregnancy tests, and intermediate weight to suggestive features such as aches and chills. Brief reflection indicates that no one metric is adequate. For example, if a patient is suspected of having the flu, sex is completely irrelevant, and should be given zero weight. In diagnosing pregnancy, on the other hand, sex is definitive, and should be given extremely high weight. Thus, the metric is dependent on the hypothesis.

The metric is calculated on the fly. The system determines how well each feature correlates with the hypothesis, and chooses weights accordingly. For example, if QUACK hypothesizes that a patient has the flu, it will compute the proportion of males and the proportion of females having the flu, and conclude that sex is not a significant predictor for influenza. Thus, it will assign a low weight to sex. On the other hand, if QUACK hypothesizes that a patient is pregnant, it will note that men are never pregnant and that females are sometimes pregnant, and will thus give sex a weight sufficiently high to exclude men from consideration.

The next step is to find the cluster of cases most similar to the current patient. This is done by applying the metric to every case in the database and extracting the cases most similar to the patient. Once the cluster of similar

Field	1	2	3	4
Eye-color	Brown	Gray	Brown	Brown
Hair-color	Blond	Brown	Black	Brown
Sex	Male	Female	Female	Male
Aches	—	—	—	—
Backache	—	—	—	—
Chills	—	—	—	—
Nausea	—	—	—	—
Period-late	N/A	1	2	N/A
Running-nose	—	—	—	—
Sneezing	—	—	—	—
Sore-throat	—	—	—	—
Fever	Mild	None	None	None
P-text	—	—	—	—
T-culture	—	—	—	—
U-text	—	—	—	—
Condition	Throat	Healthy	Pregnant	Healthy

Figure 10.5 Four items from the database.

cases has been retrieved, the proportion of cases matching the hypothesized diagnosis is calculated. This may strengthen or weaken the hypothesis. The composition of the cluster will also be used to generate additional hypotheses, which will be evaluated in their turn.

This process of hypothesis generation and evaluation is illustrated by the following sample run. The test patient was male and was experiencing a sore throat, chills, and tiredness. QUACK starts with the assumption that he is healthy; this hypothesis generally yields a good initial metric and, while the hypothesis of health is usually rejected, those cases which are unhealthy in a similar manner yield good hypotheses. In this case, several hypotheses did emerge, including the flu, non-specific throat infection, and strep throat. Each hypothesis was tested in turn; only non-specific throat infection and strep throat survived this stage of the evaluation. Figure 10.6 shows the output of this phase of the trial. Implausible hypotheses are also generated; this problem should be cleaned up, but the generation of bad hypotheses is not directly harmful because the hypothesis evaluation step is sufficiently strong to weed them out.

After the initial diagnosis has been made, several alternatives may remain, in which case a second phase of reasoning is necessary. First, QUACK computes the resolving power of various tests relative to the possibilities generated in the first phase. Second, those tests which have high resolving power are selected for use in a second phase of hypothesize-and-test. In the example we have been looking at, two tests are selected: a throat culture, and taking the patient's temperature. When these added factors are taken into account, a solid diagnosis is made. A trace of this phase is shown in Figure 10.7.

Likelihood	HEALTHY	0.0
Suggests	THROAT, STREP, FLU	
Likelihood	FLU	0.0
Suggests	PREGNANT, OV-EXER, COLD, UT-INF	
Likelihood	UT-INF	0.0
Likelihood	COLD	0.0
Likelihood	OV-EXER	0.0
Likelihood	PREGNANT	0.0
Likelihood	STREP	8.031373
Likelihood	THROAT	33.573772
THROAT	0.8070	
STREP	0.1930	

Figure 10.6 Initial diagnosis.

The example requires 8.9 seconds to run on a 16K CM-1 with a database of 4,096 examples. It was also run with a database of 32,768 examples, taking 10.7 seconds to run and generating fewer false hypotheses. Run with a database of 512 examples, it took 4.3 seconds but misdiagnosed the case (the small database contained only one case of strep). In general, increasing the size of the database improves the quality of the system's reasoning without significantly changing the time needed. For very small databases, the system's performance deteriorates somewhat, but it remains capable of handling the more common problems.

It is possible to criticize QUACK on the grounds that all its deductions are extremely obvious. But that is precisely the point: the basis of intelligent behavior *must* include a grasp of the obvious, something which is sorely lacking in rule-based systems. Remember, QUACK does not have a knowledge base, only a collection of weak methods [25] for making decisions based on memory. It has another useful property: as its base of experience grows, it gets smarter rather than stupider; the quality of its decisions improves without incurring a significant time penalty.

Tests	FEVER	T-CULTURE
Likelihood	THROAT	0.0
Likelihood	STREP	3.766667
STREP	1,000	

Figure 10.7 Diagnosis after throat culture.

6.3 Evaluation of Memory-Based Reasoning

Memory-based reasoning systems may be able to function like expert systems in domains where there are databases of situations and outcomes or actions. Potential application areas include medical diagnosis and proposal of diagnostic tests, weather forecasting, credit decision support, investment advising, and insurance risk assessment. No expert is required; a knowledge engineer only needs to identify and mark database contents according to whether they are symptoms or features of situations, outcomes or actions, or optional tests that can be ordered. Memory-based systems can form hypotheses on the basis of even a single precedent, something rule-based systems cannot do—rules are inherently summaries of regularities among classes of items. And memory-based reasoning systems "know when they don't know"; if no items in memory are closely related to the item under analysis, the system will be able to recognize this fact and tell the user.

Computer learning has generally stayed within the heuristic search/deductive reasoning paradigm; most learning algorithms try to derive a set of rules from a body of data. The approach is exemplified by the work of Winston [39], Sussman [37], and Mitchell [22,23]. Memory-based reasoning, in its pure form, differs from traditional learning methods in that no rules are ever created. This has several advantages. Firstly, the deductions made by the system are achieved without the use of rules as an intermediate representation, so there are fewer opportunities inadvertently to introduce inaccuracies, such as those that result from the combining of "confidence levels" for a chain of rules. Secondly, rule generation has to contend with a combinatorically explosive search space of possible rules, whereas memory-based systems can focus their analysis on the case at hand. It may be that, to capture a large knowledge base, so many rules are needed that the rule generation process will never finish. If this is the case, there is never any certainty that the resulting rule set is complete, and the system may fail at any moment. With a memory-based system, the data used as the basis of reasoning is always available; therefore, within limits imposed by the accuracy of the retrieval mechanism and the contents of the database, errors of omission could not occur.

Although space does not allow an extended discussion, this work is closely related to other research in parallel associative memory [15,18], semantic nets [26,11], and the case-based reasoning of Kolodner in the area of problem solving and memory for personal events [20,19].

7 BASIC OPERATIONS: SCANNING

Scan operations, which use the CM's router in a very efficient manner with highly regular communications, have proved to be valuable for a wide variety of programming tasks. For example, it is often useful to enumerate sets of items in the CM. Enumerating is the process of assigning to set elements a unique integer between 1 and n. The enumeration algorithm described below uses parallel prefix operations similar to those used in APL, and operates in log n time [2]; 64K elements can be enumerated in about 200 microseconds on the CM.

The scan operation is illustrated in Figure 10.8 for a six-element set on a hypothetical 16-processor CM:

1. In parallel, each processor tests whether its a field is nonzero; if so, it sets its b and c fields to 1; otherwise, it sets b and c to 0.

2. Each processor (except the highest-numbered one) sends and adds its value to the c field of the processor whose address is higher by one, using the (microcoded) *LISP parallel instruction *send-with-add. The send operations are illustrated by the arrows in Figure 10.8a, and the result is the configuration shown in Figure 10.8b.

3. Each processor (except the two highest-numbered ones) sends and adds its c value to the c field of the processor whose address is higher by two. The result is shown in Figure 10.8c.

4. Repeat for a total of log n ($=4$ in this case) steps, each time doubling the number of high-numbered processors that do not send their values, and also doubling the distance over which the c values are sent (see Figures 10.8d and 10.8e).

5. In each processor, multiply b and c and put the result in c.

At the end of *log n* steps, the c field of each processor will contain a unique integer. To see why this works, consider what happens to processor 12. After the first step, c is one. On the second step, the c value of processor 11 is added to it. On the next step, the current c value for processor 10 is added to it, but processor 10 already contains the sum of its own original value and that of processor 9, so processor 12 now contains the sum for processors 9–12. In the next step, the c value of processor 8 is added to processor 12; since processor 8 already contains the sum of the c values of processors 5 through 8, after this step, processor 12 holds the sum of processors 5–12. In the last step, the c value of processor 4 is added; processor 4 contains the sum of the processors 0–4, so finally processor 12 contains the sum of the c values of all the lower-numbered processors.

The scan or enumerate operation can be used as a basis for a sort program [2]. This program is included in the microcode set of the Connection Machine and can be called directly from C* or *Lisp. The sort we use is a variation of radix sort and works in the following manner. First, the lowest-order bit of the key field of each item to be sorted is tested. All the odd data items (those whose low-order bit is one) are enumerated. Next, all of the even data items (with a low-order bit of zero) are enumerated, and the total number of even items is added to each odd item. The data items are now numbered from one to the maximum number of items to be sorted, and each is routed to the correspondingly numbered processor. The result is that all the even items are stored first, starting with processor one, followed by all the odd processors. The process is

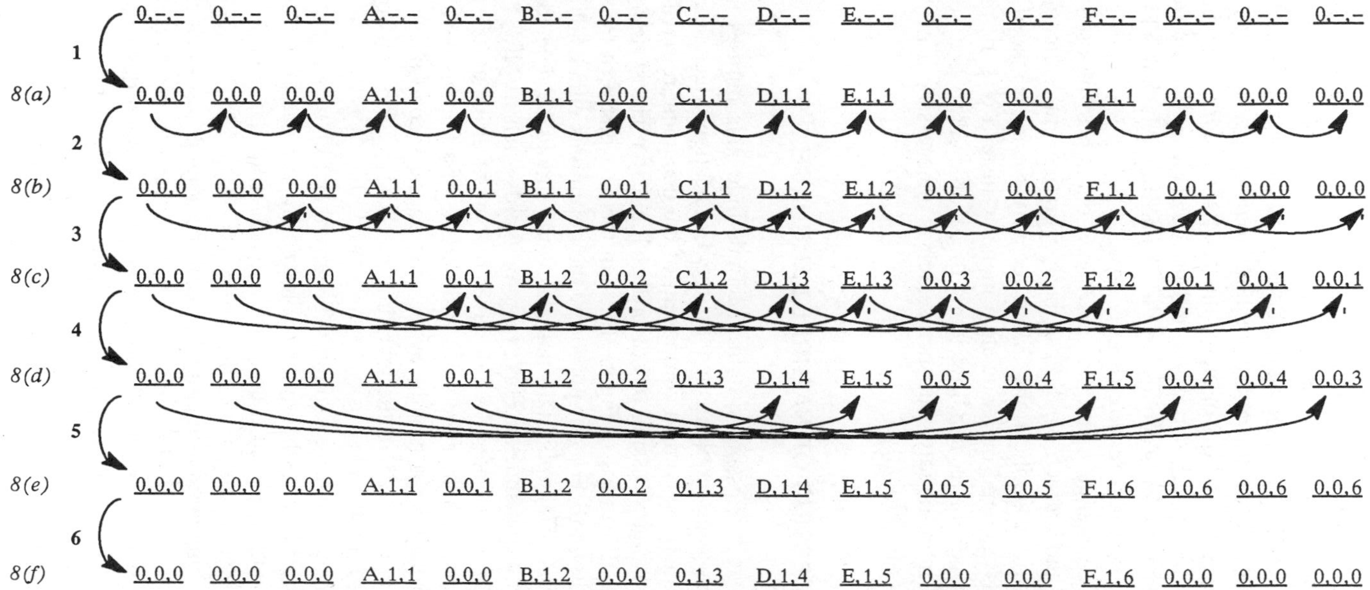

Figure 10.8 Enumeration example. Each item in a set (A, B, C, . . .) is assigned a unique integer (1, 2, 3, . . .) in log n time, where n = the size of the largest possible set. Each underlined group of letters and numbers represents the contents a, b, c of a processor. See text for explanation.

343

now repeated for each key bit, starting with the next most significant bit, until all bits in the key field have been processed. Importantly, order is preserved; for example, if two items differ only in their lowest order bits, the even item will always precede the odd item. An inductive argument shows that this will result in a total sort. The length of time required to do such a sort is proportional to $k \log n$ where k is the number of bits in the key field and n is the total number of items to be sorted. A full sort of 64,000 items, each with a key field of 16 bits, requires about 30 milliseconds on the Connection Machine. In general, if $k = \log n$, this sort will require $(\log n)^2$ time.

The sort and enumerate operations are cheap enough to be used liberally within programs and, in fact, we did use these operations rather liberally when we first began writing applications. More recently we have tended to use a mix of operations centered on associative memory instead of sort; the reason for using sort in general is to allow items to be ordered so that search can be accomplished more rapidly, but such search is unnecessary if items can be retrieved directly and globally.

8 BULK PROCESSING OF NATURAL LANGUAGE

A natural language processing system that finds indexable phrases in natural language text has been implemented on the Connection Machine. (Such a system will eventually become a component of the document retrieval system; it will be used for preprocessing or postprocessing text.) The system operates in two phases, as illustrated in Figure 10.9. In the first phase, it looks up the definition of each word. For our purposes, the definitions either hold part-of-speech information or simply state that the word should be deleted because it cannot be part of an index. (For example, words such as "if," "among," and "between," and most adverbs can never be part of an index term.) Word definitions also mark whether the word is the name of a famous person, a day of the week, the name of a month, or the like. In the second phase, a series of ATN parsers is run over all the text simultaneously; during this phase, all noun phrases are found.

Let us suppose that we have a simple ATN grammar [40] which specifies that a noun phrase consists of a determiner followed by any number of adjectives followed by a noun. Let us also assume that each word has been assigned to a (virtual) processor, and that part-of-speech information has been stored for each word. The grammar is used to parse all of the noun phrases that match the noun phrase definition in the following manner: first, an instruction is broadcast to all processors which says that if the processor contains a determiner, then the address of the processor (the starting address for the noun phrase) and a special state symbol, Q1, should be sent to the processor to the right of the given processor (the one that holds the next word of the text). In the next phase, all processors that contain symbol Q1 are selected and instructed to check the definition(s) of their word. If the word can be an adjective, then the

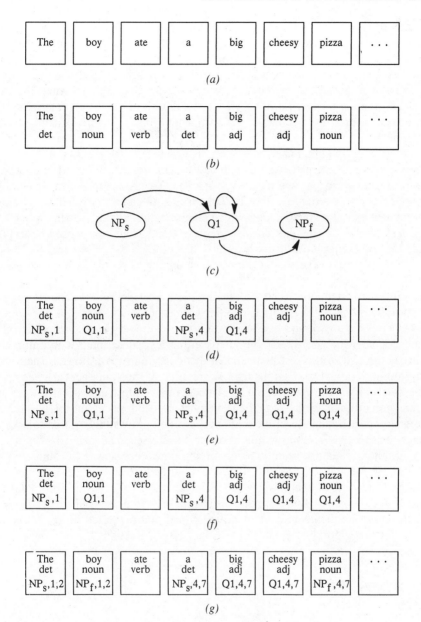

Figure 10.9 Parallel bulk processing of text: (*a*) Original text; (*b*) parts of speech assigned in parallel. See text; (*c*) Sample ATN grammar; (*d*) determiners recognized, and Q1 symbol sent; (*e*) first adjective recognized; (*f*) second (and last) adjective done; on next step no adjectives processed, so noun arc is broadcast; (*g*) noun phrase done; start and end stored in each cell.

symbol Q1 and the phrase starting address are passed to the next processor. The system then checks to see whether any cell used this rule during the last cycle. If so, the rule is repeated until no processors use it—that is to say, the longest string of adjectives preceding a noun has been parsed. The next rule can now be broadcast, instructing any processor that contains state symbol Q1 and a noun to return a noun phrase running from the initial determiner to the noun itself. The parsing of all such noun phrases in the system is now complete.

In general, the grammar will be non-deterministic; that is, noun phrases will have more than one pattern. All such patterns are broadcast, one after another, until all the noun phrases have been marked. Occasionally one noun phrase may appear to overlap another; in that case, both can be returned or a serial computer can attempt to judge which interpretation should have priority. A typical simple scheme that does not require serial arbitration would be to use the longest noun phrase if two overlap.

We have used several algorithms for assigning parts of speech to a word [30]. The one currently in use depends on the scan and sort operations and is illustrated in Figure 10.10. Both the dictionary definitions and words of text to be processed are stored, one per virtual processor. Next, all the words from both dictionary and text are sorted. The dictionary items are effectively subscripted with a zero, and the words in the text subscripted with a one, so that at the end of the sort phase, each dictionary entry is followed by all the occurrences of that word in the text. A variation of the scan operation can then be used to spread the definition to all words that follow a dictionary word in time proportional to the log of the number of occurrences of the word that appears most frequently in the text. This method allows this system to assign parts of speech to 64,000 words in under 40 milliseconds.

Another scheme currently being programmed operates according to the following method: each dictionary word is hash coded and sent to a processor. To find the parts of speech of words in a text, the hash code for each word of the text is computed, and a message is sent to the computed hash code address. Definitions are then sent back to all the words that requested them. The total time for this algorithm should be less than 10 milliseconds.

9 OTHER APPLICATIONS

Several other systems have been implemented that have application to AI research and qualitative reasoning [13,9,35].

- A SPICE-like VLSI circuit simulation system has been devised that has simulated a circuit of 25,000 transistors and could handle a much larger one. Such a simulator could be used as the basis for a system that could reason about the behavior or diagnosis of large circuits.

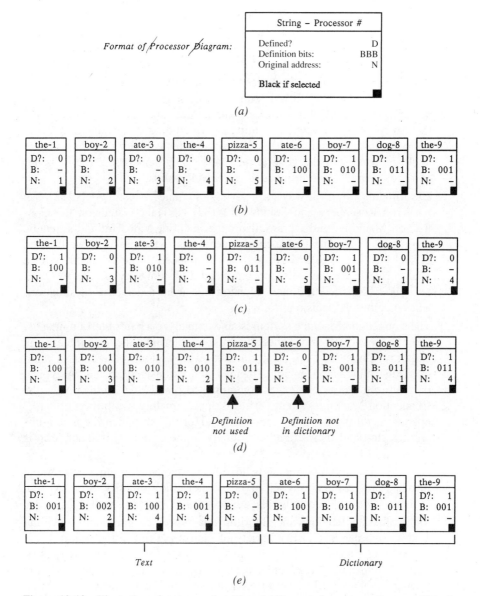

Figure 10.10 Illustration of sort-scan algorithm: (*a*) Format of processor diagram; (*b*) both the dictionary words and the text are stored in the CM; (*c*) an alphabetic sort is performed. (The dictionary is merged into the text.); (*d*) definitions are spread to all word; (*e*) definitions are sent back to original addresses.

- A VLSI layout program using simulated annealing has been devised. We are currently investigating the use of intelligent energy weighting schemes that would use a database of layout decisions made by a designer to assign priorities to proposed exchanges of cells.

- A system to model fluid dynamics has been built, using a cellular automata model [31]. The fluid flow is simulated by a 16-million-cell grid with 4 million particles moving across it. The particles interact at each cell, and their gross behavior simulates fluid flow (including vortex shedding and a periodic flow) without the use of Navier-Stokes equations. Such a system could be used to make general qualitative judgments about the behavior of fluids, liquids, and articulated or flexible solids.

- Experiments are under way for a variety of connectionist models [12]. We have run Rosenberg and Sejnowski's [34] NETtalk system on the Connection Machine, and we are also experimenting with Holland's genetic algorithms [16]. The Connection Machine is extraordinarily well suited to such tasks; a 16K CM-s has achieved a speed-up factor of 800 over a VAX 11/780. We have also programmed a system very similar to Quinlan's ID2 [28] for building a decision tree from a database of situations and outcomes or decisions.

- The Connection Machine system is able to analyze a pair of terrain images, taken from two different vantage points, to determine the terrain elevation and draw a contour map. A production level version of the contour mapping algorithm has been implemented and extensively tested. A typical program run processes images containing 512 x 512 (262,144) pixels; the Connection Machine system performs approximately two billion (2×10^9) operations during the inner loop of the algorithm in which the match-ups are detected and their alignment quality is measured. This inner loop is executed in less than two seconds.

- Dynamic programming methods have been used to program a system that matches proteins against each other to determine similarity of amino acid sequences, in order to judge the relatedness of various species [21].

- Many other applications and algorithms have been programmed, including context-free parsing [38], a variety of matrix multiplication routines, the fast Fourier transform, a maximum flow solver for directed graphs [14], minimum spanning trees and connected components of a graph [7], ray tracing graphics [10], four different algorithms for planar convex hulls, and several sorting algorithms. See Blelloch [6] for a summary of many of these applications and algorithms.

10 CONCLUSIONS

The Connection Machine has proved to be an excellent system for solving a large number of problems that we never envisioned for it at the beginning. Although for some problems the CM is of little advantage—for example trajectory calculations and the computation of pi to a million places—a much larger class of problems, generally characterized by having a large number of data elements, fit very naturally on the CM. Such problems include: finite element analysis; image, signal, and speech processing; graphics; database operations; simulations of fluid flow and aerodynamics, and simulation of molecular dynamics in crystals with defects. For all these problems, performance is typically in excess of 2.5 Gflops. The Connection Machine scales well to larger systems, both by adding more memory to each processor and by adding more processors. Other options for speed-up include replacing the bit serial processors of the CM-1 and the CM-2 with processors that operate on longer word lengths.

The Connection Machine is already cheaper than a PC on the basis of price per computational operation, and we have barely begun to come down the learning curve. While the Connection Machine is not completely general, I believe that it will establish itself as the standard across a fairly wide range of scientific, numerical, and database applications, bringing supercomputer power to a broad spectrum of new users.

11 ACKNOWLEDGMENTS

Credit is due especially to Craig Stanfill, who designed and implemented the document retrieval system and who is jointly responsible for the idea of memory-based reasoning; to Gary Sabot who implemented the system for the bulk processing of natural language; and to Brewster Kahle, Paul Mott, George Robertson, Shaun Keller, Sarah Ferguson, Jim Bailey, J. P. Massar, Guy Steele, Jr., Janet MacLaren, Danny Hillis, Donna Fritzsche, Bob Fischer, Robert Thau, Guy Blelloch, Lindsay Hotvedt, and many others at Thinking Machines, all of whom played important roles in the work reported here.

REFERENCES

1. Special issue on connectionist models and their applications, published in special issue of *Cognitive Science*, 1985.

2. S. G. Akl, *Parallel Sorting Algorithms*, Academic Press, Inc., NY, 1985.

3. A. Bawden, "Connection Graphs," *Proc. 1986 ACM Conf. on Lisp and Functional Programming,* pp. 258–265, ACM SIGPLAN/SIGACT/SIGART, Cambridge, MA, August 1986.

4. D. C. Blair and M. E. Maron, "An Evaluation of Retrieval Effectiveness for a Full-Text Document-Retrieval System," *Communications of the ACM*, 28(3):285–299, 1985.

5. G. E. Blelloch, *AFL-1: A Programming Language for Massively Concurrent Computers,* Master's thesis, Massachusetts Institute of Technology, Cambridge, MA, November 1986.

6. G. E. Blelloch, *Applications and Algorithms on the Connection Machine*, Technical Report TR87-1, Thinking Machines Corporation, Cambridge, MA, September 1987.

7. G. E. Blelloch, "Scans as Primitive Parallel Operations," *Proc. Int'l Conf. on Parallel Processing*, pp. 355–362, August 1987.

8. Thinking Machines Corporation, *Introduction to Data Level Parallelism*, Technical Report TR86-14, Cambridge, MA, April 1986.

9. J. deKleer and J. S. Brown, "Foundations of Envisioning," *Proc. AAAI-82*, AAAI, 1982.

10. H. Delaney, "Ray Tracing on a Connection Machine," *ACM/INRIA Int'l Conf. on Supercomputing*, St. Malo, France, July 1988.

11. S. E. Fahlman, *A System for Representing and Using Real-World Knowledge*, MIT Press, Cambridge, MA, 1979.

12. S. E. Fahlman and G. E. Hinton, "Connectionist Architectures for Artificial Intelligence," *IEEE Computer*, 20(1):100–109, January 1987.

13. K. Forbus, "Qualitative Process Theory," MIT AI Memo 664, February 1982.

14. A. V. Goldberg and R. E. Tarjan, "A New Approach to the Maximum Flow Problem," *Proc. ACM Symposium on Theory of Computing*, pp. 136–146, April 1986.

15. G. Hinton and J. Anderson, eds., *Parallel Models of Associative Memory*, Lawrence Erlbaum Associates, Hillsdale, NJ, 1981.

16. J. H. Holland, *Adaptation in Natural and Artificial Systems*, University of Michigan Press, Ann Arbor, MI, 1975.

17. P. Kanerva, *Self-Propagating Search: A Unified Theory of Memory*, Technical Report CSLI-84-7, Center for Language and Information, Stanford University, March 1984.

18. T. Kohonen, *Storage and Processing of Information in Distributed Associative Memory Systems*, Lawrence Erlbaum Associates, Hillsdale, NJ, 1981.

19. J. Kolodner, "Experiential Processes in Natural Problem Solving," Technical Report No. GIT-ICS/85/23. School of Information and Computer Science, Georgia Tech, Atlanta October 1985.

20. J. Kolodner, *Retrieval and Organizational Strategies*, PhD thesis, Yale University, Department of Computer Science, 1980.

21. E. Lander, J. P. Mesirov, and W. Taylor IV, "Protein Sequence Comparison on a Dataparallelcomputer," *Proc. Int'l Conf. of Parallel Processing*, 1988.

22. T. Mitchell, "Learning and Problem Solving," Computers and Thought Lecture, Eighth International Joint Conference on AI, 1983.

23. T. Mitchell, P. Utgoff, and R. Banerji, "Learning by Experimentation: Acquiring and Reforming Problem-Solving Heuristics," *Machine Learning*, Tioga, Palo Alto, CA, 1983.

24. P. Mott, D. Waltz, H. Resnikoff, and G. Robertson, *Automatic Indexing of Text*, Technical Report TR86-1, Thinking Machines Corporation, Cambridge, MA, January 1986.

25. A. Newell and H. Simon, *Human Problem Solving*, Prentice-Hall, Englewood Cliffs, NJ, 1972.

26. M. R. Quillian, *Semantic Memory*, MIT Press, Cambridge, MA, 1968.

27. J. R. Quinlan, "Discovering Rules from Large Collections of Examples: A Case Study," *Expert Systems in the Microelectronic Age*, Edinburgh University Press, Scotland, 1979.

28. J. R. Quinlan, "Induction of Decision Trees," *Machine Learning*, 1:81–106, 1986.

29. D. E. Rumelhart, D. E. Hinton, and R. J. Williams, *Learning Internal Representations by Error Propagation*, Bradford Books, Cambridge, MA, 1986.

30. G. Sabot, "Bulk Processing of Text on a Massively Parallel Computer," Twenty Fourth Annual Meeting of the Association for Computational Linguistics, June 10–13, 1986.

31. J. Salem and S. Wolfram, *Thermodynamics and Hydrodynamics with Cellular Automata*, Technical Report CA85-1, Thinking Machines Corporation, Cambridge, MA, November 1985.

32. G. Salton, *The SMART Retrieval System—Experiment in Document Processing*, Prentice-Hall, Englewood Cliffs, NJ, 1971.

33. A. Samuel, "Some Studies in Machine Learning Using the Game of Checkers," *IBM Journal of Research and Development*, 3:201–229, 1959.

34. T. J. Sejnowski and C. R. Rosenberg, *NETtalk: A Parallel Network that Learns to Read Aloud*, Technical Report EECS-86-01, The Johns Hopkins University, Baltimore, MD, 1986.

35. C. Stanfill and B. Kahle, "Parallel Free Text Search on the Connection Machine System," *Communications of the ACM 29*, 12, 1229–1239, December 1986.

36. C. Stanfill and D. L. Waltz, "Memory-Based Reasoning," *Communications of the ACM 29*, 12, 1213–1228, December 1986.

37. G. Sussman, *A Computer Model of Skill Acquisition*, Elsevier Publishing Company, NY, 1975.

38. R. Thau and S. Ferguson, *Context-Free Parsing on the Connection Machine System*, Technical Report NL87-1, Thinking Machines Corporation, Cambridge, MA, 1987.

39. P. Winston, *The Psychology of Computer Vision*, chapter: "Learning Structural Descriptions from Examples," McGraw-Hill, NY, 1975.

40. W. A. Woods, "Transition Network Grammars for Natural Language Analysis," *Communications of the ACM*, 13(10):591–606, October 1970.

CHAPTER 11

A DATABASE MACHINE BASED ON CONCATENATED CODE WORDS FOR VERY LARGE DATABASES

Soon Myoung Chung and P. Bruce Berra

1 INTRODUCTION

Knowledge-based systems consist of rules, facts, and an inference mechanism that can be utilized to respond to queries posed by users. As these systems grow, increased demands will be placed on the management of their knowledge bases. The intensional database (IDB) of rules will become large and present a formidable management task in itself. But the major management activity will be in the access, update, and control of the extensional database (EDB) of facts because the EDB is likely to be much larger than the IDB. The volume of facts is expected to be in the gigabyte range, and we can expect to have general EDBs that serve multiple inference mechanisms. In this chapter, we assume that the IDB is a set of rules expressed as logic programming clauses and the EDB is a relational database of facts. Retrieving the desired rules and facts in this context is a partial match retrieval problem where any subset of attributes can be specified in a query, and matching between terms consisting of variables and functions as well as constants should be tested as a preunification step.

   In the context of very large knowledge bases, the problem is how to obtain the desired rules and facts in the minimum amount of time. One reasonable choice of an indexing scheme to speed up the retrieval is the concatenated code word (CCW) surrogate file technique [1,7]. A CCW surrogate file is constructed by binary representations which are generated by performing well-chosen hashing functions on the original terms. The term "surrogate file" dates back to early work in information retrieval, and other terms, such as "signature file" and "descriptor file," have been used to describe similar structures [14, 23].

Suppose we have a fact type (relation) called borders given as follows:

borders (Country_1, Country_2, Body_of_Water)

To generate the CCW of a fact

borders (korea, china, yellow sea)

we first hash the attribute values to obtain binary representations:

$$h_1(\text{korea}) \quad = \quad 100...01$$
$$h_2(\text{china}) \quad = \quad 010...00$$
$$h_3(\text{yellow sea}) = \quad 110...00$$

Then the CCW of the fact is constructed by simply concatenating the binary representations of all attribute values and attaching the unique identifier (uid) of the fact as follows:

$$100...01 \mid 010...00 \mid 110...00 \mid 00...01$$

The unique identifier is also attached to the fact and serves as a link between the two. It can be converted to a pointer to the EDB block containing the fact by a dynamic hashing scheme such as Extendible Hashing [13] or Linear Hashing [19]. Thus, the CCW surrogate file of a relation is a set of CCWs, where each CCW corresponds to a fact of the relation.

The partial match retrieval process with the CCW surrogate file is as follows:

1. Given a query, obtain a query code word (QCW) by concatenating binary representations corresponding to the attribute values specified in the query. The bit positions of the QCW for the attributes which are not specified in the query are filled with don't-care bits.
2. Obtain a list of unique identifiers of the CCWs which match the QCW by comparing the QCW with all CCWs in the CCW surrogate file for that fact type.
3. Retrieve all the facts pointed to by the unique identifiers obtained in Step 2 and compare the corresponding attribute values of the facts with the query values to discard the facts not satisfying the query. These are called false drops. The facts satisfying the query are called good drops. The false drops are caused by the nonideal property of the hashing functions.
4. Return the good drops.

Compared with other full indexing schemes such as inverted lists [5], the CCW surrogate file technique yields much smaller amounts of index — usually less than 20 percent of the size of the EDB [7] while the inverted lists may be as large as the EDB. In terms of maintenance, the CCW surrogate file technique shows considerable advantage. When a new tuple is added to a relation, the CCW is generated and appended to the existing surrogate file. In the case of inverted lists, each list must be processed. Similar operations must be performed for deleting tuples from a relation. If file updates are frequent, the costs associated with updating, sorting, and garbage collection of inverted lists can be significant [25].

An important advantage of the CCW surrogate file technique is that it can be easily extended for the indexing of the rules expressed as PROLOG clauses [26], where the matching between constants, variables, and structured terms is required for unification.

An additional benefit obtained from using the CCW surrogate file approach is that relational operations can be performed on the CCW surrogate files [1]. To satisfy a query expressed in first-order logic, interrelated relational operations on the EDB are required. Thus, by performing relational operations on the CCW surrogate files first, considerable processing time can be saved. A relational operation on CCW surrogate files is a kind of relational operation algorithm using indexes [3, 20]. However, using tree-structured indexes in parallel relational operations is very difficult because the problem of synchronizing accesses to the indexes without completely serializing the actions of the processors executing in parallel has not been solved yet [2]. On the other hand, a CCW surrogate file is a set of CCWs corresponding to the facts of a relation, so it can be easily partitioned and distributed over the processors for concurrent processing.

To speed up relational operations based on the CCW surrogate file, a backend database machine is proposed. The database machine consists of a number of surrogate file processors (SFPs) and EDB processors (EDBPs) operating in SIMD mode. Each SFP has an associative memory to speed up relational operations on the CCW surrogate files. Since CCWs are quite compact and regular, they are mapped well to the associative memories. SFPs and EDBPs are connected to other processors of the same type through multistage interconnection networks.

In Section 2, the proposed backend machine is introduced. The parallel relational operation algorithms for the architecture are explained in Section 3, and Section 4 shows the performance of the proposed architecture for relational operations.

2 BACKEND DATABASE MACHINE

2.1 General Structure

The general structure of the proposed backend database machine that contains multiple processors for the management of a very large extensional database of facts is shown in Figure 11.1. We assume that there are gigabytes of data stored on the EDB disk subsystems and that there are corresponding CCW surrogate

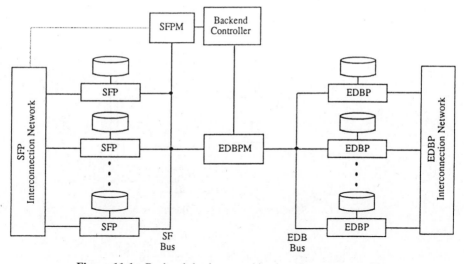

Figure 11.1 Backend database machine based on surrogate files.

files stored on the surrogate file (SF) disk subsystems. Suppose that a user is interested in retrieving fact data satisfying a partial match retrieval query from a particular relation. Then the partial match retrieval query is transferred to the backend controller from the host computer, and a query code word (QCW) is constructed in the surrogate file processor manager (SFPM) using the proper hashing functions. The QCW is then broadcasted to the surrogate file processors (SFPs) to be used as a search argument. The SFP compares the QCW with each CCW and strips off the unique identifiers of matching CCWs. Each extracted unique identifier is sent to the EDB processor manager (EDBPM) and passed on to the EDB processor (EDBP), which contains the fact with that unique identifier. The EDBP will access the EDB block containing the fact, compare the retrieved fact with the original query to check that it is a matching fact, and then send it to the host computer.

The basic idea of the proposed backend database machine is to reduce the number of EDB blocks to be transferred from the secondary storage systems by performing relational operations on the surrogate files first. To speed up relational operations, the surrogate file blocks and data blocks of each relation are evenly distributed over the surrogate file disk subsystems and the EDB disk subsystems, respectively. Thus, the SFPs can process the surrogate file concurrently, and the EDBPs can process the relation concurrently.

2.2 Surrogate File Processing System

As shown in Figure 11.2, a SFP is equipped with an associative memory unit to perform the searching operation efficiently. Associative memories are very fast because they use content addressing and parallel searching, but they are generally costly and rigid in data format. However, the format of the surrogate

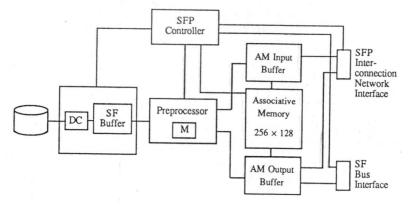

Figure 11.2 Structure of a surrogate file processor (SFP).

file is regular and maps very well into the associative memory and the cost of the associative memory hardware is decreasing as VLSI technology advances. Additionally, associative memories can be used for relational operations, such as selection and join, because associative memories can perform many associative operations such as "equal to," "not equal to," "less than (or equal to)," "greater than (or equal to)," "between limits," "outside of limits," and others depending on the structure. In our design, we used word-parallel bit-serial (WPBS) associative memory, which consists of two-dimensionally accessible memory and an array of processing elements. A word slice is a unit for memory read and write operations, and a bit slice is a unit for arithmetic and logical operations. A bit-parallel associative memory [9, 24], whose memory cells have comparison logic, is faster than a WPBS associative memory but is much more complex in structure and less flexible in functionality. The current status of associative memories and associative processors is reviewed in [30].

A surrogate file disk subsystem is attached to each SFP through the surrogate file (SF) buffer, so that the processing and accessing of the surrogate file can be overlapped. In our system, surrogate file blocks of a relation are stored consecutively within each disk subsystem, so that the surrogate file blocks can be accessed with minimum seek time. By associating a disk subsystem to each surrogate file processor, we lose some flexibility in allocating processors to a query processing, but surrogate file blocks can be processed rapidly because on-the-fly filtering is possible and there is no contention between processors for a same disk subsystem.

The surrogate file processors are connected through the SFP interconnection network. Since there are a number of surrogate file processors, flexibility and speed of the interconnection network are very important factors determining the overall performance. The mapping between SFPs will be a permutation, selective broadcasting, or broadcasting depending on the distribution of operand surrogate files among the SFPs (we consider the pair of a SFP and the associated surrogate file disk subsystem as a single unit and call it a SFP), the number of available SFPs, and the algorithms of relational operations. To handle all the mapping modes, we chose a multistage Omega network [18] implemented with 2×2 switching elements with four functions: straight, exchange, upper broadcast, and lower broadcast. Thus, any one SFP can broadcast a surrogate file block to the rest of the SFPs with uniform delay. The SFP interconnection network will operate in a circuit switching mode to facilitate the transfer of surrogate file blocks.

2.3 EDB Processing System

If a fact unique identifier is sent from a SFP to the EDB processor manager (EDBPM), the EDBPM finds the EDB processor (EDBP) containing the corresponding fact by hashing the unique identifier and then sends the unique identifier to that EDBP. The EDBPM is a general-purpose microcomputer with large memory to buffer the unique identifiers from the SFPs as well as the operation results from the EDBPs.

The structure of an EDBP is shown in Figure 11.3. In the case of fact retrieval, a fact block corresponding to the received unique identifier is accessed by the EDBP. We assume that EDB blocks are randomly distributed within a disk subsystem, and to speed up block access a disk cache is provided for each EDBP. Once the EDB block is read into the memory of the Relational Operator, the operator searches the block with the unique identifier, extracts the fact corresponding to the unique identifier, and compares the extracted fact with the query to check that it matches. The Relational Operator is a general-purpose processor that performs all the tuple-wise relational operations as well as statistical aggregate functions. Through the EDBP interconnection network, facts can be transferred from one EDBP to another. We chose to use the multistage Omega network operating in a packet switching mode to facilitate the frequent transfer of facts between EDBPs in the case of join operations.

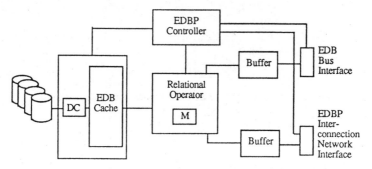

Figure 11.3 Structure of an extensional database processor (EDBP).

2.4   Processing Mode and the Multiple Backend System

The processing mode of the backend system can be SIMD or MIMD, depending on the distribution of surrogate files and relations over the processors and the assignment of the processors to a given query operation. If all the processors are working on a single operation, then it becomes a SIMD mode, but if processors are partitioned into a number of groups and each group of processors is assigned a different operation, then the processing mode is MIMD at the group level. To operate either in SIMD or MIMD mode, the interconnection network must be partitionable. A multistage Omega network of size 2^m can be partitioned into independent subnetworks of different sizes with the requirement that the addresses of all the I/O ports in a subnetwork of size 2^i agree in $(m - i)$ of their bit positions [27].

As the size of the EDB increases, the system can be upgraded to a multiple backend database machine system by adding more proposed database machines to the existing system configuration as shown in Figure 11.4. If we store the related relations and their surrogate files, which are frequently accessed together to answer queries, on a database machine, then each database machine can work on a different query and the processing mode of the system becomes multiple SIMD (MSIMD). By clustering the related relations and their surrogate files on the database machines properly, we can reduce the amount of communication between the database machines to answer queries. The inter-machine interconnections are separated from the intra-machine interconnections, and the backend controller of each machine is in charge of communication with other machines and the global backend system controller through the backend system network.

3   RELATIONAL OPERATIONS IN THE BACKEND DATABASE MACHINE

In this section, we introduce the parallel relational operation algorithms for the database machine proposed when the surrogate file blocks and the data blocks of each relation are evenly distributed over the surrogate file disk subsystems

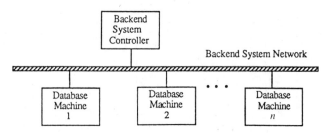

Figure 11.4   Multiple backend database machine system.

and EDB disk subsystems, respectively, and the workloads of the processors are uniform.

The problem of optimal distribution of surrogate files and data files over the processors for all possible selection and join operations is basically the same as the problem of distributing a data file over multiple disks so that the maximum parallelism can be achieved in retrieving the relative data blocks for all partial match queries. Therefore, we can adopt the Generalized Disk Modulo (GDM) method [12], which was proposed for the optimal distribution of a data file over disks for partial match retrieval, to distribute the surrogate files and data files over the processors for maximal concurrent processing.

Let's assume that we have N_{SFP} SFPs numbered from 0 to $N_{SFP} - 1$ and N_{EDBP} EDBPs numbered from 0 to $N_{EDBP} - 1$. With the GDM method, the number of the SFP allocated to the CCW $(br_1, br_2, \ldots, br_n)$ of a fact $(a_1, a_2, \ldots, a_n)$, where a_i is the ith attribute value and br_i is the binary representation obtained by hashing a_i, is

$$\left(\sum_{i=1}^{n} br_i \times PN_i \right) \mod N_{SFP} \qquad (11.1)$$

and the number of the EDBP allocated to the corresponding fact is

$$\left(\sum_{i=1}^{n} br_i \times PN_i \right) \mod N_{EDBP} \qquad (11.2)$$

where each PN_i is a positive integer relatively prime to both N_{SFP} and N_{EDBP}.

The GDM method is a heuristic algorithm and does not guarantee the optimal distribution of files for all possible selection and join operations. However, judging from its performance for partial match retrieval, it is expected to provide an optimal or near optimal distribution of files for most of the operations. Also, we need to decide the number of processors to which a file is to be distributed by estimating the steady state size of the file. If the size of the CCWs allocated to a SFP is far less than a block size, then the storage utilization and performance will be very poor. Therefore, if a file is not big, it is better to distribute the file to a subset of the processors. We can relocate CCWs and facts to balance the distribution.

3.1 Selection Operation

To select on a particular attribute, the SFPs execute a comparison, such as "equal to," "not equal to," "greater than or equal to," or "less than or equal to," between the binary representation of a CCW and the hashed selection attribute value. To retain the ordering between the binary representations of attribute values so that range queries can be performed on the CCW surrogate file, order-preserving hashing functions [15] are used to generate the CCW surrogate file.

Each SFP retrieves a block of CCWs and does projection on the binary representation of the specified attribute and the unique identifier, then loads the projected CCWs into the associative memory. The comparand register of the associative memory is loaded with the hashed selection attribute value. Those bit positions of the input mask register of the associative memory which are corresponding to the hashed selection attribute value are filled with 1's while other bit positions are filled with 0's. If there is any matching CCW, the corresponding fact unique identifier is sent to the EDBPM.

As soon as any fact unique identifier is received by the EDBPM, it finds the EDBP containing the corresponding EDB block and sends the unique identifier to that EDBP. The EDBP retrieves the EDB block by using the unique identifier, searches the block with that unique identifier to find the corresponding fact, and performs the actual selection operation on the fact. Due to the preselection operation on the surrogate file, the number of EDB blocks to be accessed from the secondary storage system is usually very small compared to the total number of EDB blocks of a relation.

3.2   Join Operation

There are three main algorithms for the join operation: sort-merge, hash-partition, and nested-loop join algorithms. The performance of the sort-merge join algorithm for the non-equijoin operation is as good as that for the equijoin operation, because once the two operand relations or subrelations are sorted, the merging step can handle the equijoin and the non-equijoin in the same way by performing the corresponding comparison operation. The database machine DELTA [16] has multiple relational database engines composed of sort-merge units and performs the sort-merge join algorithm. If a database machine has sort-merge units, the selection operation is interpreted as a join operation between a relation and a constant value specified in a query.

The hash-partition join algorithm is adopted by the database machine GRACE [17]. Each operand relation is partitioned into a number of buckets depending on the hash value of the join attribute, then matching is performed within each bucket by a processor assigned to that bucket. Usually the hash-partition join algorithm is better than the sort-merge join algorithm in the case of the equijoin operation because sorting creates a total ordering of the tuples in both relations whereas the hashing simply groups related tuples together in the same bucket [10]. However, in the case of the non-equijoin, the operation of each processor is not limited to a single bucket, and the workloads of the processors may not be uniform. One problem of the hash-partition join algorithm is the bucket overflow caused by the nonuniform distribution of the join attribute value. In this case, rehashing of the overflow bucket is necessary.

With the nested-loop join algorithm, each tuple of one operand relation is compared with all the tuples of the other operand relation. It has been shown that the nested-loop join algorithm takes advantage of different operand sizes and that the processing time is inversely proportional to the number of processors [4, 29]. In the case of the sort-merge algorithm, increasing the number of processors beyond a certain number causes very little decrease in the processing time. The reason is that the degree of parallelism is divided by two at each merge step after a certain stage [29].

Our proposed database machine adopts the nested-loop join algorithm because the associative memories in the surrogate file processors can easily perform the parallel execution of the nested-loop join operation.

If we assume that the CCW surrogate files of two operand relations, R_1 and R_2, are evenly distributed over a number of SFPs and R_1 is larger than R_2 in size, the nested-loop join algorithm is executed as follows:

1. Each SFP reads a block of CCW surrogate file of R_2 from the associated surrogate file disk subsystem, projects it on the binary representation of the join attribute and the unique identifier, and loads it into the associative memory.

2. Each SFP reads a block of CCW surrogate file of R_1 from the associated surrogate file disk subsystem, projects it on the binary representation of the join attribute and the unique identifier, and stores it in the associative processor input buffer.

3. One SFP broadcasts the projected surrogate file block from Step 2 to the rest of the SFPs, which already have a block of CCW surrogate file of R_2 in their associative memories from Step 1.

4. Each SFP searches the associative memory with the broadcasted projected CCWs as searching arguments one by one. If there is a match, the pair of unique identifiers of the two matched CCWs are read and sent to the EDBPM.

5. Repeat Steps 3 and 4 until all the projected CCW surrogate file blocks from Step 2 are broadcasted.

6. Repeat Steps 2 through 5 until all the CCW surrogate file blocks of R_1 are retrieved from the surrogate file disk subsystems.

7. Repeat Steps 1 through 6 until all the CCW surrogate file blocks of R_2 are retrieved and searched.

In Step 4, as soon as any unique identifier pair is received by the EDBPM, the EDBPM finds the EDBPs containing the corresponding facts and transfers the unique identifiers to those EDBPs. If a single EDBP contains the two operand facts, then that EDBP performs the actual join operation on the two facts retrieved, otherwise one EDBP transfers a projected fact to another EDBP containing the other operand fact, then the join is performed. To reduce the amount of communication through the EDB interconnection network, the smaller projected fact is transferred. Projection on the retrieved fact is performed on the join attribute, the attributes involved in the output relation, and the unique identifier. The prejoin operation of the SFPs is over-lapped with the actual join operation of the EDBPs.

Parallel algorithms of other relational algebra operations, such as projection, intersection, Cartesian product, union, difference, and division, for the proposed database machine are introduced in [8].

4 PERFORMANCE ANALYSIS OF THE PROPOSED DATABASE MACHINE

In this section, we analyze the performance of the proposed database machine for the selection and join operations by using deterministic analysis. We used the parameter values specified in Table 11.1, and assumed that

1. The surrogate file size is 20 percent of the relation size.
2. The surrogate file of each operand relation is evenly distributed over the surrogate file disk subsystems, as explained in Section 3. Within each surrogate file disk subsystem, the sub-surrogate file is organized as a sequential file and stored consecutively.
3. Each relation is evenly distributed over the EDB disk subsystems. Within each EDB disk subsystem, the subrelation is organized as a dynamic hashing file, such as Extendible Hashing or Linear Hashing, on the unique identifiers of facts.

TABLE 11.1 Summary of Parameter Values Used for Performance Analyses.

Parameter	Values
Average seek time of a disk	28 msec
Minimum seek time of a disk	6 msec
Rotational delay of a disk	8.3 msec
Data transfer rate of a disk	2 MB/sec
SF and EDB block size	4 KB
Effective EDB block access time	10 msec
Interconnection network transfer rate per port	10 MB/sec
SF and EDB bus bandwidth	20 MB/sec
Memory bandwidth	10 MB/sec
Projection rate	5 MB/sec
Time for loading a word into an associative memory	0.1 μsec
Associative memory searching time for k bit-slices	$(0.5 + 0.1k)\,\mu$sec
Time for reading a uid from the associative memory	0.2 μsec
Word (4 bytes) comparison time in an EDBP	2 μsec

4. The workloads of the SFPs and the EDBPs involved in a relational operation are uniform due to near optimal distribution of surrogate files and data files over the processors.
5. The storage utilization of the surrogate files and the EDB is 70 percent.
6. Disk I/O operations and the processor's operations are executed concurrently whenever possible.
7. Preoperations on the surrogate files by the SFPs and the actual operations on the relations by the EDBPs are executed concurrently whenever possible.

4.1 Selection Operation

In the case of a selection operation, the execution time is mainly determined by the surrogate file size of an operand relation, the number of SFPs and EDBPs involved in the operation, and the selectivity, which is defined as the ratio of the cardinality of the output relation to that of the operand relation.

Once a preselection instruction is broadcasted to the SFPs, each SFP reads a surrogate file block of the operand relation R_1 within

$$T_{ba} = \text{Average seek time} + \text{Rotational delay} + \text{Block transfer time} \quad (11.3)$$

Then the retrieved block is projected on the binary representation of the selection attribute and the unique identifier (uid), loaded into the associative memory (AM), and searched. The unique identifiers of the matched code words in the associative memory are read and sent to the EDBP manager (EDBPM). Thus, the time for one block processing (T_{bs}) is

$$T_{bs} = \text{Block projection time} + \text{AM loading time} + \text{AM searching time}$$

$$+ \text{uid extracting time} \quad (11.4)$$

If each SFP has m_1 surrogate file blocks and the matching code words are evenly distributed over the surrogate file blocks, the preselection time is

$$T_{\text{pre-select}} = T_{ba} + \max\!\left((m_1 - 1)\text{block transfer time},(m_1 - 1)T_{bs}\right) + T_{bs} \quad (11.5)$$

since the transfer of the last $(m_1 - 1)$ blocks can be overlapped with the processing of the first $(m_1 - 1)$ blocks.

The processing time of the EDBPM for a unique identifier received is

$$T_{\text{EDBPMS}} = \text{uid read time from buffer} + \text{uid dispatching time} \quad (11.6)$$

and the processing time of an EDBP for a unique identifier received is

$$T_{\text{EDBPS}} = \text{EDB block access time} + \text{uid searching time}$$

$$+ \text{False drop check time} + \text{Selected tuple output time} \quad (11.7)$$

We assumed that an EDBP accesses a block by using the unique identifier received and locates the corresponding fact within that block by sequentially searching the unique identifiers of the facts within that block.

If the facts to be retrieved are evenly distributed over the EDBPs, then the actual selection time on the relation is

$$T_{\text{actual select}} = \frac{\text{Number of uids extracted in preselection}}{\text{Number of EDBPs}} \times T_{\text{EDBPS}} \qquad (11.8)$$

The operation of the EDBPM is overlapped with the operations of the EDBPs and the processing time of the EDBPM is relatively very small, so we ignored the processing time of the EDBPM in estimating the actual selection time.

Since the preselection operation and the actual selection operation are performed concurrently, the total execution time of a selection operation is

$$T_{\text{select}} \approx \max(T_{\text{preselect}}, T_{\text{actual select}}) \qquad (11.9)$$

Figure 11.5 shows the total execution time of a selection operation on a relation R_1 as a function of the selectivity, the number of SFPs (N_{SFP}), and the number of EDBPs (N_{EDBP}), when

Cardinality of the operand relation $R_1 = 10^6$
Size of a tuple in $R_1 = 100$ bytes
Size of a unique identifier $= 3$ bytes
Size of a concatenated code word $= 20$ bytes
Number of the selection attribute $= 1$
Size of a selection attribute value $= 15$ bytes
Size of the binary representation of a selection attribute $= 3$ bytes
Size of an output tuple $= 100$ bytes
$$\alpha = \frac{\text{Number of false drops}}{\text{Number of good drops}} = 0.1$$

Since the hashing functions used to generate the CCW are not ideal, there are a certain number of false drops. We assumed that the total number of matched code words is $(1 + \alpha)$ times the actual number of facts satisfying a selection query.

The preselection time is mainly a function of the surrogate file size, which is proportional to the relation size, and the number of SFPs, while the actual selection time is mainly a function of the selectivity and the number of the EDBPs.

When the selectivity is low, the preselection time on the surrogate file is dominant because the number of EDB blocks to be retrieved is smaller than the total number of surrogate file blocks. In this case, the total execution time will decrease as the number of SFPs increases. As the selectivity increases, the number of EDB blocks accessed will increase, and the actual selection time

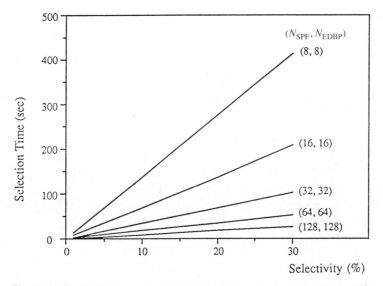

Figure 11.5 Performance of selection operations (cardinality of $R_1 = 10^6$).

on the relation becomes dominant. In this case, as the number of EDBPs increases, the total execution time decreases. Actually, as the selectivity increases, the effective EDB block access time would decrease because of the increased disk cache hit ratio. However, we assumed that the effective EDB block access time is constant in evaluating the total execution time of a selection operation.

To reduce the number of EDB block accesses when the selectivity is high, we can search an accessed EDB block with the attribute values specified in the query, instead of searching the block with a unique identifier. In this way, we can find all the desired facts in that block. Then we store the block number in memory, and whenever a unique identifier of a fact within that block is received, we discard the unique identifier since we have already retrieved that fact. The selectivity can be estimated by measuring the arrival rate of unique identifiers to the EDBPM. If the estimated selectivity is high, so that each EDB block probably contains more than one good or false drop, it is better to search the accessed EDB blocks with the attribute values specified in the query because block access time is much larger than block search time.

In Figure 11.5, we didn't apply this method in evaluating the total execution time of a selection operation.

4.2 Join Operation

In the case of a join operation, the surrogate file size of the two operand relations, the number of SFPs and EDBPs involved, and the join selectivity, which is defined as the cardinality of the output relation to the multiplication of the cardinalities of the operand relations, will mainly determine the execution time. Since an EDBP performs the join operation on two operand facts, one of which may be transferred from another EDBP, we have to consider the network delay caused by the conflict in the network. However, the size of a projected fact is usually small, so the transfer time of a projected fact is very small compared to an EDB block access time. Moreover, transfers of projected facts are overlapped with EDB block accesses. Thus, network contention is not a serious problem even when the join selectivity is very high.

Let's denote the surrogate files of the operand relations R_1 and R_2 as SF_1 and SF_2, respectively, and assume that each SFP has m_1 blocks of SF_1 and m_2 blocks of SF_2, where $m_1 > m_2$. Once a prejoin instruction is broadcasted to the SFPs, each SFP retrieves one SF_2 block, projects the block on the binary representation of the join attribute and the unique identifier, and loads the projected block into the associative memory (AM) within

$$T_1 = T_{ba} + SF_2 \text{ block projection time} + \text{AM loading time} \qquad (11.10)$$

where T_{ba} is defined in Equation (11.3). Then each SFP retrieves one SF_1 block within T_{ba} and projects the SF_1 block within

$$T_{pro} = SF_1 \text{ block projection time} \qquad (11.11)$$

One SFP broadcasts the projected SF_1 block to the rest of the SFPs through the SFP interconnection network operating in a circuit switching mode. Since only one SFP broadcasts to the other SFPs, there is no routing conflict within the SFP interconnection network. Then each SFP performs the join between the broadcasted SF_1 block and the SF_2 block loaded in the associative memory. Each projected CCW of the broadcasted SF_1 block becomes the searching argument, and if there is any matching CCW in the associative memory, then the corresponding unique identifier is read and sent to the EDBPM, paired with the unique identifier of the searching argument. Thus, one SF_2 block in each SFP is prejoined with one SF_1 block within

$$T_2 = \text{Projected } SF_1 \text{ block broadcast time} + m_s \times (\text{AM searching time}$$
$$+ \text{uid pair extracting time}) \qquad (11.12)$$

where m_s denotes the number of CCWs in the SF_1 block.

This process repeats until all the SFPs broadcast their projected SF_1 blocks. Therefore, if there are N_{SFP} SFPs, one SF_2 block in each SFP is prejoined with N_{SFP} SF_1 blocks within

$$T_3 = T_{pro} + T_2 \times N_{SFP} \qquad (11.13)$$

Since T_3 is overlapped with the time for accessing the next SF_1 blocks in each SFP, one SF_2 block in each SFP is prejoined with all the $(m_1 \times N_{SFP})$ blocks of SF_1 within

$$T_4 = T_{ba} + \max\left((m_1 - 1)SF_1 \text{ block access time}, (m_1 - 1)T_3\right) + T_3 \quad (11.14)$$

This process repeats until all the $m_2 SF_2$ blocks in each SFP are prejoined with all the $(m_1 \times N_{SFP})$ blocks of SF_1. Therefore, the total prejoin time is

$$T_{\text{prejoin}} = (T_1 + T_4) \times m_2 \quad (11.15)$$

The processing time of the EDBPM for a pair of unique identifiers received is

$$T_{\text{EDBPMJ}} = \text{uid pair read time from buffer} + \text{uid pair dispatching time} \quad (11.16)$$

If we assume that the projected tuple of R_2 is smaller than the projected tuple of R_1, then the processing time of EDBPs to join two tuples corresponding to a pair of unique identifiers received is

$$T_{\text{EDBPJ}} = P_s \times \{ R_1 \text{ tuple access time} + R_2 \text{ tuple access time}$$
$$+ \text{two-tuple join time} + \text{joined tuple output time}\}$$
$$+ (1 - P_s) \times \{ \max(R_1 \text{ tuple access time}, (R_2 \text{ tuple access time}$$
$$+ \text{network delay})) + \text{two-tuple join time}$$
$$+ \text{joined tuple output time}\} \quad (11.17)$$

where each operand tuple access time is the time for accessing and searching the EDB block containing the tuple by using the unique identifier, and P_s denotes the probability that two operand tuples are stored within the (EDB disk subsystem of the) same EDBP. If we assume that the operand tuples are distributed over the EDBPs uniformly, then

$$P_s = \frac{N_{\text{EDBP}}}{N_{\text{EDBP}}H_2} = \frac{N_{\text{EDBP}}}{(N_{\text{EDBP}}+1)C_2} = \frac{2}{N_{\text{EDBP}} + 1} \quad (11.18)$$

where $N_{\text{EDBP}}H_2$ represents the number of combinations of two out of N_{EDBP} with repetition.

When two operand tuples are stored in two EDBPs, the smaller projected tuple is transferred from one EDBP to the other EDBP through the EDBP interconnection network. Since we assumed that the projected tuple of R_2 is smaller than the projected tuple of R_1, where the projection is performed on the join attribute, output attributes, and the unique identifier, the network delay is the time to transfer a projected tuple of R_2 through the multistage interconnection network operating in a packet switching mode. The network delay of a blocking network which does not have a queue between the stages can be approximated as

$$\text{Network delay} \approx (\log_2 N_{\text{EDBP}} + M - 1) \times \frac{1}{P_A} \times (\text{Switch delay}) \quad (11.19)$$

where

$\log_2 N_{EDBP}$ = the number of stages of the network with 2×2 switching elements

M = the number of packets for a projected tuple of R_2

P_A = the probability that a transmission request by an EDBP is not blocked due to routing conflicts per network cycle when all the EDBPs are requesting simultaneously

Switch delay = time for a packet to be transmitted through a switching element

If we assume that an EDBP references the other EDBPs equiprobably and a switching element selects an input equiprobably when there is a routing conflict between the two inputs, then the nonblocking probability, P_A, can be obtained by using the following recursive equation [22]:

$$P(n) \approx 1 - \left[1 - \frac{P(n-1)}{2} \right]^2, \qquad P(0) = 1 \qquad (11.20)$$

In this equation, $P(n)$ denotes the probability that an input request is passed to the nth stage of the network implemented with 2×2 switching elements when all the processors connected to the input ports of the interconnection network are requesting simultaneously. Since there are $\log_2 N_{EDBP}$ stages within the network,

$$P_A = P(\log_2 N_{EDBP}) \qquad (11.21)$$

To reduce the network delay, we may have queues between the stages of the EDBP interconnection network. In this case, an input packet to a switching element is blocked only when the output queue is full. It has been shown that a queue size of less than three or four can significantly improve the throughput of the network compared to a network without queues by reducing the blocking probability, but queues larger than this are much less effective because the packets tend to fill up the queues and stay in the network longer [11].

Since we assumed that the operand tuples to be joined are uniformly distributed over the EDBPs, the actual join time on the relations is

$$T_{\text{actual join}} = \frac{\text{Number of uid pairs extracted in prejoin}}{\text{Number of EDBPs}(N_{EDBP})} \times T_{EDBPJ} \qquad (11.22)$$

The operation of the EDBPM is overlapped with the operations of the EDBPs, and the processing time of the EDBPM is relatively very small. Thus, we ignore the processing time of the EDBPM in estimating the actual join time.

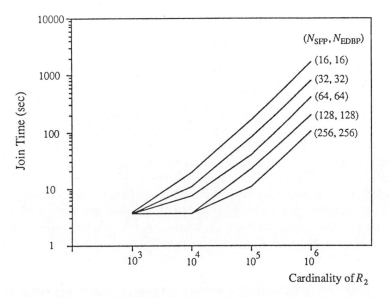

Figure 11.6 Performance of join operations between R_1 and R_2 (cardinality of $R_1 = 10^6$, cardinality of output relations = cardinality of R_2).

Since the prejoin operation and the actual join operation are performed concurrently, the total execution time of a join operation is

$$T_{join} \approx \max(T_{prejoin}, T_{actual\ join}) \qquad (11.23)$$

The execution time of a join operation on two operand relations, R_1 and R_2, is plotted in Figure 11.6 as a function of the cardinality of R_2, the number of SFPs (N_{SFP}), and the number of EDBPs (N_{EDBP}), when

Cardinality of $R_1 = 10^6$

Cardinality of the output relation = cardinality of R_2

Size of a tuple in R_1 and $R_2 = 100$ bytes

Size of a unique identifier = 3 bytes

Size of a concatenated code word = 20 bytes

Size of a join attribute value = 15 bytes

Size of the binary representation of a join attribute = 3 bytes

Contribution of each operand relation to an output tuple = 30 bytes

SFP interconnection network speed per port = 10 Mbytes/sec

Data size of a packet for EDB interconnection network = 10 bytes

EDB interconnection network switch delay = 1 μsec

$$\beta = \frac{\text{Number of unique identifier pairs extracted}}{\text{Cardinality of output relation}} = 1.1$$

Because of the nonideal hashing functions, the number of joinable CCW pairs is larger than the cardinality of the output relation, and β accounts for this effect.

As the size of the two operand relations increases, the prejoin time increases due to the increased surrogate file size. The prejoin time is dominant when the join selectivity is low because the prejoin operation is performed on every pair of surrogate file blocks whereas the actual join operation is performed on two operand facts. Therefore, when the join selectivity is low, as we increase the number of SFPs we can decrease the total join processing time. As the join selectivity increases, actual join operation time increases due to the increased number of random EDB block accesses. In this case, we can reduce the number of EDB block accesses by storing the retrieved joinable facts in the memory and reusing them whenever requested. For an equijoin operation, if the average number of tuples in R_1 (R_2) that have the same join attribute value is C_1 (C_2), then a joinable tuple of R_1, which participates in the semijoin of R_1 by R_2, can be joined with C_2 tuples of R_2 on the average. Thus, if we store that projected tuple of R_1 in memory, we can reuse it $(C_2 - 1)$ times later. For the same reason, the projected tuple of R_2 can be reused $(C_1 - 1)$ times later. However, in the results shown in Figure 11.6, we didn't apply this method; we used a constant EDB block access time independent of the join selectivity. In an actual case, as the join selectivity increases, the effective EDB block access time decreases because of the increased disk cache hit ratio. The join selectivity can be estimated by measuring the arrival rate of unique identifier pairs to the EDBPM.

4.3   Performance of the Proposed Machine with Clustered CCW Surrogate Files

In Sections 4.1 and 4.2, we analyzed the performance of the proposed backend machine assuming that a sub-surrogate file within a SFP has a sequential file structure and the whole sub-surrogate file is accessed to perform a preoperation. In this case, the preoperation time will be intolerably large when the operand relations are very large. Since the preoperation time is mainly determined by the number of surrogate file blocks to be accessed from each SFP to perform the preoperation, we can reduce the preoperation time by reducing the search space of the surrogate files. For this purpose a clustered surrogate file structure, which is a multiattribute partitioned file, has been proposed [8]. With the clustered CCW surrogate file structure, each CCW of a sub-surrogate file within a SFP is digitally hashed to a partition based on the prefix (leftmost bits) of the binary representations. Details of the clustering method and the directory structure indexing the partitions are explained in [8]. Since a sub-surrogate file within each SFP is clustered into partitions (blocks) by allocating similar CCWs (CCWs with the same prefix in one or more of the binary representations) to the same partition, only a subset of the partitions will be accessed for a preoperation.

For a selection operation, only the surrogate file partitions relative to the selection attribute values specified in the query will be retrieved and searched within the SFPs. The number of partitions to be retrieved is dependent on how many times the surrogate file is partitioned based on the binary representations corresponding to the selection attributes. If the surrogate file is partitioned based on k bits of the binary representations corresponding to the selection attributes, then the number of partitions to be retrieved is about $\frac{1}{2^k}$ of the total number of the surrogate file partitions for an equiselection operation. However, the relative partitions are accessed randomly from the disk subsystem to be searched. Therefore, the preselection operation with the clustered surrogate file structure is advantageous if the factor of reduction in search space is bigger than the factor of increase in block access time.

To perform a join operation on the backend machine with clustered surrogate files, we can use the join operation algorithm introduced in Section 3. The main difference is that we can decompose a prejoin operation into subjoins between sets of surrogate file partitions. Suppose that the surrogate file of relation $R_1(SF_1)$ is partitioned based on the leftmost k_1 bits of the binary representation of the join attribute of R_1. Then the SF_1 partitions can be grouped into 2^{k_1} sets so that there is a one-to-one correspondence between the sets of partitions and values of the k_1 bits. Similarly, if the binary representation of the join attribute of relation R_2 contributed leftmost k_2 bits for partitioning the surrogate file of relation $R_2(SF_2)$, then the SF_2 partitions are grouped into 2^{k_2} sets of partitions. If $k_1 > k_2$, each of the 2^{k_1} sets of SF_1 partitions has only one equijoin-possible set of SF_2 partitions. Therefore, for an equijoin operation, we can decompose the prejoin between SF_1 and SF_2 into 2^{k_1} subjoins between two sets of partitions. Since each sub-surrogate file is partitioned within a SFP, the partitions of a set associated with a value of $k_1(k_2)$ bits are distributed over the SFPs. Thus, we can perform a subjoin by using the join algorithm introduced in Section 3.

Let's assume that there are m_1 SF_1 partitions and m_2 SF_2 partitions within each of the N_{SFP} SFPs, and that k_1 and k_2 bits are contributed from the binary representations of the join attributes for partitioning SF_1 and SF_2, respectively. Then $(m_1 \times N_{SFP})$ SF_1 partitions are grouped into 2^{k_1} sets of partitions, and $(m_2 \times N_{SFP})$ SF_2 partitions are grouped into 2^{k_2} sets of partitions. If we represent the complexity of the prejoin by the number of joins between two partitions, the prejoin complexity of the proposed database machine with the clustered subsurrogate files is

$$\max(2^{k_1}, 2^{k_2}) \frac{m_1}{2^{k_1}} \frac{m_2}{2^{k_2}} = \frac{m_1 m_2}{\min(2^{k_1}, 2^{k_2})} \qquad (11.24)$$

if the distributions of the binary representations of the join attributes are uniform. On the other hand, the complexity of the prejoin with sequential sub-surrogate files is $(m_1 m_2)$. Thus, we can obtain a speedup of $\min(2^{k_1}, 2^{k_2})$ in prejoin by clustering the sub-surrogate files within the SFPs. For a nonequijoin operation, each set of SF_1 partitions may have many join-possible sets of SF_2 partitions, and consequently we may have less speedup in the prejoin operation compared to the case of an equijoin operation.

Figure 11.7 shows the total execution time of equijoin operations on relations R_1 and R_2 when we have the same conditions specified for the join operation evaluated in Section 4.2. Additional conditions assumed here are

1. The binary representation of the join attribute of R_1 contributes three bits for partitioning the surrogate file of R_1.
2. The binary representation of the join attribute of R_2 does not contribute a bit for partitioning the surrogate file of R_2 when the cardinality of R_2 is 10^3 and 10^4.
3. The binary representation of the join attribute of R_2 contributes one bit for partitioning the surrogate file of R_2 when the cardinality of R_2 is 10^5.
4. The binary representation of the join attribute of R_2 contributes three bits for partitioning the surrogate file of R_2 when the cardinality of R_2 is 10^6.

Comparing Figure 11.7 and Figure 11.6, the total join time with clustered surrogate files is almost the same or even larger than the total join time with unclustered surrogate files when the binary representation of the join attribute of R_2 does not contribute a bit for partitioning the surrogate file of R_2. The reason is that partitioning the surrogate files increases the average

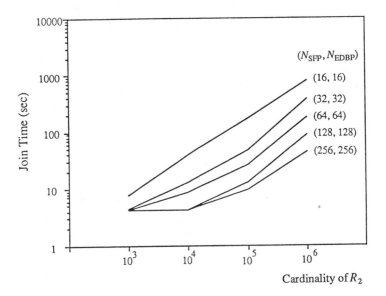

Figure 11.7 Performance of join operations between R_1 and R_2 with clustered surrogate files (cardinality of $R_1 = 10^6$, cardinality of output relation = cardinality of R_2).

block (partition) access time because of random accesses while the processing complexity remains the same. When the binary representation of the join attribute of R_2 contributes one or more bits for partitioning the surrogate file of R_2, the total join time is reduced. But the reduction ratio in total join time is not proportional to 2^3 when the cardinality of R_2 is 10^6. In this case, the speedup in the prejoin operation is proportional to 2^3, but the time for the actual join operation on the facts is not reduced and is dominant over the time for the prejoin operation on the surrogate files.

A problem of a multiattribute partitioned data file is that the number of data partitions to be retrieved to answer a partial match retrieval query is large when the data file is not sufficiently partitioned based on the attributes specified in the query. In our case, since the size of a surrogate file is usually less than 20 percent of the size of the data file, the time for searching a clustered surrogate file will be much less than the time for searching the partitioned data file. By searching the surrogate file first, we can filter out most of the data partitions which do not contain a tuple satisfying a given query. For the case of a join operation, which is much more complex than a partial match retrieval, the join on the clustered surrogate files can be performed much faster than the join on the partitioned data files. Join algorithms based on partitioned data files have been proposed in [6, 21, 28]. Furthermore, maintenance of the clustered surrogate file is very simple. When a surrogate file partition overflows, the CCWs in the partition are split into two partitions based on the value of a bit of each CCW, instead of applying numeric hashing functions on the CCWs.

5 CONCLUSION

We proposed a database machine based on the CCW surrogate files and analyzed the performance of the architecture for parallel selection and join algorithms. The basic idea of the proposed architecture is to reduce the number of EDB blocks to be transferred from the secondary storage systems by performing the relational operations on the CCW surrogate files first. It has been shown that we can obtain an almost linear speedup in executing relational algebra operations as the number of processors increases.

REFERENCES

1. P. B. Berra, S. M. Chung, and N. I. Hachem, " Computer Architecture for a Surrogate File to a Very Large Data/Knowledge Base," *IEEE Computer*, vol. 20, no. 3, pp. 25–32, 1987.

2. D. Bitten, H. Boral, et al., "Parallel Algorithms for Execution of Relational Database Operations," *ACM Trans. on Database Systems*, vol. 8, no. 3, pp. 324–353, 1983.

3. M. W. Blasgen and K. P. Eswaran, "Storage and Access in Relational Data Bases," *IBM Systems Journal*, vol. 16, no. 4, pp. 363–377, 1977.

4. K. Bratbergsengen, "Hashing Methods and Relational Algebra Operations," *Proc. VLDB*, pp. 323–333, 1984.

5. A. F. Cardenas, "Analysis and Performance of Inverted Data Base Structures," *CACM*, vol. 18, no. 5, pp. 253–263, 1975.

6. J. P. Cheiney, P. Faudemay, et al., "A Reliable Parallel Backend Using Multiattribute Clustering and Select-Join Operator," *Proc. VLDB*, pp. 220–227, 1986.

7. S. M. Chung and P. B. Berra, "A Comparison of Concatenated and Superimposed Code Word Surrogate Files for Very Large Data/Knowledge Bases," *Advances in Database Technology—EDBT '88, Proc. Int'l Conf. on Extending Database Technology*, pp. 364–387, Springer-Verlag, 1988.

8. S. M. Chung, *A Relational Algebra Machine Based on Surrogate Files for Very Large Data/Knowledge Bases*, Ph.D dissertation, Syracuse University, 1990.

9. W. A. Davis and D. L. Lee, "Fast Search Algorithms for Associative Memories," *IEEE Trans. on Computers*, vol. 35, no. 5, pp. 456–461, 1986.

10. D. J. Dewitt and R. Gerber, "Multiprocessor Hash-based Join Algorithms," *Proc. VLDB*, pp. 151–164, 1985.

11. D. M. Dias and J. R. Jump, "Analysis and Simulation of Buffered Delta Networks," *IEEE Trans. on Computers*, vol. 30, no. 4, pp. 273–283, 1981.

12. H. C. Du and J. S. Sobolewski, "Disk Allocation for Cartesian Product Files on Multiple-Disk Systems," *ACM Trans. on Database Systems*, vol. 7, no. 1, pp. 82–101, 1982.

13. R. Fagin, J. Nievergelt, et al.," Extendible Hashing: A Fast Access Method for Dynamic Files," *ACM Trans. on Database Systems*,vol. 4, no. 3, pp. 315–344, 1979.

14. C. Faloutsos and S. Christodoulakis, "Signature Files: An Access Method for Documents and Its Analytical Performance Evaluation," *ACM Trans. on Office Information Systems*, vol. 2, no. 4, pp. 267–288, 1984.

15. A. K. Garg and C. C. Gotlieb, "Order-Preserving Key Transformations," *ACM Trans. on Database Systems*, vol. 11, no. 2, pp. 213–234, 1986.

16. H. Itoh, M. Abe, et al., "Parallel Control Techniques for Dedicated Relational Database Engines," *Proc. Int'l Conf. on Data Engineering*, pp. 208–215, 1987.

17. M. Kitsuregawa, H. Tanaka, and T. Moto-Oka, "Architecture and Performance of Relational Database Machine Grace," *Proc. Int'l Conf. on Parallel Processing*, pp. 241–250, 1984.

18. D. H. Lawrie, "Access and Alignment of Data in an Array Processor," *IEEE Trans. on Computers*, vol. 24, no. 12, pp. 1145–1155, 1975.

19. W. Litwin, "Linear Hashing: A New Tool for Table and File Addressing," *Proc. VLDB*, pp. 212–223, 1980.

20. J. Menon, "Sorting and Join Algorithms for Multiprocessor Database Machine," *Database Machines: Modern Trends and Applications*, eds. A. K. Sood and A. H. Qureshi, pp. 289–322, Springer-Verlag, 1986.

21. E. A. Ozkarahan and C. H. Bozsahin, "Join Strategies Using Data Space Partitioning," *New Generation Computing*, vol. 6, pp. 19–39, 1988.

22. J. H. Patel, "Performance of Processor-Memory Interconnections for Multiprocessors," *IEEE Trans. on Computers*, vol. 30, no. 10, pp. 771–780, 1981.

23. J. L. Pfaltz, W. J. Berman, and E.M. Cagley, "Partial-Match Retrieval Using Indexed Descriptor Files," *CACM*, vol. 23, no. 9, pp. 522–528, 1980.

24. C. V. Ramamoorthy, J. L. Turner, and B. Wah, "A Design of a Fast Cellular Associative Memory for Ordered Retrieval," *IEEE Trans. on Computers*, vol. 27, no. 9, pp. 800–815, 1978.

25. C. S. Roberts, "Partial Match Retrieval via the Method of Superimposed Codes," *Proceedings of the IEEE,* vol. 67, no. 12, pp. 1624–1642, 1979.

26. D. Shin and P. B. Berra, "Surrogate File Approach to Managing First Order Terms in Secondary Storage," *Proc. SPIE Conf. on Applications of Artificial Intelligence*, pp. 1051–1062, 1989.

27. H. J. Siegel, "The Theory Underlying the Partitioning of Permutation Networks," *IEEE Trans. on Computers*, vol. 29, no. 9, pp. 791–800, 1980.

28. J. A. Thom, K. Ramamohanarao, and L. Naish, "A Superjoin Algorithm for Deductive Databases," *Proc. VLDB*, pp. 186–196, 1986.

29. P. Valduriez and G. Gardarin, "Join and Semijoin Algorithms for Multiprocessor Database Machine," *ACM Trans. on Database Systems*, vol. 9, no. 1, pp. 133–161, 1984.

30. K. Waldschmidt, "Associative Processors and Memories Overview and Current Status," *Proc. COMPEURO 87*, pp. 19–26, 1987.

CHAPTER 12

CONNECTIONIST ARCHITECTURES FOR ARTIFICIAL INTELLIGENCE

Scott E. Fahlman and Geoffrey E. Hinton

Current AI technology can do a good job of emulating many of man's higher mental functions, but some of the most fundamental aspects of human intelligence have proven elusive. AI can match the best human experts on certain narrow technical problems, but it cannot begin to approach the common sense and sensory abilities of a five-year-old child. Some important ingredients of intelligence seem to be missing, and our technology of symbolic representation and heuristic search, based on serial computers, does not seem to be closing the gap. Among the missing elements are the following:

- The human memory can store a huge quantity and variety of knowledge, and can find relevant items in this storehouse very quickly and without apparent effort. The phenomenon we call common sense is complex, but it derives in part from the ready availability of a large body of assorted knowledge about the world. Our serial machines can store large amounts of information, but it is very hard to make this knowledge an effective part of the machine's activities.

- In many domains, human recognition abilities far exceed what our machines can accomplish. Whether the domain is speech recognition, vision, or some higher-level task like medical diagnosis, the key operation seems to be an ability to locate, from among all known candidates, the

© 1987 IEEE. Reprinted, with permission, from *IEEE Computer*, vol. 20, no. 1, pp. 100–109, January 1987.

one that best matches the sample to be identified. We humans can do this even with noisy, distorted input data and faulty expectations. For us, this is quick and seemingly effortless.

- In many cases, humans seem to handle information in some form other than the symbolic assertions of traditional AI. We can all recognize an elephant, but few of us can describe its appearance symbolically in an unambiguous way. We have difficulty coming up with formal symbolic descriptions for movements, shapes, sounds, and spatial relationships, and yet people work easily in all these domains. It can be argued that we humans use internal symbolic representations that we cannot access consciously, but it seems more plausible that some other kinds of representations are in use.

The traditional AI approach to knowledge and recognition problems is to use ever more complex and clever strategies to reduce the need for excessive search and computation. An alternative is to solve such problems with a less complicated but very cycle-intensive approach, using very large numbers (millions) of very simple processors to get the job done in a reasonable time. For example, one approach to interpreting visual input is to use clever reasoning to restrict the areas in which a computationally expensive edge finder is applied. The alternative is to simply accept the cost of doing high-quality edge finding all over the image—even in places that will turn out not to be critical to later stages of interpretation.

A number of researchers have begun exploring the use of massively parallel architectures in an attempt to get around the limitations of conventional symbol processing. Many of these parallel architectures are *connectionist:* the system's collection of permanent knowledge is stored as a pattern of connections or connection strengths among the processing elements, so the knowledge directly determines how the processing elements interact rather than sitting passively in a memory, waiting to be looked at by the CPU. Some connectionist schemes use formal, symbolic representations, while others use more analog approaches. Some even develop their own internal representations after seeing examples of the patterns they are to recognize or the relationships they are to store.

Connectionism is somewhat controversial in the AI community. It is new, still unproven in large-scale practical applications, and very different in style from the traditional AI approach. We have only begun to explore the behavior and potential of connectionist networks. In this article, we describe some of the central issues and ideas of connectionism, and also some of the unsolved problems facing this approach. Part of the motivation for connectionist research is the possible similarity in function between connectionist networks and the neural networks of the human cortex, but we concentrate here on connectionism's potential as a practical technology for building intelligent systems.

1   WHAT IS CONNECTIONISM?

Jerry Feldman coined the term *connectionism* to refer to the study of a certain class of massively parallel architectures for artificial intelligence. A connectionist system uses a large number of simple processing elements or *units,* each connected to some number of other units in the system. The units have little information stored internally, typically only a few *marker bits* or a single scalar *activity-level,* used as a sort of short-term working memory. The long-term storage of information is accomplished by altering the pattern of interconnections among the units, or by modifying a quantity called the *weight* associated with each connection. This use of connections, rather than memory cells, as the principal means of storing information motivated the name *connectionism.*

The parallel processing units in a connectionist network do not follow individual programs. The units are capable of only a few simple actions, such as accepting incoming signals, performing some Boolean or arithmetic processing on the data, and sending signals out over some or all of the connections. These operations may be completely autonomous—part of a unit's built-in behavior—or they may be controlled by commands broadcast by some external controller, perhaps a serial computer of the traditional kind.

Since all of the connections can carry signals simultaneously, and all of the processing elements can act in parallel to integrate their arriving data, a connectionist system can bring a large amount of knowledge to bear simultaneously when making a decision, and can weigh many choices at once.

In some connectionist systems, the parallelism is used to implement a sort of simultaneous brute-force search through units, each representing a single item in the knowledge base. In other systems, the parallelism is used to allow richer representations; the pattern of activity in a large group of units represents an item, and different items are represented by different patterns of activity. Given some inputs and an initial state of the network, one of these patterns will emerge. Many connections, representing many small pieces of knowledge, will play a role simultaneously in determining which alternative will win. Whichever strategy is used, this ability to bring a lot of knowledge into the game at once is a major reason for the growing interest in connectionism.

The interunit signals sent through the connections are typically single-bit markers or continuous scalar values. Thus, we speak of *marker-passing* or *value-passing* parallelism. A unit can receive many of these signals at once, each arriving over a different connection. These multiple signals are combined upon receipt: multiple instances of the same marker bit are simply ORed into one, and multiple scalar values are usually just combined into a weighted sum as they arrive at the destination unit. Systems that send more complex symbolic messages from unit to unit, the so-called *message-passing* parallel systems, are not usually considered connectionist systems because they require much more complex processing units with a considerable amount of storage for the messages.

An important difference between the connectionist approach and the more conventional "modestly parallel" architectures for AI is that the connectionists are willing to assign a processing element to each tiny subtask (one element for every item of knowledge in the system, for example) and to postulate that there are enough of these simple processing elements to handle the task at hand. The more conventional approach assumes a fixed number of larger processors (typically between 2 and 1024) and tries to find ways of cutting the problem up into that many pieces, all of which can be worked on concurrently.

In a serial system, the time required to sift through a finite set of items in memory or to consider a finite set of hypotheses in a recognition task grows linearly with the size of the set. The modestly parallel approach attempts to approach N-fold speedup from N processors. The connectionist thinks in terms of performing these simple tasks in constant time, while the amount of hardware grows linearly with the number of memory items or the number of hypotheses to be considered at once. This view may seem less radical if we think of a connectionist unit not as a CPU but as a fancy kind of memory cell that stores knowledge in its connectivity or connection weights. We need enough of these memory cells to hold the system's knowledge.

Many kinds of connectionist architectures are being investigated by the small but growing community of researchers interested in this kind of parallelism. It is hard to say much more about these systems as a class, and in an article of this length it is impossible to mention all of the connectionist research going on. We will describe some of the key issues and ideas that seem important to us, and mention one or two pieces of work exemplifying each of these key ideas.

2 DISTRIBUTED REPRESENTATIONS

The simplest way to represent things in a massively parallel network is to use *local representations,* in which each concept or feature is represented by a specific piece of hardware. For example, when the system wants to work with the concept of *elephant,* it turns on the elephant unit. This kind of representation is easy to create and easy to understand. Unfortunately, if the elephant unit breaks, the system loses all of the knowledge tied to it. This creates some obvious reliability problems: in a system with millions of processing elements, not all of them will be working all the time. It also means that most of the units will be idle most of the time.

Many neuroscientists believe that the brain does not work this way, that instead it uses some sort of *distributed representation:* a concept like *elephant* is represented not by a single neuron, but by a pattern of activation over a large number of neurons. They often use the analogy of a hologram, in which each point of the image is constructed from information from all over the film; if some part of the hologram is destroyed, the total image is degraded slightly, but no part of the image is completely lost. A distributed representation in a massively parallel network (whether built from neurons or silicon) would have a similar kind of reliability: if a few of the units malfunction, the resulting pattern is imperfect but still usable. Each macroscopically important behavior of the network is implemented by many different microscopic units, so any small random subset can disappear without changing the macroscopic description of the network's behavior.

This kind of inherent fault tolerance has important implications for the construction of large-scale parallel networks. For example, wafer-scale integration becomes more feasible, since a few malfunctioning units in the wafer would simply be ignored. It is almost certainly easier to build a billion-transistor system in which only 95 percent of the circuit elements have to work than to build a million-transistor system that has to be perfect. It can also be shown [1] that in many situations it is much more efficient, in terms of the number of units required, to represent a value using a *coarse-coded* distributed scheme than to assign a single unit to represent each small interval in the range. In the coarse-coded scheme, each unit is coarsely tuned to cover a considerable range of values, and a particular value is represented by activity in a number of overlapping units.

The main disadvantage of distributed representations is that they are hard for an outside observer to understand or to modify. To add a single new piece of macroscopic knowledge to a network that uses distributed representations, it is necessary to change the interactions between many microscopic units slightly so that their joint effects implement the new knowledge. For any problem of significant size, it is nearly impossible to do this by hand, so it follows that a network using a distributed representation must employ some sort of automatic learning scheme. In the absence of an automatic learning procedure that works efficiently in large networks, distributed representations are generally too awkward to be of much practical value.

3   NETL: A CONNECTIONIST SYSTEM FOR SYMBOLIC KNOWLEDGE

The NETL system [2] is an example of a connectionist architecture that uses local representations. It was designed to store and access a large number of symbolic assertions, and to perform certain simple searches and deductions within this collection of knowledge. NETL is an implementation in hardware of a *semantic network,* a graphlike data structure in which the nodes represent noun like concepts and the links represent relationships between these concepts. Such a network is illustrated in Figure 12.1; the NETL hardware corresponding to a part of this network is illustrated in Figure 12.2.

In the NETL system, each of the nodes in the network is represented by a simple processing unit, capable of storing a few single-bit markers and of performing simple Boolean operations on the markers. Each link is also a simple processing unit, wired up to two or more node units. Link units, too,

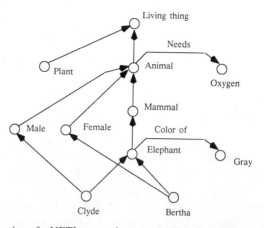

Figure 12.1 A portion of a NETL semantic network. This fragment describes the elephants Clyde and Bertha and related information. (Source: *Artificial Intelligence: An MIT Perspective*, vol. 1, P. H. Winston and R. H. Brown (eds.) MIT Press, 1979. Used by courtesy of MIT Press, publisher.)

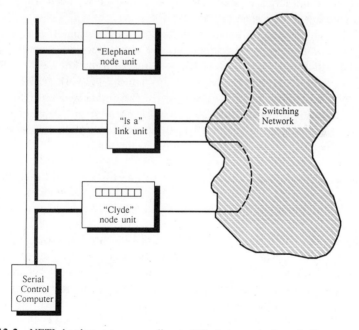

Figure 12.2 NETL hardware corresponding to "Clyde is an elephant." (Source: *Artificial Intelligence: An MIT Perspective*, vol. 1, P. H. Winston and R. H. Brown (eds.) MIT Press, 1979. Used by courtesy of MIT Press, publisher.)

can perform simple Boolean operations, which generally amount to passing a specific marker from one of the attached nodes to another. All of the nodes and links in the network can perform these operations simultaneously in response to commands broadcast by the system controller, a serial machine of the familiar sort.

Every time a new assertion is added to the system, new nodes and links must be wired into the network to represent that assertion. Since all the node-to-link connections must be capable of carrying signals simultaneously, they must be true private-line connections and not just addresses sent over a shared party-line bus. Of course, in a practical implementation, these new connections would be established in a switching system resembling a telephone exchange and not by stringing tiny wires from unit to unit.

A NETL network can perform searches and simple inferences that go beyond what can be done with a simple associative memory. For example, a very heavily used operation in most AI knowledge-based systems is *inheritance:* we want anything we know about the typical elephant to apply to the subclasses and individuals below elephant in the hierarchy of is-a links (unless that inherited information is specifically canceled for some individual). Because of branching in the is-a hierarchy, an individual node like *Clyde* might inherit information from a large number of superior nodes, so we need some very fast way of scanning a large part of the network.

In NETL, inheritance is handled by marker propagation. If the problem is to find the color of Clyde the elephant, the controller sets marker 1 on the Clyde node, then repeatedly orders any is-a link with marker 1 set on the node below it to pass it on to the node above. When the network settles, we have marker 1 on all of the nodes representing superior classes from which Clyde is supposed to inherit properties. We then command every color-of link with marker 1 on its tail to put marker 2 on its nose. Finally, we ask every node marked with marker 2 to report its identity to the controller. In this case, we get a single winner, the gray node. In a few cycles, we have followed a chain of inference to locate this information, regardless of where in the network the color information may have been attached. (The actual sequence of operations is slightly more complex than this, due to the possible presence of exceptions in the network, but the idea is the same.)

Because of its ability to do set intersections in parallel, NETL can handle recognition tasks of a certain limited kind. Suppose we are confronted with a gray four-legged mammal and we want to quickly locate any stored description that exhibits the intersection of these features. By marker propagation from the feature nodes, we can mark all of the gray things with one marker, four-legged things with a second marker, and mammals with a third. Then we simply broadcast a command that any unit with all of these markers should queue up to report its identity to the controller. The intersection step takes only a small, constant time, regardless of the size of the sets being intersected.

This approach to recognition is very strong in some ways, but weak in others. On the positive side, NETL examines all of the descriptions in memory at once, not depending on heuristics that might prematurely rule out the right answer. Any set of observed features that is sufficient to select one of the stored descriptions will suffice, and if not enough features are present, the system will still narrow the set of possibilities down as much as possible. When new descriptions are added to the network, the knowledge immediately becomes an effective part of the recognition process. We do not need to hire a programmer or knowledge engineer to specify exactly how and when each piece of knowledge is to be applied. On the negative side, NETL treats every feature as an atomic entity that is either present or absent and expects that the winning description will explain all of the observed features. Like traditional symbolic AI systems, NETL works best in clean domains, far from the noise and confusion of the low-level sensory inputs.

The NETL architecture has been simulated, but not yet implemented directly. The node and link units are so simple that several thousand of them could fit on a single chip. The hard problem is to design the switching network that implements the node-to-link interconnections. A design study by Fahlman [3] demonstrated that a NETL machine can be built for only a few times the cost of a conventional memory system able to hold the same amount of information. The Connection Machine [4] was designed in part to implement a NETL-like knowledge base, though the hardware is general enough to do other jobs as well. The current 64,000-processor version of this machine is large enough to handle a substantial body of knowledge, by the standards of current AI systems. The proposed million-processor version may be able to handle enough assorted knowledge to exhibit some degree of common sense.

4 LAYERED VALUE-PASSING NETWORKS FOR RECOGNITION

As we mentioned earlier, a marker-passing system like NETL does not gracefully handle the messy recognition problems we encounter in the real world. Whether we are dealing with vision, speech understanding, medical diagnosis, or some other real-world recognition problem, we normally face a situation where no match is ever perfect and in which few of the incoming features are noise-free and certain. Some features provide strong evidence for particular hypotheses; other features are only suggestive. Some features are clearly present; others are borderline cases or the result of low-confidence observations. Instead of the all-or-none set operations of NETL, we need a system that can combine many observations of varying quality and find hypotheses that fit well, even if they do not match perfectly.

This kind of problem can be handled by a value-passing system, in which each connection has an associated scalar weight. Each unit computes a weighted sum of the incoming values and passes this sum through a nonlinear function, with the resulting value becoming the unit's output. A value-passing network may be implemented in either digital or analog hardware, but computationally it can be viewed as a sort of analog computer.

Using these value-passing units, we can set up a layered recognition system able to handle uncertain observations and varying degrees of evidence.

Each directly observable feature is represented by an *input unit* whose value represents the probability that a discrete feature is present, the magnitude of a continuous quantity (like intensity), or the probability that a magnitude lies within a particular interval. Each of the possible hypotheses that we wish to evaluate is represented by an *output unit*.

In the simplest of these networks, the input units are connected directly to the output units by a set of connections with modifiable weights. In such networks, it is not necessary to set the connection weights by hand. Given a set of input vectors and the desired output for each, the perceptron convergence procedure [5] can be used to find a set of weights that will perform this mapping, if such a set exists.

Unfortunately, for most interesting recognition tasks there is no set of weights in a simple two-layer network that will do the job. In most cases, one cannot treat the lowest-level features as independent sources of evidence and simply add them up. It is necessary to introduce one or more intervening layers of hidden units, which combine the raw observations into higher-order features more useful in determining the output. Consider, for example, the problem of deciding whether an image contains a telephone. The fact that a single pixel in the image has high intensity cannot be used directly as evidence for or against a telephone, because telephones can be black or white. It is necessary to extract relationships between pixel intensities (such as edges) before trying to detect the telephone.

Figure 12.3 shows a simple task for which hidden units are essential. Pattern A is identical to pattern B, except that it may be rotated one step to the left or right. The task is to determine the shift for any pattern of input bits. This cannot be done without hidden units because each input bit, considered in isolation, provides no evidence whatsoever about the shift. All the information is in the joint behavior of combinations of input bits, so one or more layers of hidden units must be used to extract informative multi-unit combinations, which then can be combined into an overall answer.

Learning which of the exponentially many possible combinations are relevant for predicting the output is a hard problem. Techniques such as backpropagation (described later) can indeed discover a set of weights and hidden-unit assignments that, taken together, will solve the problem. Figure 12.4 shows some examples of hidden units and associated weights taken from one such solution.

Minsky and Papert [6] thoroughly analyzed which tasks require hidden units and demonstrated that an insightful way of categorizing tasks is in terms of how many input units must connect to each hidden unit in a three-layer network (this determines the order of the statistics that can be extracted). Unfortunately, they gave no procedure for learning appropriate hidden units, and suggested that there may not be any simple general procedure.

For many years after Minsky and Papert's book was published, AI researchers criticized the connectionist approach because the effective learning procedures were restricted to networks incapable of performing tasks like figure-ground segmentation or viewpoint-independent shape recognition. More

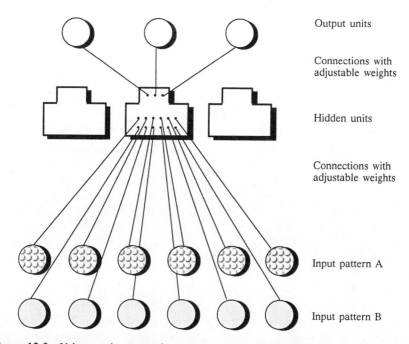

Output units

Connections with
adjustable weights

Hidden units

Connections with
adjustable weights

Input pattern A

Input pattern B

Figure 12.3 Value-passing network to compare two six-bit patterns. Input patterns A and B are identical, except that pattern A may be rotated one position to the left or right from pattern B. The three output units represent a left rotation, no rotation, and a right rotation, respectively. This task cannot be done without some hidden units. In this network we employ 12 hidden units, though only three appear in the diagram. An appropriate set of weights can be learned using the back-propagation method.

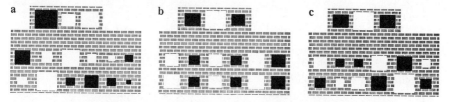

Figure 12.4 The weights learned by 3 of the 12 hidden units. The white squares are positive weights, the black squares are negative weights, and the area of a square is the magnitude of the weight. The bottom two rows in each hidden unit show the weights coming from the input units, and the top row shows the weights to the three output units that represent the three possible shifts. All the units shown learned positive thresholds (not shown here) so that they will not come on unless they receive a net positive input. Unit (*a*) detects some cases of right-shift. Unit (*b*) detects some no-shift cases. Unit (*c*) responds positively to shifts in either direction, but not to no-shift cases.

recently, a deeper appreciation of the sheer quantity of computation required and a better understanding of the nature of the computations has led to renewed interest in massively parallel approaches to perception [7].

5 LEARNING REPRESENTATIONS

Three-layered nets of the kind studied by Minsky and Papert do not have the freedom to choose their own representations, because the weights between the input units and the feature detectors in the middle layer are predetermined and do not learn. If we allow these weights to learn, the network can decide for itself what the units in the middle layer will represent. Consequently, we get a much more interesting and powerful kind of learning, especially when extended to multiple hidden layers. However, finding learning procedures that choose good representations is difficult, and because the space of possible representations is very large, any procedure that explores the space in a random way will be very slow.

One promising new procedure, called back-propagation, has been discovered independently by David Rumelhart, David Parker, and Yann Le Cun. It involves two passes each time an input vector is presented. In the forward pass, activity starts at the input units and passes through the layers to produce an output vector. In the backward pass, the derivative of the error (the difference between the actual output vector and the desired output vector) is back-propagated through the same connections but in the reverse direction. This allows the network to compute, for each weight, the gradient of the output error with respect to that weight. The weight is then changed in the direction that reduces the error. So learning works by gradient descent on an error surface in weight space.

Back-propagation has already been shown capable of learning many different kinds of interesting representations in the hidden units. It can learn optimal codes for squeezing information through narrow bandwidth channels, or sets of optimal filters for discriminating between very similar noisy signals.

Sejnowski and Rosenberg [8] showed that a back-propagation net can be trained to transform an input vector that encodes a sequence of letters into an output vector that encodes the phonemic features of the sequence; that phonemic output can then be used to drive a speech synthesis device. From the input/output examples it sees during training, this system is able to extract both the regularities exhibited by the mapping, such as the effect of a terminal *e* in changing the sound of the preceding vowel, and specific exceptions, such as the odd pronunciation of a word like *women*. This same task is performed in commercial speech-generation systems by conventional programs of considerable size and complexity.

Back-propagation can also be used to learn the semantic constraints that underlie a set of facts. One five-layer network [9] was trained on a set of 100 triples such as (Victoria has-father Christopher) or (Christopher has-wife Penelope) derived from two family trees involving people of two nationalities.

The input vector represented the first two terms of a triple and the required output vector represented the third term. As far as the network was concerned, these input vectors were just arbitrary symbols, but after extensive training the network could generalize appropriately to triples on which it had not been trained, like (Victoria has-mother ??). By recording the set of triples for which each of the hidden units became active, it was possible to show that these units had learned to represent important properties like "Italian" or "old" that were never mentioned in the input and output.

These semantic features happen to be a very good representation for the relationships the network is trained on. The hidden features and their interactions encode the underlying regularities of the domain. When viewed locally, the learning procedure just tunes the weight parameters, but the global effect is that the network does structural learning, creating new terms that allow it to express important regularities. This example also illuminates the *local versus distributed* issue. The network's internal representation of each person is a distributed pattern of active semantic features, but each of these local features captures an important underlying regularity in the domain.

Back-propagation has similarities to the Baum-Welch algorithm for tuning parameters in a stochastic, finite-state automaton. These trainable automata are widely used as generative models in practical speech recognition systems [10], and there is some hope that procedures like back-propagation will be able to improve on these models by overcoming an important limitation. In a finite-state automaton, the system's knowledge of what it has produced so far is encoded as the current node in a graph of states and transitions. So if the automaton needs to remember 20 bits of information about the partial sequence to constrain the rest of the sequence, it requires at least 2^{20} nodes. In a connectionist network, many units can be active at once, so the number of units need not grow exponentially with the amount of information that must be carried along in generating a sequence.

Another interesting new learning procedure, called A_{R-P}, was described by Andy Barto [11]. He showed how layered networks of relatively simple stochastic units can learn to cooperate in order to maximize a global payoff signal that depends on the output vector and is received by every unit. Like back-propagation, this procedure learns interesting internal representations, but it is slower because it discovers the effect of changing a weight by sampling the effects of random variation instead of by explicit computation of the gradient. The A_{R-P} model is exciting as a biological model because it does not need a separate pass in which to back-propagate detailed error information.

A major criticism of all the current multilayer learning procedures is that they are slow even for modest examples and they appear to scale poorly. The obvious problem of gradient descent—that it will get stuck in a poor local minimum—turns out to be only a minor problem in practice. The real difficulty is that simple gradient descent is very slow because we have information about only one point and no clear picture of how the surface may curve. In high-dimensional spaces the error surface usually contains ravines such that the surface is highly curved across the ravine and slopes gently downwards along the ravine. Small steps take forever to meander down a ravine, and big steps cause divergent oscillations across a ravine. Even with a large speedup from faster implementation technology, these learning techniques are too slow to handle many problems of interest. Future progress may well depend on discovering ways of partitioning large networks into relatively small modules that can learn more or less independently of each other.

6   CONSTRAINT SATISFACTION IN ITERATIVE NETWORKS

The idea that perceptual interpretation consists of transforming an input vector through successive layers of units until it becomes a categorization is rather restrictive. Most real recognition tasks require that multiple layers of features be recognized all at once and that the result of perception be a coherent, articulated structure rather than a single category. In visual recognition, for example, we might have a scene, various objects in the scene, their parts, and subparts, all constraining one another and all relating in complex ways to the low-level stimuli the system is receiving. An African context provides evidence in favor of seeing an elephant; conversely, an elephant helps us to identify an African scene, and that in turn can help us to spot a giraffe. The trunk, tusks, and ears all are features that help us to recognize an elephant. On the other hand, knowing that we're looking at an elephant helps us to recognize these parts.

In a situation like this, we want to start with a network that records the tangle of interlevel constraints and evidence relationships, put in our observations as a sort of boundary condition, and get the system to settle into the best possible solution (or set of solutions) spanning all the levels of description. Some constraints and expectations will be violated, but we want to find the best scoring constellation of decisions overall [7].

The task is similar to finding the solutions to a set of simultaneous equations. A value-passing network would seem a sensible parallel way of converging on a good solution—a sort of analog computer with links representing the constraints. Non-learning-constraint networks of this sort have been studied for a variety of problem domains, most notably at the University of Rochester [12]. In some simple cases, the network can be guaranteed to settle to the best final state no matter what values it starts with, but for more difficult problems, especially ones that involve discrete decisions, a value-passing network with loops and many nonlinear decision elements is generally not well behaved. Such networks tend to oscillate or to get trapped in uninteresting states that do not represent good solutions to the problem.

7   HOPFIELD AND BOLTZMANN NETWORKS FOR CONSTRAINT SATISFACTION

One way to guarantee that a network will settle down is to show that there exists some cost function that decreases every time one of the values changes. Hummel and Zucker [13] showed that there exists such a function for networks that pass values and have symmetrical connections. At about the same time, Hopfield [14] identified a cost function (which he called *energy*) for networks of symmetrically connected binary threshold units.

Hopfield's energy function can be interpreted in the following way: A connection between units i and j with a positive weight represents a constraint that if one of these units is on, the other one should be also. A negative connection weight says that if one of the units is on, the other should not be. The weight corresponds to the penalty to be applied if the constraint is violated. The energy of any state of the network is given by

$$ E = - \sum_{i<j} s_i s_j W_{ij} + \sum s_i \Theta_i $$

where w_{ij} is the weight on the concentration between units i and j, S_i is 1 if unit i is on and 0 otherwise, and Θ_i is a threshold for unit i.

Given Hopfield's quadratic definition of energy, the difference between the global energy of the network when unit i is off and the global energy when it is on is simply

$$ E_{i \text{ off}} - E_{i \text{ on}} = \Delta E_i = \sum_j s_j w_{ij} $$

If the energy gap is positive, the unit should turn on (or stay on) to minimize the global energy. Otherwise it should turn off (or stay off). In other words, to minimize energy it should behave exactly like a binary threshold unit.

Hopfield originally proposed his model as a theory of memory. Each local minimum corresponds to a stored vector, and the memory is content-address able because if it is started anywhere near one of the stored states it normally converges on that state. The same kind of network can also be used for perceptual interpretation by defining some of the units as input units and clamping their states on or off to represent the current perceptual input. The other units then settle into a low-energy global state consistent with these boundary conditions. This state represents a locally optimal solution given this set of inputs, but with no guarantee that it is the global optimum.

The Boltzmann machine architecture [15, 16] is essentially a Hopfield network (with hidden units) that uses a *simulated annealing* search to escape from local minima. This same general idea, with some differences of emphasis, was independently proposed by other research groups [17, 18] at about the same time.

Simulated annealing is a search technique that has been applied to a number of optimization problems [19]. The idea is that we escape from high local minima by adding a random component to the decision process of each unit. In most cases, the unit still takes a step downhill, but occasionally it will take a step uphill instead. More precisely, each unit i computes the ΔE_i value as above, then assumes the 1 state with probability p_i given the following formula:

$$P_i = \frac{1}{1 + e^{-\Delta E_i / T}}$$

The T term is a scaling factor that controls the amount of noise; it is analogous to a temperature. For large T, p is about .5 and the system assumes states at random, ignoring the constraints in the network; for $T = 0$, the random element is eliminated, and the system behaves as in the pure Hopfield net, moving downhill into the nearest local minimum. At any given T, once the system has reached thermal equilibrium, the relative probability of being in state A versus state B (see Figure 12.5) obeys a Boltzmann distribution.

$$\frac{P_A}{P_B} = e^{-(E_A - E_B)/T}$$

Thermal equilibrium does not mean that the system has settled into a particular stable state. It means that the probability distribution over states has settled down, even though the states are still changing. A deck of cards is at thermal equilibrium when it has been shuffled for long enough that the probability of finding any card in any position is 1/52, even though the cards keep moving around. The best way to reach thermal equilibrium at a given temperature is generally to start at a higher temperature, which makes it easy for the system to cross energy barriers but gives it little preference for the lower energy states. Then we gradually increase the preference for low-energy states by reducing T. If T is reduced slowly enough, there is a high probability of ending up in the best global state, or if not there, in a state not much worse than the best. This process of slow cooling is analogous to slow annealing of a metal in order to crystallize it in its lowest-energy state.

Boltzmann networks, designed by hand and running on a serial simulator, have been demonstrated successfully on a number of constraint-satisfaction problems, such as the separation of figure and ground in a two-dimensional image. Touretzky [20] is investigating ways of implementing symbolic processing, including a production system interpreter and a system that can manipulate lists and tree structures, using Boltzmann-type networks. But most of the interest in Boltzmann machines is due to a learning procedure, described later,

Figure 12.5   Energy surface with local minimum at A, global minimum at B.

that enables them to program constraints into their connections when shown examples of good solutions to a problem.

Hopfield and Tank [21] investigated a different method to avoid being trapped in local minima: they used real-valued, analog, nonlinear elements that obey the same equation as the units in a Boltzmann machine but send out a deterministic real value instead of a single stochastic bit. Reducing the temperature in a Boltzmann machine is equivalent to raising the gain (the non-linearity) of their analog units, and they have shown that they can approximate simulated annealing without actually simulating a stochastic system directly. As the gain is increased, single minima of the low-gain energy surface develop a fine structure of smaller minima.

Hopfield and Tank's approximation loses all information about the higher-order statistics. For example, they would represent a stochastic system that spent half its time in the state $(1, 1)$ and half its time in the state $(0, 0)$ by the mean values $(0.5, 0.5)$. But the very same representation would be used for a system that spends half its time in the state $(1, 0)$ and half in the state $(0, 1)$. It is not yet clear how important this loss of higher-order statistical information is.

Networks of this type have been applied to a variety of problems, the most ambitious of which is the Traveling Salesman Problem, for which the network finds moderately good solutions very fast. Hopfield's group has shown a particular enthusiasm for implementing their networks in silicon, rather than simulating them, and a number of small-scale physical implementations have appeared. At the same time, Hopfield has been working with Tank to investigate the possible connection between these networks and real neurons.

Hopfield's group has chosen not to focus on the problem of learning what to do with extra hidden units whose required behavior is not specified by the task definition. These are not needed if the experimenter decides how to represent things, but they are crucial if the network is to be given the freedom to develop its own representations. There exists an iterative form of the back-propagation learning procedure that can learn how to use hidden units in a Hopfield and Tank network and can also learn an optimal schedule for varying the gain of the units during the search process. If the same learning procedure is applied to nets with asymmetric connections, they can be trained to produce sequences [22]. The learning procedure is infeasibly slow for large nets, so the research currently divides into studies of search in larger nets, built by hand, and learning in smaller nets.

8 THE BOLTZMANN MACHINE LEARNING PROCEDURE

The Boltzmann learning scheme is surprisingly simple. It has two versions; we will describe the version in which there are input units and output units and the machine must learn to map input vectors into output vectors. We begin with a set of I/O patterns on which the network is to be trained. The goal is to adjust the weights of the network so that, when we clamp the input units into one of these patterns and anneal the network, the corresponding output pattern will appear on the output units. If we clamp the inputs into a pattern that the system has not seen before, we would like the system to generalize according to the underlying regularities in the I/O pairs that it has seen.

The learning cycle has two phases, positive and negative, followed by weight adjustment. During the positive phase, we cycle through the entire set of I/O pairs. Each of these, in turn, is clamped into the input and output units and the rest of the network is then annealed, starting with a high temperature and gradually cooling down to thermal equilibrium at a temperature of 1. Once the system is close to equilibrium, we keep running it for a few more cycles, during which time each connection keeps a record of how often the two units it joins are on at the same time. After all of the I/O pairs have been presented in this way, each connection will have recorded a value, p^+, which is the fraction of time during the positive phase in which the two connected nodes are on at the same time at thermal equilibrium.

The negative phase is identical, except that only the inputs are clamped; the output units are allowed to settle into whatever states they like. In a network that has learned perfectly, the output units will exhibit the same probability distribution over output vectors as they would if we were still clamping them along with the inputs. During the negative phase we again run for a few more cycles after reaching equilibrium. Each connection records a second value, p^-, the fraction of time in which its two units are both active during the negative phase. If the network is producing all the right answers, the output units will exhibit the same probability distribution whether or not we clamp them and p^+ will be the same as p^-. If p^+ and p^- differ for a given connection, the two probability distributions can be made to match better by changing that connection's weight. If we ignore sampling noise, the difference between p^+ and p^- is exactly equal to the gradient of an information-theoretic measure of the difference between the system's output behavior (during the negative phase) and its desired behavior (during the positive phase)[16]. So we can perform steepest descent in this difference measure by changing the weight by an amount proportional to $p^+ - p^-$.

Even though Boltzmann machines only contain pairwise connections, the learning procedure allows them to capture higher-order constraints by dedicating hidden units to represent higher-order features. The learning procedure is interesting because it decides what the hidden units should represent, and it typically chooses to use distributed representations.

Unfortunately, the Boltzmann machine learning procedure suffers from all the usual problems of gradient descent in large parameter spaces. On top of this, the estimated gradient is usually inaccurate. The proofs assume that we have a large, unbiased sample of the statistics at equilibrium in the positive and negative phases. In practice, we never quite reach equilibrium, so the statistics are systematically biased. Moreover, unless we are very patient, there is also sampling error caused by taking too few samples.

Despite these difficulties, Boltzmann machines have been used successfully for aspects of speech recognition [23]. The simplicity of the Boltzmann machine architecture makes it feasible to build VLSI chips that contain many units, and it is just possible that this will lead to sufficient speed to make this a practical scheme. Even so, the scaling properties are poor and very large networks will remain impractical unless clever scaling tricks are discovered.

Ten years ago there were no connectionist learning procedures powerful enough to build useful representations with multiple layers of hidden units. Now there are many. The current challenge is to develop faster learning schemes that can be scaled up to networks with millions of modifiable connections.

9 ACKNOWLEDGMENTS

Scott Fahlman's research was supported by the Defense Advanced Research Projects Agency (DOD), DARPA Order No. 4976, monitored by the Air Force Avionics Laboratory under contract F33615-84-K-1520. Geoffrey Hinton's research was supported by the Office of Naval Research under contract N00014-86-K-00167 and by the National Science Foundation, grant number IST-8520359. The views and conclusions contained herein are those of the authors and should not be interpreted as representing the official policies, either expressed or implied, of the sponsors or the U. S. government.

10 FURTHER READING

Parallel Models of Associative Memory, eds. J. A. Anderson and G. E. Hinton, Erlbaum, Hillsdale, NJ, 1981.

Parallel Distributed Processing: Explorations in the Microstructure of Cognition, eds. D. E. Rumelhart, J. L. McClelland, and the PDP Research Group, MIT Press, Cambridge, MA, 1986.

Cognitive Science Vol. 9, special issue on connectionist models and their applications, ed. J. A. Feldman, 1985.

REFERENCES

1. G. E. Hinton, J. L. McClelland, and D. E. Rumelhart, "Distributed Representation," *Parallel Distributed Processing: Explorations in the Microstructure of Cognition,* eds. D. E. Rumelhart, J. L. McClelland, and the PDP Research Group, MIT Press, Cambridge, MA, 1986.

2. S. E. Fahlman, *NETL: A System for Representing and Using Real-World Knowledge,* MIT Press, Cambridge, MA, 1979.

3. S. E. Fahlman, "Design Sketch for a Million-Element NETL Machine," *Proc. Nat'l Conf. AI,* Am. Assoc. for AI, Menlo Park, CA, 1980.

4. W. D. Hillis, *The Connection Machine,* MIT Press, Cambridge, MA, 1985.

5. F. Rosenblatt, *Principles of Neurodynamics,* Spartan Books, New York, NY, 1962.

6. M. Minsky and S. Papert, *Perceptrons,* MIT Press, Cambridge, MA, 1969.

7. D. H. Ballard, G. E. Hinton, and T. J. Sejnowski, "Parallel Visual Computation," *Nature,* vol. 306, pp. 21–26, November 1983.

8. T. J. Sejnowski and C. R. Rosenberg, "NETtalk: A Parallel Network that Learns to Read Aloud," Technical Report JHU/EECS-86-01, The Johns Hopkins Univ., EE and CS Technical Reports, 1986.

9. G. E. Hinton, "Learning Distributed Representations of Concepts," *Proc. 8th An. Conf. Cognitive Science Soc.,* Lawrence Erlbaum Assoc., Hillsdale, NJ, 1986.

10. L. R. Bahl, F. Jelinek, and R. L. Mercer, "A Maximum Likelihood Approach to Continuous Speech Recognition," *IEEE Trans. Pattern Analysis and Machine Intelligence,* vol. PAMI-5, no. 2, pp. 179–190, IEEE CS, Silver Spring, MD, March 1983.

11. A. G. Barto, "Learning by Statistical Cooperation of Self-Interested Neuron-Like Computing Elements," *Human Neurobiology,* vol. 4, no. 4, pp. 229–256, Springer-Verlag, New York, NY, 1985.

12. J. A. Feldman and D. H. Ballard, "Connectionist Models and Their Properties," *Cognitive Science,* vol. 6, no. 3, pp. 205–254, Ablex, Norwood, NJ, 1982.

13. R. A. Hummel and S. W. Zucker, "On the Foundations of Relaxation Labeling Processes," *IEEE Trans. Pattern Analysis and Machine Intelligence,* vol. PAMI-5, no. 3, pp. 267–287, IEEE CS, Silver Spring, MD, May 1983.

14. J. J. Hopfield, "Neural Networks and Physical Systems with Emergent Collective Computational Abilities," *Proc. Nat'l Academy of Sciences USA,* vol. 79, no. 8, pp. 2554–2558, Nat'l Academy of Sciences, Washington, DC, April 1982.

15. S. E. Fahlman, G. E. Hinton, and T. J. Sejnowski, "Massively Parallel Architecture for AI: NETL, Thistle, and Boltzmann Machines," *Proc. Nat'l Conf. AI,* Am. Assoc. for AI, Menlo Park, CA, 1983. Dist. by William Kafman, Inc., Los Altos, CA.

16. D. H. Ackley, G. E. Hinton, and T. J. Sejnowski, "A Learning Algorithm for Boltzmann Machines," *Cognitive Science,* vol. 9, no. 1, pp. 147–169, Ablex, Norwood, NJ, 1985.

17. S. Geman and D. Geman, "Stochastic Relaxation, Gibbs Distributions, and the Bayesian Restoration of Images," *IEEE Trans. Pattern Analysis and Machine Intelligence,* vol. PAMI-6, no. 6, pp. 721–741, IEEE CS, Silver Spring, MD, November 1984.

18. P. Smolensky, "Schema Selection and Stochastic Inference in Modular Environments," *Proc. Nat'l Conf. AI, AAAI-83,* pp. 109–113, Am. Assoc. for AI, Menlo Park, CA, 1983. Dist. by William Kafman, Inc., Los Altos, CA.

19. S. Kirkpatrick, C. D. Gelatt, and M. P. Vecchi, "Optimization by Simulated Annealing," *Science,* vol. 220, no. 4589, pp. 671–680, Am. Assoc. for the Advancement of Science, Washington, DC, 1983.

20. D. S. Touretzky and G. E. Hinton, "Symbols among the Neurons: Details of a Connectionist Inference Architecture," *IJCAI,* vol. 9, pp. 238–243, Morgan Kaufmann, Los Altos, CA, August 1985.

21. J. J. Hopfield and D. W. Tank, "Neural Computation of Decisions in Optimization Problems," *Biological Cybernetics,* vol. 52, no. 3, pp. 141–152, Springer-Verlag, New York, NY, 1985.

22. D. E. Rumelhart, G. E. Hinton, and R. J. Williams, "Learning Internal Representations by Error Propagation," *Parallel Distributed Processing: Explorations in the Microstructure of Cognition,* eds. D. E. Rumelhart, J. L. McClelland, and the PDP Research Group, Bradford Books, Cambridge, MA, 1986.

23. R. Prager, T. D. Harrison, and F. Fallside, "Boltzmann Machines for Speech Recognition," *Computer Speech and Language,* vol. 1, no. 1, pp. 1–20, Academic Press, London, UK, 1986.

CHAPTER 13

ARCHITECTURES FOR STRATEGY LEARNING*

Pankaj Mehra and Benjamin W. Wah

1 INTRODUCTION

Recent years have witnessed a sharp rise in the quantity and quality of knowledge that can be captured in automated systems. Artificial intelligence (AI) research in the 1980s has focused on the issue of automatic knowledge acquisition or *learning*. Learning tasks are perhaps the most complex problems that AI researchers attempt to solve. Concomitant developments in the area of artificial neural systems (ANSs) have also prompted research in machine learning—mainly by providing for learning systems an inherently parallel model of implementation, one in which generalization and constraint satisfaction are spontaneous. This chapter is an attempt to analyze and illustrate how various models (old and new) can be applied to complex learning tasks in strategic problem solving.

In his famous essay, *The Architecture of Complexity* [207], Simon outlined the nature of complexity in problem solving by comparing it to a search through a maze. Subsequently, Simon and Newell formalized the search space model of problem solving. From the very beginning of this work, it was clear that the complexity of a problem (and the associated search) was due mainly to the large number of alternatives at each step. Simon pointed out that

*This research was supported by the National Science Foundation under Grant MIP-88-10598, and by National Aeronautics and Space Administration under Grant NCC 2-481.

the trial and error [in exploring the alternatives] is not completely random or blind; it is in fact rather highly selective. . .The selectivity derives from various rules of thumb, or heuristics, that suggest which paths should be tried first and which leads are promising.

This treatment of problem solving led to work in the theory of heuristic search [166, 167], which resulted in very general models of problem solving on one hand and efficient, general algorithms for search on the other [252]. Research suggested that search algorithms draw their power from the expressiveness and efficiency of problem representation, as well as from search control knowledge encoded in heuristic rules. Work on expert systems later confirmed this assessment [85], and it was realized that overcoming the *knowledge acquisition bottleneck* was the key to designing powerful programs. With this realization, the focus in problem solving shifted from powerful search algorithms to learning of heuristic knowledge for problem solving.*

Various algorithms for learning have been proposed and implemented during the last decade and several general paradigms of learning have arisen. However, most of the algorithms were demonstrated on small, well-defined domains, such as game-playing [120, 144, 240], the blocks world [142], symbolic integration [147], and high school arithmetic [51, 206]. Furthermore, the complexity of the other applications was due more to the structure of their environments than to the dynamic variability of their parameters [148]. Our recent attempts at solving certain complex problems in resource allocation have been frustrated by the inadequate learning model assumed by existing algorithms. In order to correct this deficiency in a systematic way, we have

1. studied the origins of complexity in strategy learning, using various aspects of complexity to identify difficult problem classes,
2. analyzed the applicability of several well-known learning algorithms to various problem classes, and
3. proposed a model of learning systems that applies to a class of complex problems.

The rest of this section introduces the fundamental concepts of strategy learning systems.

1.1 Problem Solving Strategy

A problem solver has a variety of knowledge about its domain. The procedural component of this knowledge is available as primitive pieces of procedural code called *operators*. Each operator is defined by the way it transforms a problem situation (or *state*) into another. A strategy for solving a problem is

---

* The complementary question of automatic acquisition of problem representation and its relationship to the strategy acquisition process is beyond the scope of this chapter (See Section 1.7).

a body of abstract procedural knowledge stated in terms of operators. The problem itself is a generic description of several *problem instances*. It is stated in terms of parameters, constraints, and objectives. Each instance is defined by its initial situation, that is, by the initial assignment of values to parameters. A *solution* to a problem instance is typically stated as a partial order on a set of operators. When applied to the initial description, a solution generates another description that satisfies the constraints and meets the objective. A *problem-solving strategy*, in this respect, is a systematic method for generating solutions to the instances of a problem.

1.2 Strategic Knowledge

The *object level* search space of a strategy learning system is defined by the knowledge of the problem and its instances. The heuristics and strategies form the first level of *metaknowledge*. Traditionally, metaknowledge has been made available to programs directly [27] and accounts for much of their problem-solving capability. However, in some domains metaknowledge is essentially empirical and varies from one context to another; it is thus necessary to acquire this component automatically. The acquisition of metaknowledge is called *strategy learning*. Construction of strategies is accomplished using a process for learning strategic metaknowledge; the knowledge controlling the strategy-learning process can be called the *meta-metaknowledge* of the problem (Figure 13.1).

1.3 Learning

The study of learning is the study of adaptive systems, and for the last decade the modern view of learning systems has followed developments in cognitive science, artificial intelligence, and adaptive control systems theory. A learning

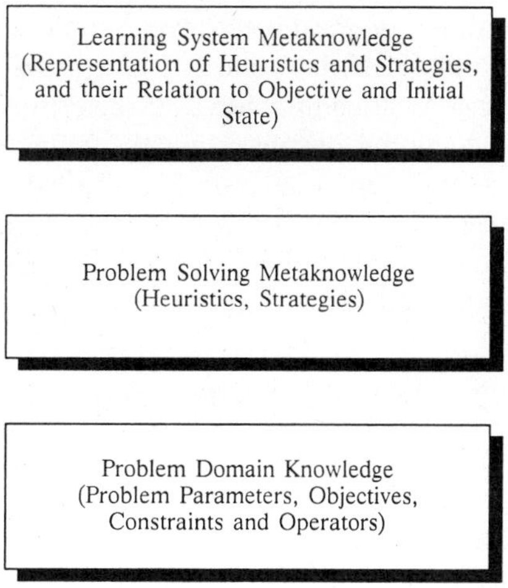

Learning System Metaknowledge
(Representation of Heuristics and Strategies, and their Relation to Objective and Initial State)

Problem Solving Metaknowledge
(Heuristics, Strategies)

Problem Domain Knowledge
(Problem Parameters, Objectives, Constraints and Operators)

Figure 13.1 Hierarchy of knowledge in strategy-learning systems.

system adapts over time by making changes that improve its performance. Under Feigenbaum's knowledge hypothesis (Knowledge is power) [61, 128], learning has become synonymous with knowledge acquisition. For insightful discussions of the fundamental issues in machine learning, the reader is referred to articles by Simon [213], Minsky [140], Schank *et al.* [204], Langley [118], and Berliner [27].

1.4 Nomenclature of Strategy-Learning Paradigms

A *strategy-learning paradigm* is a way to represent and acquire metaheuristic knowledge for problem solving. Different learning paradigms make different assumptions about the learning process; these paradigms affect the architecture of the problem solvers, which, in turn, influences the performance of the overall system. Following Dietterich, we shall attempt to analyze strategy-learning systems at the knowledge level [54, 155]. In what follows, we view strategy learning as a search process; just as the first level of metaknowledge is useful in searching for solutions at the object level, so the second level of metaknowledge is useful in searching for heuristics and strategies at the first level. Second-level search is more difficult, however, because its objectives are determined only partially by the objectives of search at the object level. The constraints at the second level include the time taken by a strategy when it is applied to a problem instance, the quality of the object-level solutions produced by the strategy, and constraints on the applicability of strategies. The operators at this level are called *strategy modifiers* or *strategy constructors*. The metaheuristic knowledge involved in the process is called the *credit assignment policy*. The strategy modifiers constitute a body of knowledge known classically as the *learning element*; the metaheuristic knowledge is implemented in a component known classically as the *critic* (see next section; also see Chapter 14, vol. 3, [21]).

It is important that the critic be sensitive both to the objectives and constraints of the object-level problem and to the constraints of the meta-level problem. If the critic is implemented procedurally, the strategy-learning process is called *analytical*. If the critic's sensitivity to the domain-level objective is acquired by observing the effect of a domain-level strategy on a reactive environment, the process is called *empirical*. Finally, if the critic is implemented so that it has background knowledge of the relationship between the objectives at the object level and the explicitly represented solutions found by strategies at the meta-level, the process is said to be *knowledge-based*.

1.5 Architectures for Strategy-Learning Systems

The components of the problem-solving and learning system and their overall arrangement constitute the architecture of a strategy-learning system. Fundamental issues in the design of such architectures have been addressed by Dietterich [52], Minsky [139], and Mitchell [144, 215]. Our research has caused us to reassess the architecture proposed by Dietterich; other architectural issues arise from recent developments in ANSs [1].

Dietterich's *classical model* has a learning element, a performance element, and a critic (see Figure 13.2). The performance element consists of the problem solver and associated bodies of knowledge, which include the basic search control rules, such as heuristics for selecting, ordering, and pruning of alternatives. At each decision point, the heuristic knowledge is invoked to determine what operator to apply to find the best path to the goal. Because the set of heuristics may be incomplete or inaccurate, the selected action may not be the best option; it is the job of the critic to evaluate either the action or, more commonly, the environment that results from its application. The evaluation is passed to the learning element, which edits the heuristic knowledge so that subsequent selections may yield better evaluations. Consequently, the learning element must know how the performance element works, including how it represents knowledge and what inference schemes it uses. This knowledge can be specified procedurally either in the form of a learning algorithm such as that used by LEX [146], or declaratively in the form of editing rules such as those in LS-1 [217], SAGE.2 [119], EURISKO [126], ODYSSEUS [246, 247], and Cupr [106]. Learning systems differ in the amounts of explicit and implicit knowledge that go into the design and interaction of these components.

A popular model of strategy learning proposed by Langley [120] is an instance of the Dietterich model. In it, the performance element is called the *behavior generation component*, the learning element is called the *behavior modification component*, and the critic is called the *behavior assessment component*. We shall refer to systems based on this model as belonging to the Dietterich model.

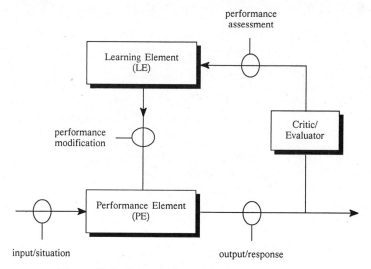

Figure 13.2   Dietterich's model of learning systems.

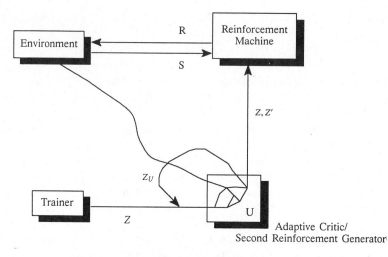

Figure 13.3   Minsky's reinforcement learning model.

An older and perhaps less restricted model of learning systems is that of Minsky [140] (see Figure 13.3).

Z is the external reinforcement; $Z = \pm 1$ for reward/extinction. The function $c_n = (2p_n - 1)$ is monotonic with the probability p_n of producing response R to stimulus S. The reinforcement machine tries to have a high value of p_n if Z_n is expected to be high, and vice versa. This can be achieved by adjusting c after every reinforced trial so that the correlation between c and Z approaches unity. The reinforcement machine adjusts c by maintaining a moving average of past reinforcements; for example,

$$c_{n+1} = (1 - \Theta) \sum_{i=0}^{n} \Theta^{n-i} Z_i, \quad \text{where } \Theta \to 0$$

which is the *exponentially decreasing average,* or

$$c_{n+1} = \left(1 - \frac{1}{N}\right) c_n + \frac{1}{N} Z_n, \quad \text{where } N \text{ varies with time}$$

which is the *uniform weighted average.* The term $\frac{1}{N}$ can be viewed as the *sensitivity* of the average to new data items.

z' is the expected reinforcement such that the probability $P(z = z'|E)$ is high, where E is the current state of the environment. Sometimes z may not follow R, in which case z' may be used. When z is available, the secondary reinforcement machine may be so adaptive that the difference between z and z' approaches zero. z_U is a positive feedback connection in the secondary reinforcement machine. It could be used to predict more than one step ahead, but positive feedback may cause instability.

The Dietterich model originated in knowledge-intensive AI systems, whereas the Minsky model originated in knowledge-lean domains. The Minsky model is distinctive because its critic is adaptive and requires only a small amount of implicit knowledge about its environment. This feature may sometimes be a drawback: if a partial internal model is available, the model has no way to use this knowledge. Samuel's Checker Player is a classic example of the Minsky model [202, 203], and other examples of the Minsky model include the REINFORCE model of Williams [248], the drive-reinforcement model of Klopf [104], and the reinforcement learning model of Barto et al. [22, 23, 226]. Even though this model of learning was proposed by Minsky, he was not convinced that such a machine could be at the heart of an intelligent system, except in the case of truly statistical environments. However, with recent advances in the ability to represent search behavior in such systems, the model has become much more applicable. We maintain that the learning paradigm in this model is more general than the learning paradigm in the classical model, even though its representations are somewhat restrictive.

In this chapter, we propose a model of learning systems that clearly defines the role played by the critic and the learning element, making explicit the knowledge they use during evaluation and editing (see Section 7). We also propose a generic scheme for training that makes it possible to acquire that knowledge automatically. Thus far, such knowledge has been programmed manually, making the resulting learning systems brittle when the training environment changes. Our model can adapt to changes in both the problem-solving objectives and the environmental reaction to the problem solver's actions.

1.6 Methods for Problem Solving

Problem solving models have become more general and powerful since the late 1960s. The first such model was proposed by Ernst and Newell [57] in their GPS program. More recently, the SOAR project [117] has put this model into a more uniform framework. In SOAR, all activity occurs as cycles of elaboration of memory, rule selection, and rule firing. Rules of preference encode all control knowledge, giving the system an ability to introspect and learn. Problem spaces and search schemes can be defined dynamically. Learning in SOAR is based on a simple, universal learning mechanism called chunking [115]. This system has been demonstrated in a wide variety of environments and on a variety of tasks: designing algorithms [222], simulating expert behavior [192], and simulating other learning mechanisms [193].

Simultaneous developments in parallel problem solving have led to the blackboard model of problem solving and learning [158]. Problem solving in this class of models (also known as BB* models) occurs through cooperative interaction among several parallel knowledge sources that communicate via a shared memory data structure called the *blackboard*. Recent extensions to the model [84] cover sources of control knowledge that make decisions at the strategic, policy, and scheduling levels, using a distinct control blackboard.

All the models described so far involve knowledge represented explicitly in a prescribed syntactic format. Connectionist problem solvers work differently [4, 11, 134, 198]. In these, problem entities are represented by a set of nodes. (Distributed, coarse-coded representations [87, 194] can provide robustness and generalization.) Connectionist systems model the search process by starting in an initial state and allowing the system to converge to some stable state [59, 88]. The long-term storage of relations among entities is encoded in the weights linking various units; the short-term state in the activation levels of individual units [196]. Typically, exploration of alternative solutions is made possible by introducing stochastic units [4, 11].

Connectionist problem solvers cannot easily represent quantifiers or predicate expressions of arbitrary complexity. One approach is to design increasingly complex ANSs for solving problems in variable binding and rule activation [50, 229, 230]. Our approach is to start with well-understood primitive networks (pattern classifier [197], autoassociator [3, 92], and bidirectional hetero-associative memory [111]) and to explore how a complex problem can be solved by an architecture composed of these basic units.

Heuristic knowledge in problem solving consists of associations between problem and solution entities. For example, if a connectionist network represents an association between the set of problem parameters and the set of operators, then it encodes heuristics for selection, ordering, and pruning of alternatives. Yet another network can learn a state evaluation function, thus supplying feedback to the action selection network [11, 12]. Note that this view of heuristic knowledge is consistent with the heuristic classification theory of problem solving [41]. Thus, we conjecture that the model used in this chapter will suffice for classification problems involving propositional variables. It has been shown by Clancey [40] that several decision problems are members of this class. For complex problems in planning and optimization, however, our results may not apply directly. In Section 5, we consider an important class of problems called dynamic decision problems, which give rise to some interesting problems in strategy learning.

1.7 Problem Representation

Choice of representation is an important determinant of problem solving performance [8, 9, 76, 98, 101]. However, a program must understand the role of representation to be able to adapt its representations automatically [127]. Because there has been no theory of representation in problem solving, it has been customary for designers to procedurally encode the knowledge of problem representation. Recently, however, researchers have tried to automate the process [66, 131, 151, 187, 188, 223]. This research stems from the work of Amarel [9, 10], who formalized the role of representation in problem solving; Davis [46], who used a methodology of metarepresentation in TEIRESIAS;

and Simon [211], whose UNDERSTAND program [209] was able to construct representations for problems stated in natural language.

Automatic acquisition of problem representation is beyond the scope of this chapter. Future developments may, however, significantly change the nature of strategy-learning systems. We consider the issue of representation complementary to the theme of this chapter.

Because we are experimenting with ANSs, a few remarks are in order regarding their powers of representation. The model assumed in this chapter is not as general as the problem space model of SOAR. We assume that ANS units collectively represent the set of problem parameters. A given network will apply to all instances of the same problem, but different networks will arise for different problems. It is possible to construct more general problem solvers by introducing an additional mapping network, but this will create complications because of unsolved problems in variable binding and unification. The simple model is sufficient to illustrate the utility of what we propose in this chapter.

There is another sense in which ANS-based models might be superior to conventional models. Conventional models assume fairly high-level primitive parameters and propositions that represent complex facts. There is little scope for partial or dynamically defined relationships between these facts. It is possible in multilayered ANSs both to construct abstract parameters of description in the internal hidden layers, and to represent partial relationships between primitive and abstract parameters [197]. This constructive element of ANSs lets them create complex descriptive parameters and form relationships between these and other variables of a problem [11].

1.8 Problem Solving Environment

A problem description covers only some of its instances. It does not include the environment of problem solving. Either the particular machine on which a problem is solved or the probability distribution of input variables may influence the optimality of a strategy. The *problem solving environment* consists of the world being modeled in the learning system, and the trainer (if there is one).

As an example of the influence that environment can have on learning, consider a computer with a two-level memory [253, 254]. It is possible that because of the high latency of accesses to the secondary memory, a depth-first search strategy (which has low memory requirement) might be preferable to a best-first search strategy (which has a high memory requirement). Without taking the memory latency into account, a best-first search may always seem preferable. The choice is obviously dependent on a context variable, namely, the amount of primary storage. In the rest of this chapter, we assume that the initial description is augmented with those context variables that affect the performance of a problem solver.

The environment plays an important role in strategy-learning systems because it reacts to the problem solver's actions. This reaction serves as feedback to guide future changes in the performance element. The nature of the feedback is an important characteristic of the environment. If the environment reacts immediately to every action, it has *immediate feedback*. Otherwise, it has *delayed feedback*, and the environment is said to be *slow reactive*. Bock [28] has suggested that feedback after a synthesized sequence of responses improves learning, when compared to systems with immediate feedback. The feedback takes the form of a desired response in a *supervised learning environment*, or of a periodic reinforcement in a *reinforcement learning environment*. In the former case, the feedback is said to be *corrective*, and in the latter, *evaluative*. In the case of delayed, evaluative feedback, the learning system needs an internal model of delay (hereafter called the *persistence model*). It uses the model to associate feedback with the actions causing that feedback.

Typically, the causal knowledge between actions and effects has an environment-dependent component. If this component is not known, the *causal model* of the system is incomplete. However, the causal model is important for proper attribution of feedback. Therefore, every learning system must have an internal model of its environment that captures the causal relationships between actions and effects, as well as the nature of the delays in the feedback process (in a persistence model).

An additional source of complexity in the organization of a strategy-learning system in a slow-reactive environment is the *asynchronous* behavior of the learning component. If the environment supplies inputs in the form of asynchronous events (which is typical of nonstationary dynamic environments), the problem solver must have an asynchronous architecture as well.

Finally, the environment controls the probability distributions of inputs. These in turn influence the applicability of various heuristic rules. This dependence can be easily represented by making the heuristic rules adaptive, and learning them through training in a specific environment.

Most of the existing learning systems have a fixed internal model of their environment. In Section 6, we examine the assumptions implicit in the design of the critic in several of these systems. In light of these assumptions, most existing systems work only in a limited range of situations. In particular, we show a class of strategy-learning problems that are hard to solve using the existing models.

1.9 Overview of This Chapter

In this chapter, we survey architectures for strategy learning. The survey is integrated with a classification of strategy-learning problems (Section 4) on the basis of several aspects of complexity introduced in Section 3. In Section 6, several architectures based on the Dieterich and Minsky models are examined. An improved model is proposed in Section 7. Unlike the Dieterich model, it does not rely on precise, complete, precoded knowledge of the environment; unlike the Minsky model, it can use whatever knowledge becomes available. It constructs dynamically an internal model of the environment. With this and other variations, our model is applicable to a certain class of problems that lies beyond the scope of both Dietterich and Minsky models.

The most crucial extensions of the classical model concern ill-defined objectives and delayed reinforcement. The classical model makes certain conceptual simplifications to eliminate these problems. We examine the deficiencies of several architectures based on this model by identifying a large class of problems that they cannot solve. We also study several systems based on Minsky's reinforcement learning model. These systems work well for small problems but require too many ad hoc decisions to work correctly on larger problems. We attempt to identify and explicitly represent the abstract bodies of knowledge acquired during learning. Our approach is illustrated with SMALL—a system that learns strategies for load balancing in multiprocessor systems [7, 24, 25]. As shown in Sections 4 and 5, load balancing belongs to a class of complex problems in our taxonomy.

2 THE NEED FOR STUDYING STRATEGY-LEARNING SYSTEMS

Even after almost two decades of prolific research in the design of intelligent machines, the applications have usually been demonstrated for simple domains. Attempts to construct complex real-world applications have been hampered by numerous factors, mainly attributed to the *knowledge acquisition bottleneck*. We examine several issues in knowledge acquisition that necessitate a reassessment of the available systems for strategy learning (Table 13.1).

Metaknowledge Not Available Strategy-learning systems have begun to require increasing amounts of metaknowledge from their designers. This knowledge is implicit in abstract parameters of representations and parameters of learning algorithms. The additional knowledge required to make a learning system work correctly has been called *bias*. The need for bias in learning systems has been recognized by several researchers (Watanabe [238, 239], Lenat [126], Davis [46], Utgoff [232, 233], Mitchell [143] and Rendell [184, 185]). These studies have quantified [183] the biases implicit in the choice of representations and in the meta-level architecture for rational setting of algorithm parameters. However, the critic has evaded similar analysis. Except for the work of Barto et al. [23], David Ackley [4], and P. J. Werbos [242], it has retained some very severe assumptions. A comprehensive treatment of the role of a critic in a learning system has been carried out by Dietterich and Buchanan [52]. Our approach is to make explicit the role of knowledge in credit assignment. Doing this enables us to analyze the critic at the knowledge level. Unlike most approaches based on the Dietterich model, we do not assume that the knowledge for credit assignment is available *a priori*.

The Need for Quick, Intelligent Decisions In real-time problem solving, strategic decisions must be made in limited time, using limited resources.

TABLE 13.1   Desiderata for Strategy Learning Systems.

Existing Systems	Proposed System
Algorithmic bias in the credit assignment process	Explicit representation of the knowledge required for doing credit assignment
Fixed strategies	Adaptive strategies
Abstract model of cost for learning and problem solving	Machine-specific model of cost for learning and problem solving
Assumption of stationary distributions	Adapting to time-varying distributions
Assumption of synchronized and/or immediate feedback	Ability to learn under asynchronous delayed feedback
TCA with implicit heuristics for persistence	Explicit, automatically-acquired model of persistence
SCA with fixed, user-supplied causal model	Dynamically varying causal model acquired automatically
Implementation on von Neuman architectures	Asynchronous, parallel connectionist architectures

The need to make quick decisions prohibits searching large search spaces, whereas the need to make intelligent decisions requires that good solutions must not be left unexplored. Moreover, the objectives of the problem may not be stated explicitly. These may have to be acquired or approximated by learning. Such constraints necessitate the inclusion of a learning component in any fast, intelligent problem solver. Learning and problem solving should be integrated so that they share processing power and other resources. Scheduling of resources is inevitable in these systems. The importance of a scheduler in a learning system was suggested by Wah and Yu [254] and in several blackboard control architectures [84, 158].

Studies of resource-constrained inference in AI have been restricted to considering some abstract performance parameters of measurement such as depth of chaining, number of states explored, the number of states memorized, the length of proposed solution, and/or the number of comparisons made. These parameters are meaningful only if they can be related to real measures of performance such as turnaround time on a task. Most researchers tend to ignore this issue because these relationships are machine-specific.

Consequently, most AI systems lack a machine-specific model of cost. Such knowledge can drastically change the evaluation of a strategy, because abstract measures of performance do not always produce a valid evaluation. An adaptive, explicit mechanism must be studied to acquire realistic performance parameters through training on the target machine. Our study in Section 7 includes the description of a system that addresses this problem. Our strategy-learning system can, therefore, make use of machine-specific information in a structured fashion.

Nonstationary Dynamic Environment   Many problems in the real world have too many parameters to be modeled exactly. Our focus is on computer architecture and performance evaluation. Traditionally, strategic problems in this area have been solved by approximation through stochastic modeling, but with the advent of large multiprocessor architectures, even approximate analytical modeling has become too cumbersome. Moreover, the environment in multiprocessor systems has been found to vary dynamically.

Recent experiments in the design of load-balancing strategies [7, 25, 163] have highlighted the need for dynamic decision making and, hence, for learning in real time. This, in turn, has posed the problems of resource-constrained and time-limited inference. Most existing learning systems do not have a real-time scheduling component and are therefore not applicable under these conditions.

Adaptive Strategies   Strategic problem solving is needed when there is a large space of alternatives to explore and no mechanism is known for generating optimal or near-optimal solutions. Therefore, unlike earlier researchers (Slagle [214], Georgeff [72, 73] and Sacerdoti [201]), we do not assume that strategies are available *a priori*. This assumption must be dropped because of the need to make intelligent decisions quickly, and because of the lack of knowledge about the problem environment. Another reason is that computing environments change continuously, and it is impossible to estimate the goodness of a decision completely at any point in time. Thus, strategies should be adaptive in order to be really efficient and generally applicable.

Slow-Reactive Environments   Learning is possible in the first place because the effect of every action becomes manifest in the state of the environment following that action. However, real environments are neither infinitely fast nor synchronous. The credit assignment process in most of the existing learning systems is based on the assumption of immediate feedback. Thus, feedback needs to be distributed only *structurally* among the various decisions causing an observed effect. In the case of delayed reinforcement, the credit assignment problem acquires an extra dimension: time! The feedback must be distributed *temporally* among several memorized decisions, before being distributed structurally among their individual causes.

Temporal Credit Assignment   Temporal credit assignment (TCA) is the mechanism by which feedback resulting from several decisions is divided among the individual decisions constituting the episode. It entails associating observed effects with a history of actions. (See Sutton's thesis [225] for a comprehensive discussion of the origins and solution techniques of this problem.) The existing approaches to TCA mainly deal with persistence of effects caused by individual actions. They employ heuristics such as recency (that the more recent an action is, the more eligibile it is to receive feedback) and frequency (the more frequently an action occurs during the episode leading to feedback, the greater its eligibility).

Although we follow the example of Sutton [11, 23, 225] in separating the temporal and structural components of credit assignment, we relax the assumptions about persistence models. In particular, the model of strategy learning we propose to study has an explicit model of persistence and retains an explicit memory of past actions. In Section 7, we give details of a mechanism for dynamically learning persistence models of various actions and effects.

Structural Credit Assignment Structural credit assignment (SCA) is the mechanism by which feedback for an individual decision is divided among the rules causing that decision. SCA has been simplified in AI systems by assuming explicit representation of cause-effect relationships, as in the preconditions and postconditions of STRIPS-like operators. In nonstationary environments, this assumption leads to a loss of accuracy. We propose an architecture that dynamically acquires the causal model relating actions and effects.

Asynchrony in the Learning Environment In real-time systems, which tend to be event-driven, and in reinforcement learning systems, which tend to be driven by reinforcement, all behavior is inherently asynchronous. Therefore, the problem solver and the learner should be able to process asynchronous inputs and asynchronous feedback, respectively. Thus, our system must have an asynchronous architecture [84, 161]. We explore ANSs because asynchrony and concurrency are inherent traits of the connectionist model.

3 A TAXONOMY OF STRATEGY-LEARNING PROBLEMS

In this section, we study the factors that make some strategy-learning problems more difficult to solve than others. Our purpose is to identify the significant characteristics of the strategy-learning task and to show relationships among the problems studied by other researchers. Assuming that problems involving similar bodies of knowledge require similar knowledge acquisition techniques, we develop a knowledge-level specification of solutions to several classes of problems. Certain strategy-learning problems are complicated by characteristics that require that more knowledge be represented and reasoned with explicitly. These characteristics, along with the considerations discussed in Section 2, allow us to recognize those problems to which existing models of learning systems do not apply.

3.1 Nature of the Objective Function

The objective function of a problem is a generic description of the set of desired solutions. During credit assignment, the critic uses this description to assess the quality of a solution. However, the specification of a problem might not include a crisp definition of its objective. We explore several possibilities for specification of problem objectives (see Table 13.2).

Single Built-In Objective The specification of the objective function is implicit in a credit assignment procedure provided by the user. Usually, a recognition procedure (as in Meta-DENDRAL [31]) is supplied for testing the extent to which a solution meets the criteria of a desired state. The strategy-learning task involves finding heuristics to generate candidate states, testing them with the given procedure, and generalizing these heuristics empirically. The generalization capabilities are implemented by the learning element. The credit assignment policy recommends the heuristics that generate acceptable states and censures those that do not. Consequently, the critic is simple to implement. This kind of system is useful in highly constrained environments where the specification of a desired solution is often too complex to be stated declaratively. However, such specification restricts the applicability of the learning system to just one problem domain. Also, some heuristic knowledge is embedded in the recognition procedure, and it cannot be modified easily because of the procedural encoding.

Assumed, General Objectives This class of learning scenarios are character-ized by the *parametric* form of their recognition procedure. A classic example

TABLE 13.2 Effect of Objective Function Specification.

Type	Examples	Credit Assignment	Strategy Learning
Built-in single objective	Meta-DENDRAL	Sufficient prior knowl-edge for procedural im-plementation of credit assignment policy	Possible to use nonadaptive strategies
Assumed, general objective	Utility-driven classification	Credit assignment policy must be sensitive to parameters computed externally	Strategies should adapt to changes in external pa-rameters
Flexible, explicit specification	Planning	Need to consider dy-namic data structures representing causal in-formation, such as goal hierarchies	Strategies must be declaratively represented and dynamically inter-preted
Abstract, implicit specification	Reinforcement learning	Need to construct an in-ternal model of the ob-jective function; also need to acquire a causal model dynamically	Strategies must be adaptively de-fined; learning of causal models in-terleaved with strategy learning

of this class of problems is utility-driven classification, wherein the assumed objective is to minimize the probability of misclassification. Yet another example is MEA (means end analysis), in which the assumed goal is to minimize the difference between current state and the goal state. Learning in this case resembles an analytic optimization procedure [58]. The critic can be implemented procedurally. It computes an evaluation based on an objective built into the body of the procedure. Credit assignment involves the relative evaluation of states before and after each move, and assignment of credit based on the change in that evaluation.

Flexible Objectives, Explicit Goal Specification   The objective in this case is defined dynamically. As a result, there can be no predefined procedure for evaluating an individual state. The critic must base its judgment on an acquired measure of goodness. The classic example of this line of problems is planning [200]. The first complication in these learning problems is that strategies have to be tagged with the preconditions of application. Each strategy is applicable to specific goals and specific problem situations. The critic tests the preconditions for overgenerality or overspecificity. It uses the causal model to identify the action(s) responsible for credit or blame. The learning element must understand the internal representations, and should be able to edit the preconditions and actions of a strategy.

Implicit Objectives, Abstract Goal Specification   The definition of objective function is implicit in a periodic reinforcement signal. This class of learning situations is complicated because the trainer evaluates a complete solution path rather than evaluate each state on that path individually. The reinforcement can be interpreted as *the net change in a hypothetical static evaluation function* from the first state on the path to the last state. Learning involves starting out with unbiased random behavior, and then biasing it towards increasing reinforcement. The critic learns relationships between actions and subsequent reinforcement. It uses a causal model of the environment, but the model must be acquired automatically. The learning of the causal model is interleaved with the learning of strategies [23]. Examples include problems in reinforcement learning, such as learning to balance a pole in a cart-pole system [43, 139]. (A negative reinforcement signal is given every time the pole is so far off balance that it falls off.) In this example, the objective is to find a sequence of balancing actions that avoids a negative reinforcement signal for the maximum length of time.

A similar class of problems arises when objective function values are available for a sufficiently large set of problem situations but the form of this function is not known. The critic must induce an internal model of the objective function using these problem situations as examples.

Learning of the objective function can precede learning of strategies. An example is the *blackbox optimization* problem discussed by Ackley [4]. This case is more complicated than the case of parametric objective functions discussed earlier, because the parameters must be *learned* while searching.

In this last class of problems, structural credit assignment (Section 2) is more difficult because both the acquired causal model and the acquired heuristics are responsible for subsequent evaluation. A scheme must be devised so that credit or blame can be assigned to the component where the acquired knowledge is in error.

3.2 Immediate Feedback versus Delayed Feedback

Three classes of learning situations can be considered (Table 13.3):

Feedback after Every Decision In this class, the feedback from the environment is made available after every decision. Examples of learning from immediate feedback are strategies employing the *greedy heuristic*—choose the locally best decision at each decision point. However, a strategy for solving a problem that is not solvable directly (in one step) involves a sequence of decisions. Immediate feedback can improve sequential behavior only by improving each decision in isolation, assuming that the net effect on the global strategy will be favorable. There is no need for TCA because there is no ambiguity about the action responsible for the feedback. Several complex problems can be reduced to this class after temporal resolution of credit or blame [52, 120]. In real-world learning situations, immediate feedback is impractical because it slows down the problem solving system. Besides, real learning environments are not reactive enough to produce feedback after every decision.

Feedback after Every Solution Path These problems involve learning *episodes*; each episode consists of the complete solution of a specific training

TABLE 13.3 Effect of Feedback Latency.

Type of Feedback	Example	Credit Assignment	Strategy Learning
Immediate	Back-propagation [196]	Each decision examined in isolation	Possible to ignore the temporal structure of an episode
Synchronous, delayed	Systems based on the classical model [147]	Using prior information about the temporal structure of learning episodes	Knowledge of temporal structure must be supplied externally
Asynchronous, delayed	Reinforcement learning [23]	Need access to a persistence model for dynamic temporal resolution of feedback	Learning of a persistence model interleaved with strategy learning

instance and the feedback. TCA uses its knowledge of a solution to extract the temporal relationships implicit in that structure. In this case *synchronous delayed feedback** is available only after a complete solution path has been found. The critic has a procedural component for TCA. Abstractly, the decisions leading down the solution path are reinforced positively while those leading away from it are reinforced negatively. This type of feedback has been investigated by Langley [120] and Sleeman et al. [215]. The critic needs to memorize the solution path and the branches leading away from it. In some cases, TCA may involve additional search [145]. It is assumed, however, that the knowledge needed for TCA is made available to the learner before any learning takes place. Thus, even though these problems require some knowledge, such knowledge does not need to be learned.

Feedback at Arbitrary Intervals   Most complex problems involve *asynchronous delayed feedback*. In these problems, there is no concept of a learning episode. A good approximation is to consider (imaginary) episodes delimited by successive reinforcements. In order to distribute the feedback the critic must memorize recent actions. However, the structure of episodic memory is not well defined. The memory consists of situation-action pairs along with the justification for each response. Because the critic has no knowledge of the internal structure of episodes, TCA is more difficult. Feedback is distributed among actions depending on their *eligibility* [139]. The eligibility of an action to receive feedback is computed as a function of the persistence of its effects. The more persistent the effects of an action are at the time of reinforcement, the greater is its eligibility to receive feedback. During learning, persistences of effects are maintained and associated with corresponding actions; these are updated in response to later actions, later events, and the passage of time. Thus, *the critic needs access to a persistence model, which maintains and updates persistences in response to events and actions, and which accounts for the decay of persistence with the passage of time*. The problem of learning strategies for pole balancing is an example of this type of feedback. In this problem, the feedback is available only when the pole falls off to one side. This feedback summarizes the evaluation of all the balancing actions since the last time such failure occurred. Moreover, this kind of failure cannot be predicted, given the knowledge of the problem.

3.3   Background Knowledge of the Environment

Background Knowledge Available   We have already seen that some problems may require the knowledge of an internal model of the environment. In *knowledge-rich domains*, for example, expert systems and planners for finite worlds, one can assume complete, consistent knowledge. Learning in such

---

* The word *synchronous* refers to the timing of the feedback relative to the timing of the decision-making process. Feedback is called *asynchronous* if one cannot predict its time of occurrence in terms of the latest decision that will be followed by feedback.

domains involves operations like macro-operator formation and construction of censors and proposers for heuristic rules. Gaps in heuristic knowledge can be detected by inference based on background knowledge and evidence. The critic may incorporate a deductive component for explaining success and failure. When prior knowledge is not assumed to be consistent and/or complete, the critic should be capable of nonmonotonic inference.

Background Knowledge Not Available   In *knowledge-lean* domains, such as control problems for physical systems, there is little background knowledge. The internal model must be constructed by the learner through observation and experimentation. The critic may need to process stochastic data when either the distribution or some parameters of distribution are unknown. Learning in such domains involves constructing causal relationships based on the statistical analysis of dependence and the observation of temporal contiguity between actions and effects. The complexity of such analysis may be controlled by prior knowledge about a causal mechanism between certain actions and certain effects. Credit assignment prefers actions that are more likely to have such effects on the environment and are frequently associated with a better evaluation.

3.4   The Nature of Feedback

Abstract Advice   If the feedback is in the form of abstract advice, then learning involves operationalizing this advice—rendering it usable for search by translating it into *concrete* advice. This may require knowledge for interpreting the abstract advice and repeatedly refining the interpretation until the advice can be stated in terms of the available heuristics. Examples include the FOO program [150]. This type of feedback requires that the trainer (advice giver) have perfect knowledge. Therefore, this type of feedback is not applicable when the environment varies dynamically.

Prescriptive Feedback   If the feedback takes the form of a description of desired output, the critic computes error by matching. The credit assignment process assigns high evaluation to actions that minimize the mismatch. Examples of this kind of learning include systems based on inductive learning. The trainer needs to know the best answer (any good answer if the learning system tolerates noise) for specific instances drawn randomly from a population of problem situations.

Explanation   If the feedback is an explanation of a solution in terms of recent behavior, the learner tries to generalize that explanation while maintaining consistency with other knowledge. Not only does the trainer need to know the correct response for randomly chosen problem situations, but also it needs to know *why* the response is correct. The learner, given an explanation in terms of its background knowledge, must isolate those components of knowledge that will be useful *in general*. Such techniques are applicable only to knowledge-rich domains.

Evaluative Feedback   If the feedback takes the form of reward or punishment, the critic prefers actions that result in the maximum reward (minimum punishment). It needs to form an internal model of the reward generation mechanism. One approach is to translate the evaluative feedback into prescriptive feedback by learning an objective function that takes high values for positive evaluation and low values for negative evaluation [5]. This type of learning occurs in several dynamic control problems because they require minimum knowledge on the part of the trainer.

3.5   Strategy Selection versus Strategy Construction

A heuristic is modular, whereas a strategy is prescriptive [189]; this implies that heuristic-learning systems can consider decisions in isolation, but strategy-learning systems must deal with a decision in its context. Thus, heuristic learning systems are classified as strategy learners only in the case of delayed feedback. If they choose between several available heuristics, they are called *strategy selection systems*, whereas if they construct new heuristics, they are called (piecewise) *strategy constructors*. The knowledge used for credit assignment in each case is described next.

Strategy Selection   Strategy selection problems are simpler than the corresponding construction problems. The selection problem requires knowledge of evaluation techniques only. If the learning problem involves evaluation of predefined operator sequences, then the critic needs to know only the extent to which the application of a sequence leads to a state that satisfies the objective.

Strategy Application   If the problem involves learning of complex preconditions for assessing the applicability and the utility of predefined operator sequences, the credit assignment process must include an analysis of violated preconditions and their eventual rectification. Since the preconditions for individual operators are available, the role of each operator in a sequence may need to be assessed by the critic.

Strategy Construction   A strategy construction problem requires knowledge of composition and evaluation of search control heuristics. If the learning task requires generation of complex sequential behavior, the critic needs the causal model and another component for constructing complex operator sequences. The latter component consists of operators for composition of operators and meta-level heuristics for the application of these composition operators.

3.6   Uncertain and Incomplete Information

If the problem knowledge is uncertain or incomplete, information must be weighted by some measure of belief, and beliefs must be revised based on evidence. Strategic knowledge in these problems depends on the state of the belief system, and needs to be revised in response to new evidence. The correctness of a strategy in the presence of uncertain parameters requires a statistical interpretation. The revision mechanism entails an explicit representation of the dependence between problem parameters, heuristic knowledge, and the net error due to uncertainty in either or both of these.

Uncertainty in Input   Uncertainty in input may arise because of noisy, incomplete, or time-varying information. Noisy inputs force strategies to have some tolerance in the checking of preconditions. Incomplete inputs require that decision making should be possible even under partial information. If the problem includes time-varying parameters, belief in the latest measured values of those parameters should decrease with the time elapsed.

Uncertainty in Heuristic Knowledge   Uncertainty in heuristic knowledge may arise due to either an inconsistent or an incomplete model of the environment. Strategies based on such knowledge have inherent uncertainty. The critic uses its error function and suggests modifications that minimize the expected error.

3.7   Resource and Time Constraints

Constraints on Object-Level Solutions   Knowledge of resource and time constraints is made available to the critic so that it can reject strategies that violate them. In the presence of such constraints, the strategies must follow the principle of *bounded rationality* [208], which asserts that in real-world problems, there is a tradeoff between the optimality of a solution and the time spent discovering it. Optimality must be sacrificed sometimes. Any acceptable strategy must be capable of generating an answer within a predefined time.

Constraints on Learning   The strategy-learning task should be completed in limited time. Learning algorithms that account for resource and time constraints require a substantial amount of scheduling knowledge and resource allocation heuristics. This knowledge may be implicit in the design of the problem solver and the strategy learner or may be made available in separate knowledge sources.

3.8   Types of Learning Techniques

A learning technique is said to be *supervised* if it involves a trainer and *unsupervised* if it does not. In between, there are techniques that reduce the role of the trainer; these are called *semisupervised*. Different learning techniques place different burdens on the learner and the trainer. In general, unsupervised learning is harder than supervised learning. Also, most of the supervised learning techniques rely on episodic learning, that is, learning under immediate or synchronous feedback.

Learning from Examples   In the case of *inductive* learning of strategies from examples, the examples may be supplied externally along with the feedback. Such learning typically leads to problems in strategy selection. However, the critic's role is minimal because the evaluation is supplied externally.

Another form of supervised learning is called *explanation-based* learning. This type of learning is common in knowledge-rich domains. The critic has sufficient knowledge and inference machinery to construct a proof of the labeling (positive or negative) of the example. This proof serves as a causal model and is generalized by the learning element to extract the sufficient preconditions of the example solution. This technique constructs the causal model by inference on its domain theory. It needs explicit assumptions about persistence to work correctly in a time-varying environment [36].

Semisupervised Learning   This type of learning uses the environmental feedback to generate more feedback internally. It is common in systems based on the Minsky model. It leads to problems of strategy construction.

Learning by experimentation is a semisupervised learning method. It typically uses a type of metaheuristic knowledge known as instance-perturbation heuristics, in order to generate examples internally. *Controlled experimentation* is a very powerful learning method in knowledge-lean domains as well as in knowledge-rich domains with inconsistent or incomplete theories [176-178]. The critic identifies candidate decisions whose ambiguity might be reduced by further experiments. It uses an explicitly represented causal model in order to detect such ambiguities. The type of feedback used is synchronous and prescriptive.

Learning in the presence of a critic (also called *reinforcement learning*) relies on evaluative feedback. The critic translates an evaluative measure into a prescriptive one by incorporating a prediction mechanism for reinforcement, which is refined after every prediction, based on a prescriptive measure: the error of prediction. The critic learns to predict reinforcement and induces a change in the direction of maximum reinforcement. The critic needs access to a causal model (in order to choose eligible candidates) and a persistence model (for temporal distribution of feedback).

Unsupervised Learning   Unsupervised learning in the form of *learning by doing* is a useful way of minimizing reliance on an external trainer. This mechanism leads to integration of learning and problem solving. However, it may not be sufficient by itself because of the training time required. The critic must be able to create and evaluate hypotheses for explanation of unexpected success or failure. The learning element integrates accepted hypotheses with the metaknowledge of the problem.

Another form of unsupervised learning, known as *self-supervised* learning, involves extra knowledge about *when* to learn. An example of this type of learning is *apprenticeship learning*, wherein learning occurs solely as a result of expectation failures.

Learning by Analogy Analogy is a general form of learning. It can be supervised or unsupervised, depending on how the strategies are generated initially. The problem solver has an explicit memory of solution schemata seen or derived in the past. It employs a reminding mechanism to retrieve a past instance in response to a new problem instance. The concept of a learning episode is crucial to this technique. If the storage and retrieval unit for the learning episodes is too large, then the memory requirements will grow and generalization abilities will drop, leading in the extreme case to a *memo-function* implementation of the problem solver. On the other hand, if the unit is too small, the retrieved solution will not have enough information to justify the overheads of a reminding mechanism. The critic's role is to detect (by matching) when a past strategy may be applicable. The learning element makes changes to the recalled strategy based on the match detected by the critic.

4 COMPLEXITY CLASSES FOR STRATEGY LEARNING

In the previous section, we saw how the complexity of strategy learning depends on some features of the problem class and the problem solving environment. Among the most important features are the nature of the objective function and the delay between an action and resulting feedback. In this section, we study several problem classes and relate them with various learning environments. This classification presents an integrated view of some well-known classes and well-researched learning environments (See Table 13.4). It allows us to assess the complexity of a strategy-learning problem in a given environment. It is also useful in studying the scope of existing learning models in terms of the problems and environments supported. We will introduce a class of complex learning problems in Section 5. The extensions to the classical learning models entailed by this class are studied in Section 7. For now, we focus on how various aspects of complexity are combined for some well-known problem classes.

As a general rule, problems requiring strategy construction subsume the corresponding problems of strategy selection; problems of learning under evaluative feedback subsume the corresponding problems of learning under prescriptive feedback; learning under resource constraints is usually more complex than learning without any constraints, and may require a radically different approach to learning and problem solving.

TABLE 13.4 Strategy-Learning Problems.

Problem Class	Simple Aspects	Difficult Aspects	Complexity of Learning
Learning strategies for combinatorial search	Well-defined objective, immediate prescriptive feedback	Large search spaces	Time-intensive knowledge-lean learning
Learning real-time scheduling	Strategy-selection, few alternatives	Complex objectives implicit in evaluative feedback, resource constraints	Dynamic learning of objective function and persistence models, only need knowledge for strategy selection
Traditional learning-to-plan problems	Background knowledge available, explicit objectives, immediate feedback	Dynamically defined objectives, delayed feedback	Knowledge-intensive, time-intensive learning; needs knowledge for strategy construction
Reactive plan revision	Explicit objectives	Dynamically defined objectives, asynchronous, delayed feedback, incomplete knowledge, resource contraints	Knowledge-intensive, time-intensive learning; needs knowledge for strategy construction; requires nonmonotonic temporal reasoning, persistence models supplied externally
Adaptive control problems	Strategy selection	Partial knowledge of complex nonlinear mappings, uncertain information, asynchronous feedback	Knowledge-lean learning; requires reasoning with uncertainty for learning nonlinear associations between time-varying, real-valued quantities

Uncertain information entails noisy or partial descriptions. It may not necessarily complicate learning, because uncertainty is inherent in learning and prediction. Hypotheses for explaining evidence naturally undergo revision in the learning process. It is possible to have a more elegant, uniform learning mechanism if data and inferences are allowed to be uncertain. Even though uncertain inputs and heuristic knowledge may require an expanded representation, they could actually reduce the complexity of learning and problem solving.

Among several possible training paradigms, learning with a critic under asynchronous delayed feedback requires the maximum knowledge. The critic has memory of several decisions not necessarily belonging to the same solution path, and it must isolate those that could have caused the feedback. Reinforcement learning with synchronous delayed feedback is less complex, because the critic has complete knowledge of temporal relationships within the sequence. As a result, the persistence model is considerably simplified. The critic does not need to resolve causal dependency between memorized actions and the feedback because all actions on the solution path are candidates for receiving credit or blame. Supervised learning with immediate feedback is substantially less complex because the learner does not need an internal model of persistence; it processes one decision at a time.

Learning Strategies for Combinatorial Search Problems Problems with well-defined objective function and immediate, prescriptive feedback are among the easiest learning problems in terms of metaheuristic knowledge.* This class includes combinatorial decision problems and combinatorial optimization problems. A decision problem involves finding any solution satisfying the problem constraints, whereas an optimization problem requires finding the best solution. Consequently, optimization problems are harder than the corresponding decision problems [69].

Real-Time Scheduling Problems Problems of learning while scheduling in real time are complicated by the presence of resource constraints. The objective function is often too complex to be stated explicitly. Instead, these problems are solved by making the feedback evaluative, so that an improvement in resource usage without significant loss of optimality is reinforced. Feedback is generated (either externally in case of supervised or reinforcement learning, or internally in the case of self-supervised learning) by taking into account the resources used by the proposed solution, as well as the extent to which scheduling constraints have been satisfied.

Planning Problems Problems in operational planning are characterized by the presence of complex logical constraints on the solution. Plans are usually discovered by inference with a causal model. Planning problems change character from constructive to selective if past planning episodes are stored and generalized. In the simplest case, an abstract plan is made available to the planner for operationalization by repeated refinement. This is a case of feedback by abstract advice. Learning in these situations has a strong deductive element and needs substantial background knowledge. Learning problems vary in complexity from those in which the background knowledge is assumed complete and consistent to those in which both these assumptions are dropped. Inconsistent knowledge requires default reasoning and nonmonotonic revision of beliefs. Learning is easier if the trainer supplies an explanation for errors

---

* In terms of heuristic knowledge required, these may well be the hardest solvable problems [132].

than if the system needs to maintain multiple possible explanations and the assumptions under which each holds.

Reactive planning, or planning in real time with resource constraints, has been the focus of recent research. Problems of this type are the most complex planning problems, because they involve time-varying parameters. Learning to plan in these domains involves constructing persistence models of various parameters. These models must account for unplanned changes in persistence caused by asynchronous external events.

Control Problems   Unlike planning problems, most problems in control involve a large number of numeric, time-varying parameters. Control problem complexity grows from the level of system identification to that of policy design. The problem of system identification roughly corresponds to supervised learning of a function from input-output behavior samples. Adaptive learning of control policy has several features of complex strategy-learning problems: uncertainty, delayed feedback, and strategy construction. Problems in control have traditionally been solved using analytical techniques. Adaptiveness is limited to changes in parameters of a general model (such as the transition probabilities in a Markov process). These analytical techniques, however, require some serious assumptions about the distribution of inputs.

It is not our goal to design a fully general system capable of solving all kinds of problems, but we would like to demonstrate a technique for solving some complex problems in strategy learning. Our focus throughout the rest of the chapter is on a class of problems called *dynamic decision problems*, which share several complexity traits with the most difficult problems in our taxonomy.

5   DYNAMIC DECISION PROBLEMS

In this section, we define characteristics, cite examples, and identify components of solution for dynamic decision problems. This category includes many practical instances of problems, stochastically modeled, especially those whose parameters are not stationary with time. Many problems that can be modeled approximately by queuing theory fall into this group. Problems in computer networks, such as flow control, routing control, and congestion control, are all based on anticipated information that may not be collected accurately. Stochastic distributions are used as an approximation. Solutions of these problems can be improved using the deterministic information acquired by learning algorithms. Many problems in computer performance evaluation, such as buffer management, disk head management, and scheduling, are modeled stochastically (Table 13.5).

Dynamic decision problems can be further subdivided into policy design problems and reactive planning problems. A *policy design problem* is characterized by the need for fast decisions to be made from a small group of stereotypical decisions. Decision making in this case is too frequent to allow

TABLE 13.5 Strategy Learning for Dynamic Decision Problems.

Dynamic Decision Problem	Corresponding Strategy-Learning Problem
Load-balancing problem • Send jobs from computers with high workload to computers with low work load • Find a method to communicate information	• Finding attributes to adaptively characterize jobs, workload, and network traffic • Learn a model to evaluate effects of balancing decisions on response time • Find decision rules for reducing the response time
Page replacement problem • Replace the page that is not going to be referenced for the longest time	• Adaptively predict page usage • Find attributes on which to base predictions • Find decision rules based on predicted usage
Instruction scheduling in pipelined computers • Prefetch the instructions for the branch that is most likely to be the target of the next conditional jump • Find a schedule that keeps the pipeline full	• Adaptively predict page usage • Find attributes on which to base predictions • Find decision rules based on branch predictions

time for searching among alternatives. In a typical scenario, any one of a few available alternatives is implemented, and performance measurement is possible only through statistics accumulated over many decision points. Examples in this class are problems of load balancing in distributed computer systems, page replacement for a virtual memory hierarchy, and prefetching for instruction level scheduling of a pipelined computer. In contrast, a *reactive planning problem* [74] involves the dynamic construction of goal-seeking behaviors in the face of a nonstationary environment, in which asynchronously generated events may violate some goal conditions achieved by past actions. Such problems are common in robotics and assembly line planning. They require more effort on the part of the problem-solver/learner because no efficient strategies are defined *a priori*, and it is not possible to completely predict the environment in which the plan is going to execute.

Since the strategies to be developed are different in each subclass, different methods may have to be applied to learn the strategies. Examples of reactive planning arise in domains in which there are multiple autonomous problem solvers and none of them has a complete internal model of the other. Further examples include the open-ended cooperative planning problem of Konolige [107], and the blocks-and-baby problem of Schoppers [205].

5.1 Characteristics of Dynamic Decision Problems

Dynamic decision problems may possess a combination of the following characteristics.

Nondeterminism and Dynamic Decision Making   The decisions involved in solving these problems may require dynamic run-time information and information about characteristics of the problem instance being solved. Attempts at these problems are known to benefit substantially from the use of adaptive strategies. The best strategy for solving the problem instance cannot always be found *a priori*.

Multiple Strategies   Typically, there is a large (and often intractable) number of alternative ways for getting a "good" solution. None of these ways can be selected *a priori* as the best one, because identification of the best strategy may depend on the problem being solved as well as on the values of certain time-varying parameters of the particular instance (possibly some environmental parameters). The enumeration of possible strategies to solve a problem under all possible conditions is generally infeasible.

Incomplete and Uncertain Information   Complete information needed for making an accurate decision may be unavailable because either the overhead of collecting this information is too high or the information is heuristic and uncertain. The solution process should be able to accommodate incomplete and uncertain information and maximize the use of information as it becomes available.

Resource Constraints   In a practical system, the available time and physical resources are limited. It is desirable that the best (as defined by the evaluation criteria) answer or strategy be found as quickly as possible. The large search space of candidate solutions and strategies prohibits an exhaustive search of all possibilities. An intelligent assignment of resources, both time and physical, must be made.

Delayed Evaluation   The exact effect of making a decision may not be known until many decision points later. This requires a history of information to be maintained in order to predict the effect of a strategy. This problem of TCA is more difficult than is usually the case in current machine learning research, primarily because of the asynchrony in delayed feedback.

Dynamic decision problems include the problem of learning strategies for real-time scheduling and reactive planning discussed in Section 4. Certain complex problems in nonlinear adaptive control, such as the pole balancing problem, are also members of this class. Because of the presence of resource constraints, this class does not include learning of strategies for optimization problems. However, for decision problems, the approach taken here is more realistic than traditional approaches such as decision theory (which considers each decision in isolation), or artificial intelligence (which usually assumes substantial prior knowledge). Also, the formulation of control problems here does not require prior knowledge of parameter value distributions. In recent past, some of these problems have been addressed by research in the area of ANSs. However, we focus on explicit consideration of delayed feedback and the role played by knowledge of the problem environment in credit assignment.

5.2 An Example of Dynamic Decision Problems

Dynamic decisions are almost inevitable in real-world problem solving. As already mentioned, the problems in policy design require different problem solving techniques than the problems in reactive planning. We restrict our attention to the former class in order to illustrate the proposed model in some detail.

As an example, consider the problem of designing a load balancing strategy. This type of strategy makes decisions about the distribution of jobs to various sites on a local area network so that some user-acceptable combination of the following objectives is optimized: overall job throughput, maximum completion time, average completion time, total communication cost, and average utilization of processors [37, 38, 156, 157, 237]. The information that can be measured includes processor utilization, number of disk requests per unit time, number of local and remote jobs that are pending, and status of communication links. These metrics are sensed periodically and used at each decision point. The set of conditions for determining the action to take at decision points constitutes the decision policy of the system. A *load balancing strategy* is, therefore, a combination of the metrics used and the decision policy [7].

A system for learning load balancing strategies should adapt to changes in its environment. This means that new combinations of low-level policy decisions and revised preconditions for applying various strategies, based on revised expressions of the measurable metrics, should be tried. Any changes should lead towards the desired goal of the system. When jobs are perfectly balanced, the interprocessor communication is at a minimum, and the overall throughput is high.

This problem belongs to the class of policy design problems. It involves runtime decision making; multiple strategies are available for solving the problem; the strategy-learning process must be executable within the time and resource constraints of the multiprocessing system; the information or metrics for the workload may not be up to date or may be uncertain; and the effect of a particular strategy does not become known until many decision points later.

We return to this problem in Section 7 following a survey of past research and a discussion of the new model of strategy-learning systems. This problem is also the subject of a prototype implementation [235] under development at the University of Illinois. However, it is only a representative problem. In general, other members of the problem class will have different objectives, decision variables, constraints, and entities. For instance, the three problems shown in Table 13.5 have similar knowledge-level learning requirements. All three have (1) a temporal aspect, which is manifest in the associated prediction problems; (2) a causal aspect, which is manifest in the construction of relevant attributes; and (3) the basic strategy selection aspect, which is manifest in the need for learning decision rules.

The methodology for the design of strategy-learning systems proposed in this chapter is motivated not so much by the specific constraints of any particular problem, but instead by the general characteristics of the entire problem class. Various aspects of complexity (Section 3) and our focus on the defining traits of dynamic decision problems (Section 5.1) underlie our survey of past research (Section 6).

6 SURVEY OF STRATEGY-LEARNING SYSTEMS

It is possible to classify learning systems along many dimensions. Three such dimensions were introduced by Michalski et al. [137]: (1) the role played by the learner (and the trainer, if there is one) and the amount of inference required; (2) the representation of acquired knowledge; and, (3) the domain of application of the performance system. Classification scheme (1) may be useful because the mechanisms of a specific learning model can be explained more concisely by examining the role of each component in a learning system, and by identifying the assumptions on which its design is based. Systems belonging to the same category under scheme (1) will employ similar mechanisms of acquisition. In strategy-learning systems, the representation of strategic knowledge exhibits substantial uniformity within a domain. Schemes (2) and (3) are equally good. The classification in this section employs scheme (3) at the top two levels and (1) at the bottom two* (See Figure 13.4)

The learning systems studied in the past have originated in domains as diverse as cognitive science, psychological decision theory, statistical decision theory, automated classifier systems, expert systems, and parallel distributed processing (or connectionism). It is interesting to note that the bodies of knowledge involved are very similar at an abstract level. Our view is that these systems are the implementations of one of the two learning models we have already introduced. When we identify a learning system or a class of learning systems as implementations of a model, we expect that the basic assessment of the model will carry over. We will, however, be careful to point out exceptions where appropriate.

Figure 13.4 shows the broad categories of learning paradigms we have studied. The hierarchy in the figure follows the general pattern

---

* Recall that the knowledge-level nomenclature introduced in Section 1.4 classifies learning paradigms primarily on the basis of the abstract acquisition mechanism employed. Specific systems specialize the abstract mechanism into an implementation scheme that is highly dependent on the representation used. Because there are fewer abstract acquisition mechanisms than representation schemes, we choose to classify by problem domains first.

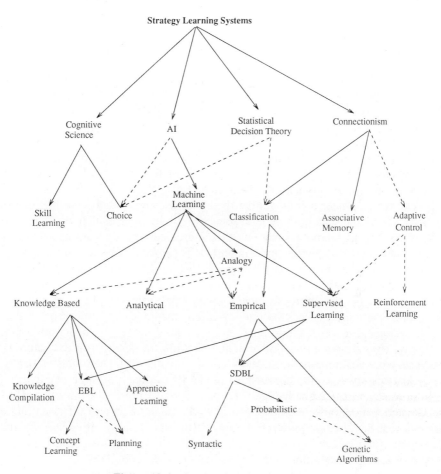

Figure 13.4 Strategy-learning paradigms

domain → problem → extensions.

The dotted lines in the figure represent approximate or indirect relationships. Some branches in the tree are short either because there is not enough variation in the abstract category, or because we are not aware of more than a few examples. In any case, this survey is meant to be representative rather than exhaustive. A quick look at the figure will, however, reveal that the strategy-learning problem has a vast scope and, therefore, a large number of techniques are available for solving it. We examine some of these techniques next and discuss their capabilities in terms of the aspects of complexity introduced earlier.

6.1 Cognitive Models of Skill Learning

The first models of problem solving behavior were motivated by a desire to understand human problem solving. Newell *et al.* [154] laid the foundation for a theory of intelligent problem solving by proposing the problem space model. The information processing theory of human problem solving [210] paved the way for later developments on the role of learning in intelligent behavior [118]. It provides common ground for comparing several strategy-learning paradigms. Even though the initial theory was built around the GPS program [57], which had a fixed general strategy of problem solving based on means-end analysis, subsequent refinements of the theory were increasingly oriented towards a production system model of learning.

The development of a cognitive theory of learning to solve problems was motivated by the search for general principles of learning, which would remain invariant across learning situations. Six ingredients of all learning situations were identified by Langley and Simon [118]: (1) *feedback* about the improvement or degradation of the performance element, (2) *behavior generation*, that is, the ability to explore alternative strategies, (3) a mechanism for *credit/blame assignment*, (4) the ability to *memorize* past performance and to *reassess* its goodness in the light of subsequent reinforcement, (5) the ability to *acquire* the examples, learning algorithms, and the causal model, by being told, and, finally, (6) continued *practice* of acquired skills. Depending on the nature of (5) above, the model can be classified as a variant of Dieterich model (if feedback is always told), a Minskian model (if feedback must be extrapolated from recent instances of being told), or neither (if the causal model and the persistence models are learned by means other than being told). These criteria show that the abstract view of learning is the same in cognitive science and artificial intelligence. Only the motivations differ. For an insightful treatment of motivation for learning, see the discussion by Simon [213].

In 1979, Anzai and Simon [17] proposed a theory of learning by doing that was centered around an adaptive production system [240]. That theory has been successfully demonstrated to acquire strategies [18] for a complex problem (steering a ship). The basic hypothesis of this theory is that problem solving proceeds by generation of problem spaces and search for goals in them. The theory applies whether or not the problem solver is capable of generating an initial problem space. In case the initial problem is ill defined, learning by doing can include understanding by doing. For example, it is possible to acquire the causal knowledge between actions and effects dynamically.

According to the theory, the basic unit of knowledge acquisition in strategy learning is the pruning heuristic, which allows the problem solver to ignore some parts of the search space during search. (For a computational approach to this kind of learning, the reader can refer to Yu's thesis [253].) The strategy-learning program is implemented as a rule discovery system much like Lenat's AM [124, 125] program, which employs heuristics for the discovery of heuristics. Learning by doing is capable of handling time lags between action and effect as well as problems with ill-defined objective functions. The theory, when augmented with the understanding-by-doing model, is capable of causal attribution and persistence modeling. Its reliance on domain knowledge and rules of discovery are a handicap in terms of speed. On the positive side, the theory is very general and applies across a variety of domains. The implementations of this theory lie outside the scope of Dietterich's model because of the understanding-by-doing component, and outside the scope of Minsky's model because of the ability to deal with metaknowledge for strategy acquisition. If implemented in full generality, following Simon's suggestion of employing recursively applicable heuristic rules for discovery, this theory can learn to solve any problem that is solvable by the production system representation. However, recent results in computational learning theory suggest that such a mechanism may not be able to learn within polynomial time [228]. In fact, such criticism applies to several other strategy-learning paradigms that are equally general (for instance, SOAR [116, 117] and ACT [14, 15, 152]).

There are several cognitive models of strategy-learning systems based on adaptive production systems with hardwired learning mechanisms in the form of procedures for composition of rules by chunking and proceduralization. These include learning by composition [13, 15], learning by knowledge compilation, the work of Lewis [129], and learning by chunking [191]. Although these systems are capable of constructive experimentation, they suffer from reliance on deep causal models of the problem domain, and the assumption of immediate feedback (for instance, p. 203 of reference [15]).

Still another class of cognitive models of strategy-learning systems are capable of learning problem solving skills from examples. In these systems, the dependence on background knowledge is minimized. The nature of feedback is prescriptive. Examples of this paradigm include the PET system for learning problem solving heuristics [103, 172, 174, 173] and the ALEX system for learning to solve textbook-algebra problems [153]. In a system based on ACT theory, Lewis and Anderson [130] consider the effect of delay on performance of human subjects. Their analysis seems to suggest that the knowledge needed for TCA is crucial to human learning, although no model of learning under delayed feedback is proposed. Most of these systems (except Anzai's learning by doing [18]) have been demonstrated on problems having a few completely specified parameters, an exact objective function, and immediate feedback for learning.

6.2 Strategy-Learning Systems for Making Choices

These systems have their origins in the psychology of decision making. Even though both psychological and statistical decision-making theories revolve around the concept of maximization of expected utility of the outcome by cost-benefit analysis, there is a major difference in their view of the learning process: statistical decision making is *normative* (how people *should* choose), whereas the psychological model is *descriptive* (how people *do* choose) [2]. Unlike the models of skill acquisition discussed above, these models take into account the uncertainty inherent in making strategic decisions. The major difference between the two models, insofar as their treatment of uncertainty is concerned, is their interpretation of probability. Psychological models use a nonfrequentistic interpretation, whereas statistical models use a frequentistic interpretation [86]. The psychological decision theory models restrict the information processing model by imposing the constraint of bounded rationality [212]. As a result, these models account for time and resource constraints. However, the temporal component of strategic behavior is played down; decisions are restricted to choices and preferences between options. Thus, these models are usually more appropriate for strategy selection and acquisition of guidance heuristics.

Einhorn and Hogarth [55, 56] identified cases in which feedback can lead to suboptimal performance. This is roughly equivalent to having no causal models. They also comment on how uncontrolled time-varying parameters make learning difficult. Both of their criticisms can be countered if a system is able to learn causal and persistence models dynamically. If there is *some* regularity in the variation of environment parameters, then the learner should be able to acquire a model of such variations by observation and controlled experimentation. While the researchers in psychological decision theory and in the psychology of learning [35] have both realized that the problem of delayed reinforcement is harder than the one with immediate feedback, there has been no attempt to construct a model that explains the temporal apportionment of feedback in delayed reinforcement learning.

Psychological models of choice under uncertainty are a rich source of possible representations of heuristic knowledge, providing models of heuristics that are invoked under resource constraints, and taking into account important variables of problem environment (see, for example, references [97, 99, 164, 199, 231, 26]). Their limited applicability in a computational environment is partly due to a lack of machinery for temporal credit assignment, and partly due to their extreme dependence on a knowledge-rich learning mechanism.

6.3 Strategy-Learning Methods in Statistical Decision Theory

As already mentioned, decision-theoretic mechanisms ignore the temporal aspects of strategies. They focus on the search for alternative means of achieving a *set* objective by ordering the alternatives according to some preference criterion [162]. In contrast, AI problem solving is concerned with problems that *cannot* be solved directly. Sequential behavior and explicit consideration of temporal aspects are central to AI problem solving. Thus, even though learning mechanisms in statistical decision theory are founded on sound learning rules, the major handicap of these approaches lies in their representation — only direct decision rules can be acquired. Decision-theoretic approaches have found their way into AI, in the form of mechanisms for assignment, fusion, and propagation of beliefs during uncertain inference, and for learning rules of selection.

The two major contenders for the use of decision-theoretic techniques in AI are the Bayesian nets of Pearl [168] and the reasoning systems based on the Dempster-Shafer theory of evidence combination [79]. These approaches are useful in learning from examples. However, the convergence and optimality of the rules depend on careful analyses of independence among the parameters. Especially, the Dempster-Shafer calculus is an excellent framework for performing resource-constrained inference [81], and can become an integral part of the performance element in a learning system. Numerous examples of AI systems using utility theory for assessing the value of a problem solving decision can be found in the literature [62, 67, 93, 195, 63].

Another class of limited strategy selection problems that can be solved using statistical decision theory techniques involve decision trees. These models assume that the decision tree is given *a priori* and that probabilities and utility values of leaf nodes are known. These techniques can tolerate some imprecision in the initial assignment of prior probabilities and utility values, and produce an optimal strategy, represented as a path in the decision tree [245].

The CART methodology of Breiman et al. [30] introduces mechanisms for automatic construction of classification trees from a learning sample. These mechanisms are nearly as general as conventional classifiers, and together with regression trees, they provide an efficient framework for managing dynamically varying representation of classes. They can be used in an adaptive action-selection network (performance element + learning element) in a data-intensive domain. Although the decision tree framework is more general than the single-decision model of choice, it requires complex mechanisms for propagation of belief. Because credit assignment can be viewed as revision of belief, these approaches can be applied to problems in which feedback is provided immediately after the last decision (leaf node) of a solution path is executed (synchronous delayed feedback).

We have criticized the decision-theory model for its lack of temporal structure, but there are a few exceptions. For instance, Kuipers [113] presents a strategy for dividing a complex (strategic) decision into a sequence of choices. Pollack et al.[171] have provided an overall architecture for integrating planning, belief revision, and choice. Decision-theoretic techniques are, therefore, promising as a model for data-intensive learning for strategy selection. Mechanisms that use explicit causal models in the form of dependency graphs (such as Pearl's Bayesian nets) appear useful for problems not involving delayed feedback. However, their neglect of temporal structure renders them unsuitable for dynamic decision problems. Recent advances in connectionist learning techniques of Sutton [227] and Werbos [241] have extended these models to include a mechanism for temporal prediction. In theory, all connectionist research can be interpreted statistically [77]. Therefore, statistical networks could be used for solving dynamic decision problems involving possibly nonlinear relationships between numerous time-varying parameters.

6.4   Strategy-Learning Methods in Artificial Intelligence

In AI, strategy-learning systems have been built for applications such as board games, problem solving, planning, and scheduling. Approaches have fallen into four broad classes:

1. *Empirical learning techniques* rely largely on syntactic similarity of examples in order to form general concepts. This approach can be further classified into (*a*) *similarity/difference-based learning* (SDBL), which works by generalization of logic expressions and structural descriptions given as examples of the concept; (*b*) *probabilistic SDBL*, which operates by forming regions in feature space through splitting and merging; and (*c*) *genetic learning* and other learning by experimentation methods, which use perturbation techniques to generate examples automatically.

2. *Analytical learning techniques* are based on extensive analysis of problem solving behavior with respect to one particular representation and a specific problem solving method.

3. *Knowledge-based approach* is used for acquisition of inference strategies in knowledge-rich domains such as expert systems, planning, and problem solving. Four learning techniques that belong to this class are: (*a*) *compilation*, which translates declarative heuristic knowledge into procedural solutions embodying the heuristics; (*b*) *explanation-based learning* (EBL), which acquires a general strategy from an explanation of a particular episode; (*c*) *plan revision*, which constructs and refines abstract descriptions of solutions; and (*d*) *apprentice* learning techniques, which employ learning by watching to fill gaps in the knowledge base.

4. *Analogical learning techniques* can be used in conjunction with any of the other learning techniques. Similarity between a new problem and an old problem, for which a solution is in memory, is exploited in order to systematically transform the recalled solution into a solution to the new problem.

In this section, we describe these learning methods in detail. Section 6.4.5 covers hybrid learning systems, which cannot be placed into any one class. Interested readers are also referred to survey papers exclusively dedicated to AI methods for strategy acquisition, by Keller [102], Mitchell [145], Sridharan and Bresina [219], and Langley [120].

6.4.1   *Empirical Learning Techniques.*
Empirical methods rely on similarity/difference-based learning techniques to acquire heuristics and strategies. The primary categories within this class are systems that learn from examples and those that learn by experimentation. The critic in the first category of systems needs only know how to interpret examples; in the second category, however, it should be able to generate its own examples. Mitchell [145, 147] discusses a *version-space* approach for learning heuristics. The LEX system incorporates a flexible representation for partially learned heuristics, in which two boundary elements describe the least and most generally consis-

tent heuristics with respect to the examples seen. The system can use learning by experimentation to restrict the version space. Its TCA employs time-bounded search, and the performance measure used is the processing time, which gives the system a machine-specific model of cost. However, the system relies on synchronous feedback. It uses procedurally encoded knowledge of the causal structure of the problem-solving episode. In dynamic decision problems, episodes do not have a predefined structure. LEX is based on the classical model of learning systems and therefore cannot be applied to problems involving asynchronous feedback.

The SAGE.2 [119] system's mechanism for credit assignment is flexible for both immediate and delayed feedback. It incorporates a framework for proposing rules for strategy learning. However, the system is very conservative in learning from examples. When a move is labeled undesirable, the system does not generate a negative example unless there is another move from the same state that has not been labeled as undesirable.

Another program that learns strategies from examples is ALEX [153], a nonfeedback learning system that relies on a form of precondition analysis. It uses a means-end analysis-based learner that only trains for differences between specific initial and final states. Credit assignment is based on whether the state is closer to the goal after applying an operator. The system has a condition creator to construct preconditions both from those pairs in successive states that cannot be explained by the current set of heuristics, and from information about the context of problem solving. Each step in the example is examined in isolation.

ALEX is also capable of learning strategies while solving problems. It learns from failure by specializing operator schemata through difference-based learning. Examples of failed episodes are examined for differences with similar situations that worked. This idea originated in the near-miss concept of Winston's ARCH program [250] and has recently been formalized by Falkenhainer [60]. Another example of this kind of learning is the discrimination learning mechanism in SAGE.2 [119].

Yu and Wah have developed TEACHER-1, a system that learns, by experimentation, the dominance relations in combinatorial searches [254]. The system generates alternatives by examining different parts of a search tree and proposes dominance relations based on positive and negative examples found. Credit assignment is based on the processing time expended, and the number of dominance heuristics found in an allocated time quantum. Dominance heuristics are hypothesized using domain-independent and domain-dependent knowledge. Because combinatorial search problems are well defined, and positive and negative examples are easily verified in the problem domain, dominance relations as good as those obtained by theoreticians have been found for a number of search problems.

Syntactical logical expressions can be generalized using Michalski's AQ program [138] and other generalization techniques [169]. These techniques have been used in various rule induction programs, such as Meta-DENDRAL [31], Poker Player [240], and UPL [160], for inducing rules from specific episodes of chaining inference. Generalization of structural descriptions can be used for learning macro-operators. Among several well-known programs for structure induction are ARCH [250] and INDUCE and its extensions [243]. Empirical techniques for learning macro-operators have been demonstrated by Whitehall [244] and Andrae [16]. Their systems generalize problem solving traces either generated internally or supplied by the user. The essential concept in this case is that of a problem solving episode. As we have already seen, in problems with asynchronous delayed feedback, the concept of an episode is very different, and so these methods do not apply directly. The same comment applies to other macro-operator learning systems, such as Macro Problem Solver [108] and FM [135].

Probabilistic methods employ numeric performance measures in making decisions. The PLS1 [180] system forms clusters of problem instances for which similar heuristics apply. Strategy learning is transformed into concept learning, and the objective function maps into the utility function of concept learning. The problem space is partitioned into rectangular regions by clustering techniques. These techniques are appropriate only when the problem-space parameters have smooth variations with respect to the applicable heuristics. (See, however, recent work on constructive induction [181, 186] as a possible counter argument.)

Zhang and Zhang [255] view search as a statistical sampling process. Learned strategies are encoded as weights on the nodes of the search space being explored. Evaluation functions of nodes that are unlikely to lead to solutions have weights added. Likewise, if a hypothesis is accepted, weights are added to all competing hypotheses. The likelihood of selection of a hypothesis varies inversely as its weight. The adjustment of weights on hypotheses amounts to learning.

Another class of search techniques use *dynamic weighting*. With heuristic estimates that are guaranteed lower bounds of the true goodness, branch-and-bound [114] methods can be used to implement learning while searching; even if the heuristics are not lower bounds, techniques such as the HPA dynamic weighting algorithm [170] and adaptive search [136] can be used to improve efficiency.

Besides being used to learn operator schemata and guidance rules, empirical methods have been also used for learning binary preference predicates [234], predicting the length of solutions [175], and for learning problem classes [19].

A class of empirical learning methods called *learning prediction* has much significance for learning in the Minskian model. The learner tries to learn a weighted evaluation function that remains invariant along the solution path, and that accurately predicts the assessment of the terminal node on the solution path [39]. The earliest example of such methods is Samuel's checker playing program [202] and its extensions [203]. Its equations have nearly the same form as the equations of the Minskian learning model in Figure 13.3. A uniform weighted average is used for maintaining the c_i's. The parameter N model is so set that oscillations are avoided during the early phases and overfitting is avoided later (N is 16 until the 32nd trial, 32 until the 64th trial, and so on, until it stabilizes at 256). Samuel's technique is quite ad hoc. Variants of this technique formalize the exponential decay factor [225] and the iterative prediction procedure. For instance, the Bayesian updating scheme is used by Lee and Mahajan [123], and linear regression is used by Rosenblatt's perceptron [190] and by Korf [110].

In subsequent research on prediction by temporal difference methods, Sutton [227] has presented a class of parameterized procedures called $TD(\lambda)$. The λ parameter is the exponential decay factor set by the user. A similar procedure had earlier been implemented in the Adaptive Critic Element [23], which can solve strategy-learning problems in knowledge-lean domains, using reinforcement learning under asynchronous delayed feedback. It includes a *horizon* parameter that defines the relative importance of making a correct prediction in the immediately following time instance versus the importance of the exponentially weighted sum of all the future predictions ad infinitum. These methods resolve the question of temporal credit assignment in a single-action machine by making the assumption that the eligibility of an action to receive feedback decreases exponentially with the time elapsed since the action was performed.

Genetic algorithms simulate nature's evolutionary mechanism to learn rules in propositional classifier systems. The classifier rules together constitute the performance element of these learning systems. This framework [90] has been augmented with learning based on genetic algorithms for discovery of new classifiers [29]. A novel TCA process called the bucket-brigade algorithm [91, 249] has been extended so that credit can be distributed structurally and temporally, using a synchronous delayed feedback paradigm of learning.

Each rule has an associated value called *strength*, and another parameter called its *specificity*, which is the fraction of the propositional literal population referred to in its antecedent. During problem solving, the *bid* of a rule is directly proportional to its strength and specificity. Among the eligible bidders (i.e., those whose preconditions are satisfied), some high bidders are selected for action. Credit is allocated either when external feedback is received from the environment, or when a rule causes another rule to be activated. Credit takes the form of increments proportional to the strength of the recommended rule. Sutton [227] notes that this algorithm results in exponentially decaying feedback, although this has not been established analytically. This algorithm is selective about the actions chosen to receive feedback, in some sense using the causal model implicit in the rules of the classifier system. It treats TCA and SCA uniformly. The eligibility of an action to receive feedback varies exponentially as its depth in the causal chain relative to the rule that draws the reinforcement. Under the assumption of infinite parallelism, the persistence model of this TCA policy is the same as that of Sutton's $TD(\lambda)$ family of procedures.

The genetic learning model is easily extended to environments with asynchronous delayed feedback [29], although the question of memorization of past decisions has not been addressed explicitly, and it is also a problem that the model's persistence model is a procedural component of the critic. This paradigm assumes a fixed objective function and a precoded causal model, although the causal model can be augmented by application of genetic operators, such as crossover, inversion, and mutation. These genetic operators are the strategy modifiers discussed in Section 1.4. The metaheuristics for the application of genetic operators are procedurally encoded in the critic.

Classifier systems share a lot of representational inadequacies with ANSs. Among them are restrictions concerning the number of propositional variables (which remains fixed throughout the learning and performance periods) and the inability to represent quantified predicates. The metaknowledge (the heuristics for classification problem solving) is expressed declaratively, and the learning technique is quite general.

Other examples of genetic learning systems for strategy learning are LS-1 [217] and PLS2 [179, 182].

6.4.2 Analytical Learning Techniques. Analytical methods are based on extensive analysis with respect to one particular problem representation and a specific problem solving technique. A typical example of this class is the DGBS system [58]. It is geared towards problems and operator representations specific to the General Problem Solver (GPS). It might be intractable on larger problems because it ends up examining all instances of an operator initially, instead of manipulating parameterized descriptions of operators. Strategy learning is reduced to automatic construction of triangular connection tables for GPS. The method takes as inputs the invariants of the procedure implementing an operator, and outputs difference-reducing connection tables. Two specific principles, namely, that the hardest difference must be reduced first, and that differences should not be reintroduced, are programmed into the strategy-learning mechanism.

An extension of DGBS [96] has been used for automatically acquiring heuristics for a robot planning problem. This method analytically derives subgoal-ordering heuristics and uses them on relaxed subproblems to get a heuristic estimate for the A* search procedure.

Other methods in this class include the heuristic generation method of Dechter and Pearl [48] for constraint satisfaction problems, the state-table analysis technique [112] for automatic completion of partial operator sequences, and the problem-relaxation technique of Gaschnig [70]. The major disadvantage of this type of method is the explicit dependence on restrictive representations. They assume that strategy construction meta-metaknowledge is understood so well that it can be implemented procedurally.

6.4.3 Knowledge-Based Methods. Knowledge-based methods are used for reasoning with explicitly represented knowledge. A strategy-learning mechanism in these environments can either result in the acquisition of new control (meta)knowledge, which can be subsequently used for controlling inference, or it can result in the re-representation of prior control knowledge so that inference becomes more efficient. The second set of methods are referred to as knowledge compilation.

Knowledge Compilation in Inference Systems   This class of methods gives a general model for improving performance with practice. These methods apply well to problems with declarative representation of heuristic knowledge. The aim of the process of *operationalization* is to reformulate this knowledge so that the resulting procedures embody the heuristic knowledge and are directly stated in terms of the problem operators.

Operationalization may be applicable in two strategy-learning scenarios: abstractly stated strategic information to be translated into solutions, and abstractly stated objective functions to be translated into methods for achieving them. Note that learning must occur in a knowledge-intensive environment, with feedback taking the form of either abstract advice or abstract specification of problem objectives.

The first set of situations occurs, for example, in rule-based systems in which strategies are stated declaratively. An example is the ACT system [14], which employs two mechanisms called composition and proceduralization. *Proceduralization* eliminates matching and retrieval from long-term memory by instantiating variables in a constrained fashion while simultaneously dropping the constraining clauses, thus reducing the amount of information needed in working memory. *Composition* works either by collapsing two productions that follow each other and eliminating the intermediate goal, or by collapsing the preconditions and actions of a number of sequenced productions into one macro-production. The original productions are not destroyed during compilation. Other examples of this technique can be found in Section 6.1. In these systems, the composition and proceduralization mechanisms are the strategy modifiers. Storage economy is maintained by empirical generalization, and validity is maintained by specialization through difference-based discrimination.

The second category of systems are called *advice-taking* systems. Mostow [150] provides a comprehensive treatment of operationalization, which refines heuristic advice expressed declaratively into a procedure that incorporates the advice. The strategy modifiers in this case consist of *reformulation operators* and metaheuristics. The metaheuristics prune those alternatives unlikely to lead to an operational solution. An example of this category of techniques is Dietterich's test incorporation theory [53]. The Incorporation Problem Solver (IPS) is a solution construction process that transforms declaratively specified tests and naive generators into constrained, procedural, intelligent generators with little or no testing. The IPS is itself a problem solver employing test incorporation operators like serialization of subgoals (called *seed growth*), and non-repetitive enumeration (called *triangle generation*, a generator implemented as a function with memory). The learning element in these systems knows the internals of the generator and performs test incorporation based on transformed heuristic advice.

Compilation techiques can be used for solving problems with abstract, implicit objective functions in knowledge-rich domains. Objectives arising out of the consideration for resource constraints usually take this form. For example, it is usually clear what can be measured (recall *metrics* in load balancing), and it is usually clear what needs to be optimized, at least abstractly (e.g., the average turnaround time of jobs in load balancing). However, the nature of the relationship between the abstract objective and the observable metrics is not clear. This theory assumes that such a relationship can be inferred by deduction. This is a useful way of incorporating a machine-specific model of cost into strategy-learning systems for knowledge-rich domains. Another system that uses compilation to operationalize abstract objectives is MetaLEX—a system for improving the performance of LEX. The metaknowledge is stated explicitly in two forms: target concept knowledge (i.e., the concept of a successful move) and performance-system knowledge. The latter includes an internalized model of the environment, an abstract description of objectives, and the performance-system internals. MetaLEX compiles this metaknowledge and memorizes those parts of the compiled description that are particularly difficult to extract from the abstract objective.

For dynamic decision problems, these mechanisms of learning cannot be applied, primarily because of a lack of sufficient background knowledge. The form of the relationship between certain objective functions and observable parameters can at best be approximated by stochastic modeling, which requires analytical, machine-specific knowledge.

Knowledge Compilation in Problem Solving Perhaps no discussion of knowledge compilation would be complete without a discussion of macro-operators. We have already seen some mechanisms such as proceduralization (in rule-based systems) and learning of procedures from traces (in empirical learning systems). Explanation-based generalization as a mechanism for macro-operator formation is discussed in the next section. In the STRIPS/MACROPS system for learning and problem solving [65], *triangle tables* map problem situations to sequences of operators. They are constructed by learning while searching: any plan constructed *ab initio* can be stored in a triangle table. Plans are generalized using a method similar to the EGGS algorithm [148]. All constants in a triangle table are replaced by variables. The (overgeneralized) precondition in this *lifted* table, which contains all uninstantiated operators, is constrained by repeating the support proof. Only *necessary* instantiations of variables occur.

A recent extension to this framework introduces hierarchy in its notion of action. The tables can be used to map situations into sequences of abstract actions, which can have further internal structure [159]. These extended tables can be used in a hierarchical, asynchronous and concurrent control architecture. Triangle tables are suitable only for primitive actions representable as STRIPS operators—using add-lists and delete-lists to represent their effects. Their applicability to dynamic decision problems is, therefore, restricted to reactive planning problems.

Another method for compilation of macro-operators while searching is due to Korf [108, 109]. It has been used for solving problems with nonserializable subgoals. In such problems the problem objective has interacting, conjunctive subgoals that cannot be achieved by a problem solver that tries to achieve them one at a time. Korf's technique uses *macro tables*, which are triangle tables compiled to learn mappings from the initial state of a variable to its final state. Each entry in the table is a macro that leaves intact (as an end result) all previously achieved subgoals, and achieves one more subgoal. It is a weak method, capable of learning all the relevant macro-operators in the same order of time as is required to solve the same problem without any heuristic knowledge. It can be used for eliminating search from finite-space, discrete-domain planning problems. In reactive planning, goals achieved previously may be affected by future changes in the problem environment. For policy design problems, this method has limited applicability because of the presence of time-varying parameters.

Explanation-Based Methods   The strategy-learning methods in this class rely on extensive domain knowledge. They work by explaining why a particular search path leads to success or failure. This explanation is generalized (sometimes specialized), and a sufficient set of preconditions inferred under which the same line of reasoning will apply. Acquired strategies, stored as schemas, are applied whenever their preconditions are satisfied. Search can be substantially reduced as learning proceeds and strategies grow in numbers and specificity.

The LEX2 system [145] uses goal regression to explain the success of problem solving episodes, and generalizes the resulting explanations. Applications of operators along the optimal solution path are the positive examples, and those leading away from the optimal path are the negative examples.

The EGGS generalization procedure [49, 148] reduces the problem of generalizing the explanation to a unification problem—that of matching the corresponding variables from adjacent rules in the explanation. EGGS can be viewed as an optimization procedure. Its goal is to maximize the number of models of its causal chain. The constraints on this procedure require that (1) the structure of the causal chain is preserved, and (2) the general explanation so obtained is valid.

The PRODIGY system [142] employs strategic information in the form of metarules (selection, rejection, and preference) to guide its search. The approach is unique in that it learns from successful solutions, failures, and goal interactions. Moreover, it learns by recursively specializing proof trees instead of generalizing ground proofs, which is the usual approach in EBL. It learns the *weakest* preconditions in which the schema represented by the particular solution is applicable. It is also able to handle disjunctive concepts (alternate paths to the goal). Carbonell and Gil [33] have incorporated learning by experimentation into the PRODIGY system. Their system is able to learn missing preconditions and postconditions of search control rules, and uses reactive experimentation to repair its strategies.

A major problem with explanation-based methods is their reliance on extensive knowledge, which is not always available. A learning system should be designed so that it does not always rely on complete and perfect information, but it should be capable of using domain-specific information when available. Most of the EBL systems fit well into the classical model of learning systems for knowledge-intensive domains with synchronized delayed feedback.

Planning Methods   Planning methods are useful when the problem objectives are not directly achievable. They are suitable for reasoning about action, anticipating the world resulting from the consequences of an action, and generating behaviors for achieving the goals. Plans are constructed from the basic operators of the problem and other plans. Three methods for construction of plans are sequencing, iteration, and recursion. The distinction between goals and preconditions is important. Goals are the conditions that the planner tries to achieve, whereas preconditions are used by the planner only for checking the applicability of a plan. Goals can be complex requirements like protecting certain conditions through the plan body, preventing certain others from becoming true, or maintaining seriality constraints on the actions of the plan. In the literature, two kinds of techniques have emerged: planning as theorem proving and planning as problem solving. The former approach, also known as the *deductive* approach [133], views planning as proving theorems about the goals (such as *there exists a sequence of actions applicable to the initial state that will achieve the goal state*). The *problem solving approach* [200] views planning as a search in the space of plans, wherein the choice points result from the large number of applicable operators.

The strategic level in a planning system tackles the problem of metaplanning [221]. It was evident from the failure of early planners to solve some problems in conjunctive subgoals (e.g., see Sussman's anomaly [224]) that planners should avoid overcommitment. Thus, one commonly found heuristic at the meta-level is the *strategy of least commitment*. This and similar heuristics can be used to limit the plan construction search space. The basic operators available to the planning process include subgoaling (setting up new goals to complete a partial plan), temporal extension (editing old plans by extending them in either direction), and specialization (by making abstract actions more concrete, typically by instantiating some variable).

The following general learning scenarios in traditional planning have been identified by Collins [42]: (1) *caching and generalization*, so that there is no need to replan in similar situations in the future; (2) *repairing failed plans*, that is, detection, characterization, and removal of failure by plan transformation; (3) *apprentice learning*, that is, noticing the *unplanned* satisfaction of certain goals during planning, possibly leading to the discovery of some new plans; and (4) *learning from observation*, wherein the planner follows causal chains resulting from environmental events or another agent's actions, and memorizes those that achieve any of its goals.

The ability to plan is inherently related to the ability to detect and construct causal links. Therefore, the causal model of the planner is the immediate target of learning mechanisms. Learning to plan involves modification of causal mechanisms that lead to an erroneous action. It can be called *strategy learning in planning*, because it creates or modifies the pieces of a strategy for plan construction.

Learning to plan in dynamic domains is still an open problem. Time-varying parameters require that the planner have a persistence model of various conditions, because conditions previously achieved may be violated in the future. Detection of failures in planning, and their attribution to causes, is more complex due to the added responsibility of TCA. The problem of maintaining persistence models for planning has recently received a lot of attention [47, 82, 149].

Planning approaches provide an excellent framework for TCA by introducing *explicit* persistence models. However, the theories of reasoning with persistence models also need heuristics in order to get accurate projections. Several such metaheuristics are based on the least commitment principle. Some examples are discussed in the following:

1. *Persistence circumscription* [100]. The approach is to define an explicit termination predicate Clip, and to assert the persistence of a fact f as follows.

$$\text{Hold}(t, f) \;\rightarrow\; \text{Hold}(t + 1, f) \;\oplus\; \text{Clip}(t + 1, f)$$

The circumscription of Clip is then solved in the theory, which includes the causal rules, the persistence axiom stated above, and the chronicle of events known to have occurred. Among all the minimal models, the one that has the longest persistence for facts is preferred.

2. *Motivated actions* [149]. A motivated action is one that is in all models of a temporal reasoning theory instance. In this approach, the models with the fewest unmotivated actions are preferred. Heuristically, this approach assumes that no action occurs unless it *must*.

3. *Probabilistic Notions of Persistence* [47]. When reasoning about the persistence of fact P, let E_P be the set of events that makes P true, and E_{-P} the set that makes P false. Letting $f(t)$ equal the probability $p(< E_P, t >)$ and $g(t)$ the probability $p(< E_{-P}, t >)$, the persistence of P at time t can be defined as:

$$p(< P, t >) = \int_{-\infty}^{t} f(z) e^{-\lambda(t-z)} \left[1 - \int_{z}^{t} g(x)\,dx \right] dz$$

This model counts some events more than once. The probability of occurrence of an event increases exponentially after the occurrence of a supporting event, and decreases similarly after an interfering event.

4. *Evidential model of persistence* [47]. This model takes into account several factors influencing the truth of a fact P at time t. These include a *natural attrition factor*, such as exponential decay, and a *causal accretion factor*, such as effects due to any element of E_P or E_{-P}. Evidential reasoning uses the *maximum entropy principle* as a heuristic for deriving a least-commitment assignment of probabilities.

In short, planning systems explicitly represent their persistence and causal models. They use meta-level heuristic knowledge to reason with these models. It is not possible to directly apply this kind of reasoning to policy design problems, but the model is suitable for reactive planning problems. Because our survey has been limited to the study of planning systems for dynamic decision problems, we may have overlooked some otherwise important aspects of planning. An excellent overview of planning techniques is given by Georgeff [75].

Learning Apprentice Systems   These systems do not require explicit training. Instead, they just watch their users interact with a knowledge-based system. The apprentice embodies certain expectations at the meta-level of the problem. Learning opportunities arise when an expectation fails due to a discrepancy between what the apprentice expects and what the user does. Such systems are also called *lazy generalizers* [174].

Apprentice learning can be combined with knowledge-based inference mechanisms in several interesting ways. The PROTOS system employs an exemplar-based categorization technique in which specific examples are retained, and these guide difference-based learning in the face of discrepancies. Another system employing this technique has recently been proposed by Wilkins [247]. The ODYSSEUS system learns by trying to complete explanations for failure of expectations. In the process of explaining a failure, it discovers gaps in its knowledge. New knowledge is hypothesized in order to fill this gap. The validity of the new knowledge is established by a confirmation theory built into the system. The apprentice learner of BB1 [83] learns new scheduling heuristics when its preferences are overridden by a human expert.

All these systems are *discrepancy-driven*. They exhibit learning with asynchronous immediate feedback in knowledge-rich domains, and derive their power from rich, explicit knowledge of abstract control metarules. They represent perhaps the most pragmatic of all knowledge-based learning paradigms, in spite of their complete reliance on generic knowledge sources that monitor the application of heuristic knowledge.

6.4.4   *Analogy-Based Methods.*
Analogical methods rely on knowledge of solutions for problems already solved. As a generic paradigm, analogy is closely related to generalization. However, in the context of knowledge-based learning systems, analogy has come to acquire a rather specialized meaning.

The method is specifically interpreted as a mechanism for the recall of past solutions, based on the similarity between the new and the old problems, including transformation of old solutions for application in the new situation. An *analogical mapping* is derived by comparing the two situations and their respective contexts of application. Part of the old solution is transferred under this mapping to the new problem situation. Analogy-based learning is practical when there is sufficient previous experience in solving similar problems.

Analogy is a very general learning mechanism applicable to almost any problem solving and learning technique. Gaschnig [70] describes an example of *analytical* analogy, Winston [251] discusses empirical structural learning by analogy, Carbonell [32] discusses planning by analogy, Davies and Russell [45] describe a deductive approach to reasoning by analogy, and Greiner [80] proposes an analogy-based approach that can be easily combined with EBL.

6.4.5 Hybrid Methods. Recently, various researchers have attempted to reduce EBL's reliance on perfect domain theory. Pazzani [165] uses the similarity of explanations to induce generalized explanation schemas, thus devising a method for *explained* empirical classification. Flann and Dietterich have proposed another extension to standard EBL, an approach that can handle multiple examples. This approach is called IOE. It involves constructing individual explanations for each example, generalizing over explanations, pruning away dissimilar subproofs, and compiling the generalized explanation into a schema.

Star [220] has proposed a method of combining EBL, similarity-based learning, and decision theory. His method, called T-BIL (Theory-Based Inductive Learning), has been applied to a reactive planning problem in robotics. Feedback in this system is in the form of explanations. The technique can handle multiple examples: after the first time, all subsequent explanations are used for generating *observation reports*. These reports are then used to update the probabilities of causal rules using a Bayesian updating mechanism.

Mitchell [147] (who called this approach *analytical* learning) proposed a method for combining EBL with an empirical approach based on the version-space method. The version-space method maintains a partial description of a heuristic in the form of two propositional descriptions S and G, which correspond to the most specific and the most general instances in a propositional Boolean lattice of the set of examples seen so far. The S and G descriptions found by the empirical learning program LEX are analogous to the necessary and sufficient conditions for application of the heuristic, as derived by the analytical learning program LEX2. Thus, the conditions derived by LEX2 could be used as positive instances for generalizing S. The resulting hybrid method was applied to simplifying integration formulae.

The UNIMEM system [122] uses similarity-based learning to infer the plausibility of causal relationships before applying EBL. Danyluk [44] has also explored the idea of combining empirical and analytical learning techniques. His technique differs from UNIMEM in that structural descriptions are matched and that inexact matching is allowed.

The idea of applying EBL as a verification step has been proposed by Golding et al. [78]. Their system learns strategies from expert advice. It works by asking an expert for advice when it cannot proceed during a problem-solving episode. The advice is followed, and the steps leading to the solution are retained, verified, and generalized by chunking.

Other methods include the precondition analysis technique used by Silver [206] in his heuristic learning system LP, and a more general method suggested by Desimone [51]. Both methods are more general than SDBL because they analyze examples by reasoning about the purpose of each step, and less powerful than EBL because they consider only the preconditions of rules rather than their interactions [148]. Silver's approach can only learn a linear sequence of rules. Desimone extends his approach to nonlinear solution trees using dependency graphs. These methods have been demonstrated on high-school algebra problems.

Rajamoney [176, 177] uses a combination of EBL and learning by experimentation to refine an initially imperfect domain theory. His technique relies on a form of causal isolation known as factoring [71]. Controlled experiments are designed from the partial description of failed instance of strategy application. One of the causal mechanisms (or processes) is allowed to dominate each experiment. Factoring simplifies the design of the critic and the learning element. This technique is applicable to knowledge-based learning for control problems.

In Section 6.4.3, we have discussed the PRODIGY system of Minton et al. [142]. Not only does this system learn from positive examples of strategy application, it also learns from explanation of failure. This technique combines EBL with compression analysis [141], a knowledge compilation technique that performs a utility-guided search through a space of plausible explanations. It results in more effective explanations, whose operationalization results in more efficient strategies.

The methods discussed in this section demonstrate that no learning technique is sufficient by itself. While empirical learning techniques exploit syntactic and numeric similarity, explanation-based learning uses similarity of causal structure. Apprentice learning techniques are adept at patching knowledge gaps, and so are learning-by-experimentation methods. Knowledge compilation improves efficiency by priming patterns of inference, and planning techniques have the constructive element necessary for creation of new strategies. Effective learning must employ a hybrid of these methods. The spectrum of learning techniques in AI is the richest, most implementable, and also the most well researched of all domains. However, existing learning mechanisms must be combined in order to design more useful and more generally applicable techniques.

6.5   Connectionist Methods

Connectionist methods also address the problem of representation and acquisition of strategic metaknowledge, but the perspective here is different. The assumed underlying model of intelligence is a massively parallel problem-solving model. All computation emerges from the collective activity of a large number of simple and richly interconnected units. Such systems for problem solving and learning are variously known as *parallel distributed processing* (PDP) systems, *neural networks*, or *artificial neural systems* (ANSs). A comprehensive review of the paradigm can be found in the references [1, 196].

What are ANSs?   A connectionist network (Figure 13.5) can be viewed as an active data structure consisting of *units* interconnected by weighted *links*. Some designated units act as sensors, so that input to the system can be supplied by influencing the states of these units. Other units act as effectors, so that the state of these units can be used to direct actions. The remaining units (appropriately called *knowledge atoms* by Smolensky [218]) capture the relationships among sensors, between sensors and effectors, and among effectors. The global short-term state of the system is captured locally by the activation of each unit. The interconnections among units allow interaction between states. The long-term memory of the system lies in weights on the links between units. Learning in these systems occurs by the modification of these weights.

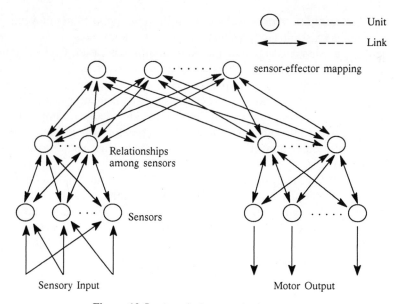

Figure 13.5   A typical connectionist network.

Connectionist Primitives   Connectionist primitive is a generic term for the various types of knowledge that can be represented using ANSs. The following primitives are discussed and used in this chapter:

1. *Pattern classifier*, which is capable of judging whether or not a pattern of inputs is a member of a class;
2. *Pattern associator*, which learns (possibly bidirectional) associations between input and output patterns;
3. *Auto-associator*, which is capable of learning how to complete partial patterns of input, based on the memory of patterns seen in the past; and
4. *Competitive activation*, which uses competing units with mutual inhibitory connections, to select the most strongly activated unit.

Learning in ANSs   A *connectionist learning paradigm* is a method for modifying weights (also called synaptic weights) in response to a pattern of changing activations. Among the several well-known paradigms for adjustment of weights with experience, we shall cite one for each kind of primitive unit discussed. Rumelhart et al. [197] have developed the *back-propagation* algorithm for learning in pattern classifiers; Kosko [111] has developed the Adaptive Bidirectional Associative Memory (ABAM) framework for learning pattern associations; Kohonen [105] has developed the Learning Vector Quantization (LVQ) framework for self-organizing auto-associative memories; and Grossberg [34] has developed the Adaptive Resonance Theory (ART2) framework for competitive learning.

ANSs for Strategy Learning   Although only a few applications of ANSs to strategy learning exist, this approach has some distinct advantages over others. In particular, decision making under partial information is easier. Connectionist networks have spontaneous generalization that results in a transfer of expertise to problems with similar values for parameters of representation. They are ideal for storing nonlinear, time-varying associations between situations and actions. The disadvantage, on the other hand, is that all knowledge is implicit and cannot be translated into other representations. Therefore, there is a tradeoff between flexibility and performance between connectionist and conventional representations.

Anderson's thesis [11] presents a comprehensive treatment of search problems using connectionist neural networks. Two separate networks are used: one for learning the state-action mapping, and the other for learning the state-evaluation function. A partial solution to the TCA problem is presented in the form of a feed-forward connection between the input layer and the state-evaluation layer. This allows the system to plan at least one move ahead of its output. The pole-balancing problem [12] is used to demonstrate the applicability of modular neural networks to complex decision problems. The system learns discriminatory features of parameter space that lead to useful representations of inputs from the perspective of strategy selection.

There are some important points to be learned from this approach. It demonstrates the feasibility of neural networks for learning state-action mappings, and more importantly, for learning continuous-valued evaluation functions using a fixed set of parameters. This means that for certain strategy-learning problems in which numeric parameters abound and the evaluation function depends in a complex way on these parameters, it might be worthwhile to sacrifice explicit knowledge of problem objectives for greater ability to learn. For problems with ill-defined objective functions, the objective function can be acquired during the course of learning through external evaluation of certain result states. Connectionist systems might be the only resort for such problems, because none of the other systems work without an explicitly stated goal.

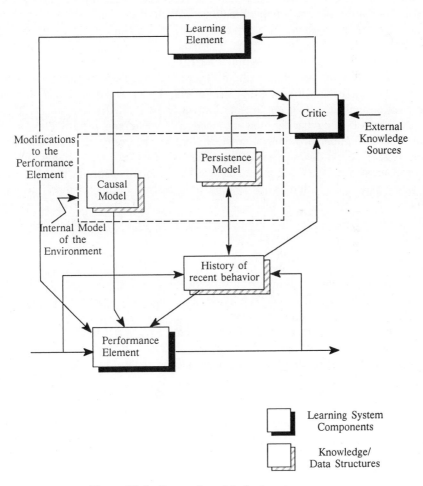

Figure 13.6 Proposed model of a learning system.

7 A PROPOSED MODEL OF LEARNING SYSTEMS

In this section we describe a learning model that solves complex problems in temporal credit assignment using explicit knowledge (see Figure 13.6). Instead of using the same network for storing past decisions, heuristics, and eligibilities of decisions, our model works with an explicit memory containing tuples that have the following abstract form:

$$< \text{problem instance, decision} >$$

The system is trained incrementally, in stages. In the first stage, a model of the reinforcer's objective function is acquired by repeatedly observing the reinforcement resulting from a single controlled action, and learning an association between the effects of that action and the feedback.* Learning in this stage involves relating an abstract specification of objectives with some observable parameters of the system. In a typical dynamic decision problem, there are so many observable parameters that (1) it is impossible to represent all the associations using just one network, and (2) it is not feasible to vary them simultaneously in a controlled fashion. Therefore, this stage is further divided into substages, each corresponding to the learning of a partial description in terms of a small group of parameters known to be mutually dependent. This is one way of using background knowledge in designing pattern classifiers with a large number of input parameters—letting the connectivity of the network reflect the top-down expectation of high covariance among related inputs. Several such partial associations are used to incrementally estimate the structure of a complete association. This scheme is appealing because it parallels the general pattern of evolution of complex, hierarchical systems [207]. Learning stays within resource and time bounds, and knowledge of complex associations can be accumulated incrementally.

In fact, this principle of layered information compression (term due to Rendell [181]) is also applied to the overall training phase. The remaining bodies of knowledge are acquired in subsequent stages. Each stage of learning a complex association is further subdivided by recursively applying the layering principle. The causal and persistence models are learned separately using controlled experimentation on the environment. The causal model is acquired first by varying the environmental control variables and performing random actions, one at a time, and acquiring associations of the form:

$$\text{Actions} \times \text{Perceived World} \rightarrow \text{Hypothesized World}.$$

The next stage in learning is to acquire the persistence model. Once again the control variables of the environment are varied over trials. First the envi-

---

* This approach can be replaced by the Sutton-Barto model [23] for acquiring the weights of the adaptive critic element, which, with small modifications, can be used to acquire the objective function by performing multiple actions at a time. However, the heuristics implicit in that model will cause problems in learning.

ronment is allowed to reach quiescence following a randomly chosen action. The action is performed and the change in the effect, as well as its average decay rate following the action, are observed. These are used to update the *causal accretion* and *natural attrition* components of the persistence model. The *spontaneous causation* component is acquired by probabilistically choosing an event, letting it occur, and observing the change in the particular effect under study. This stage results in the acquisition of associations having the form:

$$\text{Events} \times \text{Hypothesized World} \to \text{Hypothesized World},$$

$$\text{Hypothesized World} \times \text{Time} \to \text{Hypothesized World},$$

$$\text{Actions} \times \text{Hypothesized World} \to \text{Hypothesized World}.$$

The final part of the training phase involves experiments for learning search control heuristics. In this phase, the acquired objective function, the causal model, and the persistence model all remain fixed. External feedback can be supplied by comparing the heuristic performance against the best nonheuristic performance. Alternatively, some of the feedback can be generated internally using secondary reinforcement-learning algorithms. The feedback is apportioned, first temporally according to weights in the persistence model, and then structurally according to weights in the causal model. At the end of this phase, the performance element is also trained.

We have now begun to operationalize (to borrow a term from knowledge-based learning) our knowledge-level specification of the model. In this section, we have shown how the problem of learning can be reduced to a problem of acquiring complex associations. We have also proposed a *generic* layered information compression paradigm for training. Our description is still coarser than an implementation-level specification. This discussion, therefore, is applicable to almost all the implementation models of learning systems discussed in Section 6. In the following, we propose a connectionist implementation of our model, for solving a representative dynamic decision problem.

ANS-Based Implementation of the Proposed Model The implementation schematic shown in Figure 13.7 shows how the variety of associations that need to be acquired may be represented using connectionist primitives. Figure 13.8 shows the connectionist implementation of the critic. It differs from the secondary reinforcement-generation mechanism of Minskian models in that it has hooks for using explicitly represented causal and persistence models. It is different from the classical model because of its predictive reinforcement mechanism and the flexible structure of its episodic memory.

The performance element is implemented using a pattern classifier network. A competitive network is used for selecting an action given a problem instance. The causal model is realized in an ABAM (adaptive bidirectional associative memory). The persistence model and the memory of recent decisions are imple-

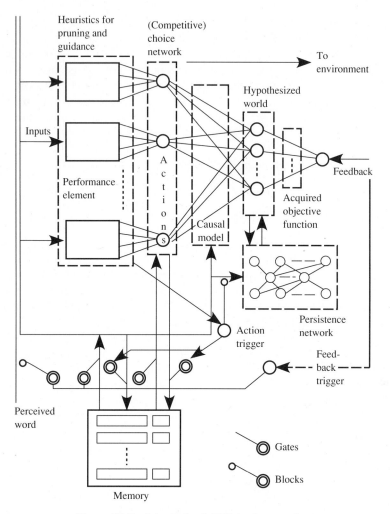

Figure 13.7 Schematic of ANS implementation.

mented using auto-associators. Numerous triggers and gates help synchronize the operation of the ANS ensemble.

Learning Strategies for Load-Balancing We view the load balancing problem as a search problem. Let migrate (t, i, j) denote the operator that migrates a task t from processor i to processor j. A balancing decision can then be seen as mapping the set of operators migrate $(t, i, \text{neighbor}(i))$ to processor i. The system makes migration decisions based on local workload information and task parameters.

The learning system for workload characterization and workload distribution is based on connectionist learning systems [89]. These have been successfully demonstrated to learn complex associations without being given any

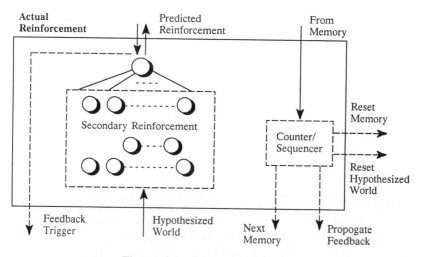

Figure 13.8 Schematic of the critic.

knowledge of the parametric form of the association. They construct abstract features of the input parameter space in their internal layers (also called hidden units). Learning and problem solving can be easily integrated in these systems, because learning proceeds by modification of weights in the problem solving component. It has been shown [23] that reinforcement-learning systems implemented as connectionist networks can withstand delays in feedback. Finally, the connectionist model of problem solving is inherently asynchronous and, once trained, can provide quick answers by parallel examination of various attributes making up the state.

The Learning System Two kinds of learning are addressed. The *search space learner* is a learning component implemented separately at each site in the distributed system that learns the workload characteristics of that site. These characteristics can be nonlinear functions of the available primitive measurements and inputs. The acquired workload parameters define an abstract search space. The *search heuristics learner* learns the relationship between workload information and the corresponding load balancing decisions. This component is essential because some of the attributes in the abstract search space are not directly available and their relationship with decision making changes dynamically.

The system operates in three distinct phases: (1) training for workload characterization, (2) training for workload distribution heuristics, and (3) application. Phases (1) and (2) involve learning.

Controlled experimentation is used for learning in phase (1). The response time of the system is observed for a wide variety of controlled tasks. For each experiment, the task remains fixed. Learning is further simplified by only learning a formula for the *relative* change in response time, rather than learning the response time function itself.

Learning in phase (2) is more complicated because a history of past actions must be maintained. Moreover, explicit justification must be maintained for each action. When external feedback is provided, it is distributed temporally and structurally. The information needed for doing credit assignment depends critically on the nature of feedback.

In our system, feedback is generated by comparing the improvement in response time for various decisions. The information for structural credit assignment is stored in the *causal model,* and that for temporal credit assignment in the *persistence model.* During training, the weights in the causal model are adjusted to reflect relationships between load balancing decisions and workloads. This is achieved by running controlled experiments—both the task and the load balancing action are control variables. The weights in the persistence model are acquired by controlled experimentation too. They represent the relationship between workload information and load balancing decisions on one side, and the time for which the effects of balancing persist on the other.

Controlled generation of problem situations in load balancing gives rise to other problems. One such problem is that of synthetic workload generation. We have developed synthetic generators for various computational resources—CPU, disk, and memory. Using these and an intelligent sampling technique, the learning program can sample the critical portions of the input parameter space.

When the inputs are gated into the heuristic networks, many action units may simultaneously try to go on. In this case, competition among these units results in the selection of some unit with strong backing from guidance heuristics and weak opposition from pruning heuristics. This action is used to update the persistence of effects it is known to cause. After this, if no action occurs, then the natural attrition component of the persistence model takes over. If we allow asynchronous external events to influence persistences, then a sudden change in environment, measured as a differential between the current state of the perceived world and the state immediately preceding, can also change the persistence of certain effects.

As soon as the competition between actions is resolved, the action trigger goes on, and the state of the perceived world and the chosen action are gated into the memory. This process continues until an asynchronous external feedback signal becomes available. The learner shuts its perception and focuses on learning rather than acting. The feedback is, meanwhile, backed up to the individual effects of the hypothesized world via the objective function model, where it stays till the end of the learning phase.

The learning phase occurs in cycles of recall and modification. The recall phase involves selecting a memory instance and gating the memorized inputs to the input lines. The network corresponding to the memorized action is selected for update. The persistence-weighted feedback is propagated back via the causal model of that action. The forward propagation of perceived inputs through the heuristic mechanism (but not the competition mechanism) is followed by a pass of the standard back-propagation method. This allows the feedback to be distributed among the weights of the heuristic mechanism. This cycle repeats until all the memories have been exhausted or the time runs out for the learner.*

During phase (3), the weights in the action selection network are used for deciding the next action. The search for the actions proceeds asynchronously and concurrently as jobs originate. Conflicts are resolved by winner-take-all [64] networks. All the weights (in the causal and persistence models, and in the action selection network) remain fixed during this phase.

Load balancing is a difficult problem in parallel search. The problem relies greatly on statistical prediction of patterns inherent in the environment. However, any predictive solution in a stochastic environment is at best approximate. Formal proofs and determination of the accuracy of approximation will be carried out in the future.

The unique feature of our system as compared to past solutions on workload characterization and distribution is that strategies are not built in. This allows the system to adapt to installation-specific and architecture-dependent parameters of workload—an activity traditionally performed by human experts. The systems reported so far, such as Engineering Computer Network [94, 95], MACH [20], Locus [235], and University of California, Los Angeles's Locus [236] and System V, depend on user-defined workload metrics for making load balancing decisions. This is not only nonadaptive, but also not portable to systems of different configuration.

We have suggested an adaptive approach to designing a system for learning load balancing strategies. The top-down development illustrates our design methodology for strategy-learning systems. The basic steps can be summarized as follows.

1. Analysis of the *knowledge-level requirements* of the learning task;
2. Study of the *knowledge-level functionality* in available learning systems;
3. Systematic mapping of various bodies of knowledge to suitable learning mechanisms;
4. Principled design of a training scheme using layered information compression.

8 CONCLUSION

In this section, we summarize our opinions on various issues pertaining to the architecture of strategy-learning systems. These conclusions are based upon an extensive survey of architectures, covering diverse domains and a variety

---

* It is possible to be more intelligent about the whole learning process if we also maintain, in the history, persistence of each action. This value should be set to its maximum value (say, 1) at the time of gating an action into the memory. Each time one of the original effects of an action is updated, the persistence should be decremented by an amount proportional to the magnitude of the change.

of techniques. Perhaps the most important issues we have identified are the effects of delayed feedback and ill-defined objectives on the design of learning systems. Such learning situations are inherent to a large class of problems called dynamic decision problems. We have identified the bodies of knowledge involved in intelligent credit assignment for such problems and illustrated how such knowledge may be acquired automatically. We have attempted to advance the field of machine learning by indicating a general methodology for learning to acquire strategies in complex domains under difficult learning situations.

We end this chapter with observations central to the architecture of strategy learning systems for complex domains.

Learning in Complex Domains   Perhaps the most important variable in real-world domains is time; it also happens to be the most neglected variable in research on learning strategies for solving problems. Considerations of delay had recently led to significant changes in learning strategy, as exhibited by Sutton's work. However, even though a few learning heuristics (such as recency and frequency) had been proposed and included in a learning algorithm, the *knowledge* for designing such heuristics for other complex problems was not explicitly available. The knowledge involved is analogous to persistence axioms being used in systems for temporal planning. This observation is being used to improve the design of learning systems by reducing the bias. This leads to systems that represent this knowledge explicitly and attempt to acquire it automatically for specific domains and/or learning environments. It also suggests a technique for implementation of temporal planning using connectionist systems similar to the ASE/ACE system studied by Barto et al. [11, 23, 227].

A Second Opinion on the Design of ANS Learning Algorithms   The trend in the design of learning algorithms for connectionist systems has been to keep them simple and local. While these goals are the crux of the argument for connectionism, they have been misinterpreted by recent research as an argument for doing all the learning in their systems through intelligent, heuristic, local, and simple algorithms. This should remind us of the classic heuristic problem solver of Slagle [214] in the beginning days of AI. The next stage in AI was its most progressive one; and the one characteristic that separated the AI systems of this age from those of its early days was the *separation of knowledge from the algorithmically implemented inference component*. Is it not obvious that if connectionist systems are to implement anything like a general purpose learning substrate, they should explicitly represent as much knowledge as possible? The human learning system, which, for now, continues to guide these developments, works in slow learning environments, in the presence of distractions, in the presence of varying amounts of delay, and in situations where goals (objectives) are set dynamically in response to asynchronous, sporadic reinforcements. Realizing that any heuristic component of learning algorithms that is associated with a dynamically varying feature of the learning environment must be flexible and available for introspective modification, we can immediately form an argument against all the learning being done algorithmically.

Our model suggests how connectionist systems may continue to use simple algorithms and still remain sufficiently flexible to be able to learn in complex environments. The major idea of our methodology is to translate complex learning into phases of knowledge acquisition. So, what should be the next stage in the design of intelligent connectionist architectures? Our answer would be: the development of *modular knowledge-level architectures* for learning. Systems employing heuristic learning algorithms are important in their own right because they establish the need for certain kinds of knowledge. However, from an architectural point of view, they only illustrate the need for an additional *module* rather than a change in a *learning algorithm*. General learning in complex situations is achievable not through more complex learning algorithms, but through more advanced interactions among dynamically acquired bodies of knowledge.

If some bodies of knowledge cannot be acquired or represented by currently existing ANS primitives, that should motivate the search for new learning algorithms and primitives. Examples of such primitives have been set forth by Touretzky et al. [230] in the domain of connectionist rule-following systems [229]. It is our conjecture that future connectionist architectures will be modular, motivated by a knowledge-level analysis of the learning scenario, and that the search for new algorithmically implemented connectionist learning procedures will only be motivated by the lack of a connectionist substrate to implement one or more types of knowledge modules. This conjecture is consistent with the fundamental tenets of connectionism as well as AI.

Reassessment of the Design of Multiprocessor Operating Systems Our experience with distributed systems in the recent past has indicated a need for automation in the management of these systems. Traditionally, complex problems in the design of multiprocessor operating systems have been solved algorithmically. Many of these problems fit into the framework of dynamic decision making. The problem of learning a load balancing strategy is representative of this class. Adaptiveness in the design of these systems is an essential step towards complete automation. It is quite plausible to think of a neural network on every processor making complex scheduling or routing decisions in response to the changing operating environment. The advent of neural networks opens exciting avenues for designing faster, more efficient multiprocessors.

The research reported here can be viewed as a marriage between learning and architecture. New developments in architecture that might lead to fast, general implementation of connectionist primitives can be used for efficient, low-overhead learning, and new developments in the field of learning can lead to the design of intelligent, efficient multiprocessor architectures that will be spending less time controlling themselves. This view has begun to be shared by other researchers in the field [68, 121, 163].

Relationship with Psychological Models of Learning We have taken an engineering approach to the design of learning systems. Psychological models, on the other hand, describe how human beings behave and learn. The complex human learning system and its methodology can guide the engineering approach. Thus, one direction in which we hope that our survey of cognitive models will benefit learning research is clear. In the opposite direction, design issues in the architecture of learning systems might shape research in psychology by providing a model and vernacular for the description of human learning.

Learning under delayed reinforcement is not a well-understood topic in psychology. Indeed, as we have indicated earlier (Section 6.1), some psychological evidence points against the incidence of such learning during simple controlled episodes of solving simple problems. However, there are complex learning phenomena that occur over prolonged periods of time and cannot be tested in a laboratory setting. Indeed, Anzai's model for skill acquisition [18] includes delay models that have not been understood and explained in a problem-independent way.

In several psychological models there is an implicit assumption that learning occurs when the learner is *prepared* to learn. This preparedness has not been characterized well. It is our hypothesis that the acquisition of causal models and persistence models sets the stage for learning of problem-solving heuristics. It is unfortunate that because of the lack of proper terminology, Anzai [18] includes both kinds of knowledge in what he calls causal knowledge. We hope that our model will help further the productive exchange of ideas in these two disciplines by providing a common vocabulary.

Issues Not Addressed Several issues need to be resolved before general purpose connectionist systems can be constructed. Among the most important issues are scaling (How well does an architecture scale up to larger and more complex problems?) and generalization (How well does the system perform on problems it has not been trained on?). These issues are the subject of ongoing research [6]. Another issue is the design of network architecture. It is obvious that there is design bias implicit in the choice of a network architecture and size. Also, the choice of an encoding for continuous-valued or integer-valued parameters leads to representational bias. The choice of a learning mechanism that incorporates more than just the internal representation used by the performance element gives rise to algorithmic bias.

We have not identified any new connectionist primitives, although, as indicated by Touretzky et al., the existing set might be inadequate for modeling all intelligent behavior. Yet another topic that needs more attention is the design of a serializing component (*scheduler*, one might say) in order to map the large number of simultaneously active bodies of knowledge onto a constrained architecture, and to achieve this mapping and management at a small overhead. Finally, the symbolic interpretation of the activities of an ANS remains an open question, and more work is needed on that front.

9 ACKNOWLEDGMENTS

We express our thanks to the members of the staff at the RII division of NASA Ames Research Center for supporting our research activities in the summer of 1987 and 1988. Thanks are also due to Dr. Katherine Baumgartner for sharing the results of her extensive study on the resource allocation problems in distributed systems, and for providing the skeleton for the load balancing testbed; and to Albert Yu for interesting discussions leading to the formulation of dynamic decision problems. This research was supported by the National Science Foundation under grant MIP-88-10598, and by the National Aeronautics and Space Administration under grant NCC 2-481. This chapter has benefitted from critical comments of Mark Gooley, Larry Rendell, Chris Matheus, Peter Haddawy, Carl Kadie, and Munidar Singh.

REFERENCES

1. *DARPA Neural Network Study*, Lincoln Laboratory, Massachusetts Institute of Technology, Lexington, MA, July 1988.

2. R. P. Abelson and A. Levi, "Decision Making and Decision Theory," *Handbook of Social Psychology*, 1983.

3. D. H. Ackley, G. E. Hinton, and T. J. Sejnowski, "A Learning Algorithm for Boltzmann Machines," *Cognitive Science*, vol. 9, pp. 147–169, 1985.

4. D. H. Ackley, *A Connectionist Machine for Genetic Hillclimbing*, Kluwer Academic Publishers, Boston, 1987.

5. D. H. Ackley, "Reinforcement Learning with Back Propagation," *Abstracts of Neural Networks for Computing*, April 1988.

6. S. Ahmad, "Scaling and Generalization in Neural Networks: A Case Study," *Proc. 1988 Connectionist Models Summer School*, eds. D. Touretzky, G. E. Hinton, and T. J. Sejnowski, Morgan Kaufmann, Palo Alto, CA, 1988.

7. R. Alonso, "The Design of Load Balancing Strategies for Distributed Systems," *Future Directions in Computer Architecture and Software Workshop*, pp. 1–6, Seabrook Island, SC, May 5–7, 1986.

8. S. Amarel, J. S. Brown, B. G. Buchanan, P. Hart, C. Kulikowski, W. Martin, and H. Pople, "Report of Panel on Applications of Artificial Intelligence," *Proc. IJCAI-77*, pp. 994–1006, 1977.

9. S. Amarel, "On Representations of Problems of Reasoning about Actions," *Readings in Artificial Intelligence*, eds. B. L. Webber and N. Nilsson, pp. 2–22, Morgan Kaufmann, Los Altos, CA, 1981.

10. S. Amarel, "Program Synthesis as a Theory Formation Task: Problem Representations and Solution Methods," *Machine Learning: An Artificial Intelligence Approach*, eds. R. S. Michalski, J. G. Carbonell, and T. M. Mitchell, Morgan Kaufmann, Los Altos, CA, 1986.

11. C. W. Anderson, "Learning and Problem Solving with Multilayer Connectionist Systems," Ph.D. thesis, University of Massachusetts, Amherst, MA, 1986.

12. C. W. Anderson, "Strategy Learning with Multilayer Connectionist Representations," *Proc. 4th Int'l Workshop on Machine Learning*, pp. 103–114, June 1987.

13. J. R. Anderson, J. G. Greeno, P. J. Kline, and D. M. Neves, "Acquisition of Problem-Solving Skill," *Cognitive Skills and Their Acquisition*, ed. J. R. Anderson, Lawrence Erlbaum Associates, 1981.

14. J. R. Anderson, "Knowledge Compilation: The General Learning Mechanism," *Machine Learning: An Artificial Intelligence Approach*, eds. R. S. Michalski, J. G. Carbonell, and T. M. Mitchell, Morgan Kaufmann, Los Altos, CA, 1986.

15. J. R. Anderson, "Skill-Acquisition: Compilation of Weak-Method Problem Solutions," *Psychological Review*, vol. 94, pp. 192–210, 1987.

16. P. M Andrae, "Constraint Limited Generalization: Acquiring Procedures from Examples," *Proc. Nat'l Conf. on Artificial Intelligence*, pp. 6–10, 1984.

17. Y. Anzai and H. Simon, "The Theory of Learning by Doing," *Psychological Review*, vol. 36, pp. 124–140, 1979.

18. Y. Anzai, "Doing, Understanding, and Learning in Problem Solving," *Production System Models of Learning and Development*, ed. Klahr et al., MIT Press, 1987.

19. A. A. Araya, "Learning Problem Classes by Means of Experimentation and Generalization," *Proc. Nat'l Conf. on Artificial Intelligence*, pp. 11–15, 1984.

20. R. V. Baron et al., *MACH Kernel Interface Manual*, Pittsburgh, PA, January 1987.

21. A. Barr and E. A. Feigenbaum, William Kaufmann, Los Altos, CA, 1981, 1982.

22. A. G. Barto and R. S. Sutton, "Landmark Learning: An Illustration of Associative Search," *Biological Cybernetics*, vol. 42, pp. 1–8, 1981.

23. A. G. Barto, R. S. Sutton, and C. W. Anderson, "Neuronlike Adaptive Elements that Can Solve Difficult Learning Control Problems," *IEEE Trans. on Systems, Man, and Cybernetics*, vol. SMC-13, pp. 834–846, 1983.

24. K. M. Baumgartner and B. W. Wah, "Load Balancing Protocols on a Local Computer System with a Multiaccess Bus," *Proc. Int'l Conf. on Parallel Processing*, pp. 851–858, University Park, PA, August 1987.

25. K. M. Baumgartner, *Resource Allocation on Distributed Computer Systems*, West Lafayette, IN, May 1988.

26. L. R. Beach and T. R. Mitchell, "A Contingency Model for the Selection of Decision Strategies," *Academy of Management Review*, vol. 3, pp. 439–449, 1978.

27. H. J. Berliner, "Some Necessary Conditions for a Master Chess Program," *Proc. 3rd Int'l Joint Conf. on Artificial Intelligence*, pp. 73–85, 1973.

28. P. Bock, "The Emergence of Artificial Intelligence: Learning to Learn," *AI Magazine*, pp. 180–190, Fall 1985.

29. L. B. Booker, D. E. Goldberg, and J. H. Holland, "Classifier Systems and Genetic Algorithms," Technical Report 8, Cognitive Science and Machine Intelligence Laboratory, University of Michigan, April 1987.

30. L. Breiman, J. H. Friedman, R. A. Olshen, and C. J. Stone, *Classification and Regression Trees*, Wadsworth International Group, Belmont, CA, 1984.

31. B. G. Buchanan and T. M. Mitchell, "Model-Directed Learning of Production Rules," *Pattern-Directed Inference Systems*, eds. D. A. Waterman and F. Hayes-Roth, Academic Press, 1978.

32. J. G. Carbonell, "Experiential Learning in Analogical Problem Solving," *Proc. Nat'l Conf. on Artificial Intelligence*, pp. 168–171, 1982.

33. J. G. Carbonell and Y. Gil, "Learning by Experimentation," *Proc. 4th Int'l Machine Learning Workshop*, pp. 256–266, 1987.

34. G. A. Carpenter and S. Grossberg, "ART 2: Self-organization of Stable Category Recognition Codes for Analog Input Patterns," *Applied Optics*, vol. 26, pp. 4919–4930, December 1987.

35. A. C. Catania, *Learning*, Prentice Hall, Englewood Cliffs, NJ, 1979.

36. S. A. Chien, "Extending Explanation-Based Learning: Failure Driven Schema Refinement," Technical Report No. UILU-ENG-87-2203, University of Illinois, 1987.

37. T. C. Chow and J. A. Abraham, "Load Balancing in Distributed Systems," *Trans. on Software Engineering*, vol. SE-8, pp. 401–412, July 1982.

38. Y. C. Chow and W. Kohler, "Models for Dynamic Load Balancing in a Heterogeneous Multiple Processor System," *Trans. on Computers*, vol. C-28, pp. 334–361, May 1979.

39. J. Christensen and R. E. Korf, "A Unified Theory of Heuristic Evaluation Functions and Its Application to Learning," *Proc. 5th Nat'l Conf. on Artificial Intelligence AAAI-86*, pp. 148–152, 1986.

40. W. Clancey, "Heuristic Classification," in *Artificial Intelligence*, vol. 27, pp. 289–350, Amsterdam, 1985.

41. W. J. Clancey, "Classification Problem Solving," STAN-CS-84-1018, Stanford University, CA, July 1984.

42. G. C. Collins, "Plan Creation: Using Strategies as Blueprints," Ph.D. Thesis, Yale University, 1987.

43. M. E. Connell and P. E. Utgoff, "Learning to Control a Dynamic Physical System," *Proc. 6th Nat'l Conf. on Artificial Intelligence*, pp. 456–60, Seattle, WA, June 1987.

44. A. P. Danyluk, "The Use of Explanations for Similarity-Based Learning," *Proc. 10th Int'l Joint Conf. on Artificial Intelligence*, pp. 274–276, Milan, Italy, 1987.

45. T. R. Davies and S. J. Russell, "A Logical Approach to Reasoning by Analogy," *Proc. 10th Int'l Joint Conf. on Artificial Intelligence*, pp. 264–270, 1987.

46. R. Davis and R. B. Lenat, *Knowledge-Based Systems in Artificial Intelligence*, McGraw-Hill, New York, 1982.

47. T. Dean and K. Kanazawa, "Probabilistic Temporal Reasoning," *Proc. Nat'l Conf. on Artificial Intelligence AAAI-88*, pp. 524–528, 1988.

48. R. Dechter and J. Pearl, "Network-Based Heuristics for Constraint-Satisfaction Problems," *Artificial Intelligence*, vol. 34, pp. 1–38, 1988.

49. G. F. DeJong and R. J. Mooney, "Explanation-Based Learning: An Alternative View," *Machine Learning*, vol. 1, pp. 145–176, 1986.

50. M. Derthick, "Counterfactual Reasoning with Direct Models," *Proc. Nat'l Conf. on Artificial Intelligence AAAI-87*, pp. 346–351, July 1987.

51. R. Desimone, "Learning Control Knowledge within an Explanation-Based Learning Framework," *Progress in Machine Learning*, eds. I. Bratko and N. Lavrac, pp. 107–119, Sigma Press, Cheshire, U.K., 1987.

52. T. G. Dietterich and B. G. Buchanan, "The Role of Critic in Learning Systems," Technical Report No. STAN-CS—81-891, Stanford University, CA, December 1981.

53. T. G. Dietterich and J. S. Bennett, "The Test Incorporation Theory of Problem Solving," *Proc. Workshop on Knowledge Compilation*, Oregon State University, Department of Computer Science, pp. 145–159, September 1986.

54. T. G. Dietterich, "Learning at the Knowledge Level," *Machine Learning*, vol. 1, pp. 287–316, Boston, 1986.

55. H. J. Einhorn, "Learning from Experience and Suboptimal Rules in Decision Making," *Cognitive Processes in Choice and Decision Behavior*, ed. T. Wallsten, Erlbaum, Hillsdale, NJ, 1980.

56. H. J. Einhorn and R. M. Hogarth, "Behavioral Decision Theory: Processes of Judgment and Choice," *Annual Review of Psychology*, vol. 32, pp. 53–88, 1981.

57. G. Ernst and A. Newell, *GPS: A Case Study in Generality and Problem Solving*, Academic Press, 1969.

58. G. W. Ernst and M. M. Goldstein, "Mechanical Discovery of Classes of Problem-Solving Strategies," *J. of the ACM*, vol. 29, pp. 1–23, January 1982.

59. S. E. Fahlman, G. E. Hinton, and T. J. Sejnowski, "Massively Parallel Architectures for AI: NETL, Thistle, and Boltzmann Machines," *Proc. Nat'l Conf. on Artificial Intelligence*, pp. 109–113, 1983.

60. B. Falkenhainer, "The Utility of Difference-Based Reasoning," *Proc. Nat'l Conf. on Artificial Intelligence AAAI-88*, Ann Arbor, Michigan, 1988.

61. E. A. Feigenbaum, "The Art of Artificial Intelligence–Themes and Case Studies of Knowledge Engineering," *Proc. 5th Int'l Joint Conf. on Artificial Intelligence*, pp. 1014–1029, Los Altos, CA, August 1977.

62. J. A. Feldman and Y. Yakimovsky, "Decision Theory and Artificial Intelligence: I. A Semantics-Based Region Analyzer," *Artificial Intelligence*, vol. 5, pp. 349–371, 1974.

63. J. A. Feldman and R. F. Sproull, "Decision Theory and Artificial Intelligence II: The Hungry Monkey," *Cognitive Science*, vol. 1, pp. 158–192, Norwood, NJ, 1977.

64. J. A. Feldman, "Dynamic Connections in Neural Networks," *Biological Cybernetics*, vol. 46, pp. 27–39, 1982.

65. R. E. Fikes, P. E. Hart, and N. J. Nilsson, "Learning and Executing Generalized Robot Plans," *Readings in Artificial Intelligence*, eds. B. L. Webber and N. Nilsson, pp. 231–249, Morgan Kaufmann, 1981.

66. N. S. Flann, "Improving Problem Solving Performance by Example Guided Reformulation of Knowledge," *Proc. 1st Workshop on Change of Representation*, 1988.

67. L. M. Fu and B. G. Buchanan, "Enhancing Performance of Expert Systems by Automated Discovery of Metarules," Technical Report No. HPP-84-38, Stanford University, CA, 1984.

68. S. Fujita, "Self-Organization in Distributed Operating System," *Abstracts 1st Annual INNS Meeting, (Neural Networks)*, vol. 1 (supplement 1), p. 93, 1988.

69. M. R. Garey and D. S. Johnson, *Computer and Intractability: A Guide to the Theory of NP-completeness*, San Francisco, CA, 1979.

70. J. Gaschnig, "A Problem Similarity Approach to Devising Heuristics: First Results," *Readings in Artificial Intelligence*, eds. B. L. Webber and N. Nilsson. pp. 21–29, Morgan Kaufmann, Los Altos, CA, 1981.

71. M. R. Genesereth and N. J. Nilsson, *Logical Foundations of Artificial Intelligence*, Morgan Kaufmann, Los Altos, CA, 1987.

72. M. P. Georgeff, "Search Methods Using Heuristic Strategies," *Proc. Int'l Joint Conf. on Artificial Intelligence IJCAI-81*, pp. 563–568, 1981.

73. M. P. Georgeff, "Strategies in Heuristic Search," *Artificial Intelligence*, vol. 20, pp. 393–425, 1983.

74. M. P. Georgeff and A. L. Lansky, "Reactive Reasoning and Planning," *Proc. 6th Nat'l Conf. on Artificial Intelligence*, pp. 677–82, Seattle, WA, June 1987.

75. M. P. Georgeff, "Planning," *Annual Review of Computer Science*, pp. 359–400, Annual Reviews Inc., Palo Alto, CA, 1987.

76. A. Ginsberg, "Representation and Problem Solving: Theoretical Foundations," Technical Report No. CBM-TR-141, Rutgers University, October 1984.

77. R. M. Golden, "A Unified Framework for Connectionist Systems," *Biological Cybernetics*, vol. 58, no. 2, p. 109, 1988.

78. A. Golding, P. S. Rosenbloom, and J. E. Laird, "Learning General Search Control Rules from Outside Guidance," *Proc. 10th Int'l Joint Conf. on Artificial Intelligence*, pp. 334–337, Milan, Italy, 1987.

79. J. Gordon and E. H. Shortliffe, "A Method for Managing Evidential Reasoning in a Hierarchical Hypothesis Space," *Artificial Intelligence*, vol. 26, pp. 323–357, 1985.

80. R. Greiner, "Learning by Understanding Analogies," Technical Report CSRI-188, University of Toronto, August 1986.

81. P. Haddawy, "A Variable Precision Logic Inference System Employing the Dempster-Shafer Uncertainty Calculus," Technical Report No. UIUCDCS-F-86-959, Department of Computer Science, University of Illinois, Urbana, IL, 1986.

82. S. Hanks, "Representing and Computing Temporally Scoped Beliefs," *Proc. Nat'l Conf. on Artificial Intelligence AAAI-88*, pp. 501–505, 1988.

83. B. Hayes-Roth and M. Hewett, "Learning Control Heuristics in BB1," Technical Report No. KSL 85-02, Knowledge Systems Lab., Stanford University, CA, January 1985.

84. B. Hayes-Roth, "A Blackboard Architecture for Control," *Artificial Intelligence*, vol. 26, pp. 251–321, July 1985.

85. F. Hayes-Roth, D. A. Waterman, and D. B. Lenat, *Building Expert Systems*, Addison-Wesley, Reading, MA, 1983.

86. L. Hendricks, H. Oppewal, and C. Vlek, "Relative Importance of Scenario Information versus Frequency Information in the Judgment of Risk," Technical Report, University of Gronigen, The Netherlands, 1987.

87. G. E. Hinton, "Distributed Representations," Technical Report No. 84-157, Department of Computer Science, Carnegie-Mellon University, October 1984.

88. G. E. Hinton, T. J. Sejnowski, and D. H. Ackley, *Boltzmann Machine: Constraint Satisfaction Network that Learns*, Carnegie-Mellon University, Pittsburgh, PA, 1984.

89. G. E. Hinton, "Connectionist Learning Procedures," *Artificial Intelligence*, vol. 40, pp. 185–234, 1989.

90. J. H. Holland and J. S. Reitman, "Cognitive Systems Based on Adaptive Algorithms," *Pattern Directed Inference Systems*, eds. D. A. Waterman and F. Hayes-Roth, Academic Press, 1978.

91. J. H. Holland, "Escaping Brittleness: The Possibilities of General-Purpose Learning Algorithms," *Machine Learning II*, eds. R. S. Michalski, J. G. Carbonell, and T. M. Mitchell. pp. 593–623, Morgan Kaufmann, 1986.

92. J. J. Hopfield and D. W. Tank, "Computing with Neural Circuits: A Model," *Science*, pp. 625–633, August 1986.

93. E. J. Horvitz, "Reasoning about Beliefs and Actions under Computational Resource Constraints," *Proc. 3rd AAAI Workshop on Uncertainty in Artificial Intelligence*, Seattle, WA, July 1987.

94. K. Hwang, B. W. Wah, and F. A. Briggs, "Engineering Computer Network (ECN): A Hardwired Network of UNIX Computer Systems," *Proc. Nat'l Computer Conf.*, pp. 191–201, May 1981.

95. K. Hwang, W. J. Croft, G. H. Goble, B. W. Wah, F. A. Briggs, W. R. Simmons, and C. L. Coates, "A UNIX-based Local Computer Network with Load Balancing," *Computer*, vol. 15, pp. 55–66, April 1982. Also in *Tutorial: Computer Architecture*, eds. D. D. Gajski, V. M. Milutinovic, H. J. Siegel, and B. P. Furht, pp. 541–552, IEEE Computer Society, 1987.

96. K. B. Irani and J. Cheng, "Subgoal Ordering and Goal Augmentation for Heuristic Problem Solving," *Proc. 10th Int'l Joint Conf. on Artificial Intelligence*, pp. 1018–1024, Milan, Italy, 1987.

97. E. J. Johnson and J. W. Payne, "Effort and Accuracy in Choice," *Management Science*, vol. 31, April 1985.

98. B. Julish, F. Klix, R. Klein, W. Krause, and F. Kukla, "Some Experimental Results Regarding the Effect of Different Representations in Human Problem Solving," *Human and Artificial Intelligence*, ed. F. Klix, North-Holland, 1979.

99. D. Kahneman and A. Tversky, "Choices, Values, and Frames," *American Psychologist*, vol. 39, pp. 341–350, 1984.

100. H. A. Kautz, "The Logic of Persistence," *Proc. Nat'l Conf. on Artificial Intelligence AAAI-86*, p. 401, 1986.

101. D. S. Kay and J. B. Black, "The Evolution of Knowledge Representations with Increasing Expertise in Using Systems," *Proc. Cognitive Science Society Annual Meeting*, University of California, Irvine, 1985.

102. R. M. Keller, "A Survey of Research in Strategy Acquisition," Technical Report DCS-TR-115, Department of Computer Science, Rutgers University, May 1982.

103. D. Kibler and B. W. Porter, "Episodic Learning," *Proc. Nat'l Conf. on Artificial Intelligence AAAI-83*, pp. 191–196, 1983.

104. A. H. Klopf, "Drive-Reinforcement Learning: A Real-Time Mechanism for Classical Conditioning," *Proc. ICNN*, pp. II-441–II-445, 1987.

105. T. Kohonen, *Self-Organization and Associative Memory*, 2nd ed., Springer-Verlag, 1988.

106. M. M. Kokar and W. W. Zadrozny, "A Logical Model of Machine Learning," *Proc. 1st Workshop on Change of Representation*, 1988.

107. K. G. Konolige, "Experimental Robot Psychology," Technical Report No. 363, SRI International, Menlo Park, CA, November 1985.

108. R. E. Korf, *Learning to Solve Problems by Searching for Macro-Operators*, Pitman, Boston, 1985.

109. R. E. Korf, "Macro-Operators: A Weak Method for Learning," *Artificial Intelligence*, vol. 26, pp. 35–77, 1985.

110. R. E. Korf, "Heuristics as Invariants and its Application to Learning," *Machine Learning: A Guide to Current Research*, pp. 161–165, eds. T. M. Mitchell, J. G. Carbonell, and R. S. Michalski, Kluwer Academic, Boston, 1986.

111. B. Kosko, "Adaptive Bidirectional Associative Memories," *Applied Optics*, vol. 26, pp. 4947–4960, December 1987.

112. T. Kramer, "Automated Analysis of Operators on State Tables: A Technique for Intelligent Search," *Journal of Automated Reasoning*, vol. 2, pp. 127–151, 1986.

113. B. J. Kuipers, A. J. Moskowitz, and J. P. Kassirer, "Critical Decisions under Uncertainty," AI TR87-61, University of Texas at Austin, August 1987.

114. V. Kumar, "Branch-and-Bound Search," Technical Report AI TR85-11, University of Texas at Austin, August 1985.

115. J. E. Laird, P. S. Rosenbloom, and A. Newell, "Towards Chunking as a General Learning Mechanism," *Proc. Nat'l Conf. on Artificial Intelligence AAAI-84*, pp. 188–192, 1984.

116. J. E. Laird, P. S. Rosenbloom, and A. Newell, "Chunking in Soar: The Anatomy of a General Learning Mechanism," *Machine Learning*, vol. 1, pp. 11–46, 1986.

117. J. E. Laird, P. S. Rosenbloom, and A. Newell, "Soar: An Architecture for General Intelligence," *Artificial Intelligence*, vol. 33, pp. 1–64, 1987.

118. P. Langley and H. A. Simon, "The Central Role of Learning in Cognition," *Cognitive Skills and Their Acquisition*, ed. J. R. Anderson, Lawrence Erlbaum Associates, 1981.

119. P. Langley, "Learning Effective Search Heuristics," *Proc. 8th Int'l Joint Conf. on Artificial Intelligence*, pp. 419–421, Los Altos, CA, 1983.

120. P. Langley, "Learning to Search: From Weak Methods to Domain-Specific Heuristics," *Cognitive Science*, vol. 9, pp. 217–260, 1985.

121. D. Lawson and B. Williams, "A Neural Network Implementation of a Page-Swapping Algorithm," *Abstracts of the 1st Annual INNS Meeting (Neural Networks)*, vol. 1 (supplement 1), p. 451, 1988.

122. M. Lebowitz, "Integrated Learning: Controlling Explanation," *Cognitive Science*, pp. 219–240, 1986.

123. K. F. Lee and S. Mahajan, "A Pattern Classification Approach to Evaluation Function Learning," *Artificial Intelligence*, vol. 36, pp. 1–25, 1988.

124. D. B. Lenat, "HEURETICS: Theoretical and Experimental Study of Heuristic Rules," *Proc. Nat'l Conf. on Artificial Intelligence AAAI-82*, pp. 159–163, 1982.

125. D. B. Lenat, "The Role of Heuristics in Learning by Discovery: Three Case Studies," *Machine Learning: An Artificial Intelligence Approach*, eds. R. S. Michalski, J. G. Carbonell, and T. M. Mitchell, pp. 243–306, Morgan Kaufmann, Los Altos, CA, 1983.

126. D. B. Lenat, "Theory Formation by Heuristic Search; The Nature of Heuristics II: Background and Examples," *Artificial Intelligence*, vol. 21, pp. 31–59, 1983.

127. D. B. Lenat and J. S. Brown, "Why AM and EURISKO Appear to Work," *Artificial Intelligence*, vol. 23, pp. 269–294, 1984.

128. D. B. Lenat and E. A. Feigenbaum, "On the Thresholds of Knowledge," *Proc. 10th Int'l Joint Conf. on Artificial Intelligence*, pp. 1173–1182, 1987.

129. C. Lewis, "Composition of Productions," *Production System Models of Learning and Development*, ed. Klahr et al. , MIT Press, 1987.

130. M. W. Lewis and J. R. Anderson, "Discrimination of Operator Schemata in Problem Solving: Learning from Examples," *Cognitive Psychology*, vol. 17, pp. 26–65, 1985.

131. M. R. Lowry, "The Logic of Problem Reformulation," *Proc. Workshop on Knowledge Compilation*, ed. T. G. Dietterich, 1986.

132. F. Maffioli, "The Complexity of Combinatorial Optimization Algorithms and the Challenge of Heuristics," *Combinatorial Optimization*, pp. 107–128, 1979.

133. Z. Manna and R. Waldinger, "A Theory of Plans," *Reasoning about Actions and Plans*, eds. M. P. Georgeff and A. L. Lansky, pp. 11–45, 1987.

134. J. L. McClelland, "The Programmable Blackboard Model of Reading," *Parallel Distributed Processing: Psychological and Biological Models*, eds. J. L. McClelland and D. E. Rumelhart, pp. 122–169, MIT Press, 1986.

135. T. L. McCluskey, "Combining Weak Learning Heuristics in General Problem Solvers," *Proc. 10th Int'l Joint Conf. on Artificial Intelligence*, pp. 331-333, Milan, Italy, 1987.

136. L. Mero, "A Heuristic Search Algorithm with Modifiable Estimate," *Artificial Intelligence*, vol. 23, pp. 13–27, 1984.

137. R. S. Michalski, J. G. Carbonell, and T. M. Mitchell, *Machine Learning: An Artificial Intelligence Approach*, 1983.

138. R. S. Michalski, "A Theory and Methodology of Inductive Learning," *Machine Learning*, ed. R. S. Michalski, J. G. Carbonell, and T. M. Mitchell, Tioga, 1983.

139. D. Michie and R. Chambers, "BOXES: An Experiment in Adaptive Control," *Machine Intelligence 2*, pp. 137–152, eds. E. Dale and D. Michie, Oliver and Boyd, Edinburgh, Scotland.

140. M. Minsky, "Steps toward Artificial Intelligence," *Computers and Thought*, eds. E. A. Feigenbaum and J. Feldman, McGraw-Hill, New York, 1963.

141. S. Minton, J. G. Carbonell, C. A. Knoblock, D. Kuokka, and H. Nordin, "Improving the Effectiveness of Explanation Based Learning," *Proc. Workshop on Knowledge Compilation*, pp. 77–87, 1986.

142. S. Minton and J. G. Carbonell, "Strategies for Learning Search Control Rules: An Explanation-Based Approach," *Proc. 10th Int'l Joint Conf. on Artificial Intelligence*, pp. 334–337, Milan, Italy, August 1987.

143. T. M. Mitchell, "The Need for Biases in Learning Generalizations," CBM-TR-117, Rutgers University, Computer Science Department, 1980.

144. T. M. Mitchell, P. E. Utgoff, B. Nudel, and R. Benerji, "Learning Problem-Solving Heuristics through Practice," *Proc. 7th Int'l Joint Conf. on Artificial Intelligence*, pp. 127–134, Los Altos, CA, 1981.

145. T. M. Mitchell, "Learning and Problem Solving," *Proc. 8th Int'l Joint Conf. on Artificial Intelligence*, pp. 1139-1151, Los Altos, CA, August 1983.

146. T. M. Mitchell, P. E. Utgoff, and R. B. Banerji, "Learning by Experimentation: Acquiring and Refining Problem-Solving Heuristics," *Machine Learning*, eds. R. S. Michalski, J. G. Carbonell, and T. M. Mitchell, Tioga, 1983.

147. T. M. Mitchell, "Toward Combining Empirical and Analytical Methods for Inferring," *Artificial and Human Intelligence*, pp. 81–103, eds. R. B. Banerji and T. Elithorn, Elsevier, 1984.

148. R. J. Mooney, "A General Explanation-Based Learning Mechanism and Its Application to Narrative Understanding," Technical Report UILU-ENG-87-2269, University of Illinois, Urbana-Champaign, December 1987.

149. L. Morgenstern and L. A. Stein, "Why Things Go Wrong: A Formal Theory of Causal Reasoning," *Proc. Nat'l Conf. on Artificial Intelligence AAAI-88*, pp. 518–523, 1988.

150. D. J. Mostow, "Machine Transformation of Advice into a Heuristic Search Procedure," *Machine Learning: An Artificial Intelligence Approach*, eds. R. S. Michalski, J. G. Carbonell, and T. M. Mitchell, pp. 367–404, Morgan Kaufmann, Los Altos, CA, 1983.

151. B. A. Nadel, "Representation Selection for Constraint Satisfaction Problems: A Case Study Using n-Queens," DCS-TR-208, Laboratory for Computer Science Research, Rutgers University, NJ, 1987.

152. D. M. Neves and J. R. Anderson, "Knowledge Compilation: Mechanisms for the Automatization of Cognitive Skills," *Cognitive Skills and Their Acquisition*, ed. J. R. Anderson, Lawrence Erlbaum Associates, 1981.

153. D. M. Neves, "Learning Procedures from Examples and by Doing," *Proc. Int'l Joint Conf. on Artificial Intelligence*, pp. 624–630, 1983.

154. A. Newell and H. A. Simon, *Human Problem Solving*, Prentice Hall, Englewood Cliffs, NJ, 1972.

155. A. Newell, "The Knowledge Level," *Artificial Intelligence*, vol. 18, pp. 87–127, 1982.

156. L. M. Ni and K. Hwang, "Optimal Load Balancing Strategies for a Multiple Processor System," *Proc. 10th Int'l Conf. on Parallel Processing*, pp. 352–357, August 1981.

157. L. M. Ni, C.-M. Xu, and T. B. Gendreau, "A Distributed Drafting Algorithm for Load Balancing," *Trans. on Software Engineering*, vol. SE-11, pp. 1153–1161, October 1985.

158. H. P. Nii, "Blackboard Systems, Blackboard Application Systems, Blackboard Systems from a Knowledge Engineering Perspective," *AI Magazine*, pp. 82–106, August 1986.

159. N. J. Nilsson, "Triangle Tables: A Proposal for a Robot Programming Language," Technical Note 347, SRI International, Menlo Park, CA, February 1985.

160. S. Ohlsson, "A Constrained Mechanism for Procedural Learning," *Proc. Int'l Joint Conf. on Artificial Intelligence*, pp. 426–428, Los Altos, CA, 1983.

161. P. S. Ow, S. F. Smith, and A. Thiriez, "Reactive Plan Revision," *Proc. 10th Nat'l Conf. on Artificial Intelligence AAAI-88*, vol. 1, pp. 77–82, Saint Paul, MN, 1988.

162. V. M. Ozernoi and M. G. Gaft, "Methods for the Best Solutions Search in Multi-objective Decision Problems," *Proc. Int'l Joint Conf. on Artificial Intelligence IJCAI-75*, pp. 357–361, Moscow, USSR, 1975.

163. J. C. Pasquale, "Intelligent Decentralized Control in Large Distributed Computer Systems," Ph.D. Thesis, University of California, Berkeley, April 1988.

164. J. W. Payne, J. R. Bettman, E. J. Johnson, and E. Coupey, "Selection of Heuristics for Choice: An Effort and Accuracy Perspective," *Illinois Interdisciplinary Workshop on Decision Making*, 1988.

165. M. J. Pazzani, "Explanation and Generalization Based Memory," *Proc. Annual Meeting of the Cognitive Science Society*, pp. 323–328, Irvine, CA, 1985.

166. J. Pearl, "Some Recent Results in Heuristic Search Theory," *Trans. on Pattern Analysis and Machine Intelligence*, vol. PAMI-6, pp. 1–13, January 1984.

167. J. Pearl, *Heuristics—Intelligent Search Strategies for Computer Problem Solving*, Addison Wesley, Reading, MA, 1984.

168. J. Pearl, "Bayesian Networks: A Model of Self-Activated Memory for Evidential Reasoning," *Proc. 7th Annual Conf. of Cognitive Science Society*, pp. 329–334, August 1985.

169. G. D. Plotkin, "A Note on Inductive Generalization," *Machine Intelligence*, ed. Meltzer and Michie, pp. 153–163, Edinburgh University Press, Edinburgh, Scotland, 1970.

170. I. Pohl, "The Avoidance of (Relative) Catastrophe, Heuristic Competence, Genuine Dynamic Weighting, and Computational Issues in Heuristic Problem Solving," *Proc. of IJCAI–73*, pp. 12–17, Stanford, CA, 1973.

171. M. E. Pollack, D. J. Israel, and M. E. Bratman, "Toward an Architecture for Resource-Bounded Agents," CSLI-87-104, Center for the Study of Language and Information, Menlo Park, CA, August 1987.

172. B. Porter and D. Kibler, "Experimental Goal Regression: A Technique for Learning Heuristics," Technical Report AITR 86-20, Department of Computer Science, University of Texas, Austin, 1986.

173. B. W. Porter and D. F. Kibler, "Learning Operator Transformations," *Proc. Nat'l Conf. on Artificial Intelligence AAAI–84*, pp. 278–282, 1984.

174. B. W. Porter and R. E. Bareiss, "PROTOS: An Experiment in Knowledge Acquisition for Heuristic Classification Tasks," Technical Report AI TR-86-35, A.I. Lab., University of Texas at Austin, September 1986.

175. J. R. Quinlan, "Predicting the Length of Solutions to Problems," *Proc. Int'l Joint Conf. on Artificial Intelligence IJCAI-75*, pp. 363–369, 1975.

176. S. Rajamoney, G. DeJong, and B. Faltings, "Towards a Model of Conceptual Knowledge Acquisition through Directed Experimentation," Working Paper 68, Artificial Intelligence Research Group, Coordinated Science Lab., University of Illinois, 1985.

177. S. A. Rajamoney and G. F. DeJong, "Active Ambiguity Reduction: An Experiment Design Approach to Tractable Qualitative Reasoning," Technical Report No. UILU-ENG-87- 2225, University of Illinois, April 1987.

178. S. A. Rajamoney, "Exemplar-Based Theory Rejection: An Approach to the Experience Consistency Problem," *Machine Learning*, pp. 284–289, Boston, Kluwer, 1989.

179. L. Rendell, "Conceptual Knowledge Acquisition in Search," *Computational Models of Learning*, ed. L. Bolc, Springer Verlag, 1987.

180. L. A. Rendell, "A New Basis for State-Space Learning Systems and a Successful Implementation," *Artificial Intelligence*, vol. 20, pp. 369-392, 1983.

181. L. A. Rendell, "Substantial Constructive Induction Using Layered Information Compression: Tractable Feature Formation in Search," *Proc. Int'l Joint Conf. on Artificial Intelligence IJCAI-85*, p. 65, 1985.

182. L. A. Rendell, "Genetic Plans and the Probabilistic Reasoning System: Synthesis and Results," UIUCDCS-R-85-1217, Department of Computer Science, University of Illinois, 1985.

183. L. A. Rendell, "A Framework for Induction and a Study of Selective Induction," *Machine Learning*, vol. 1, June 1986.

184. L. A. Rendell, "Representations and Models for Concept Learning," UIUCDCS-R-87-1324, Department of Computer Science, University of Illinois, 1987.

185. L. A. Rendell, R. Seshu, and D. Tcheng, "Layered Concept Learning and Dynamically Variable Bias Management," *Proc. Int'l Joint Conf. on Artificial Intelligence IJCAI-87*, Milan, Italy, 1987.

186. L. A. Rendell, "Learning Hard Concepts," *European Workshop in Learning (EWSL-88)*, 1988.

187. P. J. Riddle, "An Overview of Problem Reduction: A Shift of Problem Representation," *Proc. Workshop on Knowledge Compilation*, ed. T. G. Dietterich, 1986.

188. P. J. Riddle, "An Approach for Learning Problem Reduction Schemas and Iterative Macro-Operators," *Proc. 1st Workshop on Change of Representation*, 1988.

189. M. H. Romanycia and F. J. Pelletier, "What Is a Heuristic?," *J. Computational Intelligence*, vol. 1, pp. 47–58, Toronto, 1985.

190. F. Rosenblatt, *Principles of Neurodynamics*, Spartan Books, New York, 1962.

191. P. Rosenbloom and A. Newell, "Learning by Chunking: A Production System Model of Practice," *Production System Models of Learning and Development*, ed. Klahr et al., MIT Press, 1987.

192. P. S. Rosenbloom, J. E. Laird, J. McDermott, A. Newell, and E. Orciuch, "R1-Soar: An Experiment in Knowledge-Intensive Programming in a Problem Solving Architecture," *IEEE Trans. on Pattern Analysis and Machine Intelligence*, vol. PAMI-7, pp. 561–569, 1985.

193. P. S. Rosenbloom and J. E. Laird, "Mapping Explanation-Based Generalization onto SOAR," STAN-CS-86-1111 (also KSL-86-46), Department of Computer Science, Stanford University, 1986.

194. R. Rosenfeld and D. S. Touretzky, "Four Capacity Models for Coarse-Coded Symbol Memories," Technical Report No. CMU-CS-87-182, Computer Science Department, Carnegie-Mellon University, Pittsburgh, PA, December 1987.

195. J. Rosenschein and V. Singh, "The Utility of Metalevel Effort," HPP-83-20, Stanford University, CA, March 1983.

196. D. D. Rumelhart, G. Hinton, and J. L. McClelland, "A General Framework for Parallel Distributed Processing," *Parallel Distributed Processing: Explorations in the Microstructure of Cognition*, eds. D. E. Rumelhart, J. L McClelland, and the PDP Research Group, MIT Press, Cambridge, MA, 1986.

197. D. E. Rumelhart, G. E. Hinton, and R. J. Williams, *Learning Internal Representations by Error Propagation*, Institute for Cognitive Science Report 8506, UCSD, September 1985.

198. D. E. Rumelhart, P. Smolensky, J. L. McClelland, and G. Hinton, "Schemata and Sequential Thought in PDP Models," *Parallel Distributed Processing: Psychological and Biological Models*, MIT Press, 1986.

199. J. E. Russo and B. A. Dosher, "Strategies for Multiattribute Binary Choice," *Journal of Experimental Psychology: Learning, Memory, and Cognition*, vol. 9, pp. 676–696, 1983.

200. E. D. Sacerdoti, *A Structure for Plans and Behavior*, Elsevier, New York, 1977.

201. E. D. Sacerdoti, "Problem Solving Tactics," *Proc. IJCAI-79*, pp. 1077–1085, 1979.

202. A. L. Samuel, "Some Studies in Machine Learning Using the Game of Checkers," *IBM J. Research and Development*, vol. 3, pp. 210–229, 1959.

203. A. L. Samuel, "Some Studies in Machine Learning Using the Game of Checkers II—Recent Progress," *J. of Research and Development*, vol. 11, pp. 601–617, 1967.

204. R. C. Schank, G. C. Collins, and L. Hunter, "Transcending Inductive Category Formation in Learning," *Behavioral and Brain Sciences*, vol. 9, pp. 639–686, 1986.

205. M. J. Schoppers, "Representation and Automatic Synthesis of Reaction Plans," forthcoming Ph.D. thesis, Department of Computer Science, University of Illinois, Urbana, 1988.

206. B. Silver, *Meta-Level Inference: Representing and Learning Control Information in Artificial Intelligence, Studies in CS and AI*, New York: North-Holland, 1986.

207. H. A. Simon, "The Architecture of Complexity," *The Sciences of the Artificial*, pp. 193–230, MIT Press, Cambridge, MA, 1969.

208. H. A. Simon, "Economic Rationality: Adaptive Artifice," *The Sciences of the Artificial*, p. 46, MIT Press, Cambridge, MA, 1969.

209. H. A. Simon, "Artificial Intelligence Systems that Understand," *Proc. 5th Int'l Joint Conf. on Artificial Intelligence*, vol. 2, pp. 1059–1073, Cambridge, MA, 1977.

210. H. A. Simon, "Information-Processing Theory of Human Problem Solving," *Handbook of Learning and Cognitive Processes*, Erlbaum, 1978.

211. H. A. Simon, "What the Knower Knows: Alternative Strategies for Problem Solving Tasks," *Human and Artificial Intelligence*, ed. F. Klix, North-Holland, 1979.

212. H. A. Simon, *The Science of the Artificial*, 2nd edition, MIT Press, Cambridge, MA, 1981.

213. H. A. Simon, "Why Should Machines Learn?" *Machine Learning: An Artificial Intelligence Approach*, eds. R. S. Michalski, J. G. Carbonell, and T. M. Mitchell, pp. 25–37, Morgan Kaufmann, Los Altos, CA, 1983.

214. J. R. Slagle, "A Heuristic Program that Solves Symbolic Integration Problems in Freshman Calculus," *Computers and Thought*, eds., E. A. Feigenbaum and J. Feldman, McGraw-Hill, New York, 1963.

215. D. Sleeman, P. Langley, and T. M. Mitchell, "Learning from Solution Paths: An Approach to the Credit Assignment Problem," *AI Magazine*, vol. 3, pp. 48–52, 1982.

216. R. G. Smith, T. M. Mitchell, R. A. Chestek, and B. G. Buchanan, "A Model for Learning Systems," *Proc. 5th Int'l Joint Conf. on Artificial Intelligence*, pp. 338–343, Los Altos, CA, August 1977.

217. S. F. Smith, "Flexible Learning of Problem Solving Heuristics through Adaptive Search," *Proc. Int'l Joint Conf. on Artificial Intelligence*, pp. 422–425, 1983.

218. P. Smolensky, "Information Processing in Dynamical Systems: Foundations of Harmony Theory," *Parallel Distributed Processing: Foundations*, pp. 194–281, MIT Press, 1986.

219. N. S. Sridharan and J. L. Bresina, "Exploration of Problem Reformulation and Strategy Acquisition—A Proposal," Laboratory for Computer Science Research, Rutgers University, March 1984.

220. S. Star, "Theory-Based Inductive Learning: An Integration of Symbolic and Quantitative Methods," *Uncertainty in Artificial Intelligence Workshop*, pp. 237–248, Seattle, WA, 1987.

221. M. Stefik, "Planning and Meta-Planning (MOLGEN: Part 2)," *Readings in Artificial Intelligence*, pp. 272–286, eds. B. L. Webber and N. Nilsson, Morgan Kaufmann, 1981.

222. D. Steier, "CYPRESS-Soar: A Case Study in Search and Learning in Algorithm Design," *Proc. 10th Int'l Joint Conf. on Artificial Intelligence IJCAI-87*, pp. 327–330, Milan, Italy, 1987.

223. D. Subramanian and M. R. Genesereth, "Reformulation," *Proc. Workshop on Knowledge Compilation*, ed., T. G. Dietterich, 1986.

224. G. J. Sussman, "A Computational Model of Skill Acquisition," Ph.D. thesis, Artificial Intelligence Lab., MIT, 1973.

225. R. S. Sutton, "Temporal Credit Assignment in Reinforcement Learning," Ph.D. thesis, University of Massachusetts, Amherst, February 1984.

226. R. S. Sutton and A. G. Barto, "Toward a Modern Theory of Adaptive Networks: Expectation and Prediction," *Psychological Review*, vol. 88, pp. 135–170, 1984.

227. R. S. Sutton, "Learning to Predict by the Methods of Temporal Differences," *Machine Learning*, vol. 3, pp. 9–44, 1988.

228. C. Swart and D. Richards, "On the Inference of Strategies," TR 86-20-3, Oregon State University Computer Science Department, 1986.

229. D. S. Touretzky and G. E. Hinton, "A Distributed Connectionist Production System," CMU-CS-86-172, Department of Computer Science, Carnegie-Mellon University, 1986.

230. D. S. Touretzky, "Beyond Associative Memory: Connectionists Must Search for Other Cognitive Primitives," *Proc. AAAI Spring Symp. Series: Parallel Models of Intelligence: How Can Slow Components Think So Fast?*, Stanford, CA, 1988.

231. A. Tversky and D. Kahneman, "Judgment under Uncertainty: Heuristics and Biases," *Science*, vol. 185, pp. 1124–1131, 1974.

232. P. E. Utgoff and T. M. Mitchell, "Acquisition of Appropriate Bias for Inductive Concept Learning," *Proc. Nat'l Conf. on Artificial Intelligence*, pp. 414–417, 1982.

233. P. E. Utgoff, "Adjusting Bias in Concept Learning," *Proc. Int'l Joint Conf. on Artificial Intelligence*, pp. 447–449, Los Altos, CA, 1983.

234. P.E. Utgoff and S. Saxena, "Learning a Preference Predicate," *Proc. Int'l Machine Learning Workshop*, pp. 115–121, Irvine, CA, June 1987.

235. B. W. Wah and P. Mehra, "Learning Parallel Search in Load Balancing," *Workshop on Parallel Algorithms for Machine Intelligence and Pattern Recognition*, Minneapolis, MN, August 21, 1988.

236. B. Walker, G. Popek, R. English, C. Kline, and G. Thiel, "The LOCUS Distributed Operating System," *Proc. 9th Symp. on Operating System Principles*, pp. 49–70, 1983.

237. Y. T. Wang and J. T. Morris, "Load Sharing in Distributed Systems," *Trans. on Computers*, vol. C-34, pp. 204–217, March 1985.

238. S. Watanabe, *Pattern Recognition: Human and Mechanical*, Wiley Interscience, 1985.

239. S. Watanabe, "Inductive Ambiguity and the Limits of Artificial Intelligence," *Computational Intelligence*, vol. 3, pp. 304–309, 1987.

240. D. A. Waterman, "Generalization Learning Techniques for Automating the Learning of Heuristics," *Artificial Intelligence*, vol. 1, pp. 121–170, 1970.

241. P. J. Werbos, "Learning How the World Works: Specifications for Predictive Networks in Robots and Brains," *Proc. IEEE Int'l Conf. on Systems, Man, and Cybernetics*, vol. 1, pp. 302–310, October 1987.

242. P. J. Werbos, "Backpropagation: Past and Future," *Proc. Int'l Conf. on Neural Networks ICNN-88*, San Diego, CA, July 1988.

243. B. L. Whitehall, "Incremental Learning with INDUCE/NE," Working Paper 84, AI Group, Coordinated Science Lab, University of Illinois, 1986.

244. B. L. Whitehall, "Substructure Discovery in Executed Action Sequences," Technical Report UILU-ENG-87-2256, University of Illinois, College of Engineering, September 1987.

245. C. C. White III and S. Dozono, "A Generalized Model of Sequential Decision-Making under Risk," *European Journal of Operational Research*, vol. 18, pp. 19–26, 1984.

246. D. C. Wilkins, "Knowledge Base Refinement by Monitoring Abstract Control Knowledge," *Knowledge Acquisition for Knowledge Based Systems*, ed. Boose et al., Academic Press, New York, 1987.

247. D. C. Wilkins, "Knowledge Base Refinement Using Apprenticeship Learning Techniques," *Proc. Nat'l Conf. on Artificial Intelligence AAAI-88*, 1988.

248. R. J. Williams, "On the Use of Backpropagation in Associative Reinforcement Learning," *Proc. 2nd Int'l Conf. on Neural Networks ICNN-88*, vol. I, pp. 263–270, San Diego, CA, July 1988.

249. S. W. Wilson, "Hierarchical Credit Allocation in Classifier Systems," *Genetic Algorithms and Simulated Annealing*, ed. L. Davis, Pitman, London, 1987.

250. P. H. Winston, "Learning Structural Descriptions from Examples," *The Psychology of Computer Vision*, ed. P. H. Winston, McGraw-Hill, New York, 1975.

251. P. H. Winston, T. O. Binford, B. Katz, and M. R. Lowry, "Learning Physical Descriptions from Functional Definitions, Examples, and Precedents," *Proc. AAAI-83*, pp. 433–439, 1983.

252. P. H. Winston, *Artificial Intelligence*, Addison-Wesley, Reading, MA, 1984.

253. C. F. Yu, *Efficient Combinatorial Search Algorithms*, Ph.D. thesis, School of Electrical Engineering, Purdue University, West Lafayette, IN, December 1986.

254. C. F. Yu and B. W. Wah, "Learning Dominance Relations in Combinatorial Search Problems," *Trans. on Software Engineering*, vol. SE-14, no. 8, August 1988.

255. B. Zhang and L. Zhang, "The Statistical Inference Method in Heuristic Search Techniques," *Proc. Int'l Joint Conf. on Artificial Intelligence*, pp. 757–759, 1983.

CHAPTER 14

AI AND SOFTWARE ENGINEERING: A CLASH OF CULTURES?*

Wei-Tek Tsai, K. Heisler, D. Volovik, and I. A. Zualkernan

1 INTRODUCTION

Herbert Simon has had a profound impact on the development of artificial intelligence (AI). His paper, "Whether Software Engineering Needs to Be Artificially Intelligent" [82] and a reaction to that paper provide a good example of an apparent cultural clash between AI and software engineering (SE). Simon's paper appeared in the July 1986 issue of *IEEE Transactions on Software Engineering*. In the October 1986 issue of *ACM Software Engineering Notes*, Peter Neuman made the following observations [60]:

"In the July 1986 issue of the *IEEE Transactions on Software Engineering* (vol SE-12, page 720), Herb Simon offers a paper with the following abstract:

> This paper is entitled 'Whether Software Engineering Needs to Be Artificially Intelligent.' I will not hold you in suspense; of course, I would not have selected that title, unless I thought the answer were yes. Otherwise, I could just give a simple answer.

"Other authors might counter with a paper bearing the same title and this abstract:

> This paper is entitled 'Whether Software Engineering Needs to Be Artificially Intelligent.' I will not hold you in suspense; of course, I would not have selected that title,

---

*See the Appendix to this chapter beginning on page 510 for a discussion of recent publications in this area.

unless I thought the answer were yes. Otherwise, I could just give a simple answer. What it most needs is real intelligence.

"But an even more important issue arises in connection with the risks of using artificially intelligent programs in life-critical and other sensitive applications. This suggests an alternative paper:

> This paper is entitled 'Whether Artificial Intelligence Needs to Be Software Engineered.' I will not hold you in suspense; of course the answer is yes. Unfortunately, most of the allegedly production quality AI systems I have seen never heard of software engineering. How often have you heard something like this: 'Don't give me any of the junk about software engineering, requirements, specifications, or even formal verification—I just write the code bottom up and it works fine.'

"Well, then along comes a remarkable book edited by Charles Rich and Richard C. Waters of the MIT AI Lab, entitled 'Readings in Artificial Intelligence and Software Engineering.' Most of the book is devoted to what might be paraphrased as 'Ask not what software engineering can do for AI, ask what artificial intelligence can do for software engineering.' I hope that developers of real-world AI systems, particularly systems for use in critical environments, will take software engineering more seriously. A fluid exploratory programming style may well be appropriate in AI research, but should be tempered with a more rigorous approach as AI begins to be applied to real-world systems."

In these remarks, Neuman implies that SE needs real intelligence and not artificial intelligence. Parnas [63] makes a similar observation about the application of expert systems (a subset of AI) to the problem of Strategic Defense Initiative (SDI) software:

> SDI presents a problem that may be more difficult than those being tackled in AI-1* and expert systems. Workers working in those areas attack problems that now require human expertise. Some of the problems in SDI are in areas where we now have no human experts. Do we now have humans who can, with high reliability and confidence, look at missiles in ballistic flight and distinguish warheads from decoys?

Neuman and Parnas are both correct in their observations that AI cannot reproduce knowledge that does not exist, but Simon's paper is not in conflict with the view that SE needs real intelligence. The point that Simon is making about the application of AI to SE becomes clear if one looks at his definition of AI:

> The question before us when we talk about artificial intelligence is, "How can we do problem solving when we do not have . . . a systematic, direct algorithm?"

*Parnas defines AI-1 as "the use of computers to solve problems that previously could be solved only by applying human intelligence."

This question can be applied to the problem solving that we do when we go through the process of SE. There is real intelligence in the way SE is conducted today. In the past 20 years, several extremely large and complex programs have been successfully developed (for example, the space shuttle computer programs [88, 51]) using traditional SE techniques. These programs not only work but are also reliable enough to be operational in critical environments. Parnas would probably agree that there are no known algorithms for solving the programming-in-the-large problem. Therefore, at the outset it seems that AI is exactly what is needed to understand and automate this real intelligence in the SE process.

In his remarks, Neuman also states that people in AI should take SE more seriously. Simon's paper is also in perfect agreement with this view. For example, he states:

> The recent history of expert systems shows that tasks comparable in technology and complexity to the tasks you perform in building software can be automated. I do not want to get into an argument as to whether doctors are more professional than software engineers or software engineers are more professional than doctors. These are both clearly professional level tasks which require intelligence and knowledge. We know today how much knowledge it takes and what kinds of intelligence it takes to do credible jobs such as medical diagnosis. I think we can take a great deal of heart from that.

Simon also says:

> Most of the ideas and information which are imbedded today in artificial intelligence and the lore of artificial intelligence were borrowed more or less by watching how human beings perform.

The importance of these comments is that Simon is not proposing to ignore SE. On the contrary, just as progress in medical AI was achieved by working closely with physicians [68], AI can achieve progress in SE by working closely with software engineers. Ignoring software engineers, who have been developing production-quality software for years, would be like ignoring physicians during the course of medical AI work.

Neuman criticizes the collection of papers by Rich and Waters [73] as being based on the premise "Ask not what software engineering can do for AI, ask what artificial intelligence can do for software engineering." This criticism is justified to a degree but misplaced. The premise articulated by Neuman is a reflection of the views of Rich and Waters that "the ultimate goal of artificial intelligence applied to software engineering is automatic programming" [74].

Rich and Waters and the automatic programming school have a definite view of how programming should be done and hence what the SE process should be [5]. Being a part of the computer science community, they are completely justified in their approach, which should be judged entirely on the basis of its success or failure. This is not very different from people within the SE community proposing, for example, how requirements should be written. However, it is important to distinguish this type of research (automatic programming paradigm) from the more typical type of AI research, wherein a domain of problem solving (in this case the SE process), is studied and attempts are made to understand and improve it using AI techniques.

Neuman's comment on AI being useful for exploratory techniques and not so much for production-quality software (at least in its current stage) warrants attention. However, it is dangerous to underestimate the power of exploratory programming techniques. For example, Parnas [63] makes the following comment:

> Lately we have heard a great deal about the success of a particular class of rule-based systems known as expert systems. Every discussion cites one example of such a system that is being used by people other than its developers. That example is always the same—a program designed to find configurations for VAX computers. To many of us, that does not sound like a difficult problem; it sounds like the kind of problem that is amenable to algorithmic solution because VAX systems are constructed from well-understood, well-designed components. Recently, I read a paper that reported that this program had become a maintenance nightmare. It was poorly understood, badly structured, and hence hard to change. I have good reason to believe that it could be replaced by a better program written using good software engineering techniques instead of heuristic techniques.

Even if what Parnas says about rewriting R1 (as the mentioned program is known) using "good software engineering techniques" is true, it should be realized that several traditional approaches to the problem R1 solves were tried and abandoned before the R1 project was started [54]. The programming of R1 using "heuristic" techniques was exactly what led to the understanding of the way R1 solves the configuration problem.

As Parnas [63] points out elsewhere, most real-world programming has a large exploratory component:

> We can write software requirements documents that are complete and precise. We understand the mathematical model behind such documents and can follow a systematic approach to document all necessary requirement decisions. Unfortunately, it is hard to make the decisions that must be made to write such a document. We often do not know how to make those decisions until *we can play with the system*. Only when we have built similar systems before is it easy to determine the requirements in advance. It is worth doing, but it is not easy.

Neuman's last remarks imply that a precondition for application of AI to SE is that AI's program development methodologies be made more rigorous. No one would argue with this position; however, AI is more than a collection of programming techniques.

Parnas refers to one of the reasons why conventional software development does not produce reliable programs:

It should be clear that writing and understanding very large real-time programs by "thinking like a computer" will be beyond our intellectual capabilities. How can it be that we have so much software that is reliable enough for us to use it? The answer is simple; programming is a trial and error craft. People write programs without any expectations that they will be right the first time. They spend at least as much time testing and correcting errors as they spent writing the initial programs.

What Parnas is referring to is the fact that people (including programmers) have what Simon [81] has termed "bounded rationality"; any systems including people as a component should be tolerant of this fact. Simon reflects this view that we should not "postulate a new man." A substantial portion of research in AI deals with how effective problem solving can be done under the conditions of bounded rationality (resource-limited reasoning). Clearly, an understanding of conditions for effective problem solving in such situations will help in determining which SE environments and languages are easier and more effective to use. This research is independent of whether AI programming techniques are judged as structured or not according to SE criteria.

The cultural clash between AI and SE has not deterred a continuing interest in the use of AI techniques for SE [7, 73, 82, 9, 74, 75, 93]. There is also considerable interest in the use of SE technology for AI program development [31, 94]. However, in cross-pollination of different fields, it is very important to communicate at the correct level.

Fields of study are defined and held together by common objectives. For example, SE is concerned with studying, modeling, and improving the software development process. Various approaches arise to achieve these common objectives, and specific solutions are developed based on these approaches. The waterfall model, which decomposes the SE process into distinct subproblems, is an example of an approach to SE objectives. The use of specifications is a solution to the maintenance problem within this approach, because it cuts down the number of errors that need to be fixed at later stages of software development [72].

The correct level of communication between fields is the level of objectives and approaches rather than specific solutions. One danger of communication at the level of solutions is that there is a tendency to fit a solution from one field to a problem in another, without realizing what the solution was developed for and where its weaknesses lie. Research must be problem-oriented rather than solution-oriented. The temptation to force-fit a problem into the mold of an elegant solution should be avoided.

AI as a field is changing, and there is no consensus on a single definition [57]. For example, early definitions distinguished AI as

that part of computer science concerned with designing intelligent computer systems, that is, systems that exhibit the characteristics we associate with intelligence in human behavior. This includes language understanding, learning, reasoning and solving ill-structured problems [7].

Waterman [98], on the other hand, defined AI programs as those that exhibit intelligent behavior by skillful application of heuristics. He considers knowledge-based systems as a subset of AI programs and characterizes them as programs that make domain knowledge explicit and separate from general problem-solving knowledge. Expert systems then are defined as a subset of knowledge-based systems that apply human expert knowledge to difficult, real-world problems.

More recently Schank [79] gives probably the most accurate description of AI: "[AI's] primary goal is to build an intelligent machine. The second goal is to find out about the nature of intelligence."

SE, on the other hand, is an engineering discipline whose objective is to develop and maintain production-quality software most efficiently. It includes methods, tools, and procedures that enable a software engineer to orchestrate the building and maintenance of high-quality software in a disciplined manner [70].

Some of the major methodological challenges facing SE can be characterized as issues of programming in the large [10], which involves structuring large collections of modules to form an efficient and effective system [28, 62]. Its counterpart, programming in the small, relates to the design and implementation of specific algorithms and data structures. Programming in the small is considered a distinct and different intellectual activity from programming in the large.

SE is also concerned with the development and maintenance of production-quality software, and therefore has concentrated on the development of disciplined methodologies [11, 29]. Tools are developed to support methodologies. Examples of such tools are REVS [2], which supports Software Requirement Engineering Methodology (SREM), and graphical interfaces involving display and manipulation of large amounts of complex information. SE has achieved considerable success in some domains, such as data processing. Current challenges to SE are to develop very complex, mission-critical, real-time, distributed applications such as the Strategic Defense Initiative (SDI) [63] and nuclear reactor management [71].

It is clear that both AI and SE involve the development of software, but their aims and objectives are different. AI's main concerns have been developing intelligent programs and understanding intelligence, while SE is concerned with disciplined approaches to developing reliable software. By its nature, AI deals with problems for which solution methods are not well understood. An exploratory programming paradigm is often used in approaching these types of problems [66], and this paradigm has given rise to a number of powerful programming tools and techniques [80].

In this chapter we discuss seven important areas of research related to application of AI approaches to SE and vice versa. The purpose of this discussion is to present both perspectives (AI and SE) of research in these areas and dispel any misunderstanding caused by the cultural difference. The perspectives presented are our interpretation of previous work as it conforms to either a typical AI or SE viewpoint. The perspectives are not based on the alliance of individuals with a particular area (SE or AI).

2 AI FOR SE

This section discusses five issues arising from applications of AI approaches to SE problems.

2.1 The Role of Automatic Programming

An AI Perspective Hammer [40] wrote the following about the types of problems that automatic programming addresses:

> To some extent, all automatic programming is an artificial intelligence problem, since its goal is to produce a computer system that performs activities that are currently unstructured and imperfectly understood. These activities range from resolving ambiguous specifications to designing program organizations to choosing efficient representations for abstract data structures.

Talking about the scope of automatic programming, Rich and Waters [76] wrote:

> Most of the productivity improvements introduced by automatic programming will almost certainly be used to attack applications that are enormous rather than merely huge.

There are three major aspects to the automatic programming process:

1. The existence of a formal software specification is a major assumption of the automatic programming paradigm. Emphasis is often placed on a language that is as close as possible to the user domain and on the ability of the language to deal with ambiguous and incomplete requirements from the user. High-level requirements "describe desired behavior without describing mechanisms for generating this behavior" [46]. Automatic programming systems should be able to use knowledge of the application domain to disambiguate and, in some sense, to complete requirements by evolving a formal specification from them [76].

2. Once the formal system specification has been derived, an automatic programming system should, through a sequence of steps, be able to transform the specification (automatically or semi-automatically) into a form suitable for direct compilation. The transformations are usually required to be invariant with respect to correctness (defined in some suitable logic) or sometimes some other properties.

3. Finally, automatic programming systems should have the ability to replay the transformation process as changes are being made to the original specification. This should enable program evolution and maintenance based on the formal system specification alone.

A number of long-standing research projects have involved automatic programming [6, 76]. These have mainly focused on producing tools that perform (or assist many aspects of) the SE process within the overall automatic programming paradigm. These projects have met with only limited success. However, they continue to provide us with a better understanding of the requirements specification, design, and maintenance processes within SE.

Early work on automatic programming by Green and Barstow [7] included PSI and CHI. The CHI knowledge-based programming environment included a wide-spectrum language called V for specifying both programs and programming knowledge [84].

Early work of Balzer included Specification Acquisition From Experts (SAFE) and Transformational Implementation (TI). This work led to the development of the Gist language for specification within the automatic programming paradigm [6, 46]. The work also lead to what Balzer calls a new paradigm for software technology in the 1990s [5].

The Programmer's Apprentice Project [99, 75, 77] has a long-term goal of developing a theory of programming (i.e., how expert programmers understand, design, implement, test, verify, modify, and document programs) and of automating the programming process. However, the intermediate goal has been to build an intelligent assistant for programmers, emphasizing division of labor between human and machine. Recently, the emphasis has shifted towards automating the acquisition and formalization of specifications from the user [75].

Most knowledge-based investigations of the algorithmic design process focus upon selecting or constructing solutions from known alternatives. Selections are made on the basis of feasibility and further narrowed down to a single solution using some established criteria for determining the best alternative. Goldberg [39], for example, discusses in some detail the nature of the knowledge needed to carry out such design decisions as finite differencing, loop fusion, data structure selection, storage versus computation, converting logic assertions to procedural interpretation, and parameter selection in algorithm design. Another approach, taken by Kant [48], is directed toward capturing the expert process of creative design. Kant studied how people design algorithms in limited highly formal domains (like geometry).

An SE Perspective   The SE view is best summarized by Brooks, Parnas, and Boehm. In particular, Brooks [21] quotes Parnas as follows:

> Many people expect advances in artificial intelligence to provide the revolutionary breakthrough that will give order-of-magnitude gains in software productivity and

quality. I do not. . . . The term [automatic programming] is used for glamor, not for semantic content. In short, automatic programming always has been a euphemism for programming with a higher-level language than was presently available to the programmer.

Parnas [63] also remarks that

the claims that have been made for automatic programming systems are greatly exaggerated. Automatic programming in a way that is substantially different from what we do today is not likely to become a practical tool for real-time systems like the SDI battle-management system. Moreover, one of the basic problems with SDI is that we do not have the information to write a specification that we can trust. In such a situation, automatic programming is no help at all.

Finally, Boehm [16] noted a number of difficulties with the transformation step of the automatic programming process. First, automatic transformation capabilities are available for only small products in a few limited areas. Second, the transform model shares some of the difficulties of the evolutionary development model, such as the assumption that users' operational systems will be flexible enough to support unplanned evolution paths. Third, he points to the formidable knowledge-base maintenance problem in light of the rapidly increasing and evolving supply of reusable software components and commercial software products.

The SE view of automatic programming is also shared by some within the AI community. For example, Partridge [66] discusses three problems with the automatic programming paradigm. First, the specification must be of an operational nature, and it is not clear how users can translate desired changes in program behavior into changes in operational specification. Second, completely automated implementation is not a real possibility because of the wide gap between high-level specification languages and implementations. Finally, under the automatic programming paradigm, the formal specification replaces the implementation as the fundamental object, and this places demanding constraints on suitable specification languages.

Barstow [9] also notes that no demonstrable experience of automatic programming exists and that most efforts have concentrated on programming in the small. Goldberg [39] notes that the transformational approach

promises attractive solutions to most of the problems plaguing software development, but to date there has been no system created which demonstrates that, for a reasonably general class of problems, software can be produced more cheaply using this rather than traditional methods.

He also notes that replay, which is an essential part of evolution and maintenance, is not a well-understood methodology.

Finally, Balzer [6] states that after 15 years of intensive research it is still difficult to compile the specification.

The need to base automatic programming paradigms on models of human problem solving (including specification, design, implementation, and maintenance, as well as validation and verification) is now being recognized in the automatic programming community [34, 76].

2.2 The Role of Expert Systems

An SE Perspective   Expert systems (ES) are AI programs that take advantage of knowledge in semantically rich domains to do problem solving. These programs have been successful in such domains as computer configuration [54] and medical diagnosis [69]. However, if expert systems are to become a part of the SE process, one expects them to satisfy some of the basic requirements of the field. Since SE is concerned with the development of production software, it demands that any of its subprocesses have a strong developmental methodology that yields similar results regardless of the implementer. It also demands that the subprocesses themselves be maintainable and that their development be carried out through standardized tools.

It should also be realized that SE tackles problems in programming in the large, which require large-scale solutions. However, current literature in ES shows that despite their success, large-scale expert systems are expensive to develop, lack standardized development methodologies, and face severe maintainence problems. Furthermore, it is not clear whether parts of the SE process are amenable to an expert system solution.

Expert systems developed to date are hard to maintain [14, 33, 67]. As Partridge points out, the two major problems facing expert systems are the need for explanation facilities and the problem of incrementally upgrading an incomplete knowledge base. Explanation is used to understand what is in the knowledge base, and incremental upgrading involves maintenance or redevelopment. In PUFF, an expert system that interprets data from pulmonary function tests, the rules had to be increased from 100 to 400 just to get a 10 percent improvement in performance [98].

In a recent survey of successes and failures of current expert systems, Ernest points out [33]:

> As a system grows in size to, say, 8–10 thousand rules, there begins to be a loss of control on part of the developers. The system is too large to be fully comprehended, and our means for representing knowledge are inadequate in terms of providing tools or structures to maintain a manageable perspective and level of understanding of exactly what is happening. . .

Bobrow [14] further points out that

> The maintenance cost for expert systems is substantial. Once the initial thrill of a prototype system and a fancy interface wears off, some projects come abruptly to an end as the expense of developing them further and maintaining them is assessed all too belatedly.

Many researchers have described the maintenance of R1 (now called XCON), an expert system to configure VAX architectures. R1 is one of the most successful expert systems in practical use. It is an important case study because its development and maintenance have been extensively documented and discussed. It is interesting to trace the evolution of R1 with respect to maintenance.

At the beginning [54]:

... during the course of developing R1 a great deal of knowledge was added that R1 was unprepared for. . . . it resulted in a considerable amount of redundancy. . . . knowledge required to perform two subtasks was not represented in the form of rules . . . .

Given these problems, it appeared that a serious re-implementation effort would be likely to have the effect of making clearer precisely what knowledge R1 had, and thus make the task of maintaining and extending R1 easier.

I was surprized by the length of time the reimplementation took.

Four years later [4]:

In the early days, when R1 was small, people who joined the project were able, reasonably quickly, . . . to become competent developers. But now that R1 has grown substantially, its sheer magnitude seems to serve as a barrier to the would-be developer.

In another two years [22]:

One has to know R1 well to modify its behavior in some desired fashion. . . . Since R1 has so much knowledge, gaining such familiarity has become more and more formidable.

And one year later [84]:

XCON started as a relatively small, rule-based system (about 700 rules). It has grown to over 6,200 rules to meet the needs of DEC. Frankly, there is no end in sight: XCON will continue to expand and change. Unfortunately, the problems of continually updating such a large system do not grow linearly; moving from 700 rules to 6,200 rules, with 50% of the rules changing every year, makes for an exceedingly difficult software enhancement problem.

According to Brug [22], R1 is very difficult to maintain because virtually no structures exist in R1 software. A small version of R1, Rime, is claimed to be easy to maintain because it contains only 10 percent of the knowledge of R1, and most importantly because it has an explicit formulation of different roles of software such as apply-operator, recognize-failure, and so on. The latter is actually a variant of SE technique. Thus, Brug is in fact advocating the use of some SE techniques in expert system development.

Langley [50] indicates that many AI systems that use production systems (a prevalent architecture for expert systems) are not modular, in spite of claims to the contrary. Adding or removing rules from a production system usually results in errors. This phenomenon is known as a ripple effect in SE [101]. Rules are easy to add to an expert system, but knowledge is difficult to add. This difficulty arises from the ad-hoc mixing of different stages of development. Although modularity is observed at the stage of implementation or design, that is, via rule teams and so on, that does not guarantee modularity at the requirement specification. The current knowledge base systems tend to be a collection of tightly coupled knowledge. Two types of knowledge coupling can be observed in expert systems: (1) domain rules in the knowledge base coupled with one another, and (2) the knowledge base tightly coupled with the inference engine.

The cost of developing expert systems is comparable to the cost of developing large-scale software. For example, Bobrow [14] says:

> For large-scale systems with knowledge spanning a wide domain, the time needed to develop a system that can be put in the field can be measured in years not months, and tens of worker-years, not worker-months.

Martins [53] also points out:

> Expert systems development costs are high, development times seem unusually long, and the resulting programs put a heavy burden on computing resources.

This is also confirmed by historical evidence; the earlier expert systems required 5 to 30 person-years to build [98].

Martins presents interesting hypotheses about why the current expert programs (large-scale) are successful. One of the reasons is brilliant programmers; as an example he points out that the person who created DENDRAL is Joshaua Lederberg, a Nobel Prize–winning chemist and the president of New York's Rockefeller University. Some other reasons he cites are lots of time and generous funding. He also points out that in the course of these projects developers developed their own custom-built tools rather than using commercial expert systems tools.

One way to assess the applicability of expert systems to SE is to look at the types of problems that have been solved by using expert systems. In Bobrow [14], the problems cited as good candidates for expert system solutions are problems that have a good set of test cases. Also, a suitable problem is probably one that takes an expert an hour or two. Problems that require common sense knowledge, English language understanding, recoginition of complex causal or temporal relations, or understanding of human intentions are probably beyond expert system treatment. Most problems in programming in the large are unsuitable in one or more of these ways. See Zualkernan [104] for a discussion of these criteria.

Both historical evidence and current opinions of people working in AI attest that expert systems for large-scale problems do not have a well-defined methodology and get progressively more difficult to maintain as they get bigger in size. Even though there is active research in the area of expert system development and maintenance (e.g., see Neches [59]), expert systems for any realistic problem have required just as much effort as large-scale software does. This raises the obvious question from an SE perspective of how a solution (expert systems) that itself fails the basic objectives of SE (strong methodology and easy maintenance) will fit into the SE process.

An AI Perspective

> The expression expert systems has been used as a buzz word for funding and a flag to wave for all sorts of projects. . . . Expert systems are neither the answer to all questions nor the answer to none. One can build expert systems for appropriate problems— ones that are valued, bounded, routine and knowledge intensive—provided experts are available who can articulate [14].

The preceding quote by itself should be enough to dispel the belief that expert systems are the key to all SE problems. However, one has to realize that expert systems are a solution generated by AI to automate complex problem solving in domains with rich semantic problem solving knowledge. Why should they even be considered as a solution to the problem of programming in the large when we have only a very shallow understanding of the processes required to develop large-scale software? Are there any techniques in traditional SE that even come close to providing a solution? Probably not. The correct question to ask is, Do we understand the nature of knowledge in the task of programming in the large? Only when the answer is yes can we begin to consider expert systems as a solution.

Even if we do not expect to be able soon to develop production-level expert system tools to support the software development process, experimenting with them for different phases of the software life cycle will help us organize the knowledge applicable to these phases. This is the first step towards automation or semi-automation.

We have to avoid saying that application of AI representations and expert systems alone will solve SE problems. Rather, these representations are solutions to problems in AI. Within the AI community there is general agreement with McCarthy's view as presented in Israel [43]: "Before we spend too much time worrying about the adequacy of a particular representational formalism, we should have some better idea of what we want to represent."

Given this direction, the first task that needs to be tackled in SE is a characterization of the knowledge that is used in the SE process. The concept of knowledge level [61] has had a considerable effect on knowledge representation research in AI and provides a useful conceptual tool for this purpose [45]. Although knowledge levels were originally proposed to talk about the knowledge of intelligent agents, they are now used to specify the nature of knowledge that needs to be represented in solving a particular problem. As Brachman [19] states,

> The knowledge level view urges us to look at the logical content of a knowledge base—which we expect to represent something about the world—and understand its implications independent of its implementation. It is central to the enterprise of knowledge representation to assess the "knowledge" of a system independent of how that knowledge is realized in data structures and algorithms.

Chandrasekaran [23] also makes an important point that

> the available paradigms [in expert system development] often force us to fit the problems to tools rather than fashion tools to reflect the structure of the problem. This situation is caused by a failure to distinguish between what we might call the information processing level (or the knowledge level in Allen Newell's words) and the implementation language level. Most available languages, be they rules-, frame-, or logic-based, are more like assembly languages of the field than programming languages with constructs essential for capturing the essence of the information processing phenomenon.

The knowledge level descriptions proposed by Chandrasekaran involve recognizing the types of problems that exist in different domains and classifying them according to a task specification, the specific forms and organization of domain knowledge needed for the task, and a family of control regimes appropriate for the task. Clancey [25] has also shown that the concept of knowledge level can be used to construct specifications for expert systems after they are built. These specifications serve the useful purpose of making the knowledge embedded in the program explicit. Zualkernan et al. [105] have also shown the utility of the knowledge of the specification for understanding and maintaining expert systems. They present the view that a knowledge base can be considered a *symbol-level* specification, while its specification is a *knowledge-level* specification. Using GALEN [92] as an example, Zualkernan et al. show how a knowledge-level specification can be extracted from the knowledge base and how it can be analyzed.

In conclusion, in applying AI techniques (specifically expert systems technology) to problems in SE, we should not re-invent the wheel by an ad-hoc application of AI techniques to SE problems but recognize and understand what the nature of problem solving in SE is like. Only then can we either use AI representations or develop new ones to fit the problems at hand.

2.3 The Role of AI Languages and Environments

An AI Perspective AI has contributed to the development of powerful programming language concepts [80]. Examples of languages based on these concepts are Flavors [100], Loops [12], Smalltalk [38], Common Lisp [89], and

Prolog [90]. These and other languages are designed to support a number of programming paradigms, including object-oriented programming, logic programming, frame-based programming, incremental program development and compilation, and interpretation interleaved with compilation.

Partridge [66] suggests that a powerful programming environment can solve a number of problems in the development of AI programs. Two examples of AI programming environments are Interlisp [91] and Genera [96]. These environments support the exploratory style of programming, wherein different versions of a solution can be rapidly built and discarded. Code can be reused, easily transferred among programs, and incorporated into applications. These environments also have support for incremental compilation, graphical tools for code browsing and debugging, word processing capabilities, version control, and code history tracing. However, they are relatively expensive to build and are hard to port from one machine to another.

The Genera environment has such tracing facilities as inspect, describe, and break functions. When *inspect* is called, the browser window appears on the screen. The browser allows a programmer to inspect objects in the program and the environment. For example, when a list of elements is inspected, each element of the list can be seen in the appropriate order. When compound objects are inspected, pointing the mouse to the component and clicking the appropriate button allows recursive inspection of subobjects. *Describe* is similar to *inspect* but offers much faster performance, although it is not as sophisticated. *Break* allows the programmer to call a window-oriented debugger. The debugger allows inspection of the program stack, including all pending function returns. It allows changing a definition of any function, including functions whose calls have not yet been completed, and then allows the computation to continue from the point where the break was called.

Sophisticated AI languages often support good SE practices on the code and design levels. For example, most AI languages support object-oriented programming. The object-oriented paradigm presents a programmer with a special view of the world in which both the system down to the smallest submodule and the environment are modeled as abstract entities exchanging messages. Properties defining behavior or constraining internal states can be defined for objects. These properties are passed down or inherited by subobjects, or submodules. Since information can be exchanged between objects only by sending messages through interfaces, modularity, information hiding, and data abstraction are enforced and high module cohesion and low coupling are encouraged.

Prolog is another advanced AI language that supports SE principles on the code level. Prolog programs are ordered sets of inference rules, which are executed by a Prolog virtual machine (Prolog interpreter) in a manner similar to the way a theorem is proved in mathematics, by exhaustive, depth-first trial and error. The programmer's art is to guide the proof, that is, the execution of a Prolog program, to avoid as many blind alleys as possible. Prolog allows specification of a module by asserting its properties without elaborating on implementation details. Unfortunately, the resulting Prolog programs are excessively inefficient. A compromise is for a programmer to specify the module along with a set of rules limiting the combinatorial space in which the proof takes place and thus increasing the efficiency of a program. Prolog allows a limited form of declarative programming, in contrast to purely procedural programming languages. Prolog can be used as a wide-spectrum language throughout the software life cycle.

Structured editors, or syntax-directed editors, are powerful productivity-enhancing tools typical of those supported by AI environments [27]. Structured editors allow programmers to create syntactically correct structures and to develop programs in a disciplined top-down fashion. They achieve this through built-in syntax and simple semantic checks. Structured editors produce syntactically correct programs and support learning of new programming languages.

Code browsers, debuggers, and tracers used in AI environments allow programmers to dynamically peek into program stacks and data heaps. They allow a programmer to suspend a program at any point, investigate every data structure, and observe the value of any variable. They also allow a programmer to modify the values of variables and to resume execution after modifications, or to change control flow during execution [96].

However, Fischer [35] points out that programmers often do not use their systems efficiently because they do not know what tools exist, how to use them, when to use them, what they do, and how to combine, adapt, or modify them.

An SE Perspective   AI environments are among the most powerful SE environments developed to date. However, they cannot solve SE problems unless used in conjunction with disciplined SE methodologies. AI uses language-centered programming environments as a tool for AI software development. These environments provide sophisticated debugging, tracing, browsing, and conditional breakpoint facilities [27]. However, the more power you have vested in your tools, the greater is your need for discipline and your dependence on methodological development. Many prototyping tools are available for building expert systems [96, 58], such as Advisor, ART, HPRL, KEE, and KES; however, we still need systematic processes to use them for production software.

Ideas from AI environments are gaining acceptance in the SE community. Several SE environments under development incorporate most of the best features of Genera, or Interlisp, and other AI environments. For example, an intelligent assistant called Marvin [47] is being developed to automate mundane tasks such as recompilation and reconfiguration. Marvin uses the object-oriented software paradigm and stores its knowledge in the form of rules. Each rule has a precondition that must hold for the action of the rule to apply, and a postcondition describing the state of the system after action has been completed. Another recent development is the MicroScope system [3]. MicroScope's goal is to facilitate code maintenance by using graphic interfaces for viewing code from different perspectives, code browsers, code slicers, bug models, and impact (ripple effect [100]) analysis. MicroScope has been demonstrated on a 50,000-line code.

Despite the great promise of high-level languages, there is a limit to their usefulness. Brooks observes that most of the progress in programmer productivity has been achieved through gradually moving from low-level programming languages to high-level ones. The advances are mostly concentrated in the coding (or implementation) stage. Even though the physical amount of code produced daily by an average programmer stays fairly constant, code quality and productivity are greatly influenced by the programming language used for coding [20]. However, once a language becomes overly sophisticated and voluminous, its more advanced features are not exploited by programmers [21].

Many studies in the software process, both in conventional SE [63, 21] and AI [13], have indicated that programming languages themselves cannot make the impact required for radical improvement in the software process. It is how they are used that makes the difference. In other words, methodology is crucial to improving productivity.

2.4   The Role of Automated Software Design

In this section we will discuss the current applications of AI to software design and some of the SE reactions to these AI approaches.

An AI Perspective   Goldberg [39] has pointed out that software design is a knowledge-intensive activity. Programmers know the alternatives for representing problem domain objects and processes in terms of a programming language. They have an intuitive feel for the time and space tradeoffs for alternative representations. They also know how to analyze these tradeoffs and have a great deal of heuristic knowledge of what works best and what to avoid in a wide range of situations.

Goldberg believes that current AI technology can provide automatic knowledge-based software assistance that goes beyond merely performing the clerical tasks associated with the design process. Automatic programming, knowledge-based design assistance, and knowledge-based algorithmic design are three examples.

Automatic Programming   Software design work within AI has been largely carried out in conjunction with automatic programming (see Section 2.1). The ultimate goals of automatic programming are a requirements specification language and a methodology that allows a software requirement to be automatically converted to optimized code suitable for running on a target system. In this approach maintenance is performed directly on the requirement specification [5]. The specification language should be able to represent software requirements and at the same time be suitable for input to a knowledge-based automatic optimizing compiler. Current research toward this end focuses on the concept of a wide-spectrum language and knowledge-based transformations within this language to move from very high-level specifications to low-level specifications suitable for automated compilation. Each transformation represents a design decision. This transformation process is far from automatic and depends on human intervention, interaction, and evaluation.

Examples of high-level transformation strategies include divide and conquer, selecting more efficient lower-level representations for higher-level data objects, storing the values used from a function rather than doing repeated calculation, and finite differencing.

Many lower-level transformation techniques are used in conjunction with functional specifications. For example, Goldberg defines [39]

1. Definition: introduce a new recursion relation.
2. Instantiation: substitute specific values for a variable in an equation.
3. Unfolding: replace a function call by its body.
4. Folding: replace a function body by a call.
5. Abstraction: give a variable name to an expression.
6. Algebraic transformation: apply the laws of algebra.

A typical approach to automatic programming is presented in [6]. The steps of this methodology are as follows:

1. Manual conversion of informal natural language requirements into formal high-level requirements specification using the wide-spectrum specification language (like Gist [46]). This process is currently manual. Experiments in the early 1970's, with automatic natural language translation met with some limited success in several small but highly informal specifications.

2. Verification of high-level requirement specification against the informal requirements. Typical methods for carrying out this step include theorem proving [7], executable specifications, and symbolic evaluation. The work reported by Balzer focuses upon a semisymbolic evaluation methodology that allows test cases to be partially specified.

3. Use of a semi-automatic interactive translation methodology to produce a low-level optimized specification. This step corresponds most directly to the design phase in the waterfall model. According to Balzer, optimization cannot be fully automated; therefore, this step must include some interactive direction. To achieve this end the specification language Gist was defined to "minimize the difference between the way we think about processes and the way we write about them" [6]. However, this makes the gap wider between the level of requirements specification and a level suitable for automatic compilation.

4. Automatic compilation of low-level specification into optimized executable code.

Research points to the following observations on automatic programming and design [7, 6, 99, 39, 77];

1. A wide-spectrum language is suitable to go from very high-level specifications to an automatic compiler.
2. Some high-level automated design issues are really high-level formal requirements specification issues.
3. The ability to move low-level optimizing design decisions into the compiler is slowly improving.
4. There remains a significant gap between formal high-level specifications and specifications suitable for input to an automated optimizing compiler. The majority of traditional software design currently falls into this gap. This gap is lessening but is unlikely to be completely closed except in very narrow domains in the near future [8].

Feasibility and verification of a design is guaranteed by the transformation processes, provided the transformation algorithms are fully tested [39]. Since the automatic transformations preserve correctness [5], testing of the final product is claimed to be unnecessary. As an alternative, tradeoff optimization is carried out in a restricted domain by a chosen sequence of transformations. Hence, the greatest impact of these techniques is on the algorithmic and data structure issues involved in module design and less on the issues of decomposition.

Knowledge-Based Design and Specification Assistance   The Programmer's Apprentice [7, 99, 75] is an example of knowledge-based design assistance (see also Section 2.1). It emphasizes division of labor between humans and machines. The current focus is on the task of program construction. For this purpose a knowledge-based editor, KBEmacs, has been developed. The knowledge is organized into clichés and plans. Cliché refers to a standard method for dealing with a task. A cliché consists of a set of roles embedded in an underlying matrix. The plan formalism used by KBEmacs represents the structure of programs and knowledge about clichés. The structure of a program is expressed in a flowchart form; data flow as well as control flow are represented by explicit arcs. To represent clichés, added support is provided for representing roles and constraints.

The short-term utility of systems like KBEmacs is to provide an environment that assists designers with presentation of preprogrammed design alternatives, focuses attention on selected issues, records design decisions, and offers assistance in the analysis of complete or partial designs. The current experimental implementation contains a limited number of clichés and requires relatively long execution times. Waters [99] estimates that the number of clichés needs to be increased by a factor of 100 and execution times improved by a factor of 100 to make a practical tool. These formidable targets could perhaps be reduced somewhat by narrowing the domain of application.

Knowledge-Based Algorithmic Design   Mostow [55] has noted that since design is a complex task and AI addresses the automation of complex, knowledge-intensive tasks, it is a natural discipline for the study of design. At the same time he pointed out that

> the study [of design] is still at an early stage in that much of the design process is still poorly understood, let alone automated. We are still developing our models of the design process. Developing better models is the key research problem in AI design today.

An SE Perspective   Within transformational approaches associated with automatic programming, high-level architectural decomposition remains a human activity and is a part of the specification process. However, the choice of design methodology and decomposition criteria is significantly narrowed, since the initial decomposition is dictated by the higher-level abstractions provided by the wide-spectrum languages used.

Traditional SE has approached the design task through the use of modularity, information hiding, and abstract data types, as well as by developing reliable empirical software methodologies. These empirical methodologies (like Jackson's [44]) emphasize formalization of the development process by introducing appropriate models for system specification and design. Textbooks like Pressman [70] and Birrell [11] discuss some modern design methodologies and SE management techniques.

Improved software design methodology will only come by way of a deeper understanding of the tasks involved in real design, the knowledge needed to carry out each task, and the ways in which that knowledge is used within each task. It is not clear how study of small design tasks can be scaled up to large-scale software development. The real process of design within existing and established SE methodologies (e.g., SREM and structured design) has been mostly ignored by AI researchers. However, AI can provide a wide array of insights, methodologies, and tools to aid in understanding the nature and use of knowledge in such situations.

2.5   The Role of Rapid Prototyping

AI has developed a wide spectrum of tools and environments to support its basic prototyping approach to problem exploration and program development. It is tempting to apply these tools to prototyping problems within SE.

An AI Perspective   AI programming developed, in part, to deal with constantly evolving programs with incomplete specifications. The exploratory prototyping paradigm of software development used in AI calls for sophisticated support from the machine to relieve programmers of mundane low-level operations [80]. AI introduced the concept of language-centered programming environments. These environments provide sophisticated debugging, tracing, browsing, and conditional breakpoint facilities [27]. Waterman provides an overview of many of the basic expert system prototype development tools in use today [98].

Prototyping within the expert systems development domain is mainly used for evolution, but prototyping for exploration is also important. Exploratory prototyping concepts are of potential use in the requirements specification stage of SE, since that process is also exploratory in nature [26]. The developing knowledge representation and knowledge acquisition techniques from AI are potentially useful for future improvements to software specification methodology. Corresponding inference techniques can aid in understanding, analyzing and developing specifications. However, several issues should be considered before AI techniques are applied (see Section 2.2 and [106]).

Waterman formalized the rapid prototyping paradigm as applied to expert systems development. His steps are similar to the steps of the waterfall model, but they are designed to be cycled through many times in an evolutionary development spiral, and there is no provision for verification and validation between successive steps of Waterman's model.

Waterman's five steps are as follows:

1. Identification. The identification step identifies the important features of the problem, problem type, problem scope, participants in the development, and required resources.

2. Conceptualization. During conceptualization the knowledge engineer and expert decide on the concepts, relationships, and control mechanisms needed to describe the problem and its solution.

3. Formalization. In the formalization stage, key concepts and relations are expressed in some formal way. Often this is done using the syntax of some formal expert systems development tool.

4. Implementation. During implementation, the knowledge engineer turns the formalized knowledge into a working computer program. This is often done by using an expert systems prototyping tool.

5. Testing. Testing involves evaluating the performance and utility of the prototype program and revising it by revisiting steps 1 to 5 as necessary.

An SE Perspective   Rapid prototyping is the dominant development strategy within the expert systems domain [41]. This approach is used to deal with uncertain and rapidly changing requirements. Such requirements are also common in large-scale software. Unfortunately, within the SE community, there seems to be little consensus on how to build and maintain prototypes in general or how to most effectively use prototypes as part of an integrated SE approach [1].

The rapid prototyping method of expert systems development is only one of several ways in which the rapid prototyping paradigm is used within SE.

Other SE prototyping paradigms have applicability within AI as well. For example, Figure 14.1 shows the prototyping method used to test and perfect a requirements specification prior to full implementation. The specification postulates a solution to the requirement, and the prototype helps to test the validity and adequacy of that solution. The prototype also helps in the process of refining the specification toward a more adequate solution by enabling many rapid cycles through the specification/prototype loop.

Tests are often focused on selected facets of the specification rather than the full specification; for example:

1. Functional simulation without concern for efficiency or error handling.
2. Simulating user interfaces without functional details.
3. Simulating selected features of the system that appear unclear or ambiguous in the user domain.
4. Simulating a portion of the system for which feasibility may be in question.

Prototyping for testing specifications is part of the requirements analysis process and the specification verification and validation (V & V) process [15].

Figure 14.2 shows an application of the prototype in situations where requirements are unknown or difficult to establish. Examples include the following:

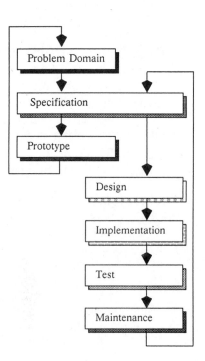

Figure 14.1 A prototyping method for testing specifications.

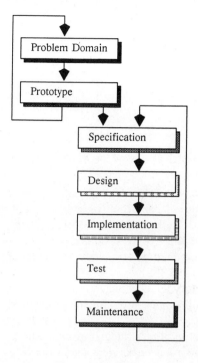

Figure 14.2 A prototyping method for generating specifications.

1. An expert system is being developed for which neither the expert nor the knowledge engineer is able to describe the underlying processes being used by the expert.
2. Requirements are in dispute. Multiple users place conflicting demands on the system and compromise is required.
3. The problem domain is specialized and difficult to understand. There is great risk of miscommunication between user and software engineer.
4. Feasibility is in question.
5. Management demands demonstrable progress throughout the development process.

In these situations, a specification is extracted from the prototype once an acceptable prototype has been completed. This process can be difficult. The prototype can intermingle problem requirements and implementation requirements in a way that is difficult to untangle. Furthermore, the prototype represents only a single solution to the requirement, and thus a specification based on it may be far too restrictive. This may lead to the elimination of many important implementation options.

The prototyping scheme in Figure 14.3 represents a common reaction to the problem of extracting a specification from the prototype. The prototype is used as the production system, at least in some limited sense. The success of this solution depends upon the existence of powerful prototyping tools, very high

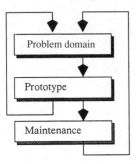

Figure 14.3 A prototyping method for system evolution.

level languages, or fourth-generation languages. The prototyping approach depicted in Figure 14.3 is the currently preferred paradigm for expert system development. There are many expert system shells and other tools available to support this prototyping paradigm within the expert system development domain [98]. Major problems with this solution include efficiency of execution and maintenance over the lifetime of the product [85].

The use of a specification to derive the prototype, as depicted in Figure 14.4, has been suggested as a solution to the maintenance problem for production prototypes [94].

Some of the highly developed rapid prototyping strategies used with AI have potential within SE. However, it is important to be clear about the type of prototyping to be done. It is also important to know the limitations of the AI prototyping paradigm and to recognize that the principal usefulness is in the prototyping concepts and not in the specific tools themselves.

Floyd [36] divides prototyping in SE into three classes: exploration, experimentation, and evolution.

Prototyping for exploration is used to clarify the requirements and test various solution strategies in cases where the developers do not have a very good idea of the domain of application. The prototypes are usually thrown away. Exploratory prototypes can also be considered rapid specification prototypes [49].

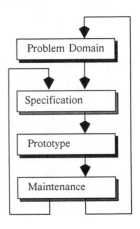

Figure 14.4 A prototyping method supported by a specification.

Prototyping for experimentation tests whether a solution presented by the prototype is adequate. Several strategies can be used in this process: full functional simulation, human/machine interface simulation, skeleton programming, base machine construction, and partial functional simulation.

Prototyping for evolution is used for gradually adapting the system to changing requirements. Prototyping for evolution breaks the linear ordering of development steps into development phases. It can take two forms: *Incremental system development* slowly develops the system at the implementation level. It is based on an overall design. *Evolutionary system development* views development as a sequence of cycles:

1. (re)design
2. (re)implement
3. (re)evaluate

Evolution can take place at any phase of the software development.

Floyd's categories of prototypes for experimentation, exploration, and evolution correspond to the uses of prototyping depicted in Figures 14.1, 14.2, and 14.3 respectively.

The testing phase in the prototyping paradigm needs to be improved when used for experimentation and exploration. Since the process of exploration or discovery is not well understood, strong testing techniques are essential to rule out inadequate explorations.

Maintenance of a prototype can become a serious problem as the prototype grows in size. The use of a separate specification as part of the prototype process can reduce this problem [105].

There is a tendency for a prototype to become large and complex when used for evolution. When this happens, it is common to use the prototype as a requirement for more formal development [98]. However, the true requirements tend to become intertwined with the prototype's implementation features so that the two cannot be separated. Maintaining a separate specification along with the prototype is a solution to this problem.

3 SE FOR AI

This section explores two common AI attitudes toward SE. These attitudes are (1) that SE is predominantly the use of the waterfall model, which is not suitable for AI; and (2) that expert systems cannot be specified, and hence SE approaches do not apply.

3.1. The Role of the Waterfall Model

SE is trying to find practical methods (within the current state of the art of hardware and software) for controlling (managing, predicting) software processes.

Therefore the methodologies of practical large-scale software development play a central role in SE. The waterfall model of the software development process has been a particularly convenient vehicle for communicating ideas in SE. However, the popularity of the waterfall model and the simplicity of the language often used to describe it have led to some unfortunate misunderstandings.

An AI Perspective AI is exploratory in nature and must use methodologies that allow cycling through a rapid succession of exploratory implementations. The waterfall model from SE is viewed as a long linear process of requirements analysis, specification, design, implementation, and testing. This model of development is far too rigorous and lengthy for the needs of AI. For example, Balzer [6] writes,

> Software is produced via some form of the "waterfall" model, in which a linear progression of phases is expected to yield correct results. Rework and maintenance are afterthoughts.

Partridge [66] discusses methodological issues in development of AI software. He argues that software must be developed in an evolutionary manner using a Run-Debug-Edit (RUDE) paradigm of development, because AI problems are so ill structured that specifications are not available until the code is implemented and tested, that is, prototyped.

An SE Perspective SE is not synonymous with a sequential waterfall model of software development. It is a misconception that each stage of development has to be completely finished before going on to the next stage. SE calls for disciplined development, but not for step-by-step forward-only sequential development. Boehm [16] points out that the original waterfall model of software development was proposed by Royce [78]. Royce pioneered the waterfall model (which should be properly called waterfall with feedbacks), but also explicitly included prototyping. In particular, he wrote,

> If the computer program in question is being developed for the first time, arrange matters so that the version finally delivered to the customer for operational deployment is actually the second version in so far as critical design/operations areas are concerned.

The waterfall model of software development is a useful idealization for studying different software development paradigms [72]. No software development paradigm matches the waterfall model exactly, but every software development process either implicitly or explicitly goes through the different stages of this model.

The conventional SE life cycle or waterfall model is shown in Figure 14.5. It defines specific steps through which each software development project must pass [72, 11]. The model includes verification, which is a well-defined process of feedback from one step in the waterfall to its preceding step to insure

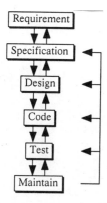

Figure 14.5 Waterfall.

compatibility between the two and in particular to ensure the correctness of any step in terms of its predecessor.

Following Charette [24], the waterfall steps can be described as follows:

Requirements Specification Extracting and translating user needs and requirements into a precise description. This is a complete description of the problem that needs to be solved.

Design Creation of a model of a software system that is consistent with the requirements specification and provides a reasonable description for implementation. Design is divided into architectural design and module design. Architectural design identifies the software modules needed to implement the requirements specification and gives precise black-box descriptions of their behavior, including their interactions with all other black-box elements. Module design identifies the specific data structures and algorithms to be used to implement each module.

Implementation Creating an operational software system that implements the design.

Testing Validation that the implementation satisfies the requirements using techniques and criteria established as part of the requirements specification.

Maintenance Tracing and fixing errors in any of the previous steps as errors are revealed by use of the system, changing the system to adapt to changed operating conditions, and changing the system to incorporate new or modified requirements. Incorporating new requirements involves updating all steps of the waterfall model. The other maintenance activities involve as many of the waterfall steps as are affected by the change.

The following paragraphs show how the waterfall model can be used to analyze various important paradigms of development used within SE.

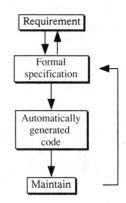

Figure 14.6   Automatic programming.

Automatic Programming   The automatic programming paradigm [76] is characterized in Figure 14.6. As in the waterfall model, initial user requirements are analyzed and a formal specification is created through a process that is predominantly manual. This specification is typically written in an executable, wide-spectrum language [39]. The conventional steps of design and implementation are no longer explicit, but become implicit within an automatic (or semi-automatic) process that converts the specification to a compilable form. This automates much of the design knowledge.

Creation of the formal specification corresponds to the waterfall step of requirements specification. Some aspects of the waterfall design and coding steps are also implicit in this step. The process of using the executable specification to validate against user requirements corresponds to the waterfall steps of specification validation and testing. The (semi-)automatic step of mechanical optimization implicitly contains the waterfall steps of design and coding. The process of maintaining the formal specification and replaying the mechanical optimization corresponds to the waterfall maintenance step.

Operational Approach   The operational approach [102, 103, 1] is depicted in Figure 14.7. This methodology arises principally in response to the need to extend SE into the realm of real-time, embedded, and distributed applications.

Requirements specification is in the form of an operational specification, which specifies the system in terms of implementation-independent structures in an executable format. Internal structure is explicit from the start. External behavior is implicit in the specification; it can be observed and analyzed by executing the specification.

The specification is subjected to transformations that preserve its external behavior but alter or augment the mechanisms by which that behavior is produced so as to yield an implementation-oriented specification. Transformations are based on balancing performance and various implementation resources available.

In the operational approach, all functional mechanisms have been chosen by the time the operational specification is complete. Optimizations of all types

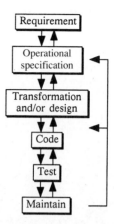

Figure 14.7   Operational approach.

of mechanisms may be interleaved during the transformation phase. A coding phase follows the transformation phase. The transformed specification becomes the detailed requirement for coding in the chosen implementation language. Unlike very high level languages, (see following), the operational approach is not focused on a narrow application domain or a fixed run-time environment. The waterfall design step is embedded in and distributed over the specification and transformation processes.

Maintenance is carried out on the operational specification. The operational specification is optimized for maintainability while the transformations, which are replayed as part of the maintenance cycle, break the specification structure down in favor of something more efficient.

Rapid Prototyping   In the rapid prototyping methodology [11], the software engineer creates a working system that displays selected aspects of the required external behavior. Under this paradigm, the user very quickly gets a working model that can be used to analyze and modify requirements.

The rapid prototyping paradigm is characterized in Figure 14.8. The waterfall steps of specification, design, code and test for the demonstration prototype are all combined within the prototype development cycle. Maintenance is a loop back through the prototype cycle to perfect the demonstration prototype. Rapid prototyping is a large number of repetitions of the waterfall model in miniature [36].

Once the demonstration prototype is perfected, it serves as the requirement specification for a complete waterfall development. Exploratory prototyping is one of the major paradigms used within AI.

Very High Level Languages   Very high level languages (VHLLs) represent another powerful SE paradigm [40]. The term *fourth-generation language* (or 4GL) is commonly used for this approach within the data processing application domain [52]. Under this paradigm, diagrammed in Figure 14.9, the waterfall steps are each carried out, but much of the detailed design and

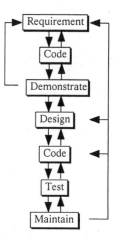

Figure 14.8 Rapid prototyping.

implementation becomes implicit within the VHLL compiler. Requirements specification and design are still needed, but design is aimed toward the VHLL as an implementation language and is generally easier, because most VHLLs are specific to the application domain and thus not too different in concept and structure from the requirement.

Reusable Modules   Reusability is another emerging approach to SE [72]. Major functions can be pre-engineered and then used where applicable in a parameterized fashion. Some examples are an existing data base manager integrated into a major application and parameterized sort routines. Reusability is currently considered an important methodology by AI tool builders [32].

This paradigm includes a dual use of the waterfall process as shown in Figure 14.10. The first track analyzes requirements on a continuous basis within an entire application domain. This analysis is similar to an architectural decomposition and leads to the specification of requirements for individual reusable modules. Each of these modules can be developed by means of the conventional waterfall steps.

Along the second track, an individual application is specified in terms of a configuration of reusable modules. This corresponds to the waterfall steps of specification and architectural design. The modules are customized to the

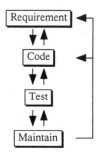

Figure 14.9   Very high level languages (VHLL).

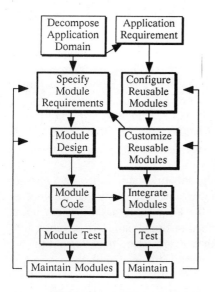

Figure 14.10 Reusable modules.

particular application in a step that corresponds to module design within the waterfall model. This activity places additional requirements on individual module design and modification, which are transferred to the first track of the dual process.

Specific module code comes from module libraries that are maintained along the first track. This corresponds to the waterfall coding step. The waterfall steps of testing and maintenance are carried out as shown in the figure.

We have demonstrated how the elements of the waterfall model are present in each of the SE paradigms. Waterfall steps are commonly used as a framework for discussing SE methods [11] therefore, any observations made about the properties of a waterfall model apply to all software development paradigms.

3.2 The Role of Specification

AI researchers [30, 57, 66] argue that AI programming, particularly expert system development, is exploratory in nature, and thus conventional software development methodologies do not apply. Powerful AI prototyping tools are the solution to this problem. In this section we argue that precisely due to the nature of exploratory environments, SE techniques, such as an explicit statement of requirements and specifications, are necessary and desirable for production AI software. Support for this position is growing even within the AI community.

We first give a brief synopsis of some current AI views that tend to support some form of specification process. We then shift to the SE perspective and summarize a recent SE study focused on the issue of a specification process for expert system development.

An AI Perspective   Chandrasekaran [23] has investigated several high-level building blocks for expert system design. For example, he identifies hierarchical classifications, hypothesis matching or assessment, knowledge-directed information passing, abductive assembly, and others as basic components of an expert system. Such generic tasks can be indentified in AI software, including expert systems. Generic tasks are one way to characterize a software system.

In a series of discussions, Mostow [55, 56] and Partridge [64, 65] have explored various AI design issues. Partridge advocates an incremental design approach, while Mostow emphasizes the need for design specifications and a formal design process.

Brachman [17, 18] indicates that semantic networks and frames, two popular knowledge representations, are ambiguous. He even suggests that it is necessary to have a functional specification of the knowledge-based system under development. KRYPTON, a system described in Brachman [18], severely limits the interface between a user and a knowledge base, thus separating the knowledge base and the implementation details. This is essentially a common SE design technique, namely, separating software specifications from their implementations.

Partridge [66] argues that because AI problems are ill structured, software must be developed in an evolutionary manner using the RUDE (Run-debug-edit) paradigm. *The implementation is in fact a specification.* Even though AI problems are ill structured, *solution techniques* in a given context are always well structured. Thus, it is always possible to write the specification for an AI program. Furthermore, the *process* of developing an AI program should and can be well structured. Making the software development ill structured will aggravate the problems of AI program development.

To aid the software evolution process, Partridge suggests that *abstraction* should be applied during each RUDE cycle. In other words, during each cycle, AI programmers should abstract the underlying theory from the code. These abstractions can have multiple levels as the multiple decompositions in the software specifications. Both regular software specifications and the abstractions advocated by Partridge are useful for software evolution.

However, there are some important differences between software specifications and software abstractions. Software specifications are developed before any code is written, and they are changed as software evolves. The changes may be driven from the top, when a change is made in a higher-level specification, or from the bottom, when an error is found in design, implementation, or testing. Software specification is an integrated part of a software methodology. On the contrary, abstraction of the underlying theory from code is an after-the-fact activity, and it is a much harder task than software specification because it requires reverse engineering, deducing the intentions of the designer from code. For example, Parnas and his team attempted to document and formalize (i.e., abstract) the intent for the real-time software controlling a military aircraft [42]. The major lessons learned from that experiment are that it is not easy to do and that all the documentation should be done during the development process. Software abstractions can be considered a part of software specifications, as they mainly deal with the process from the bottom up. If AI software can be abstracted, it can be specified as well.

An SE Perspective   Development of explicit specifications can aid in cutting down errors [72] and facilitate maintenance. An immediate question is whether it is possible to write a nontrivial specification for, say, an expert system. A pilot study has shown that a separate specification is feasible and desirable as part of the expert system development process [95].

The pilot study compared a standard expert system prototyping approach (the control group) with three specification-based approaches: a conceptual-structures approach, a conventional object-oriented specification approach, and a knowledge model–based object-oriented approach. All four approaches were analyzed on the basis of the resulting product (an expert system requirements specification or a prototype) and the process (methodologies) used to obtain, maintain, and use the product. Both the product and the process were analyzed by a domain expert and a software engineer.

This study suggested that it is feasible to specify an expert system. In particular:

1. The three specification-based approaches were successful in creating a specification for the target expert system.
2. The three specification-based approaches compared favorably with the control group.
3. The ability of the expert to understand the specification and work with the developer is significantly better for the knowledge model approach than for any of the others.
4. The comparative experimental technique is useful for providing valuable insight into the feasibility and relative merits of competing approaches to expert system development.

A brief description of the study follows:

The Task   The domain of expertise considered for this problem was auditing. The particular procedure chosen was the "going concern" audit judgment made during a concurring partner review. The fact that this task is typically assigned to senior audit partners suggests that a great deal of expertise is required to assess a client's going concern status. The assessment consists of a high-level review of the reasonableness of the company's financial statements.

Specification Techniques   The four competing teams were made up of graduate students in SE and AI. All team members had similar experience with the basic techniques used by their team.

Team 1   Team 1 (the control group) followed a standard expert system rapid prototyping development approach. This approach was based on Waterman's model of the expert system development process [98].

The specific prototyping tool used was AGNESS [83]. AGNESS is a network-based expert system shell using forward reasoning when propagating probabilities, and backward reasoning when explaining.

Team 2   Team 2 adopted a *software-engineered* version of the standard expert system development methodology. This approach follows the Waterman model with the following additions:

1. A class-oriented requirements specification technique was used for conceptualization rather than some existing expert system building tool.

2. The process includes a perfective loop from prototypical test back through reconceptualization, reformalization, reimplementation, and retesting, as in the Waterman model. There is an additional inner loop from formalization back through reconceptualization and reformalization. This loop provides verification and validation of the specification prior to prototype implementation and testing.

3. The resulting specification serves as the basis for implementing a detailed prototype using existing expert system tools and/or formal software-engineered system development. The resulting specification also serves as a basis for future expansion and maintenance.

Team 3   Team 3 adopted a conceptual-structures methodology for creating a specification of an expert system.

The conceptual-structures representation is a flexible and powerful knowledge representation that provides guidance to the knowledge engineer. The representation is described in detail in Sowa [86, 87], and Gardiner [37].

This methodology uses six types of conceptual structures:

1. Type definitions
2. Schemata
3. Implications
4. Facts
5. Type lattices
6. Canonical graphs

The conceptual-structures approach allows rapid prototyping and iterative development of an expert system. The methodology consists of five phases:

1. Requirements analysis
2. Knowledge acquisition
3. Knowledge specification
4. Verification
5. Validation

Team 4 Team 4 adopted a knowledge model methodology. The knowledge model methodology distinguishes between Newell's knowledge-level specification and the symbol-(program-) level specification [61]. The knowledge model methodology is built around goal/subgoal structures and concepts used by the expert.

The knowledge model methodology is significantly different from other methodologies because it stresses specification of a problem rather than specification of a system to solve the problem. The methodology is also different in that it concentrates on the conceptual adequacy of the specification, as opposed to concentrating on the blueprint adequacy (implementability), as most specifications do.

Summary of Evaluation Table 14.1 provides a summary of the evaluation of the four approaches by the domain expert (user) and the software engineer. The numbers in each category are normalized, with 100 as a perfect score in each case.

We conclude that the experimental methodology borrowed from SE provides insights to the feasibility and relative merits of competing approaches to difficult problems in expert systems development. This pilot study is only one sample, but it demonstrates the value of the experimental technique. The study seems to indicate the feasibility of a requirements specification for expert systems. Three very different specification approaches drawn from SE all compare favorably in most respects with the conventional prototyping approach to expert system development.

The ability of the expert to understand the specification product and the expert's confidence in the specification process (i.e., the contractual adequacy of the process) is best for the knowledge model approach (Team 4). This correlates well with the fact that the knowledge model approach separates knowledge issues from implementation issues to a much greater extent than the other approaches. The apparent advantages of the knowledge model approach are offset to a certain degree by some perceived inadequacy of its requirements specification in the SE sense.

This study shows the potential usefulness of SE-related specification techniques in facilitating dialogue among knowledge engineers, software engineers, and experts during the early stages of expert systems development. Table 14.1

TABLE 14.1 Comparison of Specification Techniques for Expert Systems Development.

Approach	Understandability Rated by User	Overall Rating Made by User	Overall Rating by Software Engineer
1. Standard prototype	50	41	34
2. Object-oriented	68	63	74
3. Conceptual-structure	35	54	49
4. Knowledge-based	75	69	52

indicates the quality of this dialogue was generally better with the three approaches using explicit specifications than with the prototyping approach.

4 SUMMARY

We need to overcome the cultural clash between AI and SE before we can make effective progress. There is much to be learned from each side. SE scientists must understand the knowledge involved in the real SE process and make it explicit. Conversely, AI software does need an SE approach as it enters the commercial marketplace. The cultural clash discussed in this paper is summarized below.

Automatic Programming

AI: The automatic programming approach to software development is the solution to SE problems.

SE: Automatic programming has limited scope and is only one of many proposed approaches to SE.

Expert Systems

AI: Regardless of maintainability problems, expert systems can be used as an exploratory tool to codify and formalize knowledge in SE processes.

SE: Expert systems cannot help in SE processes, because they are difficult to maintain.

AI Environments

AI: AI environments are the key to many SE problems.

SE: AI environments are powerful, but they need to be augmented with methodologies of use.

Software Design

AI: Knowledge-based techniques can be used to automate parts of software design.

SE: So far, AI can tackle small-scale design only; we need to look at the real software design domains.

Prototyping

AI: Rapid prototyping is uniquely used in AI to deal with evolving programs with incomplete specifications.

SE: Many types of rapid prototyping are used to carry out SE tasks, including specification development.

Waterfall Model

AI: SE follows a linear waterfall model.

SE: Waterfall model is an *idealized* stepwise iterative model.

AI Software

AI: AI programs cannot be specified.

SE: Any large-scale production software must be specified.

We end this discussion of the cultural difference by the following quote from Simon [82], which reflects the spirit of this paper:

> It is not a question of whether we want to automate more of this [SE] process, and since the part of the system we want to now automate is a highly unstructured part of the process, it is not really a question of whether we want to use artificial intelligence methods in software engineering: it is a question of whether artificial intelligence is powerful enough, whether we yet know enough about artificial intelligence, of whether it is advanced enough to really help us. Of course, if it is not, we should not just sit here wringing our hands; we just join the AI community, which has always been shorthanded, and help with the task of advancing not only software engineering, but simultaneously with the task of advancing artificial intelligence.

REFERENCES

1. W. W. Agresti, "Tutorial: New Paradigms for Software Development," *IEEE Computer Society,* June 1986.

2. M. W. Alford, "SREM at the Age of Eight: The Distributed Computing Design System," *IEEE Computer,* vol. 18, no. 4, pp. 36–36, April 1985.

3. J. Ambros and V. O'Day, "MicroScope: A Knowledge-Based Programming Environment," *IEEE Software,* vol. 5, no. 3, pp. 50–58, May 1988.

4. J. Bachant, "R1 Revisited: Four Years in the Trenches," *AI Magazine,* pp. 21–32, Fall 1984 .

5. R. Balzer, T. E. Cheatham, and C. Green, "Software Technology in the 1990's: Using a New Paradigm," *IEEE Computer,* vol. 16, no. 11, pp. 39–45, November 1983.

6. R. Balzer, "A 15 Years Perspective on Automatic Programming," *IEEE Trans. Software Engineering,* vol. SE-11, no. 11, pp. 1257–1268, November 1985.

7. A. Barr and E. A. Feigenbaum, eds., *The Handbook of AI,* vol. 2, Addison-Wesley, 1982.

8. D. R. Barstow, "Domain-Specific Automatic Programming," *IEEE Trans. Software Engineering,* vol. SE-11, no. 11, pp. 1321–1336, November 1985.

9. D. R. Barstow, "Artificial Intelligence and Software Engineering," *Proc. Int'l. Conf. on Software Engineering,* pp. 200–211, 1987.

10. L. A. Belady and M. M. Lehman, "The Characteristics of Large Systems," *Research Directions in Software Technology,* MIT Press, ed. P. Wegner, pp. 106–138, 1979.

11. N. B. Birrell and M. A. Ould, *A Practical Handbook for Software Development,* Cambridge University Press, 1985.

12. D. G. Bobrow and M. Stefik, *The Loops Manual,* Xerox Corp., December 1983.

13. D. G. Bobrow, "If Prolog Is the Answer, What Is the Question? or What It Takes to Support AI Programming Paradigms," *IEEE Trans. Software Engineering,* vol. SE-11, no. 11, pp. 1401–1408, November 1985.

14. D. G. Bobrow, S. Mittal, and M. J. Stefik, "Expert Systems: Perils and Promise," *Comm. of the ACM,* vol. 29, no. 9, pp. 880–894, September 1986.

15. B. W. Boehm, "Verifying and Validating Software Requirements and Design Specification," *IEEE Software,* vol. 1, no. 1, pp. 75–88, January 1984.

16. B. W. Boehm, "A Spiral Model of Software Development and Enhancement," *IEEE Computer,* vol. 21, no. 5, pp. 30–36, May 1988.

17. R. J. Brachman, "What IS-A Is and Isn't: An Analysis of Taxonomic Links in Semantic Networks," *IEEE Computer,* vol. 16, no. 10, October 1983, pp. 30–36.

18. R. J. Brachman, R. E. Fikes, and H. J. Levesque, "KRYPTON: A Functional Approach to Knowledge Representation," *IEEE Computer,* vol. 16, no. 10, pp. 67–73, October 1983.

19. R. J. Brachman and H. J. Levesque, "Tales from the Far Side of KRYPTON," *Proc. 1st Int'l Conf. on Expert Database Systems,* pp. 3–43, ed. L. Kerschberg, Benjamin/Cummings, 1987.

20. F. P. Brooks, Jr., *The Mythical Man-Month,* Addison-Wesley, Reading, MA, 1975.

21. F. P. Brooks, Jr., "No Silver Bullet: Essence and Accidents of Software Engineering," *IEEE Computer,* vol. 20, no. 4, pp. 10–19, April 1987.

22. A. van de Brug, J. Bachant, and J. McDermott, "The Taming of R1," *IEEE Expert,* vol. 1, no. 3, pp. 33–39, Fall 1986.

23. B. Chandrasekaran, "Generic Tasks in Knowledge-Based Reasoning: High-Level Building Blocks for Expert Systems," *IEEE Expert,* vol. 1, no. 3, pp. 23–30, Fall 1986.

24. R. N. Charette, *Software Engineering Environments, Concepts, and Technology,* McGraw-Hill, 1986.

25. W. J. Clancey, "Heuristic Classification," *Artificial Intelligence,* vol. 27, pp. 289–350, 1985.

26. B. Cohen, W. T. Harwood, and M. I. Jackson, *The Specification of Complex Systems,* Addison-Wesley, pp. 10-11, 1986.

27. S. A. Dart, R. J. Ellison, P. H. Feiler, and A. N. Habermann, "Software Development Environments," *IEEE Computer,* vol. 20, no. 11, pp. 18–28, November 1987.

28. F. DeRemer and H. H. Kron, "Programming-in-the-large Versus Programming-in-the-Small," *IEEE Trans. Software Engineering,* vol. SE-2, no. 6, pp. 80–86, June 1976.

29. E. W. Dijkstra, *A Discipline of Programming,* Prentice-Hall, Englewood Cliffs, NJ, 1976.

30. J. Doyle, "Expert Systems and the 'Myth' of Symbolic Reasoning," *IEEE Trans. Software Engineering,* vol. SE-11, no. 11, pp. 1386–1390, November 1985.

31. R. O. Duda, P. E. Hart, R. Regoh, J. Reiter, and T. Risch, "Syntel: Using a Functional Language for Financial Risk Assessment," *IEEE Expert,* vol. 2, no. 3, pp. 18–31, Fall 1987.

32. L. D. Erman, J. S. Lark, and F. Hayes-Roth, "Engineering Intelligent Systems: Progress Report on ABE," *Proc. Expert Systems Workshop,* Defense Advanced Research Projects Agency, Information Processing Techniques Office, SAIC Report Number SAIC-86/1701, April, 1986.

33. M. L. Ernest, and H. E. Ojha, "Business Applications for Artificial Intelligence KBS's," *Future Generation Computer Systems,* vol. 2, pp. 173–186, 1986.

34. S. Fickas and P. Nagarajan, "Being Suspicious: Critiquing Problem Specifications," *Proc. 7th Nat'l Conf. on Artificial Intelligence,* pp. 19–24, August 21–26, 1988, Saint Paul, MN.

35. G. Fischer, "Cognitive View of Reuse and Redesign," *IEEE Software,* vol. 4, no. 4, July 1987, pp. 60–72.

36. C. Floyd, "A Systematic Look at Prototyping," *Approaches to Prototyping,* eds. Buddie et al., Springer-Verlag, 1984.

37. D. A. Gardiner, J. R. Slagle, J. W. Esch, and T. E. Nagle, "Selectional Constraints in Knowledge Representation," Technical Report TR 88-15, Institute of Technology, University of Minnesota, April 1988.

38. A. Goldberg and D. Robinson, *Smalltalk-80: The Language and Its Implementation,* Reading, MA, 1983.

39. A. T. Goldberg, "Knowledge-Based Programming: A Survey of Program Design and Construction Techniques," *IEEE Trans. Software Engineering,* vol. SE-12, no. 7, pp. 752–768, July 1986.

40. M. Hammer and G. Ruth, "Automating the Software Design Process," *Research Directions in Software Technology,* pp. 767–790, ed. P. Wegner, MIT Press, 1979.

41. F. Hayes-Roth et al., *Building Expert Systems,* Addison-Wesley, Reading, MA, 1983.

42. K. L. Heninger, "Specifying Software Requirements for Complex Systems: New Techniques and Their Applications," *IEEE Trans. Software Engineering,* vol. SE-6, pp. 2–13, January 1980.

43. D. J. Israel, "Some Remarks on the Place of Logic in Knowledge Representation," *The Knowledge Frontier,* eds. N. Cercone and G. McCalla, Springer-Verlag, 1987.

44. M. Jackson, *System Development,* Prentice-Hall, Englewood Cliffs, NJ, 1983.

45. P. E. Johnson, I. Zualkernan, and S. Garber, "Specification of Expertise," *Int'l Journal of Man-Machine Studies,* vol. 26, pp. 161–181, 1987.

46. W. L. Johnson, "Deriving Specifications from Requirements," *Proc. 10th Int'l Conf. on Software Engineering,* pp. 428–438, Singapore, April 11–15, 1988.

47. G. E. Kaiser, P. H. Feiler, and S. S. Popovich, "Intelligent Assistance for Software Development and Maintenance," *IEEE Software,* vol. 5, no. 3, pp. 40–49, May 1988.

48. E. Kant, "Understanding and Automating Algorithm Design," *IEEE Trans. Software Engineering,* vol. SE-11, no. 11, pp. 1361–1374, November 1985.

49. H. E. Keus, " Prototyping: A More Reasonable Approach to System Development," *ACM SIGSOFT Software Engineering Notes,* vol. 7, no. 5, 1982.

50. P. Langley, "Representational Issues in Learning Systems," *IEEE Computer,* vol. 16, no. 10, pp. 47–51, October 1983.

51. W. A. Madden and K. Y. Rone, "Design, Development, Integration: Space Shuttle Primary Flight Software System," *Comm. of the ACM,* vol. 27, no. 9, pp. 914–925, September 1984.

52. J. A. Martin, *Fourth Generation Languages,* Prentice-Hall, 1985.

53. G. R. Martins, "The Overselling of Expert Systems," *Datamation,* vol. 30, no. 18, pp. 76–80, 1984.

54. J. McDermott, "R1's Formative Years," *AI Magazine,* vol. 2, no. 2, pp. 21–29, Summer 1981.

55. J. Mostow, "Toward Better Models of Design Process," *AI Magazine,* vol. 6, no. 1, pp. 44–57, Spring 1985.

56. J. Mostow, "Response to Derek Partridge," *AI Magazine,* vol. 6, no. 3, pp. 51–52, Fall 1985.

57. J. Mostow, "Forward: What Is AI? and What Does It Have to Do With Software Engineering?" *IEEE Trans. Software Engineering,* vol. SE-11, no. 11, pp. 1253–1255, November 1985.

58. W. Myers, "Introduction to Expert Systems," *IEEE Expert,* vol. 1, no. 1, pp. 100–109, Spring 1986.

59. R. Neches, W. R. Swartout, and J. Moore, "Enhanced Maintenance and Explanation of Expert Systems through Explicit Models of Their Development," *IEEE Trans. Software Engineering,* vol. SE-11, no. 11, pp. 1337–1351, November 1985.

60. P. G. Neuman, "Software Engineering and Artificial Intelligence," ACM Software Engineering Notes, vol. 11, no. 5, p. 2, October 1986.

61. A. Newell, "The Knowledge Level," *AI Magazine,* vol 2., no. 2, pp. 1–20, Summer 1981.

62. D. L. Parnas, "Research Problems in Programming Methodology," *Research Directions in Software Technology,* pp. 352–264, eds. P. Wagner et al., MIT Press, 1979.

63. D. L. Parnas, "Software Aspects of Strategic Defense Systems," *Comm. of the ACM,* vol. 28, no. 12, pp. 1326–1335, December 1985.

64. D. Partridge, "Letter to Editor," *AI Magazine,* vol. 6, no. 3, pp. 48–51, Fall 1985.

65. D. Partridge, "Rude vs. Courteous," *AI Magazine,* vol. 6, no. 3, pp. 28–29, Fall 1985.

66. D. Partridge, *Artificial Intelligence: Applications in the Future of Software Engineering,* Addison-Wesley, Reading, MA, 1986.

67. D. Partridge, "The Scope and Limitations of First Generation Expert Systems," *Future Generation Computer Systems,* vol. 3, pp. 1–10, 1987.

68. R. S. Patil, "A Case Study on Evolution of System Building Expertise: Medical Diagnosis," *AI in the 1980s and Beyond,* eds. W. Eric, L. Grimson and R. S. Patil, The MIT Press, Cambridge, MA, London, England, 1987.

69. H. E. Pople, "CADUCEUS: An Experimental Expert System for Medical Diagnosis," *The AI for Business,* eds. P. Winston and K. Prendergast, MIT Press, Cambridge, MA, 1984.

70. R. S. Pressman, *Software Engineering: A Practitioner's Approach,* 2nd edition, McGraw-Hill, 1987.

71. C. V. Ramamoorthy, Y. K. Mok, F. Bastani, G. Chin, and K. Suzuji, "Application of a Methodology for the Development and Validation of Reliable Process Control Software," *IEEE Trans. Software Engineering,* vol. SE-7, no. 6, pp. 537–555, November 1981.

72. C. V. Ramamoorthy et al., "Software Engineering: Problems and Perspective," *IEEE Computer,* vol. 17, no. 10, pp. 191–209, October 1984.

73. C. Rich and R. C. Waters, eds., *Readings in Artificial Intelligence and Software Engineering,* Morgan Kaufmann, 1986.

74. C. Rich and R. C. Waters, "Artificial Intelligence and Software Engineering," *AI in the 1980s and Beyond,* eds. W. Eric, L. Grimson, and R. S. Patil, The MIT Press, Cambridge, MA, London, England, 1987.

75. C. Rich, R. C. Waters, and H. B. Reubenstein, "Toward a Requirements Apprentice," *Proc. 4th Int'l Workshop on Software Specification and Design,* April 3-4, 1987, pp. 79–86, Monterey, CA.

76. C. Rich and R. C. Waters, "Automatic Programming: Myths and Prospects," *IEEE Computer,* vol. 21, no. 8, pp. 40–51, August 1988.

77. C. Rich and R. C. Waters, "The Programmer's Apprentice: A Research Overview," in *IEEE Computer,* vol. 21, no. 11, pp. 10–25, November 1988.

78. W. W. Royce, "Managing the Development of Large Software Systems: Concepts and Techniques," *Proc. WESCON,* pp. 1–9, 1970.

79. R. C. Schank, "What Is AI, Anyway?" *AI Magazine,* vol. 6, no. 4, pp. 59–65, Winter 1985.

80. B. Sheil, "Power Tools for Programmers," *Interactive Programming Environments,* eds. D. Barstow, H. Shrobe, and E. Sandewall, McGraw-Hill, New York, 1984.

81. H. A. Simon, *The Sciences of the Artificial,* 2d ed., MIT Press, Cambridge, MA, 1982.

82. H. A. Simon, "Whether Software Engineering Needs to Be Artificially Intelligent," *IEEE Trans. Software Engineering,* vol. SE-12, no. 7, pp. 726–732, July 1986.

83. J. R. Slagle, M. R. Wick, and M. O. Poliac, "AGNESS: A Generalized Network-Based Expert System Shell," *Proc. Nat'l Conf. on Artificial Intelligence,* AAAI, pp. 996–1002, August 1986.

84. D. R. Smith, G. B. Kotik, and S. J. Westfold, "Research on Knowledge-Based Software Environments at Kestrel Institute," *IEEE Trans. Software Engineering,* vol. SE-11, no. 11, pp. 1278–1295, November 1985.

85. E. Soloway, J. Bachant, and K. Jensen, "Assessing the Maintainability of XCON-in-RIME: Coping with the Problems of a Very Large Rule Base," *Proc. 6th Nat'l Conf. on Artificial Intelligence,* vol. 2, pp. 824–829, July 1987.

86. J. F. Sowa, *Conceptual Structures: Information Processing in Mind and Machine,* Addison-Wesley, Reading, MA, 1984.

87. J. F. Sowa and E. C. Way, "Implementing a Semantic Interpreter Using Conceptual Graphs," *IBM J. Res. Develop.,* vol. 30, no. 1, pp. 57–69, 1986.

88. A. Spector and D. Clifford, "The Space Shuttle Primary Computer Systems," *Comm. of the ACM,* vol. 27, no. 9, pp. 874–900, September 1984.

89. G. L. Steele, Jr., *Common Lisp,* Digital Press, 1984.

90. L. Sterling and E. Shapiro, *The Art of Prolog,* MIT Press, Cambridge, MA, 1986.

91. W. Teitelman and L. Masinter, "The Interlisp Programming Environment," *IEEE Computer,* vol. 14, no. 4, pp. 25–33, April 1981.

92. B. Thompson et al., "Knowledge-Based Diagnostic Reasoning," *IJCAI* 8, August 1983.

93. W. F. Tichy, "What Can Software Engineers Learn from Artificial Intelligence?" *IEEE Computer,* vol. 20, no. 11, pp. 43–54, November 1987.

94. W. T. Tsai, I. Zualkernan, D. Volovik, and K. Heisler, "Artificial Intelligence Techniques for Software Engineering," TR 87-60, Computer Science Department, University of Minnesota, Minneapolis, MN, 1987.

95. W. T. Tsai et al., "Requirement Specification for Expert Systems? A Case Study," *Technical Report,* Computer Science Department, University of Minnesota, Minneapolis, MN, 1988.

96. H. J. Walker, D. A. Moon, D. L. Weinreb, and M. McMahon, "The Symbolics Genera Programming Environment," *IEEE Software,* vol. 4, no. 6, pp. 36–45, November 1987.

97. D. A. Waterman, "An Investigation of Tools for Building Expert Systems," *Building Expert Systems,* pp. 169–215, eds. B. Hayes-Roth et al., Addison-Wesley, Reading, MA, 1983.

98. D. A. Waterman, *A Guide to Expert Systems,* Addison-Wesley, Reading, MA, 1986.

99. R. C. Waters, "The Programmer's Apprentice: A Session with KBEmacs," *IEEE Trans. Software Engineering,* vol. SE-11, no. 11, pp. 1296–1320, November 1985, .

100. D. Weinreb and D. Moon, "Objects, Message Passing, and Flavors," *Lisp Machine Manual,* Symbolics, Inc., pp. 279–313, July 1981.

101. S. S. Yau, J. S. Collofello, and T. M. MacGregor, "Ripple Effect Analysis of Software Maintenance," *Proc. COMPSAC 78,* pp. 60–65.

102. R. T. Yeh, P. Zave, A. P. Conn, and G. E. Cole, Jr., "Software Requirements: New Directions and Perspectives," *Handbook of Software Engineering,* eds. C. R. Vick and C. V. Ramamoorthy, Van Nostrand Reinhold, pp. 519–543, 1984.

103. P. Zave, "The Operational versus the Conventional Approach to Software Development," *Comm. of the ACM,* vol. 27, no. 2, pp. 104–118, February 1984.

104. I. Zualkernan, W. T. Tsai, and D. Volovik, "Expert System and Software Engineering: Ready for Marriage?" *IEEE Expert,* vol. 1, no. 4, pp. 24–31, Winter 1986.

105. I. Zualkernan, W. T. Tsai, P. E. Johnson, and J. H. Moller, "Utility of Knowledge Level Specifications," *Proc. of 4th Annual Artificial Intelligence & Advanced Computer Technology Conference,* pp. 79–85, 1988.

106. I. Zualkernan and W. T. Tsai, "Are Knowledge Representations the Answer to Requirement Analysis?" *Proc. IEEE Computer Languages,* pp. 437–443, 1988.

APPENDIX TO CHAPTER 14: RECENT DEVELOPMENTS

Many papers related to the interaction between AI and software engineering have appeared since this paper was prepared. Both fields have advanced. For example, in software engineering, object-oriented programming and databases are getting widespread acceptance [A7, A16]; the software development process has received more attention [A1, A5]; existing techniques such as requirements [A3] are also improved. Recently, Knowledge Systems Institute has sponsored a conference dedicated to the interaction between AI and software engineering; that is, how knowledge engineering methods can be applied to software engineering and how software engineering can be applied to knowledge engineering.

AI for Software Engineering

The application of AI techniques for part of the software engineering process has continued at a rapid rate [A8, A9]. We have also analyzed the expertise used in large-scale software maintenance tasks [A6] and built a software prototype based on the expertise [A13]. We have used AI protocol analysis techniques to analyze two problem specification processes [A15]. We have also proposed a research framework for applying AI to software engineering [A12].

Software Engineering for AI

Software engineering is also being used extensively in building AI software [A2]. In a recent article [A4], Fox classifies several beliefs about AI software as *myths*, which are "perceptions not based on any fact," *legends,* which are "perceptions, once based on fact, that have been blown out of proportion," and *facts,* which are "perceptions that have a real basis in fact." He considers the following as legends:

- AI systems are easy to build.
- Rapid prototyping leads more quickly to final solutions.
- AI systems can be easily verified and validated.
- AI systems are easy to maintain.

He considers the following as myths:

- Small prototypes can be scaled up into full-scale solutions.
- Managing AI systems differs from conventional project management.

These viewpoints are consistent with ours in this paper. Fox even says:

Due in part to academic ignorance of requirements for building productional-level systems, an incorrect belief prevails that managing AI system engineering should differ from managing conventional system engineering.

Rapid prototyping is an important means for acquiring problem requirements and specifications, and for eliciting and verifying knowledge maps. This is not meant to circumvent the need for creating requirements, specification, test code, and the like. The success of rapid prototyping indicates that the waterfall model of software development is inappropriate for many software development projects. Instead, the spiral model is the most suitable for managing the construction of conventional and AI-based systems alike.

This is probably the most open statement from an AI researcher describing the need of software engineering for AI systems; it conforms well to Neuman's view that AI systems need to be software engineered [60] and the position of this paper.

The wider acceptance of software engineering techniques in AI software also comes from the fact that many AI systems must be verified and validated before they can be used in mission-critical environments such as a nuclear power plant [A2, A10, A17]. The need of a specification for expert systems is even greater because without a specification [95, 105] many testing techniques cannot be carried out [A14]. We have also analyzed the life cycle of expert systems [A11].

References

A1. V. R. Basili, "Software Development: A Paradigm for the Future," *Proc. of IEEE COMPSAC*, pp. 471–485, 1989.

A2. C. L. Chang, R. A. Stachowitz, and J. B. Combs, "Testing Integrated Knowledge-Based Systems," *Proc. of IEEE Workshop on Tools for AI*, pp. 12–18, 1989.

A3. A. Davis, *Software Requirements: Principle and Design*, Prentice Hall, Englewood Cliffs, NJ, 1990.

A4. M. S. Fox, "AI and Expert System Myth, Legends, and Facts," *IEEE Expert*, vol. 5, no. 1, pp. 8–20, 1990.

A5. W. S. Humphrey, *Managing the Software Process*, Addison-Wesley, Reading, MA, 1989.

A6. P. E. Johnson, W. T. Tsai, and K. Heisler, "Mental Model for Software Maintenance," *Proc. of Naval Surface Warfare Center Re-Engineering Workshop*, Silver Spring, MD, February 1990.

A7. W. Kim and F. H. Lochovsky, eds., *Object-Oriented Concepts, Databases, and Applications*, Addison-Wesley, Reading, MA, 1989.

A8. M. Lowry and R. Duran, "Knowledge-Based Software Engineering," in *Handbook of Artificial Intelligence*, vol. IV, edited by A. Barr, P. R. Cohen, and E. A. Feigenbaum, Addison-Wesley, Reading, MA, 1989.

A9. C. Rich and R. C. Waters, *The Programming Apprentice*, Addison-Wesley, Reading, MA, 1990.

A10. C. Y. Suen, P. D. Grogono, R. Shinghal, and F. Coalier, "Evaluation of Expert Systems," Technology Transfer Series, Institute for Industrial Technology Transfer, Cournay-sur-marne, France, 1989.

A11. W. T. Tsai, K. G. Heisler, D. Volovik, and I. A. Zualkernan, "An Analysis of Expert System Life Cycle," *Proc. of Int'l Conf. on New Generation Computer Systems*, pp. 389–405, 1989.

A12. W. T. Tsai and I. A. Zualkernan, "Towards a Framework for Normative Software Engineering Research," *Proc. of IEEE Workshop on Tools for AI*, pp. 296–304, 1989.

A13. W. T. Tsai, P. E. Johnson, K. G. Heisler, and Y. Kasho, "Computational Theory for Software Maintenance," *Proc. of Naval Surface Warfare Center Re-Engineering Workshop*, Silver Spring, MD, February 1990.

A14. W. T. Tsai and I. A. Zualkernan, "Towards a Unified Framework for Testing Expert Systems," to appear in *Proc. of Software Engineering and Knowledge Engineering*, 1990.

A15. D. Volovik, R. Mojdehbakhsh, and W. T. Tsai, "What Software Engineering Can Learn from Practitioners," to appear in *Proc. of Software Engineering and Knowledge Engineering*, 1990.

A16. A. B. Zdonik and D. Maier, eds., *Readings in Object-Oriented Database Systems*, Morgan Kaufmann, San Mateo, CA, 1990.

A17. D. Zhang and D. Nguyen, "A Technique for Knowledge Base Verification," *Proc. of IEEE Workshop on Tools for AI*, pp. 399-406, 1989.

CHAPTER 15

DEVELOPMENT SUPPORT FOR AI PROGRAMS

C. V. Ramamoorthy, Shashi Shekhar, and Vijay Garg

1 INTRODUCTION

Artificial intelligence (AI) is a growing branch of computer science that studies ways of enabling computers to do tasks that seem to require human intelligence. These tasks include game playing, expert problem solving, natural language understanding, and theorem proving. The AI programs that have been developed perform such tasks with varying degrees of success. Whereas a program that can understand natural language is still a dream, many chess-playing programs can beat expert human players. More successful AI programs include knowledge-based expert systems that are being applied to a wide spectrum of real-life problems from airline catering to oil exploration. These systems can acquire knowledge about a domain from a human expert and then use it to solve routine problems in that area. For example, expert systems can perform almost as human experts in the diagnosis of infectious diseases (MYCIN), finding the structure of chemical compounds (DENDRAL), and performing mathematical symbol manipulations (MACSYMA). These AI systems offer new capabilities to tackle several problems that have been difficult to solve using the conventional algorithmic approach.

However, there are some difficulties in using AI techniques for solving real-world problems. The AI programming style is quite different from that of conventional languages such as Fortran or Pascal. Because of this, there is a shortage of AI programmers and experts. The problem is aggravated by the fact that each area of AI has its own programming style, which is often unsuitable for other applications of AI. Also, AI typically requires a software development environment that supports all phases of the software life cycle: requirements specification, design, implementation, testing, and maintenance. The existing AI development environments support just the implementation phase, which results in programs that are unreliable and hard to maintain.

In this chapter, we discuss improved software development environments for AI programs. The chapter is organized as follows: Section 2 contrasts AI techniques with conventional algorithmic techniques. Section 3 introduces popular AI languages and describes how their features have been combined in expert system shells. Section 4 presents the main theme of the chapter. It starts with an overview of the evolution of development support for AI programming. The support methods divide into five classes: discrete tools, toolbox, life cycle support, knowledge-based tools, and intelligent life cycle support environment. After a discussion of the classification and associated terminology, we take up examples from each class. Discrete tools from Prolog systems are described. The Interlisp environment is introduced as an example of the toolbox method. Since almost no existing systems exemplify other classes, we simply summarize the main ideas behind them. The present commercial AI program development support environments fall into the toolbox class, but they are likely to have knowledge-based tools and life cycle support for AI programs in the near future.

2   INSIDE AN AI PROGRAM

AI programs differ from conventional algorithmic programs. The latter often have a fixed sequence of steps defined precisely by the programmer, whereas AI programs, like human beings, often take a trial-and-error approach. Table 15.1 lists some of the differences between conventional programs and AI programs.

The essence of AI lies in encoding knowledge about a domain and then using it to solve problems in that domain. This task is usually broken into three parts—*knowledge representation* for encoding information about the domain,

TABLE 15.1   Differences between AI and Conventional Programming.

Feature	AI Programming	Conventional Programming
Processing type	Symbolic	Numeric
Technique	Heuristic search	Algorithmic
Definition of solution steps	Not explicit	Precise
Answers sought	Satisfactory	Optimal
Control/data separation	Separate	Intermingled
Knowledge	Imprecise	Precise
Modification	Frequent	Rare

pattern matching for retrieving knowledge relevant to the problem, and *searching* for a satisfactory solution in a large number of solution possibilities.

Knowledge Representation   The amount of knowledge needed for any interesting AI application tends to be very large, and therefore to access, update, and maintain this knowledge, AI programs need good knowledge representation schemes. AI researchers have used Prolog clauses, semantic networks, frames (see Figure 15.1), and many other schemes for representing knowledge, each

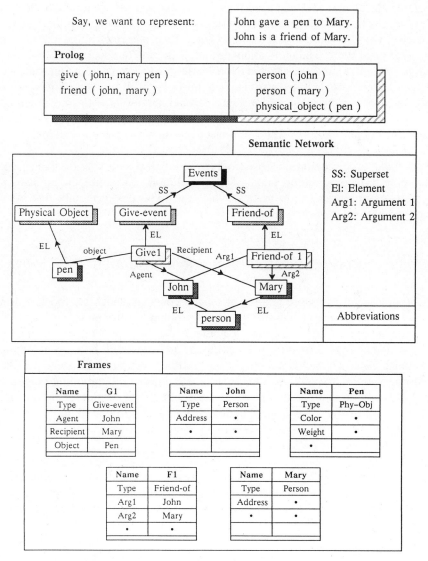

Figure 15.1   Knowledge representation schemes.

having its advantages and disadvantages. For example, a semantic network keeps the related information together but has difficulty inferring from it. Prolog, on the other hand, has mechanisms for logical deduction but does not keep related information together. Most traditional AI schemes, however, are suitable only when all the knowledge is in main memory. If the knowledge base grows too large to fit in the main memory, then disk-based databases have to be used. The design of efficient databases to store rules, objects, frames, and so on is still in the research stage.

Pattern Matching   AI language interpreters use sophisticated pattern matching for selecting relevant rules at any stage of problem solving. Prolog systems match (unify) the current goal against various facts and rules in the database to find pertinent rules. The unification algorithm checks for equality of atoms and bound variables, and tries to bind free variables to the value being matched against. More sophisticated matching, as in OPS-5, allows condition evaluation to check whether the current state of global memory satisfies the specified condition. The conditional expressions can include relational operators such as *equality* and *less-than,* as well as connectives such as *and, not,* and *or.*

Searching   AI programs choose from the set of all possibilities (search space) the most promising one to pursue for the solution. To reduce the amount of search, AI programs use heuristics, which are like rules of thumb, that often succeed but may fail occasionally.

3   AI PROGRAMMING PARADIGMS AND LANGUAGES

A programming paradigm is a set of related concepts which can be useful in conceptualizing, analyzing, and solving a problem; a programming language is a medium in which one or more paradigms can be expressed. For example, object-oriented programming is a paradigm, and Smalltalk is a programming language based on that paradigm. In this section, we summarize some popular AI programming paradigms by presenting, for each, a programming language that is based on that paradigm. We also discuss efforts to provide a homogeneous combination of multiple paradigms in a single programming language. The section ends with a discussion of some new developments that will affect the design of future AI languages.

Lisp: Functional Paradigm   A Lisp program consists of independent functions each of which is a mapping rule that translates its arguments to a result. These functions are created by composing system functions or user-defined functions using control structures such as *if* and *recursion.* System functions include list-processing primitives such as *car, cdr, member,* and *append;* mathematical primitives such as *add, times,* and *division;* and boolean predicates such as *and, negation,* and *equal.*

Another underlying structure used in Lisp is *list*. Lisp provides functions to concatenate, build, extend, and dissect lists. Some examples of lists are (ON TABLE BOOK) and (ADD 3 (ADD 4 8)). Another example of a Lisp function is given in Figure 15.2*a*. Because both data and functions are represented as lists, one can write Lisp functions to manipulate other functions.

A number of versions of Lisp are in use today. MacLisp, developed at MIT, is noted for efficiency, and Interlisp of Xerox Corporation has the most sophisticated program development environment. Common Lisp and Franz Lisp are other popular versions. We will take a look at the Interlisp environment in Section 4.

Prolog: Logic Programming Logic programming is usually based on a theorem prover, which can tell us whether a given set of logical formulas has any contradiction. A Prolog program may be viewed as a collection of formulas in propositional logic with a theorem (query) to be proved. Formulas are restricted to implications, where each side of an implication is a restricted logical expression. They are used to represent various facts (data) and rules of implication. An example of facts and rules is given in Figure 15.2*a*.

Prolog provides a database facility to store facts and rules for efficient manipulation. A Prolog interpreter uses pattern matching (called unification) to select relevant rules from the database to answer a particular query. The user does not have to specify the exact sequence of steps to carry out a computation. This mechanism helps in programming natural language interfaces, because rules of language grammar fit easily into Prolog syntax. Prolog is younger than Lisp, and none of its program development environments is yet as good as that of Interlisp.

3.1 OPS-5: Production Systems An OPS-5 program consists of two parts: a global memory and a set of rules. The global memory can store data structures, which are referenced and manipulated by rules. Each rule has a left-hand side composed of a set of condition elements and a right-hand side that specifies actions (see Figure 15.2*b*). If all the condition elements of a rule match with the contents of global memory, then the rule is applicable. The OPS-5 interpreter selects, in a *recognition-action* loop, all those rules that are applicable and uses a criterion, known as the *conflict resolution strategy,* to chose one rule out of them. This strategy often depends on execution-time parameters, making it difficult to predict the behavior of a program, but it has the advantage of making unplanned responses possible.

Production system languages such as OPS-5 are generally unsuitable for coding algorithms, since it is difficult to do iteration and recursion in them [6]. Also, OPS-5 rules are independent of each other, and cannot be grouped together as a set of related rules in a module, making the development of large programs difficult.

Smalltalk: Object-Oriented Style In this paradigm, a program consists of objects and messages. This way of programming implements modularity by

Lisp: functional paradigm

```
(define is-ancestor (person1 person2)
   (cond ((is-parent person1 person2)      true)
      ((either (is-ancestor person1 (father-of person2))
            (is-ancestor person1 (mother-of person2))   )
         true)))
```

Prolog: Logic programming

```
parent (john, mary).            ; a fact
                       ; rules
Is_Ancestor (Person1, Person2)     :-     Parent (Person1, Person2).
Is_Ancestor (Person1, Person2)   :-
   Parent (Person3, Person2) , Is_Ancestor (Person1, Person3).
```
(a)

OPS-5: Production Systems

```
global memory elements
      (time
            ^date; day, month, and year
            ^hours     ) hour of the day
      (walkman-sales
            ^make     ; manufacturer
            ^model ;   model of the walkman
            ^figure)  sales-figure for the model
- - - - - - - - - - - - - - - - - - - - - - - - - - - - - - - - - - -
productions (rules) :

(p sales_report
   (time ^date  TODAY. ^hours  5pm)
   (walkman-sales ^make sony ^model<model> ^figure <sales>)
   --->
   (write (crlf) The sales figure for sony model <model> is <sales>)
)
```
(b)

Figure 15.2 Glimpses of various programming languages: (*a*) The is-ancestor function takes two arguments and returns true if Person1, the first argument, is an ancestor of Person2, the second argument. The Prolog program for the same purpose would do something extra. It can also find out all ancestors of Person2, or all descendants of Person1; (*b*) This sales report rule will be activated at 5 pm every day and will print the sales figure for all models of Sony Walkman. The names surrounded by angle brackets [< and >] are variables, and the names prefixed by ^ are slot-names.

data abstraction and is specially suitable to situations that have a clear hierarchical classification of objects.

Smalltalk introduces two new concepts in programming: data abstraction and inheritance. Data abstraction is the encapsulation of data-structure definitions with procedures that can directly manipulate the representation of the data.

All other procedures have to send a message to this object to access or update the data stored in it. The concept of inheritance is useful in representing knowledge about hierarchical classification. For example, in the taxonomy used by biologists for living things, the class of humans is a subclass of mammals, and hence it will suffice to store the properties related to mammals at one place with the understanding that all instances of humans will inherit the properties of the super class mammals. This representation will also avoid inconsistencies arising out of duplication of the same information at several places.

Others　　Several other programming models are used by researchers in AI. The ATN (Augmented Transition Network) parser is used in natural language research to implement grammar rules for the English language. ATN parsers are top-down parsers with facilities of special variables, conditions, and actions to aid in programming. A program for this system consists of a set of grammar rules and associated statements to be carried out when a grammar rule is applied. *Frame-based systems* have some features of the object-oriented style and of production systems. Each frame represents knowledge about a stereotypical situation. During execution, the current state can be matched against one part of a frame to check if that frame is applicable. The frame consists of an action to be taken and other conditional checks to see that the action has been performed correctly. The paradigm has been used successfully in understanding stories about certain stereotypical situations.

Towards an Amalgam: KEE, LOOPS, and ART　　There are several programming languages in AI. Each of them was designed with a set of specific problems in mind, and each features some interesting and useful constructs for solving those problems. For example, Prolog and ATN are more convenient for developing *parsers* for a subset of English than Lisp. This multiplicity of languages, however, poses several problems. It makes it difficult for people to share their work and development tools. People also have to learn a new language whenever they try to explore a new problem domain. In view of this, three commercial systems, KEE, ART and LOOPS, have recently tried to combine some features from objects, frames, rules, and so on. They provide an integrated set of tools to allow multiple paradigms in the same program. These systems are called *expert system shells* and they concentrate on the programming needs for building knowledge-based expert systems.

The Knowledge Engineering Environment (KEE) is a frame-based system built on top of Lisp. Frames are used to represent both objects and rules and can be linked together to group them in a hierarchical manner. Frames consist of slots that describe attributes of objects, constraints on their values, sources of inherited values, procedures to be invoked when these attributes are accessed or updated, and rules that are affected by those objects. Lisp functions are used to represent procedures, and predicate calculus is used to represent production rules and relations between various objects. Figure 15.3 shows the knowledge base for a particular process control application. This hierarchically organized knowledge base has a mix of rules, frames, and so on.

A deductive retrieval facility named TellAndAsk [16] provides a Prologlike interface for accessing the knowledge base. Prologlike logic programming is also used for rules interpretation. Rules systems provides some tools for debugging and explanation. A *rule class* graph shows conflicting rules that can be invoked for the same goal. A *how* graph shows the derivation tree for any conclusion.

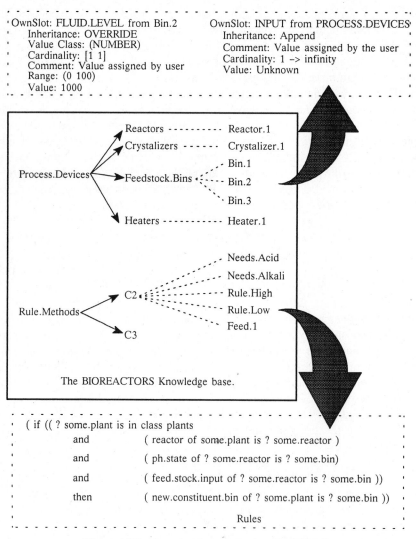

OwnSlot: FLUID.LEVEL from Bin.2
  Inheritance: OVERRIDE
  Value Class: (NUMBER)
  Cardinality: [1 1]
  Comment: Value assigned by user
  Range: (0 100)
  Value: 1000

OwnSlot: INPUT from PROCESS.DEVICES
  Inheritance: Append
  Comment: Value assigned by the user
  Cardinality: 1 –> infinity
  Value: Unknown

Process.Devices
- Reactors --------- Reactor.1
- Crystalizers ------ Crystalizer.1
- Feedstock.Bins
  - Bin.1
  - Bin.2
  - Bin.3
- Heaters ---------- Heater.1

Rule.Methods
- C2
  - Needs.Acid
  - Needs.Alkali
  - Rule.High
  - Rule.Low
  - Feed.1
- C3

The BIOREACTORS Knowledge base.

```
( if (( ? some.plant is in class plants
    and            ( reactor of some.plant is ? some.reactor )
    and            ( ph.state of ? some.reactor is ? some.bin)
    and            ( feed.stock.input of ? some.reactor is ? some.bin ))
    then           ( new.constituent.bin of ? some.plant is ? some.bin ))
```
Rules

Figure 15.3 A typical knowledge base in KEE85b.

LOOPS [24], a system built on top of Interlisp-D, combines the functional paradigm of Lisp, the object-oriented style, and rules [3]. LOOPS represents data structures and rules as objects. It introduces the concept of composition of rules by allowing a rule to call other rules directly or by sending messages. Similarly, objects can be combined or grouped together to form higher-level objects. Procedures defined inside an object can be rules or Lisp functions. Interlisp tools have been upgraded to provide a uniform set of tools for the mixed paradigm.

The ART system [2] combines frames, rules, and processes. It represents facts by logical assertions or frames. For activities involving inferences and consequences, it uses a rule-based paradigm. ART also introduces the concept of a process, which can incorporate rules and procedures with internal control information for doing independent computations.

The Future   The present efforts towards integrating useful concepts of all paradigms can be useful only as short-term measures. These amalgams will need to incorporate other important developments in AI, including representation of belief, cooperating distributed AI systems, and disk-based AI systems [8]. The representation of belief helps in decision-making situations where some of the evidence is in favor of a hypothesis and some against. These uncertain situations are quite common in real-world problems such as medical diagnosis (MYCIN) and oil exploration (PROSPECTOR).

It is difficult for a tightly integrated system to incorporate all the new features. We would rather see an open architecture, which can accommodate newer paradigms, as the choice of the future.

4   DEVELOPMENT SUPPORT FOR AI PROGRAMS

The purpose of development support is to facilitate the creative and intellectual step from the conception of an idea to its description in a computer-understandable form. Peculiarities of tools, programming languages, and user interfaces often present a formidable problem to the user: the user has to spend more time translating an idea into a computer language than polishing the idea to improve it. Development support should relieve the user from as many details as possible and let him concentrate on the idea.

The evolution of development support technology can be traced through five approaches as shown in Figure 15.4.

Tools   Development tools have been around for a long time. Translators of programming languages and debuggers to monitor the execution-time state of programs were some of the first tools. Debuggers, along with screen editors, remain the most useful tools even now [14]. Other popular tools include pretty printers, configuration control support, program cross referencers, and execution profilers.

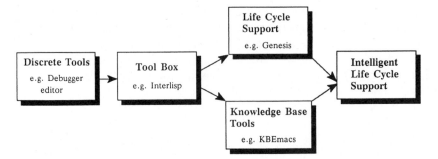

Figure 15.4   Classification of development environments.

These tools were at first discrete and independent: the programmer could access only one at a time. For example, during debugging one had to use the debugger to identify a bug and then terminate debugging to invoke the editor for fixing the bug. After this he would have to compile his program and call the debugger for further debugging. This edit-compile-run cycle forced the programmer to type in extra commands and waste time starting and terminating system tool execution. Interrupting the debugging process to edit, recompile, and rerun the program severely hampered the thinking process of the programmer.

Toolbox   The next development in support environments was the provision of an integrated set of tools, or *toolbox*. Each tool was designed keeping others in mind, and therefore the system could provide access to other tools from inside the executing tool. Interlisp allowed a programmer to run a function (program) right after it was typed into the system. On error, the system would invoke the debugger to allow the programmer to investigate the cause of trouble; the programmer then could edit the function and continue execution. This approach reduced the time spent fixing minor errors during the exploratory (let us try and see) programming commonly used by the AI community. The Unix system provided a loosely integrated set of tools for a wider range of functionality. The editor could be called from inside the debugger, but the programmer still had to recompile and rerun the program. Unix also presented a wide array of tools for data manipulation (*grep, wc, head, more, awk, join,* etc.), rapid prototyping (*csh, awk,* piping, I/O redirection), configuration management (*make, ar, awk*), and many other such routine operations.

Life Cycle Support (Software Engineering)   The design of a large software system is complex and difficult. Its division into several phases—requirement analysis, specifications, design, implementation, testing and debugging, operation, and maintenance [5, 9, 22]—is aimed at reducing the complexity by isolating and ordering the important tasks in the development life cycle.

Unix and Interlisp support only the implementation and debugging phase. Studies show that the largest cost of the life cycle comes from the maintenance phases. The requirement and specification phase are also important, since early errors may be very difficult and expensive to correct in later stages. Life cycle considerations are important for any piece of software that is going to be used for an appreciable length of time by end users. Although the life cycle patterns for most experimental AI programs are different, similar issues are usually involved.

Very few programming environments provide support for the entire life cycle. GANDALF [12], Genesis [21], and SAGA [18] are representative efforts in this direction.

Intelligent Tools   The recent success of AI techniques in several areas motivated their use in the development of software. The Programmer's Assistant project at MIT and the Psi project at Stanford University are two representative systems. These projects model the knowledge a programmer uses in understanding, designing, implementing, and maintaining a program. This knowledge can be used by expert systems to partially automate the program development process. AI techniques are also being used to automate certain optimization decisions like data structure selection. We will describe this approach in more detail later in this section.

Future Intelligent Life Cycle Support   Program development environments of the future will provide knowledge-based tools for life cycle support. AI techniques will support different phases of program development. The RADC report [11] describes knowledge-based tools for developing and validating requirements for any user program. The Fifth Generation Project [19] includes plans to provide intelligent life cycle support.

We now follow our abstract view of support systems with a look at reality. Section 4.1 introduces some *discrete tools* from various Prolog environments. Section 4.2 describes the Interlisp environment, which is an example of a *toolbox*. Section 4.3 discusses some issues in *life cycle support*. Finally, section 4.4 presents some of the existing *knowledge-based tools*.

4.1   Tools from Prolog Environments

This section describes some of the development tools available for Prolog programming. The tools have been taken from various versions of Prolog, including the Prolog-10 family [10], POPLOG [15], Micro-Prolog [7], and APES [13].

Explanation   The explanation of any conclusion (result) derived by the system is the sequence of rules and facts that was used to reach the conclusion. The explanation is like an argument in favor of the conclusion, and of course the user may or may not agree with the argument. This can be used effectively for debugging the knowledge base of rules and facts. The explanation facility presents a good interface for a doctor using an expert system for the diagnosis of a disease. The expert system explains its conclusions, and the doctor remains in control and can accept or overrule them. *Explanation* of APES and *Why* of Micro-Prolog are examples of such facilities.

In some sense, an explanation is like the execution trace of the program computing a result, but it is quite useful if the number of rules and facts applied is relatively small. APES displays an edited form of execution trace showing how rules and facts were used.

Tracing and Debugging   The Prolog-10 family provides a trace package based on the box model, which views the execution of a Prolog program as a sequence of *call, exit, fail,* and *redo* events. Prolog program execution starts when the user types in a query. The Prolog interpreter selects a rule, which can answer the query or break the query into simpler queries. The selection and invocation of a rule is a *call*. If the query is answered, then the invoked rule *exits* successfully. Otherwise the invoked rule *fails* to satisfy the query and other applicable rules are tried (*redo* the query). The user can place watchdog points at the four events (call, exit, fail, and redo) to trace the execution.

Editor and Filing System   POPLOG provides a screen-oriented editor with limited windowing facilities. The editor is integrated with the debugger, and either of them can be accessed from inside the other tool. The editor also has a documentation facility for user programs. The documentation is stored at three levels—*help, teach,* and *reference.* The POPLOG system also has an associated program library, which can be automatically linked to the user programs. It can also be accessed and modified by the user.

Micro-Prolog provides an English-like syntax for entering the Prolog facts and rules to the system. For example, mother (mary, sue) can be paraphrased as *Mary is the mother of Sue.* Many versions of Prolog provide metavariables. A meta-level Prolog program is capable of processing other Prolog programs. This facility makes it easier for the user to write his own development tools in Prolog itself. Small tools such as pretty printers and experimental Prolog interpreters with various chaining/evaluation strategies can be written in Prolog itself.

4.2   The Interlisp Environment

Interlisp, or Interactive Lisp, is an integrated set of tools, or *toolbox,* for developing programs in the Lisp language. It was developed around 1980 and provided one of the most sophisticated forms of development support at that time. It included uniform error handling, automatic error correction, a debugger, a filing system, and a screen-oriented editor [23, 25]. Lisp was enriched with new constructs to access these tools via function calls. The tools are integrated in two ways. (*a*) All the tools are at the programmer's disposal in almost every state of computation. For example, the programmer can access the editor from inside the debugger and then resume debugging. (*b*) A single tool is used for one purpose in all states. For example, the automatic error correction module is used in all other tools for error correction.

The tools are now described individually.

Do What I Mean (DWIM)   The command interpreter of most tools are intolerant of trivial errors by its user. Often a small spelling mistake requires the user to type in the entire command again. DWIM is an automatic error correction facility that is invoked by the system whenever the interpreter detects an error. The DWIM tries to guess what the user might have intended and fixes the mistake. It tries to seek the closest match to the misspelled word in the collection of recently used words by the user. It can take care of spelling mistakes in the command languages of the editor, the debugger, and the file system. DWIM also takes care of many other syntactic errors. The corrections are not always right, and this can cause execution of undesired commands.

Masterscope   During the debugging process, the programmer often wants to look at definitions of some particular functions and variables in his program. If the program is small, a hard-copy printout or the editor window may be adequate, but if it is large (say 10,000 lines scattered over several files and directories, with scores of functions and variables), then retrieval of particular definitions requires a substantial effort. To make this task easier, Interlisp has a tool called Masterscope, which consists of a database of program-related information, a program syntax analyzer (parser) to build the database, and a query language for the database. Masterscope lets the user retrieve the declaration of any program object, such as a function, a variable, or a type, with little effort.

Programmer's Assistant   The interactive environment offers ways to provide a more friendly and helpful design environment to users. A user working on a new program often tries an exploratory approach to design or debug a module. He might modify a file and later want to revoke the decision if the modifications do not yield desirable results. The Programmer's Assistant provides a facility for this kind of exploratory programming. It maintains a *history list* of all the commands issued by the programmer with the side effects (changes in functions, directory, etc.) they caused. The list is used to revoke (undo) a previous command, if this is requested by the user. Similarly, the Programmer's Assistant lets the user repeat (redo) a sequence of past commands. Old commands can be composed together and then modified to form a new command, which saves the user's typing effort and time.

File System, Editor, and Debugger   On any modification, the file system loads, recompiles, or creates backups of relevant files. It gives the user the view that his entire program resides in a database, and functions, variables, and so on can be accessed conveniently via a powerful query language. The editor can be called by a user command or from inside an executing function. The users can add new commands by adding their own functions to the editor module. The debugger is invoked automatically when the executing user program has a major error. The debugger enters a *read-eval-print* loop and lets the user execute arbitrary functions as debugging commands. The functions can look at back-trace or the present state of variables, change the present state, invoke the editor to change a function, and resume execution.

Windows and Graphics   The Xerox 1108 Interlisp-D System uses a large-format display and a mouse pointing device to allow better computer-user interaction. A window system lets the user divide the workstation screen into several logical screens, or windows. Each window is like an independent screen to view a file or to execute a command, and it can display text, graphic images, line drawings, or a combination of all these. A *mouse* is used to select one window out of the existing windows to work in.

4.3   Life Cycle Support for AI Programming

As expert systems are being used increasingly in commercial applications, they should be made reliable and robust. Software engineering has provided concepts for developing reliable programs in conventional applications. It divides the life of a program into phases of requirement analysis, specification, design and implementation, usage, and maintenance. This reduces the complexity of design by grouping and ordering the main tasks of program development. Requirement analysis is aimed at getting a complete, consistent, and unambiguous idea of the needs of the user, which can also serve as a common agreement between the customer and the developer. This is used to chalk out the specification of the software system, against which the software product can be developed and tested. The implementation, testing, and debugging phases need little elaboration. The maintenance phase comprises software updates as well as fixing newly discovered bugs. We will discuss each phase of the life cycle in this section.

Requirements Engineering   AI program development should begin with writing complete, consistent, and unambiguous requirements of the system. The main advantages of writing requirements are as follows:

1. It helps in early detection of errors. Studies [3] have shown that 30 percent of errors are due to faulty specification of requirements, and that errors are 10–100 times more expensive to fix at the implementation stage than at the requirements stage.
2. It helps in the detection of misunderstanding among the customer, the developer, and the user.

3. If requirements are written formally, then expert system programmers can check that all the requirements of the system are met. This is difficult to do otherwise.

4. It helps in generating good test cases and judging the quality of the used test cases.

Design Concepts   A system should be designed using software engineering principles such as information hiding, separation of concerns, layering, and modularity. The principle of *information hiding* can guide us in designing rules. It suggests that rules and their clusters should be designed so that they hide some internal details about the system. For example, if a queue is being used in the system, its priority mechanism should be hidden from the rest of the system. In designing rules, care must be taken that different functionalities are separated and different rules implement these functionalities. For example, rules governing a pressure and temperature control in a system should be different from rules governing security of the system. This is the principle of *separation of concerns*. The principle of *layering* suggests that a system should be designed as a composition of layers. Any layer is aware only of the layers underneath it. Thus activation of rules at higher level due to activation of rules at lower level should be minimized. Such a design simplifies not only implementation but also testing. The principle of *modularity* suggests that rules that access common information and implement related functionality should be clustered into one module for ease in implementation and maintenance.

Testing by Prototyping   Testing of AI systems is different from that of conventional systems. Firstly, AI systems often show nondeterministic behavior, because the *conflict resolution strategy* can depend on some execution-time parameters. This makes the behavior unrepeatable and therefore more difficult to debug. Secondly, there is no precise input-output relationship for rules as in the case of procedures in conventional software. This makes it difficult to use input-output analysis for testing. Thirdly, the number of ways rules can be activated is too large to use branch and path coverage tools.

Prototyping is the only effective way to test an AI program. User experience with a prototype can reveal vital problems in the design. However, prototyping of a large system can take substantial effort and time.

Maintenance   Maintenance and modification are an important part of expert systems development. Rules evolve with the experience of their use and are therefore modified more often than algorithms. In the case of inexact reasoning, the belief support of rules varies with the feedback from previous support numbers. Rules can also be time-dependent, and hence their validity can change over time. Unfortunately, not much attention has been paid to maintenance in expert system programming, though many useful tools for maintenance of conventional programs can be adapted for expert system programming. These tools include version management systems, configuration management systems, and modification request systems. A *version management system* is responsible for storage and easy retrieval of various versions of a program. A *configuration management system* is responsible for keeping the information about various modules of a program and generating the program from these modules. A *modification request system* helps in estimating the impact of a change in the program. Besides the use of these tools, it is also necessary that the program be designed keeping maintenance in mind. This involves proper documentation of all design decisions and the constraints that affected these decisions.

Traceability   Traceability refers to the ability to locate within a database (of program-related information) objects of current interest. For example, if the user changes a particular requirement, traceability enables him to get all specification and code modules that need modification. This can be as useful for large expert systems as it is for conventional software. For example, if the user changes a particular rule, it lets him retrieve all rules that may possibly conflict with the new rule.

Traceability can be provided using database technology: useful relations can be defined to link rules with each other and useful attributes can be defined to classify rules, and then rules can be retrieved using attribute or relation qualification. Examples of relations are *can_activate, semantically_related,* and *mutually_exclusive.* Examples of attributes are *author, date, relevant_requirement_section,* and *version.*

4.4   Knowledge-Based Tools

AI techniques deal with inherently difficult problems that cannot be solved without human assistance. Programming has been one of these tasks requiring a high degree of human intellectual involvement. Naturally, there has been a considerable effort to apply AI techniques to the programming process. People have worked on three different approaches to this problem—the transformational approach to program generation, modeling of the programming process, and using AI techniques to improve other tools.

The transformational approach aims at generating software directly from its specification. This method can be viewed as an extension of work in providing high-level language as a substitute for assembly-level programming. High-level languages provide relatively concise and more understandable notation for programs. They also delegate many lower-level decisions (e.g., register allocation, type checking) and implementation details (storage allocation, recursion) to a compiler and allow programmers to concentrate on the design process. The transformational approach can create a high-level specification language to relieve programmers of intermediate-level decisions (e.g., data structure selections, coding well-known algorithms) for various optimizations. KBEmacs, a knowledge-based editor, illustrates some of these concepts [26]. KBEmacs provides *clichés* and *plans* for creating Lisp or Ada programs with very few commands. A *cliché* is a generic parametrized procedure for performing a certain task (e.g., searching). The user can invoke a cliché and fill in parameters to use it as a part of his program. A good set of clichés can provide a vocabulary of relevant intermediate and high-level concepts for effective realigning and communication. A *plan* provides convenient structures to represent the entire program for various kinds of manipulations by other tools.

One can separate the knowledge about a program into two parts—software engineering knowledge and application-specific domain knowledge. The software engineering knowledge is related to experience in software development (e.g., life cycle models). There have been several efforts to develop a theory of software engineering knowledge—that is, how expert programmers understand, design, implement, test, verify, modify, and document programs [1, 26]. This research can be used to automate some aspects of the program development process. Application-specific domain knowledge is particular to a problem. For example, the laws of mechanics and gravitation will be a part of the domain knowledge of a program guiding the space shuttle. This domain knowledge can be used for automatic checking of various programming decisions for conformance to the physical laws of the application domain. We can also use this knowledge to check the feasibility of requirements, and for validation and testing of the design.

Various AI techniques can add new power to existing development tools. The RADC report describes some applications of AI techniques to acquisition, verification, and validation of requirements [20]. Often a user does not know precisely the features and capabilities he wants from software, so the development of the user requirements for future software resembles the knowledge acquisition process. The knowledge acquisition tools can be used to get the requirements from the user. Applications of AI techniques to software validation include the use of theorem provers that can add the power of deduction to the database storing the program. The rule-based paradigm can make the prototyping of products quick and easy.

The Japanese Fifth Generation Project refers to knowledge-based development support as *systematization technology* [19]. This project tries to use all of the three approaches discussed here to using AI for development support. The project aims to make it possible to synthesize programs from user requirements with automatic algorithm selection from an algorithm bank. It plans to achieve this by developing a theory for specification and verification of programs for automatic synthesis. Other goals of the project include the development of a consultant system for program design and many other tools to maintain, improve, and manage programs. It is the most ambitious project in this direction.

5 CONCLUSIONS

AI programs are no longer mere experimental objects for a small group of researchers in universities; they are now being used for commercial purposes. Hence, development support tools for producing reliable programs in a cost-effective manner are more necessary than ever.

The large number of programming paradigms in use presents a difficulty in developing a standard set of development tools. It also poses a challenge for creating a single environment that provides powerful features of all of the paradigms. Developments like KEE, ART, and LOOPS are a step in this direction. But several issues are still open in designing an open extendible architecture, which could incorporate newer paradigms and features when needed. Presenting a simple and uniform semantic view of such a system to all programmers is a difficult design problem.

AI programming has seldom been concerned with reliability and maintainability of the software, although these are the most important considerations for software that is used for a long time. To make the products more reliable and maintainable, software engineering techniques must be used during development. These software engineering techniques include formal requirements specification, support for maintenance, and enforcement of a software life cycle development methodology. The use of these techniques in the context of AI will become increasingly important in the future.

6 ACKNOWLEDGMENTS

We would like to thank our Berkeley colleagues, namely Prof. Bastani, Abhijit Sahay, Andrew Guest, Atul Prakash, Jaideep Srivastava, Keshav, Moon Kim, and Yih-Farn Chen, for their careful reviews and comments on various drafts of this paper, which improved the readability of the article a great deal. We are extremely grateful to referees, who went through each and every line of the original draft to improve the presentation style and technical soundness.

REFERENCES

1. G. Arrango et al., "Modeling Knowledge for Software Development," *IEEE 3rd Int'l Workshop on Software Specifications and Design*, 1985.

2. ART User's Manual, Inference System Inc., 1984.

3. V. R. Basili and B. T. Pericone, "Software Errors and Complexity: An Empirical Investigation," *CACM*, January 1984.

4. D. G. Bobrow and M. Stefik, "The Loops Manual," Memo KB–VLSI–81–13, August 1984.

5. B. Boehm, "A Spiral Model of Software Development and Enhancement," *Proc. IEEE 2nd Software Process Workshop*, 1986.

6. L. Brownston et al., *Programming Expert Systems in OPS-5: An Introduction to Rule Based Programming*, Addison-Wesley, 1985.

7. K. L. Clark, *Micro-Prolog*, Prentice-Hall, 1984.

8. A. S. Cromaty, "What Are Current Expert System Tools Missing?" *Proc. AAAI Conf.*, 1985.

9. R. Davis et al., "The Dipmeter Advisor: Interpretation of Geological Signals," *Proc. 7th Int'l Joint Conf. on Artificial Intelligence*, August 1981.

10. M. Eisendadt, "PTP: A Prolog Trace Package for the Prolog-10 Family," Technical Report #12, Human Cognition Research Laboratory, The Open University, Milten Keynes, U. K., 1984.

11. K. A. Frenkel, "Towards Automating the Software Development Cycle," *CACM*, vol. 28, no. 6, June 1985.

12. A. N. Haberman, "Special Issue on the Gandalf Project", *The Journal of Systems and Software,* vol. 5, no. 2, May 1985.

13. P. Hammond, "APES (A Prolog Expert System Shell): A Users Manual," Department of Computer Science, Imperial College, London, Doc. Report #82/9, 1982.

14. S. J. Hansen and R. R. Rosinski, "Programmer Perception of Productivity and Programming Tools," *CACM*, vol. 28, no. 2, February 1985.

15. S. Hardy, "A New Software Environment for List Processing and Logic Programming," *Artificial Intelligence: Tools, Techniques, and Applications*, eds. T. O'Shea and M. Eisentadt, Harper and Row, New York, 1984.

16. KEE Software Development System User's Manual, IntelliCorp Inc., January 1985.

17. "The Knowledge Engineering Environment," IntelliCorp Inc., January 1985.

18. P. A. Kirlis et al., "The SAGA Approach to Large Program Development in an Integrated Modular Environment," *Proc. GTE Workshop on Software Engineering Environments for Programming-in-the-Large*, June 1985.

19. T. Moto-oka, "Preliminary Report on Fifth Generation Computer Systems," Keynote Speech, *Proc. Int'l Conf. on 5th Generation Computer Systems*, Tokyo, Japan, 1981.

20. Rome Air Development Center Report on Knowledge Based Software Assistant, 1983.

21. C. V. Ramamoorthy, Y. Usuda, W. T. Tsai, and A. Prakash, "Genesis — An Integrated Environment for Development and Evolution of Software," *COMPSAC*, 1985.

22. C. V. Ramamoorthy, V. Garg, and A. Prakash, "Programming in the Large," *IEEE Trans. on Software Eng.*, July 1986.

23. B. Sheil, "Power Tools for Programmers," *Datamation*, February 1983.

24. M. Stefik, D. G. Bobrow et al., "Knowledge Programming in LOOPS: Report on Experimental Course," *The AI Magazine,* vol. 4, no. 3, Fall 1983.

25. W. Teitelmen and L. Manister, "The Interlisp Programming Environment," *IEEE Computer,* vol. 14, no. 4, April 1984.

26. R. C. Waters, "The Programmer's Apprentice: A Session with KBEmacs," *Trans. on Software Eng.*, vol. SE-11, no. 11, November 1985.

CHAPTER 16

RELIABILITY OF AI PROGRAMS

Farokh B. Bastani

1 INTRODUCTION

AI programs and techniques are increasingly being used for developing intelligent embedded computer systems for factory automation, aircraft control, robot control, and so on. The failure of such real-time systems can result in the loss of life and property. Hence, as these programs shift from an advisory role to being autonomous components in the control loop, it becomes important to determine their reliability prior to putting them into operational use.

Various methods have been proposed for estimating hardware and software reliabilities [4, 16, 31, 34]. Procedures for estimating the reliability of hardware systems are well developed and are used routinely in industry, while the assessment of software reliability has proved to be more difficult and controversial. Several competing techniques and models for determining software reliability are still being refined and subjected to experimental validation.

The failure process of hardware components can be modeled using a bathtub curve for the failure rate, namely, a decreasing failure rate during the burn-in period, followed by a constant failure rate for an interval constituting the useful life of the component, and finally, an increasing failure rate in the terminal phase of the life of the component. The reliability of a hardware system consisting of many components can be computed assuming parallel, serial, and other structures for the system [4]. We can also determine which component has the most significant impact on the reliability of the system. These and some related hardware reliability topics are reviewed in Section 2.

Software does not fail the way hardware does, since there is no physical deterioration such as wear and tear. Instead, software failures are due to residual specification, design, or implementation faults remaining in the program due to inadequate validation and verification. Assuming that inputs are chosen randomly according to some distribution, these faults cause failures that can be modeled as Poisson processes. There are several models for estimating software reliability, including software reliability growth models and sampling models [16, 31, 34]. The reliability of a software system can be computed from those of its components by using a semi-Markov model based on the transition probability from one component to another and the length of stay in a component [8, 20]. Some of these methods are reviewed in Section 3.

Most software reliability models have been developed to be applicable to all types of programs, so that, in particular, they can also be applied to AI programs. However, as we shall see in Section 4, some characteristics of AI programs need to be considered explicitly in the reliability model. These include the following:

1. The correctness of the outputs of some AI programs is a fuzzy [40, 41] rather than a binary quantity, in the sense that we cannot categorically state whether it is correct or not. For example, a program controlling an automated factory may assemble acceptable products though not in the best possible way. This problem is compounded by the fact that deviations from the optimum point are usually not easily discernible.
2. Some of the underlying techniques used in AI programs do not work all the time. For example, the resolution method for proving theorems in first-order predicate calculus terminates only when the theorem is false. Similarly, even the best heuristics may fail to work for certain cases.
3. In planning, there is often a strong correlation between the time spent in forming a strategy and the probability that the strategy is correct. For example, the deeper the depth of a search tree, the more likely that the strategy selected is an appropriate one.
4. The response time and, perhaps, the correctness of the output of AI programs incorporating learning mechanisms is strongly dependent on all prior inputs encountered by them. This history dependency greatly complicates the task of validating such programs.

In Section 5 we propose two modifications to existing software reliability models to take care of points 1 and 2. Firstly, we allow for the possibility that failures may have a certain distribution over the interval [0,1] with 0 indicating that the output is acceptable (e.g., when a summarization program generates a summary that is both concise and captures the full meaning of the input text) and 1 indicating that the output is definitely incorrect (e.g., a summary that is larger than the input text). Secondly, we recognize that for some AI programs there will be a nonzero failure rate even after an infinite amount of testing and debugging, because of deficiencies in the underlying principles used in implementing the program.

In Section 6 we discuss why response time is sometimes a central factor in computing the overall system reliability of embedded AI systems. The main reason is that for embedded computer systems the software determines the mission time. We discuss two cases, namely, (1) when the state of the system can change during the time spent by the controller in deciding what action to take, and (2) when the controller can achieve its task using strategies having different reliabilities. Finally, Section 7 summarizes this chapter.

2 REVIEW OF HARDWARE RELIABILITY

Hardware failures can be attributed to wear-out due to physical deterioration of components, residual design faults, and shortcomings in the manufacturing process. Residual design faults include the existence of inadequately cooled areas (hot spots) and the use of incorrect components, while failures attributable to the manufacturing process include poor soldering and the use of low-reliability components [14]. The reliability of a hardware system can be defined as the probability that the system functions according to its specification in a specified environment for a specified duration called the *mission time*.

Let a continuous random variable T denote the life of a fresh copy of a certain component. Then, by definition, the reliability of the component for a mission time t is $R(t) = P\{T > t\} = 1 - F(t)$, where F is the life distribution of the unit. Now,

$$P\{t < T < t + dt | T > t\} = \frac{P\{t < T < t + dt, T > t\}}{P\{T > t\}}$$

$$= \frac{P\{t < T < t + dt\}}{R(t)} \approx \frac{f(t)dt}{1 - F(t)} = r(t)dt$$

where f is the density of F, and $r(t)$ represents the conditional failure rate or the hazard rate of a t-year-old component. Since $f(t)dt = dF(t) = -dR(t)$, or

$$f(t) = \frac{dF(t)}{dt} = \frac{-dR(t)}{dt}$$

we have

$$\frac{dF(t)}{1 - F(t)} = \frac{-dR(t)}{R(t)} = r(t)dt$$

F is an *increasing failure rate* (IFR) distribution if $r(t)$ is an increasing function of t while it is a *decreasing failure rate* (DFR) distribution if $r(t)$ is a decreasing function of t. The *exponential* distribution with parameter λ, that is, $F(t) = 1 - e^{-\lambda t}$ has a *constant* failure rate $(r(t) = \lambda)$ and is conventionally considered to be both IFR and DFR.

The failure rate function of a hardware component often has a *bathtub* shape, that is, the failure rate initially decreases during the *infant mortality* phase, stays constant during the *useful life* phase, and increases during the *wear-out* phase. Infant mortality is critical for electronic devices since they have a nonzero probability of being nonfunctional before they are used or of failing upon the initial surge of current. Such devices go through a *burn-in* period prior to being put into operational use. On the other hand, mechanical devices are more susceptible to wear-out failures due to friction or increasing brittleness.

Consider a *system* of n components characterized by a *state vector* $\mathbf{x} = (x_1, x_2, \ldots x_n)$, where x_i is an indicator variable defined by

$$x_i = \begin{cases} 1 \text{ if the } i\text{th component is functioning} \\ 0 \text{ if the } i\text{th component has failed} \end{cases}$$

The functioning of the system is indicated by a *structure function* $\phi(x)$, which is 1 if the system is functioning when the state vector is $\mathbf{x}$ and 0 otherwise. For a series structure

$$\phi(\mathbf{x}) = \min(x_1, x_2, \ldots, x_n) = \prod_{i=1}^{n} x_i$$

since the system fails if even one of its components fails. For a parallel structure

$$\phi(\mathbf{x}) = \max(x_1, x_2, \ldots, x_n) = 1 - \prod_{i=1}^{n} (1 - x_i)$$

since the system fails only if all its components fail. Similarly, for a k-out-of-n structure

$$\phi(\mathbf{x}) \text{ is 1 if } \sum_{i=1}^{n} x_i \geq k \text{ and 0 otherwise}$$

that is, the system continues to operate so long as at least k components are functioning. Note that a series system is an n-out-of-n structure while a parallel system is an 1-out-of-n structure.

Let $X_i(t)$ be a random variable indicating the state of the ith component at time t. We shall only consider systems with *permanent* faults [3], so that $X_i(t') = 0 \Rightarrow X_i(t) = 0$ for all $t \geq t'$. The reliability of the ith component for a mission of duration t is $R_i(t) = P\{X_i(t) = 1\} = 1 - F_i(t)$, *where F_i is* the distribution of the life of the ith component. The overall reliability of a system with structure function ϕ is $R(t) = P\{\phi(\mathbf{X}) = 1\} = E[\phi(\mathbf{X})]$, where $E[.]$ denotes mathematical expectation. If the components are independent, then $R(t)$ can be expressed as a function of the component reliabilities, $R(t) = h(\mathbf{R}(t))$, where $\mathbf{R}(t) = (R_1(t), R_2(t), \ldots, R_n(t))$. Thus, for a series $R(t) = \prod_{i=1}^{n} R_i(t)$, for a parallel system $R(t) = 1 - \prod_{i=1}^{n}(1 - R_i(t))$, and for a k-out-of-n system where all components are identical (i.e., $R_i(t) = R_1(t), 1 \leq i \leq n$), $R(t) = \sum_{i=k}^{n} \binom{n}{i}(R_1(t))^i(1 - R_1(t))^{n-i}$. The latter is just the binomial probability that at least k out of n trials are successful.

The reliability importance [4] $I_h(j)$ of component j is defined as

$$I_h(j) = \frac{\partial h(\mathbf{p})}{\partial p_j}$$

where $\mathbf{p} = \mathbf{R}(t)$ and $p_j = R_j(t)$. Assume that we have a system of n independent components, which have been labeled so that $p_i \leq p_{i+1}$ for $1 \leq i \leq n$. Then for a series system

$$I_h(j) = \prod_{i \neq j} p_i, \text{ so that } I_h(1) \geq I_h(2) \geq \ldots \geq I_h(n)$$

Thus, as expected, for a series system the component with the lowest reliability is the most important to the system. Similarly, for a parallel system

$$I_h(j) = \prod_{i \neq j}(1 - p_j), \text{ so that } I_h(1) \leq I_h(2) \leq \ldots \leq I_h(n)$$

Hence, the component with the highest reliability is the most important to the system. For a k-out-of-n system, the most important component depends on the specific values of $p_i, 1 \leq i \leq n$.

Reliability importance can be used to determine which component should be improved in reliability in order to improve the system reliability. In general, other factors, such as the cost of improving the reliability of a component, should also be considered. A reasonable assumption is that this cost increases as the reliability of the component increases. For a series system the reliability can be improved by improving the reliability of its most important component which, being the most unreliable component, also has the least reliability improvement cost. However, for other structures, the selected component may not be the one with the highest reliability importance or the lowest reliability improvement cost. For example, consider a 1-out-of-3 structure with $p_1 = 0$, $p_2 = 0.5$, and $p_3 = 0.9$. Assume that the cost of improving the reliability of component 1 by 0.1 = cost of improving the reliability of component 2 by 0.06 = cost of improving the reliability of component 3 by 0.01. Component 3 has the highest reliability importance and component 1 has the lowest reliability; however, in this case it is more cost effective to improve the reliability of the system by improving the reliability of component 2 since for the same cost this results in a system reliability of 0.956 while improvement of components 1 or 3 yield a system reliability of 0.955. A relevant result for this purpose states that redundancy at the component level is more effective than redundancy at the system level [4]. Formally, $h(\mathbf{1} - (\mathbf{1} - \mathbf{p})(\mathbf{1} - \mathbf{p}')) \geq \mathbf{1} - (\mathbf{1} - h(\mathbf{p}))(\mathbf{1} - h(\mathbf{p}'))$, where $\mathbf{1}$ is a vector of 1's and $-$ is vector subtraction.

One interesting question is whether a system of IFR components is IFR. The answer is no, since it is possible to construct simple systems which are not IFR even though all their components are IFR [4]. An important result in reliability theory states that if component distributions are *increasing failure rate on the average* (IFRA), then the system reliability is also IFRA. Since the average failure rate over time t is $1/t \int_0^t r(t)dt$, that is, $\frac{-1}{t}\log(1 - F(t))$, a distribution F is said to be IFRA if $\frac{-1}{t}\log(1 - F(t))$ is increasing in $t \geq 0$. Similarly, F has a *decreasing failure rate on the average* (DFRA) if $\frac{-1}{t}\log(1 - F(t))$ is decreasing in $t \geq 0$. It can be shown via counter examples that systems of only DFR components or only DFRA components are not necessarily DFR or DFRA, respectively.

So far we have assumed that life distributions are not changed during the life of a component or a system. However, in the development phase of complex systems, the reliability of a system (generally) increases during the development tests as a result of testing the system, finding faults, and fixing them. The Duane model [11] and a modified version developed by Finkelstein [14] can be used to predict the reliability of a system based on its development history. The MTBF (Mean Time Between Failures) of the modified Duane model is given by

$$\frac{1}{\text{MTBF}(t)} = \frac{1}{\theta_L} + \frac{1}{(\mu/\beta)(t/\mu)^{1-\beta}}$$

where θ_L is the limiting MTBF and μ and β are other parameters. Given that the times of uncorrectable faults are $(x_1, x_2, \ldots, x_m, t^*)$ while those of removable faults are $(y_1, y_2, \ldots, y_v, t^*)$ where t^* is the final observed time, the maximum likelihood estimates of the parameters of the model are as follows [14]:

$$\hat{\theta}_L = \frac{t^*}{m}$$

$$\hat{\beta} = \frac{v}{v \log t^{v*} - \sum_{i=1}^{v} \log y_i}$$

$$\hat{\mu} = \frac{t^*}{v^{1/\beta}}$$

Other topics dealt with in hardware reliability studies are systems whose components may have related (as opposed to independent) failure processes due to common stresses, various shock models, and analysis of maintenance policies. Two excellent references are [4] and [3]; the latter contains an overview of hardware reliability, while the former presents an in-depth treatment. Some recent works appear in [35].

3 REVIEW OF SOFTWARE RELIABILITY

The main focus of hardware reliability is on the estimation of the system reliability given the reliability of the individual components. The reliability of a hardware component can be determined by observing the behavior of a random sample of copies of the component under normal operating conditions. This method, however, cannot be used to determine the reliability of a software component, since there is only one copy of each component (or module). In this section we discuss three methods of estimating software reliability. The first approach attempts to predict the reliability of a program on the basis of its failure history, and is similar (in spirit) to Duane's hardware reliability growth model. The second method estimates the reliability of a program on the basis of its behavior for a random sample of points taken from its input domain. In the third technique a number of faults are inserted into the program at the start of the debugging phase. The correctness of the program at the conclusion of this phase is determined on the basis of the number of seeded and actual (unseeded) faults detected by the test team.

3.1 Software Reliability Growth Models

After a program has been coded it enters a testing and debugging phase. During this phase, the implemented software is tested till a failure is encountered. Then the fault causing the failure is located and removed from the program. The *failure history* of the program is defined to be the realization of the sequence of random variables $T_1, T_2, \ldots, T_n$, where T_i denotes the CPU time spent in testing the program after the fault causing the $(i-1)$th failure has been removed till the ith failure is detected. One class of software reliability models attempts to predict the reliability of a program on the basis of its failure history. It is frequently assumed that the correction of a fault does not introduce any new faults into the program. Hence, the reliability of the program increases, and therefore such models are called *software reliability growth* models. These models can be further classified according to whether they express the reliability in terms of the number of faults remaining in the program or not. These constitute fault-counting and non–fault-counting models, respectively.

The failure history of a program depends strongly on the testing process. For example, consider a program that has three paths, thus partitioning the input domain into three disjoint subsets. If the testing is random, then initially failures occur frequently. As the faults responsible for these failures are removed, the interval between failure detection increases, since fewer faults remain. If, on the other hand, one path is tested "well" before testing another path, then whenever a switch is made to a new path the failure detection rate increases. Similarly, if we switch from random testing to boundary value testing, the failure detection rate can increase. Nevertheless, all software reliability growth models assume that inputs are selected randomly and independently from the input domain according to the operational distribution.

Fault-counting models attempt to estimate software reliability in terms of the estimated number of faults remaining in the program. They assume that the failure rates of the faults remaining in the program are independently, identically distributed random variables and that the program failure rate is the sum of the individual failure rates. As an illustration, consider the general Poisson model (GPM) discussed in [2]. It assumes that the failure rate, $r_j(t)$, after the faults causing the $(j-1)$th failure have been removed is proportional to the number of faults remaining in the program and a power of the elapsed CPU time, that is,

$$r_j(t) = \phi(N - M_j)\alpha\, t^{\alpha-1}$$

where N is the number of faults originally present in the program, $M_j = \sum_{i=1}^{j} m_i$, m_i is the number of faults removed following the ith failure, and α and ϕ are constants. Hence, the reliability of the program after the jth failure is given by

$$R_j(t) = e^{-\phi(N-M_j)t\alpha}$$

Given $m_1, m_2, \ldots, m_n$ and $t_1, t_2, \ldots, t_n$, where t_j is the CPU time required to detect the jth failure after the faults causing the $(j-1)$th failure have been removed, the maximum likelihood estimates of the parameters N, ϕ, and α can be obtained by solving the following equations:

$$\sum_{j=1}^{n} \frac{1}{\hat{N} - M_{j-1}} - \sum_{j=1}^{n} \hat{\phi} j^{\hat{\alpha}} = 0$$

$$\frac{n}{\hat{a}} + \sum_{j=1}^{n} \log t_j - \sum_{j=1}^{n} \hat{\phi}(\hat{N} - M_{j-1})t_j^{\hat{a}} \log t_j = 0$$

$$\frac{n}{\hat{\phi}} - \sum_{j=1}^{n} (\hat{N} - M_{j-1})t_j^{\hat{a}} = 0$$

These are discussed further in [2].

Non–fault-counting models consider the effect of a debugging action on the failure rate without concern as to the number of failures detected at a time. An example of a model in this category is the Musa-Okumoto logarithmic model [23]. The inputs to the model are $t_1, t_2, \ldots, t_n$ where t_j is the CPU *time* (not interval as in the GPM model) at which the jth failure was observed. The failure rate is given by

$$r(t) = \frac{\lambda_0}{\lambda_0 \theta t + 1}$$

Thus, the model assumes that the failure rate decreases continuously over the testing and debugging phase, rather than at discrete points corresponding to failure detection and removal times. Further, the rate of decrease in $r(t)$ itself decreases with time, thus modeling the decrease in the size of errors detected as debugging proceeds [21]. The reliability during the jth interval is given by

$$R_j(t) = \left(\frac{\lambda_0 \theta t_j + 1}{\lambda_0 \theta (t_j + t) + 1} \right)^{1/\theta}$$

The maximum likelihood estimates of λ_0 and θ can be obtained by solving the following equations:

$$\frac{n}{\hat{\lambda}_0} - \hat{\theta} \sum_{j=1}^{n} \frac{t_j}{\hat{\lambda}_0 \hat{\theta} t_j + 1} - \frac{t_n}{\hat{\lambda}_0 \hat{\theta} t_n + 1} = 0$$

$$-\hat{\lambda}_0 \sum_{j=1}^{n} \frac{t_j}{\hat{\lambda}_0 \hat{\theta} t_j + 1} + \frac{1}{\hat{\theta}^2} log(\hat{\lambda}_0 \hat{\theta} t_n + 1) - \frac{\hat{\lambda}_0 t_n}{\hat{\theta}(\hat{\lambda}_0 \hat{\theta} t_n + 1)} = 0$$

Further discussions concerning this model appear in [23].

3.2 Sampling Models

This method is similar to the sampling technique used to determine the reliability of hardware components, except that instead of selecting a random sample of components and subjecting them to operational use, here the program is tested with a random sample of points from its input domain. Faults discovered in this process are not removed. If we observe n_f failures out of n runs, then the estimate of the reliability of the program for a single run is

$$\hat{R} = 1 - \frac{n_f}{n}$$

Assuming that inputs are selected independently according to the same probability distribution used to choose the random sample, the reliability of the program over i runs is given by

$$\hat{R}(i) = (\hat{R})^i$$

This method of estimating software reliability is the basis of the Nelson model [24, 37].

While the theoretical foundations of the Nelson model are sound, it suffers from a practical drawback, namely, the need to select a very large number of random test cases in order to have a high confidence in the reliability estimate. The input domain–based model discussed in [30] overcomes this (and other) objections to the Nelson model. It was developed for assessing the reliability of critical real-time process control programs for which no failures should be detected during the reliability estimation phase, so that the reliability estimate is one. Hence, the important metric of concern is the confidence in the reliability estimate. This model provides an estimate of the conditional probability that the program is correct for all possible inputs given that it is correct for a given set of inputs. The basic assumption is that the outcome of each test case provides at least some stochastic information about the behavior of the program for points which are close to the test point. The main result of the model developed in [30] is

$$P\{\text{program is correct for all points in } [a, a + V] \,|$$
$$\text{it is correct for all test cases having successive}$$
$$\text{distances } x_j, j = 1, \ldots, n - 1\}$$

$$= e^{-\lambda V} \prod_{j=1}^{n-1} \frac{2}{1 + e^{-\lambda x_j}}$$

where λ is a parameter that is deduced from some measure of the complexity of the source code.

The foregoing equation is derived under the assumption that the correctness for an input given that the program works for all its neighbors depends only on its distance from its nearest neighbor. More general assumptions, such as the influence of boundary value test cases, result in mathematically intractable derivations. An alternative approach based on fuzzy set theory [40] is proposed in [5, 6, 7] for incorporating more general assumptions. The program is decomposed into equicomputation (or equivalence) classes $E_1, E_2, \ldots$, where the inputs in E_i are subjected to similar computations. The computational correctness possibility of E is defined as

$$CC(E) = \min \left\{ 1 - \left(\frac{D_i}{M_i} \right)^{t_i}, 1 \leq i \leq k \right\}$$

where k is the number of output variables, and for the ith output variable D_i is the degree of computation, M_i is the potential input volume, and t_i is the number of *random* test cases. In order to compute the *control flow correctness possibility* of an equivalence class E, we associate a fuzzy set $F_{E,x}$ with membership function $\mu_{E,x}$ with each test case $x = (x_1, x_2, \ldots, x_m)$ contained in E. The *control flow correctness possibility* of E for a set of test cases $X = \{\, x1, x2, \ldots, xn\}$ is defined as

$$CFC(E, X) = \min \{\mu_{E,X}(y), y \text{ is in } E\}$$

where $\mu_{E,X}(y) = P\{y \text{ is in at least one } F_{E,xk}, 1 \leq k \leq n\}$

$$= \max \{\mu_{E,xk}(y), 1 \leq k \leq n\}$$

$$\text{and } \mu_{E,x}(y) = \prod_{j=1}^{m} \left(1 - \prod_{k=1}^{j} d_k\right)$$

in which $d_1, d_2, \ldots, d_m$ is a descending permutation of $(|\mu_{E,xk} - \mu_{E,yk}| 1 \leq k \leq m)$ and $\mu_{E,xk} = a^{xk - x0k}$ where $x0$ is a boundary point. This is used to determine the optimal distribution of test cases. Further details and examples appear in [5, 6, 7].

3.3   Fault Seeding

Fault seeding is an experimental approach to predicting the number of faults in a program. It has been proposed and used by Mills and Basin [22, 34]. Artificial faults are inserted into the program without the knowledge of the test team. Assuming that the seeded faults have the same distribution as the original faults in the program, an approximate estimate for the number of faults remaining in the program given that m_a actual faults and m_s seeded faults have been detected is $m_a(M_s - m_s)/m_s$, where M_s is the total number of faults artifically created in the program.

The problem with the fault seeding approach is that there is no way to ensure that the seeded faults have the same distribution as the original faults. Simple changes to the source code, such as deleting some statements or modifying some expressions, do not adequately reflect subtle design and requirements specification problems. An alternative approach is to use two test teams and compare the set of faults detected by each team. Suppose that the first team finds m_{a2} faults and m_{a12} of these were also found by the first team. Then, an estimate for M_a, the total number of faults in the original program, is $m_{a1}m_{a2}/m_{a12}$. The number of faults remaining in the program is $M_a - m_{a1} - m_{a2} + m_{a12}$. This approach does not have to deal with the issue of selecting artificial faults. However, a problem here is that the faults detected by the two teams are likely to be correlated since faults with a high failure rate have a high probability of occurring in both the sets. Finally, there is no way to make reliability statements about a program given just the number of faults remaining in it. This number, however, may be useful in estimating the resources needed for future maintenance of the program. Some additional results appear in [12].

3.4 System Reliability

Here we briefly consider the problem of estimating the reliability of a *software* system given the reliabilities of the modules constituting the software. The estimation of the reliability of a hardware/software system is discussed in Section 6.

Assuming that only one component is active at a given time, then an approach proposed independently by Littlewood [20] and Cheung [8] can be used to assess the reliability of the software system. The system is assumed to consist of j components, among which control is switched randomly according to a semi-Markov process [33]. Assuming that the failure process of each component is a Poisson process with the ith component having parameter λ_i, and that no failures occur during the transition from one module to another, the overall system failure process can be approximated as a Poisson process with failure rate

$$\lambda = \sum_{i=1}^{k} \sum_{j=1}^{k} \pi_i p_{ij} \mu_{ij} \lambda_i \bigg/ \sum_{i=1}^{k} \sum_{j=1}^{k} \pi_i p_{ij} \mu_{ij}$$

where π_i is the steady-state probability that component i is active and equals $\sum_{j=1}^{k} \pi_j p_{ji}$ subject to $\sum_{j=1}^{k} \pi_j = 1$, where p_{ij} is the transition probability from component i to component j, and μ_{ij} is the mean CPU time spent in component i before switching to component j (the *sojourn* time).

3.5 Discussion

During the first decade of software reliability research the major emphasis was on developing models based on various assumptions. This resulted in a proliferation of models, most of which were neither used nor validated. Currently the consensus appears to be that perhaps there is no single model that can be applied to all types of projects. Hence, one active research area is to investigate whether a set of models can be combined so as to achieve more accurate reliability estimates for various situations. Other research topics include (1) developing methods of analyzing the confidence in the predictions of a model, and (2) using software reliability theory to assist with the management of a project throughout its life cycle.

4 CHARACTERISTICS OF AI PROGRAMS

Most software reliability models have been developed to be applicable for all types of programs. However, in this section we demonstrate that AI programs possess certain features that require modification of conventional reliability models. Conventional programs have a well-defined specification (such as to sort a file) and a well-defined algorithm for solving the problem. If failures are detected, then the fault can generally be attributed to certain sections of the code and debugging actions can eliminate the fault. AI programs, on the other hand, do not have a well-defined *efficient* algorithm and, usually, do not have a well-defined problem specification. Examples of problems that can be completely specified but are difficult to solve efficiently include tasks such as moving an object from position A to position B without any collisions and the traveling salesman's tour. Mathematically provable solutions to these problems are generally NP-complete. The AI approach here is to use heuristics-guided search procedures. Examples of problems that are difficult to specify completely include summarizing a story and understanding natural language. For AI programs, faults may be due to deficiencies in the underlying approach, so that it may not be possible to fix one fault without introducing another. For example, heuristics usually involve tradeoffs and hence will not work in all cases. Similarly, some learning procedures may lose important information prematurely (e.g., induction-based systems) while others may cause the database to overflow with trivial knowledge (e.g., memory-based systems).

Software reliability models for AI programs must take three special considerations into account, namely, the correctness of the underlying approach, the difficulty in recognizing failures, and history dependency when learning is involved. In the following we discuss each of these aspects along with the type of modifications that are required.

There are several competing and controversial methods of developing AI programs. A typical AI program addresses three problem-independent issues, namely, the problem solving method, the knowledge representation method, and the knowledge acquisition method. Some problem solving methods that have been used in practice are goal decomposition, constraint propagation, various search methods, the resolution principle for first-order calculus, generate and test, and forward/backward chaining [32, 39]. All these involve heuristics, such as which subgoals to select, which branch to search next, which clauses to resolve, which rule to apply, and which solution to generate next. The effect of these decisions ranges from variation in the amount of time spent in finding a solution to the correctness of the solution when heuristic pruning is involved.

Knowledge representation methods include procedural semantics, various forms of logic, semantic networks, frames, and production systems [10, 29]. There is a question of which methods (or combinations thereof) are most suitable for a given application, and furthermore, they all have deficiencies in ensuring the completeness of the knowledge base and in representing fuzzy information. The latter requires methods of evaluating the fuzziness of conclusions drawn from fuzzy premises. Some controversial approaches are reviewed in [39].

Finally, two extreme methods of knowledge acquisition are to store the raw data or to generate rules [13, 36, 38]. The former runs into the problem of having to cope with the sheer volume of information, while the latter involves the use of heuristics that do not work perfectly in all the cases.

In summary, the underlying approach affects the duration of the mission time and splits faults into two classes, namely, those that can be corrected (e.g., incomplete or faulty knowledge) and those that cannot (e.g., the use of heuristics). In case of the latter, the failure rate of the software will be nonzero even after infinite testing, as in the case of hardware systems [14].

For conventional programs each run can be viewed as a Bernoulli trial; that is, the outcome is either a success or a failure. However, this sort of binary classification is not possible for certain AI programs. For example, consider a program that summarizes a story, or analyzes the style of an essay, or plans a strategy. We cannot specify whether a particular output is correct or incorrect, since there can be a range of acceptable summaries, writing styles, or strategies, as the case may be. This aspect can be modeled by associating a fuzzy membership function or probability of correctness with each output. Another factor that needs to be considered, given that current AI programs are devoid of common sense, is the severity of failures. For example, an expert system that is excellent on the average but that fails catastrophically occasionally will not be acceptable for real-time critical process control applications. One way of dealing with this is to classify failures into different categories and determine the reliability of the program for each class. This, however, reduces the sample size for each class and thus reduces the confidence in the reliability estimate.

The failure process of AI programs involving knowledge acquisition or learning behavior depends on the history of inputs to the program. This makes the reliability analysis of the program impractical since failure data must be gathered for a large number of different histories, a potentially expensive and time-consuming process. For example, consider a real-time system incorporating AI control. If it learns (discovers) a good strategy under noncritical circumstances, then it can survive under a critical situation by merely recalling the strategy. However, if the inputs are such that it does not get the opportunity to learn and store the strategy, then it can fail disastrously in a critical case [15]. Methods of dealing with this are beyond the scope of reliability analysis alone and are not considered further in this chapter.

In this section we have seen that software reliability models must be modified in order to be applicable to AI programs. In the next section we illustrate this by modifying the Musa-Okumoto logarithmic model for this purpose. The changes model the facts that not all faults in the program can be corrected, since they may be due to the fundamental technique used in implementing the program, and that failures have a distribution over the interval $[0,1]$, with 1 representing a definitely unacceptable output and 0 representing a definitely acceptable output. In Section 6 we consider the effect of computation time on embedded process control systems incorporating AI programming techniques.

5 SOFTWARE RELIABILITY MODEL

All software reliability growth models assume that after an infinite amount of testing the reliability of the program will be one. However, as in the case of hardware [14], this is not true for AI programs, since there can be faults that are an intrinsic part of the approach used in implementing the program and therefore cannot be fixed easily. For example, consider the *hill climbing* search heuristic used in some AI programs. It is susceptible to the following problems [32, 39]:

1. *Foothill problem:* The search may return a local optimum instead of the desired global optimum.
2. *Plateau problem:* The search procedure may not be able to make any improvements in the results obtained so far by probing reasonably nearby points.
3. *Ridge problem:* Inspection of most nearby points shows that the current point is an optimum point even though it is neither a local nor a global optimum.

Another problem is the possibility of hardware limitations for AI subsystems implemented directly in hardware. An illustration is a novel optical system developed for image recognition [1] that requires that images presented to it be properly oriented.

The possibility of there being incorrigible as well as corrigible faults can be modeled by splitting the failure process into two independent Poisson processes, one of which has a decreasing failure rate reflecting the growth in the reliability as a result of faults that are detected and removed, while the other has a constant failure rate and is due to faults that are not removed. For the Musa-Okumoto model, we have

$$\lambda_1(t) = \frac{\lambda_{10}}{\lambda_{10}\theta t + 1}, \lambda_2(t) = \lambda_2$$

During the operational phase, no faults are removed, so that $\lambda_1(t)$ is also constant, say λ_1. Thus the reliability is given by

$$R(t) = e^{-(\lambda_1 + \lambda_2)t}$$

The next step is to model the distribution of failures over $[0, 1]$, with 0 denoting a benign (i.e., no) failure and 1 denoting a definite failure. We assume that all correctable faults are definite failures since they are generally due to design or coding errors. Faults in the underlying approach are assumed to have a distribution G, with $G(0) = 0$ and $G(1) = 1$. The *level of service* provided by the program can be measured in several ways, some of which are discussed in the following.

One approach is to consider the system as having failed if the sum of the failures encountered exceeds some limit, say x. Let x_i be a random variable indicating the level of failure of the ith fault. Then,

$P\{\text{system is alive at } t\}$

$= P\{\text{accumulated failures } \leq x\}$

$= \sum_{m=0}^{|x|} P\{\text{accumulated uncorrectable failures} \leq x - m \,|\, m \text{ correctable failures}\}$

$\times P\{m \text{ correctable failures}\}$

$= \sum_{m=0}^{|x|} \sum_{n=0}^{\infty} P\{X_1 + \ldots + X_n \leq x - m\} P\{n \text{ uncorrectable failures}\} \times \frac{e^{-\lambda_1 t}(\lambda_1 t)^m}{m!}$

$= \sum_{m=0}^{|x|} \frac{e^{-\lambda_1 t}(\lambda_1 t)^m}{m!} \sum_{n=0}^{\infty} \frac{e^{-\lambda_2 t}(\lambda_2 t)^n}{n!} G^{(n)}(x - m)$

where $G^{(n)}(x)$ denotes the n-fold convolution of G, defined as

$$G^{(n)}(x) = \begin{cases} 1 & \text{if } n = 0 \\ G(x) & \text{if } n = 1 \\ \int_0^x G^{(n-1)}(x - y)dG(y) & \text{if } n > 1 \end{cases}$$

If we cannot tolerate unexpected coding/design faults (i.e., correctable faults), then

$$P\{\text{system is alive at } t\} = e^{-\lambda_1 t} \sum_{n=0}^{\infty} \frac{e^{-\lambda_2 t}(\lambda_2 t)^n}{n!} G^{(n)}(x)$$

An alternative service requirement is that the maximum failure level should not exceed some specified limit $\epsilon \leq 1$. For $\epsilon = 1$ the survival probability is obviously 1. Hence, we consider $\epsilon < 1$, so that no correctable failures can occur.

$P\{$ system is alive at $t\}$

$= P\{$no correctable failures and maximum uncorrectable failure $\leq \epsilon < 1\}$

$= P\{$no correctable failures$\}P\{$maximum uncorrectable failure $\leq \epsilon < 1\}$

$= e^{-\lambda_1 t} \sum_{n=0}^{\infty} P\{n \text{ uncorrectable failures}\} P\{X_1 \leq \epsilon < 1, \ldots, X_n \leq \epsilon < 1\}$

$= e^{-\lambda_1 t} \sum_{n=0}^{\infty} \frac{e^{-\lambda_2 t}(\lambda_2 t)^n}{n!} [G(\epsilon)]^n$

$= e^{-\lambda_1 t} e^{-\lambda_2 (1 - G(\epsilon))t}$

We can combine the above two approaches by limiting the level of each failure (given by $\epsilon < 1$) as well as the accumulated level of all faults (given by x). Then the survival probability for time t is given by

$$e^{-\lambda_1 t} \sum_{n=0}^{\infty} \frac{e^{-\lambda_2 t}(\lambda_2 t)^n}{n!} G'^{[n]}(x, \epsilon)$$

$$\text{where } G'^{[n]}(x, \epsilon) = \begin{cases} 1 & \text{if } n = 0 \\ G(min(x, \epsilon)) & \text{if } n = 1 \\ \int_0^{min(x, \epsilon)} G'^{[n-1]}(x - y, \epsilon)dG(y) & \text{if } n > 1 \end{cases}$$

For $x > n\epsilon$, $G'^{[n]}(x, \epsilon) = [G(\epsilon)]^n$.

Three sets of data are required in order to estimate the parameters of the model. These are $(t_{u1}, t_{u2}, \ldots, t_{ur})$, $(t_{c1}, t_{c2}, \ldots, t_{cs})$, and $(f_1, f_2, \ldots, f_r)$ where t_{ui} is the CPU time required to find the ith uncorrectable failure, t_{ci} is the CPU time required to find the ith correctable failure, and f_i is the failure level of the ith uncorrectable failure.

The failure probability density of the uncorrectable failures is

$$PDF_{ui}(t) = \lambda_2 e^{-\lambda_2 t}$$

Hence, the likelihood function is given by

$$L = \prod_{i=1}^{r} \lambda_2 e^{-\lambda_2 t_{ui}}$$

Taking the logarithm of the likelihood function, we get

$$log\ L = r\ log\ \lambda_2 - \lambda_2 \sum_{i-1}^{r} t_{ui}$$

Setting $\partial log\ L/\partial \lambda_2$ equal to 0 yields the maximum likelihood estimate of λ_2, that is, $\hat{\lambda}_2$:

$$\hat{\lambda}_2 = \frac{r}{\sum_{i=1}^{r} t_{ui}}$$

The maximum likelihood estimates of λ_{10} and θ can be obtained by solving the equations given in Section 3 for the Musa-Okumoto model with $\lambda_0 = \lambda_{10}$, $n = s$, and $t_j = \sum_{i=1}^{j} t_{ci}$.

A reasonable model for G is the Beta (α, β) distribution [17] with density

$$g(x) = \begin{cases} \dfrac{\Gamma(\alpha + \beta)}{\Gamma(\alpha)\Gamma(\beta)} x^{\alpha-1}(1 - x)^{\beta-1} & \text{if } 0 \leq x \leq 1 \\ 0 & \text{if } x < 0 \text{ or } x > 1 \end{cases}$$

The maximum likelihood estimates of α and β can be obtained by numerically solving the following equations:

$$r\frac{\dfrac{\partial \Gamma(\alpha + \beta)}{\partial \alpha}}{\Gamma(\alpha + \beta)} - r\frac{\dfrac{\partial \Gamma(\alpha)}{\partial \alpha}}{\Gamma(\alpha)} + \sum_{i=1}^{r} log\ f_{ci} = 0$$

$$r\frac{\dfrac{\partial \Gamma(\alpha + \beta)}{\partial \beta}}{\Gamma(\alpha + \beta)} - r\frac{\dfrac{\partial \Gamma(\beta)}{\partial \beta}}{\Gamma(\alpha)} + \sum_{i=1}^{r} log(1 - f_{ci}) = 0$$

$$\text{where } \frac{\partial \Gamma(\alpha + \beta)}{\partial \alpha} = \int_0^\infty (log\ x) x^{\alpha+\beta-1} e^{-x} dx$$

A less general, though simpler, model is to consider a single-parameter beta distribution with α equal to 1. In this case, the density is $\beta(1 - x)^{\beta-1}$ for $0 \leq x \leq 1$ and 0 otherwise. The maximum likelihood estimate of β is

$$\hat{\beta} = \frac{r}{\sum_{i=1}^{r} log\ 1/(1 - f_{ci})}$$

6 SYSTEM RELIABILITY

Embedded systems incorporating AI programs include robots, automated manufacturing systems, and some defense systems. These are embedded computer systems consisting of both hardware and software components. A simple way of estimating the reliability of an embedded AI system is to view it as a series connection of hardware and software components, so that

$$R_{\text{system}}(t) = R_{\text{hardware}}(t) \times R_{\text{software}}(t)$$

However, this equation ignores the fact that the length of the mission time is frequently determined by the software. For example, the time taken by a robot to complete a certain task depends on the strategy selected by the program. Since exhaustive search is usually not practical, the optimality of the strategy depends on the heuristics used to guide the search procedure. Further, there is often a tradeoff in the optimality of the strategy and the time spent in computing the response. For example, in game playing situations, the greater the depth of the evaluation tree (and hence the greater the computation time) for determining the next move, the more likely that the move is correct.

In this section we focus on process control applications incorporating artificial intelligence techniques. Process control programs are real-time programs for applications ranging from controlling the gasoline intake in automobiles to landing an aircraft. The programs can be characterized by a predicate, $G(.)$, on the state of the system, this being their control *task* or *goal*. To achieve its goal, a control program monitors its *sensors* to determine the current state of the environment. Then it computes an appropriate response, and finally it issues commands to *actuators,* which make the desired changes in the state of the environment. These steps are repeated till the control task is achieved. The problem is how much processing the controller should do before deciding how to react to a situation. The more the processing time, the greater is the probability that the decision is correct. But longer processing also means that the situation is more likely to have changed so that the decision is no longer valid. It could be the case that a more reliable response takes a longer time to implement. For example, consider the action of crossing a road. If a pothole is encountered, it is more reliable to move around it than to jump over it. However, if cars are approaching rapidly on the road, then the *system* reliability may be improved by jumping over the pothole since the mission time is smaller than moving around it.

Section 6.1 discusses the system reliability for several search techniques, such as *generate and test* and *heuristic pruning* [32, 39] and derives the amount of time to be spent in the *compute response* phase to optimize the reliability of the system for a particular example. During this phase, the more time spent in deciding what the response should be, the more likely it is that the response is correct. However, for a real-time system, longer processing also means that the situation is more likely to have changed so that the response is no longer valid. Section 6.2 discusses the *response execution* phase, when there are several possible responses for a particular situation, each with a different probability of success and time to achieve the goal. If the more reliable responses take a longer time, then a tradeoff is involved in computing the system reliability. For example, as discussed above, jumping over a pothole encountered while crossing a road may be less reliable than going around it, but it may be preferable in some cases since it is faster. Another example is when responses require different numbers of actuators. For example, a reliability tradeoff is involved when a faster response requires two actuators (a series connection) while a slower response can use either of the actuators (a parallel connection).

6.1 Response Computation Time

Normally, in a real-time process-control system, there is a stringent real-time constraint that must be satisfied. When a real-time situation arises, there is a response computation period in which an optimal or near-optimal strategy must be formulated. This is followed by a response execution time that activates the underlying hardware mechanisms and carries out the strategy. Very frequently, if the response computation period for formulating an optimal strategy is too long, there is a high risk that the response may not be completed within the real-time constraint since there is not enough time left for response execution. On the other hand, if a non-optimal strategy is selected in order to meet the real-time constraint, the reliability of the resulting strategy may not be acceptable because of the poor hardware reliability associated with the strategy selection. Thus, there is a tradeoff between response computation time and response execution time under a real-time situation. In this subsection, we illustrate this tradeoff by investigating some AI search procedures. Specifically, we investigate the employment of A^*, which is known to be optimal [26], with some other search heuristics, which we show may provide a better overall system hardware/software reliability under certain conditions. Our intention is not to explore findings of heuristics that would lead to better real-time performance of search procedures [27]. Instead, we are interested in identifying the conditions under which a search strategy may provide a better system reliability than the others. Also, we restrict our analysis to a very simple problem space so as to obtain meaningful results.

6.1.1 Generate and Test Procedure. As our first example, consider a generate-and-test search procedure, wherein each step takes T time units to process and has a probability p of passing the test. Suppose that this procedure is part of a hardware system where the hardware has a constant failure rate λ. Then,

$P\{$Search procedure terminates and the system is alive$\}$

$$= \sum_{i=1}^{\infty} p(1-p)^{i-1}e^{-\lambda iT}$$

$$= pe^{-\lambda T}\sum_{j=0}^{\infty}[(1-p)e^{-\lambda T}]^{j}$$

$$= \frac{pe^{-\lambda T}}{1-(1-p)e^{-\lambda T}}$$

6.1.2 Branch and Bound Procedure.   The generate-and-test technique is not suitable for cases where optimal or reasonably optimal solutions are required, since all solutions would have to be inspected. One alternative is to use the branch-and-bound method along with heuristics to limit the search [32, 39]. Let the branching factor of the resulting tree be b and the maximum depth of the tree be D. Let the probability that the heuristic results in a correct answer (the reliability of the heuristic) be r. Also, assume that the probability that a correct answer is found when the depth of the search tree is d (independently of the heuristics used) is d/D. Now, if the depth of the search tree is limited to d, then between $2b^{d/2}$ and b^{d} nodes have to be examined [39]. Let the probability that i nodes are examined be $h(i)$. Finally, assume that this is a part of a hardware system having failure rate λ. Then, if T time units are required to analyze each node, the probability that the system will complete the search successfully is

$$R(r,d) = r\frac{d}{D}\sum_{i=2b^{d/2}}^{b^{d}} e^{-\lambda iT}h(i)$$

For the case where h is the uniform distribution over $[2b^{d/2}, b^{d}]$, we obtain

$$R(r,d) = \frac{rd}{D}\frac{1}{b^{d}-2b^{d/2}+1}\sum_{i=2b^{d/2}}^{b^{d}} e^{-\lambda iT}$$

$$= \frac{rd}{D}\frac{e^{-\lambda(2b^{d/2})T}-e^{-\lambda(b^{d}+1)T}}{(b^{d}-2b^{d/2}+1)(1-e^{-\lambda T})}$$

The optimal value of d can be obtained by solving $\partial R(r,d)/\partial d = 0$.

An alternative approach is to examine each node up to depth d without using any heuristics. In this case $r = 1$, and the reliability of the system is

$$R(d) = \frac{d}{D} e^{-\lambda b^d T}$$

In this case the optimal value of d is obtained by solving

$$d\lambda T (\log\ b) b^d = 1$$

It is interesting to note that the optimal value of d is independent of the maximum depth of the tree D (of course, $1 \le d \le D$). Comparing this reliability with that of heuristic search, we observe that clearly there is a value of r below which a complete search results in a more reliable system.

6.1.3. A* with Unique Solution Path [9].

It is well known that in the context of a graph problem, $A*$ (or its corresponding algorithms such as IDA* [19]) is optimal when its admissibility condition is satisfied. This is, it is guaranteed that $A*$ will always find a minimal cost (optimal) path to a goal when $h(n) \le h*(n)$ for all nodes n, where $h*(n)$ is the minimum path cost from a node n to a goal node and $h(n)$ is an estimate of $h*(n)$. However, in real-world situations, it is not practical to rely on the statement that a heuristic will always statisfy the inequality condition other than for trivial cases such as when $h(n)$ is always zero. The worst case for A*—when $(h*(n) - \epsilon) \le (h*(n) + \epsilon)$, where ϵ is a nonnegative quantity—has been analyzed [28]. In the extremely simple problem space of infinite binary trees with unit cost on all the arcs of the search graph, it was concluded that in the worst case $k2^{\epsilon} + 1$ nodes have to be visited before the unique goal node, which resides at level k of the tree, can be located. When this result is compared with the worst case of other search strategies such as pure branch-and-bound, which require $2^{k+1} - 1$, we can perform a simple tradeoff analysis for the two search strategies as follows.

Let T_C be a random variable with density function $f_C(.)$ denoting the execution time required when the cost of a solution path is C. Then

$$T_C \le T_{C'} \quad \text{if } C \le C'$$

If the hardware reliability is $g(t)$, then the system reliability is given by

$$r = P\{\text{mission is completed successfully}\} = \int_0^{\infty} g(t) f_C(t)\ dt$$

where we have assumed the software reliability of the solution is 1. Since the whole system fails when the response cannot satisfy a real-time constraint, say t_R, the above equation is modified to

$$r = \int_0^{t_R - t_S} g(t) f_C(t) \, dt$$

where t_S is the response computation time for finding the solution path. For example, in the infinite binary tree problem space when A^* is employed to search for the solution path, t_S is equal to $(k2^\epsilon + 1)T$, where T is the number of time units required to analyze each node.

As a particular example, consider $g(t) = e^{-\lambda}t$ and $f_C(t) = \delta(K_C C_0 - t)$, where K_C is a contrast and $\delta(t)$ is the impulse function defined in Section 6.1.5. This models a system with a hardware failure rate of λ and a constant time for executing the response that increases as the cost associated with a solution path increases. Hence, the system reliability is

$$r = e^{-\lambda K_C C_0} \qquad \text{if } t_R > (k2^\epsilon + 1)T + K_C C_0$$

$$r = 0 \qquad \text{if } t_R \leq (k2^\epsilon + 1)T + K_C C_0$$

The interpretation of the above result is intuitively clear. Clearly, ϵ must be bounded by

$$\frac{t_R - K_C C_0 - T}{kT} > \epsilon$$

in order to satisfy the real-time constraint, t_R. On the other hand, if the pure branch-and-bound heuristic is used to guide the search, then in the worst case the real-time constraint is met when

$$\ln\left(\frac{t_R + T - K_C C_0}{T}\right) - 1 > k$$

and the system reliability is the same as that obtained from A^* when the inequality is true.

From this we see that strategies that do not use heuristic functions are not necessarily worse than those that use heuristics, especially if certain information is given a priori. For example, if the depth of the unique goal state is known, then the last inequality can be used as a criterion to judge whether the former should be employed to enhance the reliability of the system. This is particularly the case when the heuristic error, ϵ, is unknown.

6.1.4 A* with Multiple Solution Paths [9].

In the previous subsection, we assumed that there is a unique solution path. In this subsection, we consider the possibility of multiple paths, all of which could lead to the same goal state. Certainly, of these multiple solution paths, some are optimal solutions, whereas others are not. Therefore, there is a difference in the *quality* of the solution. Generally, the more time one invests in finding a solution, the more likely the solution that is found is close to optimal. However, this may not be permitted in real time and a near-optimal solution may be desired as a compromise between the real-time constraint and the maximum system reliability. This subsection illustrates this by comparing two search strategies using two different heuristics, namely $A*$ and hill climbing [39].

$A*$ is well known in the domain of AI search. Hill climbing represents an extreme case where search efficiency rather than quality of the solution, as used by $A*$, guides the search for a solution path. As in the analysis performed in the previous subsection, our search space is an infinite binary tree, except that arcs are no longer of the same weight. Unlike $A*$, which uses weight of the arc as the search heuristic, hill climbing uses the remaining number of *nodes* to guide the search for a solution path. Furthermore, to obtain maximum search efficiency at the expense of the quality of the solution path, search is streamlined from level to level without checking whether there are other nodes at the same level that may lead to a more optimal solution path. Also, to ensure that there is at least one solution under hill climbing in this situation, we assume that our search tree has monotonic property, namely that the admissibility property is always observed for both $A*$ and hill climbing [19]. After this problem space is defined, it is clear that the number of nodes that need to be visited for the hill climbing case is only $k + 1$ nodes in the worst case (k is the depth of an optimal solution path in the tree) and two in the best case. The average number of nodes to be visited before a solution is found is therefore $(k + 3)/2$, assuming a uniform distribution over $[2,(k + 3)/2]$. On the other hand, the number of nodes to be visited for $A*$ is between $k + 1$ and $2^{k+1} - 1$. The worst case upper bound occurs when the probability that the relative error exceeds some fixed positive quantity is greater than .5 [18]. The average number of nodes to be visited for $A*$, also assuming a uniform distribution, is therefore equal to $(2^{k+1} + k)/2$ in our binary tree space.

The gain in search efficiency in hill climbing is associated with the decline in the quality of the solution. Let C_{hc} and C_{A*} denote the cost of the solution path associated with hill climbing and $A*$, respectively. Clearly, $C_{hc} \geq C_{A*}$. We can rewrite this relationship as follows.

$$C_{hc} = (1 + \epsilon_{hc})C_{A*} \qquad \epsilon_{hc} \geq 0$$

where ϵ_{hc} is a nonnegative quantity indicating the degradation of the solution quality. Using the equation

$$r = \int_0^{t_R - t_s} g(t) f_C(t)\, dt$$

and modeling $g(t) = e^{-\lambda t}$ and $f_C(t) = \delta(K_C(1 + \epsilon_{hc})C_{A*} - t)$, we obtain

$$r(A^*) = e^{-\lambda K_c C_{A^*}} \qquad \text{if } t_R > \left(\frac{2^{k+1} + k}{2}\right) T + K_c C_{A^*}$$

$$= 0 \qquad \text{if } t_R \le \left(\frac{2^{k+1} + k}{2}\right) T + K_c C_{A^*}$$

and

$$r(\text{hill climbing}) = e^{-\lambda K_c (1 + \epsilon_{hc}) C_{A^*}} \qquad \text{if } t_R > \left(\frac{k + 3}{2}\right) T + K_c (1 + \epsilon_{hc}) C_{A^*}$$

$$= 0 \qquad \text{if } t_R \le \left(\frac{k + 3}{2}\right) T + K_c (1 + \epsilon_{hc}) C_{A^*}$$

Several conclusions can be drawn from the above result: (a) if the real-time constraint can be satisfied for A^*, the resulting system reliability will be better than that obtained from hill climbing; that is, $e^{-\lambda K_c C_{A^*}} \ge e^{-\lambda K_c (1 + \epsilon_{hc}) C_{A^*}}$ because $\epsilon_{hc} \ge 0$; (b) the advantage of the hill climbing strategy is that it has a higher probability of satisfying the real-time constraint, as can be seen by the weaker condition required by it; and (c) the advantage of hill climbing disappears as ϵ_{hc} becomes bigger, that is, when

$$\epsilon_{hc} > \frac{t_R - [(K + 3)/2]T}{K_c C_{A^*}} - 1$$

From these results, we see that there is a tradeoff between system reliability and the satisfaction of the real-time constraint, especially when the constraint, t_R, is stringent. Perhaps the best way to satisfy both requirements is to employ a strategy that combines the merits of hill climbing and A^*, as in the A_ϵ^* algorithm [25].

6.1.5 Planning in a Changing Environment.
In the previous sections we assumed that the state of the system does not change while the response is being computed. While this is true for some cases, such as playing chess, it is not true for a real-time system. In this section we illustrate the tradeoff between the length of the computation period (and, hence, the reliability of the response) and the validity of the sensor inputs on which the computation is based. Consider the problem of hitting a target moving along a line between points a and b. The position of the target is observed at time 0, its probable position at time t is computed (assuming that it takes a finite time for the actuator to be aimed and activated and for the projectile to reach the target), and then the response is executed. We assume that the target is hit if the projectile lands within a distance d of the target, $0 \le d \le (b - a)/2$. If the computation considers a larger knowledge base, such as the velocity and acceleration of the target, the wind speed, and so on, then the response is more

accurate, though the computation will take a longer time. The position of the target becomes more and more unpredictable as t increases.

This problem can be modeled by the following two probability density functions:

$$\text{pdf(target is at } y \text{ at time } t) = p_1 \delta(x - y) + \frac{p_2}{b - a}$$

where $p_1 = e^{-\lambda t}$, $p_2 = 1 - e^{-\lambda t}$, and $\delta(t)$ is the impulse function, that is,

$$\int_{-\infty}^{x} \delta(t)\,dt = 0 \text{ for all } x < 0, \int_{-\infty}^{0+} \delta(t)\,dt = 1, \int_{-\infty}^{x} \delta(t)\,dt = 1 \text{ for all } x > 0$$

Thus, the target is at x at time 0 with probability 1, while at time ∞ it can be at any position with equal probability. This models the fact that its position can be predicted with less and less certainty as time goes by. Similarly, we assume that

$$\text{pdf(projectile is at } y \text{ at time } t) = p_2 \delta(x - y) + \frac{p_1}{b - a}$$

This models the situation where at time 0 the mechanism delivers the projectile at any position with equal probability since no computation has been done. As time goes by, the positioning of the mechanism becomes more and more accurate.

We are interested in the reliability of the system at time t, $r(t)$, defined as follows:

$$r(t) = \text{Prob\{target is hit at time } t\}$$

Conditioning on the position of the target at time t, we get

$$r(t) = \int_{a}^{b} \text{pdf(target is hit at } t \mid \text{it is at } y) \text{ pdf(it is at } y \text{ at } t)\,dy$$

Substituting for these two density functions and simplifying the expression, we get

$$r(t) = (e^{-t} - e^{-2t}) \frac{2d(b - a - 2d)}{(b - a)^2}$$

$$+ \left[(e^{-t} - e^{-2t})\left(1 + \frac{4d^2}{(b - a)^2}\right) + (1 - 2e^{-t} + 2e^{-2t})\frac{2d}{b - a} \right]$$

The optimal computation time can be determined by solving $\partial r(t)\partial t = 0$ for

t. This yields $t = \ln 2$. Hence, the optimal reliability is $0.25 + [1.5d/(b - a)]$ while $r(0) = r(\infty) = 2d/(b - a) \leq 0.25 + [1.5d/(b - a)]$—recall that $0 \leq d \leq (b - a)/2$.

6.2   Response Execution Time

When a control program can achieve its goal in several different ways, it should select the one that optimizes the system reliability. The strategy that can tolerate the maximum number of failures may not be optimal if the time required to execute it is much larger than that required by the other strategies. To see this, consider a case involving two identical hardware components and two strategies, S_1 and S_2. Strategy S_1 requires both the components and achieves the goal in T_1 units, while strategy S_2 requires only one of the components and achieves the goal in T_2 units, where $T_2 > T_1$. We also assume that once S_1 is partially executed we cannot switch to S_2 in case one of the components fails.

Assuming that the failure rate of the components is λ, the reliability of the system when strategy S_1 is used is

$$r_1 = e^{-2\lambda T_1}$$

Similarly, the reliability when S_2 is used is

$$r_2 = 2e^{-\lambda T_2} - e^{-2\lambda T_2}$$

If $T_2 > 2T_1 + \ln 2/\lambda$, then strategy S_1 is definitely better than S_2. However, if $T_2 < 2T_1$, then S_2 is definitely better than S_1.

Another tradeoff occurs when the time required to execute the response increases with its effectiveness. Let T_r be a random variable with density function $f_r(.)$ denoting the execution time required when the software reliability is r. Then

$$T_r \underset{\text{st}}{\leq} T_{r'} \text{ if } r \leq r'$$

If the hardware reliability is $g(t)$, then the system reliability is given by

$$s(r) = P\{\text{mission is completed successfully}\} = r \int_0^\infty g(t)f_r(t)dt$$

As a particular example, consider

$$g(t) = e^{-\lambda t}$$

and

$$f_r(t) = \delta\left(\frac{r}{1-r} - t\right)$$

This models a system with a hardware failure rate of λ and a constant time for executing the response, which increases as the software reliability increases. Hence, the system reliability is

$$s(r) = re^{-\lambda[r/(1-r)]}$$

From this we see that $s(0) = s(1) = 0$. In fact, the optimal system reliability is obtained by solving $\partial s(r)/\partial r = 0$, yielding

$$s_{\text{opt}}(r) = s(r_{\text{opt}}), \text{ where } r_{\text{opt}} = 1 + 0.5\lambda - \sqrt{\lambda + 0.25\lambda^2}$$

To illustrate the case where the distribution of T_r is not constant, consider the aforementioned system with $f_r(t) = \alpha e^{-\alpha t}$, where $\alpha = (1 - r)/r$. In this case the system reliability is

$$s(r) = \frac{r(1-r)}{\lambda r + 1 - r}$$

Again, we observe that $s(0) = s(1) = 0$, and the optimal system reliability is obtained when r is $(1 - \sqrt{\lambda})/\lambda$.

7 SUMMARY

In this chapter, we first reviewed current hardware and software reliability assessment techniques. Then we discussed why these methods have to be modified for estimating the reliability of AI systems. Finally, we developed an approach for evaluating the reliability of both AI programs and embedded AI systems.

For the software components of an AI system, we explicitly modeled the possibility that there are unremovable faults (or shortcomings) in some basic AI methodologies, such as the use of heuristics in knowledge acquisition and various problem solving methods. We also considered cases where the output of some AI programs cannot be classified as being simply correct or incorrect since this may be a fuzzy quantity.

For embedded AI systems consisting of both hardware and software we modeled the facts that (1) the mission time is often determined by the software, (2) the system state can change while the response is being computed, (3) more reliable responses require more processing time, and (4) there can be varying strategies of hardware reliability requirements. In particular, we looked at the influence of the response computation time and the response execution time on the system reliability. In the former case, the tradeoff is between a quick response time and the correctness (reliability) of the response, which generally improves with an increase in computation time, especially when exhaustive search techniques have to be used. In the second case, the choice is between the reliability of a strategy and the time required to execute it. For example, a faster strategy may use more components and hence be more unreliable than a slower strategy. We computed the system reliability for some cases and determined the optimal reliability in each case.

8 ACKNOWLEDGMENTS

The author wishes to thank Ing-Ray Chen and Atul Prakash for their extensive comments and discussions, which have greatly improved the quality of the presentation.

REFERENCES

1. Y. S. Abu-Mostafa and D. Psaltis, "Optical Neural Computers," *Scientific American,* vol. 256, no. 3, pp. 88–95, March 1987.

2. J. E. Angus, R. E. Schafer, and A. Sukert, "Software Reliability Model Validation," *Proc. Annual Reliability and Maintainability Symposium,* pp. 191–199, San Francisco, CA, January 1980.

3. A. Avizienis and J.-C. Laprie, "Dependable Computing: From Concepts to Design Diversity," *Proc. IEEE,* vol. 74, no. 5, pp. 629–638, May 1986.

4. R. E. Barlow and F. Proschan, *Statistical Theory of Reliability and Life Testing,* Holt, Rinehart and Winston, Inc., New York, 1975.

5. F. B. Bastani and C. V. Ramamoorthy, "A Methodology for Assessing the Correctness of Control Programs," *Computers and Electrical Engineering,* vol. 11, no. 3, Pergamon Press, 1984.

6. F. B. Bastani, "On the Uncertainty in the Correctness of Computer Programs," *IEEE Trans. Software Engineering,* vol. SE-11, no. 9, September 1985.

7. F. B. Bastani and C. V. Ramamoorthy, "Input Domain Based Models for Estimating the Correctness of Process-Control Programs," *Theory of Reliability,* eds. A. Serra and R. E. Barlow, North-Holland, 1986.

8. R. C. Cheung, "A User-Oriented Software Reliability Model," *Proc. COMPSAC '78,* pp. 565–570, Chicago, IL, November 1978.

9. I.-R. Chen, "An AI Based Architecture of Self-Stabilizing Fault Tolerant Process Control Programs and its Analysis," Ph.D. dissertation, Department of Computer Science, University of Houston - University Park, Houston, TX, 1988.

10. *Computer* (IEEE), special issue on knowledge representation, vol. 16, no. 10, October 1983.

11. J. T. Duane, "Learning Curve Approach to Reliability Monitoring," *IEEE Trans. Aerospace,* vol. 2, pp. 563–566, 1964.

12. J. W. Duran and J. J. Wiorkowski, "Capture-Recapture Sampling for Estimating Software Error Content," *IEEE Trans. Software Engineering,* vol. SE-7, no. 1, January 1981.

13. S. E. Fahlman and G. E. Hinton, "Connectionist Architectures for Artificial Intelligence," *Computer* (IEEE), vol. 20, no. 1, pp. 100–109, January 1987.

14. J. M. Finkelstein, "Starting and Limiting Values for Reliability Growth," *IEEE Trans. Reliability,* vol. R-28, pp. 111–114, June 1979.

15. M. P. Georgeff and A. L. Lansky, "Procedural Knowledge," *Proc. IEEE,* vol. 74, no. 10, pp. 1383–1398, October 1986.

16. A. L. Goel, "Software Reliability Models: Assumptions, Limitations, and Applicability," *IEEE Trans. Software Engineering,* vol. SE-11, no. 12, pp. 1411–1423, December 1985.

17. P. G. Hoel, S. C. Port, and C. J. Stone, *Introduction to Probability Theory,* Houghton Mifflin Co., Boston, 1971.

18. N. Huyn, R. Dechter, and J. Pearl, "Probabilistic Analysis of the Complexity of A*," *Artificial Intelligence,* vol. 15, pp. 241–254, 1980.

19. R. E. Korf, "Depth-First Iterative Deepening: An Optimal Admissable Tree Search," *Artificial intelligence,* vol. 27, pp. 97–109, 1985.

20. B. Littlewood, "A Reliability Model for Markov Structured Software," *Proc. 1975 Int'l Conf. on Reliability Software,* pp. 204–207, Los Angeles, CA.

21. B. Littlewood, "A Bayesian Differential Debugging Model for Software Reliability," *Proc. COMPSAC '80,* pp. 511–519, Chicago, IL., 1980.

22. H. D. Mills, "On the Development of Large Reliable Software," *Rec. IEEE Symposium Computer Software Reliability,* pp. 155–159, New York, May 1973.

23. J. D. Musa and K. Okumoto, "A Logarithmic Poisson Execution Time Model for Software Reliability Measurement," *Proc. 7th Int'l Conf. on Software Engineering,* pp. 230–237, Orlando, FL, March 1984.

24. E. Nelson, "Estimating Software Reliability from Test Data," *Microelectronics and Reliability,* vol. 17, pp. 67–74, Pergamon Press, 1978.

25. J. Pearl and J. H. Kim, "Studies in Semi-Admissible Heuristics," *IEEE Trans. Pattern Analysis and Machine Intelligence,* vol. PAMI-4, no. 4, pp. 392–399, July 1982.

26. J. Pearl, "Some Recent Results in Heuristic Search Theory," *IEEE Trans. Pattern Analysis and Machine Intelligence,* vol. PAMI-6, no. 1, pp. 1–12, January 1984.

27. J. Pearl, *Heuristics,* Addison-Wesley, Reading, MA, 1984.

28. I. Pohl, "First Results on the Effect of Error in Heuristic Search," *Machine Intelligence,* vol. 5, pp. 219–236, 1970.

29. *Proc. IEEE,* special issue on knowledge representation, vol. 74, no. 10, October 1986.

30. C. V. Ramamoorthy and F. B. Bastani, "An Input Domain Based Approach to the Quantitative Estimation of Software Reliability," *Proc. Taipei Seminar on Software Engineering,* Taipei, 1979.

31. C. V. Ramamoorthy and F. B. Bastani, "Software Reliability—Status and Perspectives," *IEEE Trans. Software Engineering,* vol. SE-8, no. 4, pp. 354–371, July 1982.

32. E. Rich, *Artificial Intelligence,* McGraw-Hill, 1983.

33. S. M. Ross, *Introduction to Probability Models,* Academic Press, New York, 1985.

34. G. J. Schick and R. W. Wolverton, "An Analysis of Competing Software Reliability Models," *IEEE Trans. Software Engineering,* vol. SE-4, no. 2, pp. 104–120, March 1978.

35. A. Serra and R. E. Barlow, eds., *Theory of Reliability,* North-Holland, Amsterdam, 1986.

36. C. Stanfill and D. Waltz, "Toward Memory-Based Reasoning," *Comm. ACM,* vol. 29, no. 12, pp. 1213–1228, December 1986.

37. TRW Defense and Space Systems Group, Software Reliability Study, Report No. 76-2260.1-9-5, TRW, Redondo Beach, CA, 1976.

38. D. L. Waltz, "Application of the Connection Machine," *Computer* (IEEE), vol. 20, no. 1, pp. 85–97, January 1987.

39. P. H. Winston, *Artificial Intelligence,* 2nd ed., Addison–Wesley, 1984.

40. L. A. Zadeh, "Fuzzy Sets and Information Granularity," *Advances in Fuzzy Set Theory and Application*, eds. M. M. Gupta, R. D. Ragade, and R. R. Yager, North-Holland, 1979.

41. L. A. Zadeh, "Is Probability Theory Sufficient for Dealing with Uncertainty in AI: A Negative View," *Uncertainty in Artificial Intelligence,* pp. 103–116, eds. L. N. Kanal and J. F. Lemmer, North-Holland, Amsterdam, 1986.

INDEX